Functional Supramolecular Materials
From Surfaces to MOFs

Monographs in Supramolecular Chemistry

Series Editors:
Philip Gale, *The University of Sydney, Australia*
Jonathan Steed, *Durham University, UK*

Titles in this Series:
1: Cyclophanes
2: Calixarenes
3: Crown Ethers and Cryptands
4: Container Molecules and Their Guests
5: Membranes and Molecular Assemblies: The Synkinetic Approach
6: Calixarenes Revisited
7: Self-assembly in Supramolecular Systems
8: Anion Receptor Chemistry
9: Boronic Acids in Saccharide Recognition
10: Calixarenes: An Introduction, 2nd Edition
11: Polymeric and Self Assembled Hydrogels: From Fundamental Understanding to Applications
12: Molecular Logic-based Computation
13: Supramolecular Systems in Biomedical Fields
14: Synthetic Receptors for Biomolecules: Design Principles and Applications
15: Polyrotaxane and Slide-Ring Materials
16: Boron: Sensing, Synthesis and Supramolecular Self-Assembly
17: Porous Polymers: Design, Synthesis and Applications
18: Pillararenes
19: Supramolecular Chemistry at Surfaces
20: Aromatic Interactions: Frontiers in Knowledge and Application
21: Naphthalenediimide and its Congeners: From Molecules to Materials
22: Functional Supramolecular Materials: From Surfaces to MOFs

How to obtain future titles on publication:
A standing order plan is available for this series. A standing order will bring delivery of each new volume immediately on publication.

For further information please contact:
Book Sales Department, Royal Society of Chemistry, Thomas Graham House, Science Park, Milton Road, Cambridge, CB4 0WF, UK
Telephone: +44 (0)1223 420066, Fax: +44 (0)1223 420247
Email: booksales@rsc.org
Visit our website at http://www.rsc.org/Shop/Books/

Functional Supramolecular Materials
From Surfaces to MOFs

Edited by

Rahul Banerjee
CSIR National Chemical Laboratory, Pune, India
Email: r.banerjee@ncl.res.in

ROYAL SOCIETY OF CHEMISTRY

Monographs in Supramolecular Chemistry No. 22

Print ISBN: 978-1-78262-540-7
PDF eISBN: 978-1-78801-027-6
EPUB eISBN: 978-1-78801-141-9
ISSN: 1368-8642

A catalogue record for this book is available from the British Library

The Royal Society of Chemistry is a charity, registered in England and Wales, Number 207890, and a company incorporated in England by Royal Charter (Registered No. RC000524), registered office: Burlington House, Piccadilly, London W1J 0BA, UK, Telephone: +44 (0) 207 4378 6556.

Visit our website at www.rsc.org/books

Printed in the United Kingdom by CPI Group (UK) Ltd, Croydon, CR0 4YY, UK

Preface

Supramolecular chemistry and crystal engineering are two extremely well known, as well as relevant, subjects in modern day science. Supramolecular materials have great potential in materials science, as the reversibility of intermolecular interactions can create new functions that are unavailable in covalently bonded materials. This book highlights multiple aspects of structure and function in organic as well as metal organic supramolecular materials. It also highlights the chemical functions of supramolecular materials. The chapters of this book will provide an overview of organic and metal organic supramolecular materials (from surfaces to MOFs) and their assembly at various length scales (including the formation of 2D-polymers and molecular cages) and how this determines their usage for novel applications in molecular recognition, catalysis, light harvesting and environmental remediation. This book will also provide an idea about the role of intermolecular interactions in the solid state and solution state supramolecular self-assembly. Another highlight of this book will be the role of chemical bonding on the molecular and supramolecular architectures.

This book consists of 12 chapters that have been contributed by leading experts in the field of structural chemistry, covering the areas of surface self-assembly, soft supramolecular materials, supramolecular receptors, electron transfer reactions, morphology control and solid-state photo-reactivity, new devices with tunable electrical and optical properties, bio-materials, supramolecular polyhedral capsules and cages and metal organic frameworks. As the development of new materials is key to the continuous improvement of the performance of organic and metal organic supramolecular materials, a significant portion of this book have been dedicated to discussing the design concept and synthesis of novel organic and metal organic supramolecular materials for diverse applications.

Monographs in Supramolecular Chemistry No. 22
Functional Supramolecular Materials: From Surfaces to MOFs
Edited by Rahul Banerjee

Published by the Royal Society of Chemistry, www.rsc.org

Champness and Yang describe how intermolecular interactions can be exploited to control molecular organisation on surfaces, demonstrating the use of hydrogen bonds, van der Waals interactions, metal–ligand coordination and covalent coupling to construct low-dimensional structures on surfaces. The idea behind the design of soft supramolecular materials using intermolecular interactions and guest binding within the supramolecular capsular assemblies are covered in detail by Dastidar and co-workers and Ghosh *et al.* in Chapters 2 and 3 respectively. Photochemistry features heavily in the next two chapters by Mukhopadhyay *et al.*, and Quah and Vittal on supramolecular materials for electron transfer reactions and photo-reactivity in solid-state supramolecular materials. Nidamanuri and Saha in Chapter 6 and D'Alessandro *et al.* in Chapter 7 describe framework materials for electrical, optical and photoconduction. Ma *et al.* in Chapter 8 describe the structure-based evolution of porous biomaterials. Comparing and contrasting the controlled release from these porous biomaterials is an area of active debate. This aspect of "Intrinsic *vs.* Extrinsic Porosity" has been discussed by Green and Lloyd in Chapter 9. Research on metal organic frameworks (MOFs) and covalent organic frameworks (COFs) continues to evolve rapidly and has recently seen new applications in areas such as storage and separation, conduction and catalysis. In the last three chapters of this book, Clegg *et al.*, Arab and El-Kaderi and Maji *et al.* discuss many new examples from their own research and the literature.

I would like to thank Professor Jonathan W. Steed and Professor Philip Gale for their invitation to contemplate a book project on the topic of "Functional Supramolecular Materials: From Surface to MOFs" for the Royal Society of Chemistry series Monographs in Supramolecular Chemistry, and Rowan Frame, Katie Morrey, Rachel Wood and Harriet Manning at the Royal Society of Chemistry for their support throughout the production of the book. I sincerely thank all the authors of the chapters for their time and sincere efforts as well as their commitment to this project. I sincerely hope that this book will be of interest among undergraduate and graduate students, research scholars, academicians (including technical service providers) and industrial chemists interested in supramolecular chemistry and materials science.

I dedicate this book to Professor Gautam R. Desiraju on the occasion of his 65th birthday.

Rahul Banerjee
CSIR-National Chemical Laboratory
India

Contents

Monographs in Supramolecular Chemistry No. 22
Functional Supramolecular Materials: From Surfaces to MOFs
Edited by Rahul Banerjee

Published by the Royal Society of Chemistry, www.rsc.org

CHAPTER 1

Surface Self-assembly of Functional Supramolecular Networks

LIXU YANG AND NEIL R. CHAMPNESS*

School of Chemistry, University of Nottingham, Nottingham NG7 2RD, UK
*Email: neil.champness@nottingham.ac.uk

1.1 Introduction

The use of self-assembly processes to prepare nanoscale structures lies at the very heart of supramolecular chemistry.[1] Indeed, through the use of self-assembly it is possible to prepare highly complex molecular architectures with targeted properties that can avoid lengthy, multi-step synthetic pathways. In the initial stages of research in the field of supramolecular chemistry the majority of studies were performed in the solution phase[2] and more recently solid-state supramolecular chemistry has been developed combined with the wider field of crystal engineering.[3] However, more recently the field of surface-based supramolecular chemistry has begun to be explored which has led to a significant interest in developing low-dimensional, predominantly two-dimensional, structures.[4–10] The developments in this area are the focus of this chapter, including the synthetic approaches to such structures and the detailed characterisation of the resulting two-dimensional arrays that can be achieved using scanning probe microscopies.

Surface-based chemistry has long been utilised for assembling arrays of molecules. Perhaps the most widely studied area of such chemistry is the

Monographs in Supramolecular Chemistry No. 22
Functional Supramolecular Materials: From Surfaces to MOFs
Edited by Rahul Banerjee

Published by the Royal Society of Chemistry, www.rsc.org

development of self-assembled monolayers (SAMs) of thiolate molecules adsorbed onto Au(111) substrates.[11] The development of SAMs has been expansive and lies at the heart of many studies that attach molecules and, in some cases supramolecular arrays, to surfaces.[12] Recent studies have demonstrated the use of supramolecular interactions between thiolate species to control the SAM formation.[13] In the case of Au-thiolate SAMs and in many other instances deposition of molecules onto surfaces often leads to close-packed arrays that have well-defined arrangements; such arrays typically rely on van der Waals interactions and simple geometric preferences. However, the concepts of supramolecular chemistry open up the possibility of using stronger intermolecular interactions to create more complex and potentially porous structures. The logical extension of this approach is to use stronger interactions, ultimately leading to covalent coupling reactions to create robust low-dimensional arrays.

The emphasis of this chapter will be on studies where molecules are specifically *designed* in an attempt to control relative molecular organisation. In particular, the chapter will focus on the use of stronger supramolecular interactions, such as hydrogen-bonding, van der Waals interactions and lastly covalent coupling, to form extended arrays that propagate in two dimensions, parallel to the surface. Two-dimensional self-assembly of molecules on surfaces, when combined with scanning probe techniques, provides direct evidence of the potential of this approach. In some instances, examples show that such structures can be used to trap diffusing species as guests, in a similar fashion to porous architectures constructed in three-dimensional solids such as metal-organic frameworks (MOFs)[14] and covalent-organic frameworks (COFs).[15]

The other main difference between solution-phase supramolecular chemistry and surface-based systems is of course the surface itself. Firstly, the surface defines a two-dimensional boundary upon which the self-assembly process is developed and secondly, the surface is far from innocent in the reaction process. Suitable surfaces often studied include Au(111),[16] Ag-Si(111)$\sqrt{3} \times \sqrt{3}$R30° [Ag(Si(111))][17] and highly-oriented pyrolytic graphite (HOPG).[18] However recent studies have started to explore other substrates including graphene[19] and boron-nitride monolayers.[20] The choice of substrate is typically driven by their tendency to adopt weak interactions with organic molecules, which are commonly the focus of self-assembly studies. The surface organisation needs to be strongly influenced by intermolecular, supramolecular interactions, rather than surface effects to allow ready design of the resulting structure. Additionally, the requirements of specific scanning probe microscopies heavily influence the choice of substrate, *i.e.* STM requires conducting surfaces for imaging, and this is in many instances the determining feature. Recent improvements in resolution of atomic force microscopy (AFM)[21] and the development of dynamic force microscopy (DFM)[22] have led to a wider scope for characterisation and therefore the choice of substrate is no longer restricted to (semi)conducting materials and insulating substrates are now realistic targets. Lastly, notably in the case of covalently coupled structures, the choice of substrate is extremely important with the substrate often taking an active role in promoting coupling reactions.

The chapter is subdivided into four sections, systems assembled using (i) hydrogen bonds, (ii) van der Waals interactions, (iii) covalent bonds and lastly a section (iv) of self-assembled systems with unusual ordering that illustrate the power of the approach and particularly the molecular-level characterisation of such systems.

1.2 Two-dimensional Arrays Assembled Using Hydrogen-bonding

The use of hydrogen bonds to create supramolecular structures goes back to the origins of the field.[23] Indeed it is relatively straightforward to use a simple molecule capable of self-recognition to create extended framework structures. Indeed, an early example of surface-based supramolecular chemistry—the formation of a unimolecular supramolecular structure mediated by hydrogen-bonding—was reported by Griessl *et al.*[24] The structure was formed *via* deposition of trimesic acid onto a HOPG substrate in UHV conditions and imaging using STM (Figure 1.1a,b) reveals precise details of the self-assembled molecular structure. The carboxylic acid–carboxylic acid hydrogen bonds, which adopt the classic $R^2_2(8)$ intermolecular arrangement,[25] ensure that an open structure is adopted in preference to a hypothetical, alternative close-packed arrangement, confirming that hydrogen-bonding interactions are the dominant force in producing these structures. The observed structure is the expected 'chicken-wire', or honeycomb arrangement leading to a porous network structure (Figure 1.1a,c). In addition to the honeycomb structure a secondary so-called 'flower' arrangement is also observed by STM measurements (Figure 1.1d). This alternative self-assembled structure results from the adoption of $R^3_3(12)$ supramolecular synthons formed by three carboxylic acid moieties from separate trimesic acid molecules. The 'flower' structure may form due to a higher molecular density on the surface, confirming that surface coverage also influences the final self-assembled structure by maximising the energy gained through adsorbate-substrate interactions.

A subsequent study of an elongated analogue of trimesic acid, 1,3,5-tris(4-carboxyphenyl)benzene, deposited on HOPG, but deposited from a range of alkyloic acids rather than by sublimation in UHV, also results in the formation of a honeycomb lattice at low temperatures, but a more densely packed phase at higher temperatures.[26] The phase transition and the transition temperature between honeycomb and denser structures were found to depend on the nature of the solvent and molecular concentration. The authors suggested that the co-adsorption of solvent molecules within the honeycomb structure stabilises the nominally porous structure at low temperatures, but upon elevation of the temperature the weakly bound

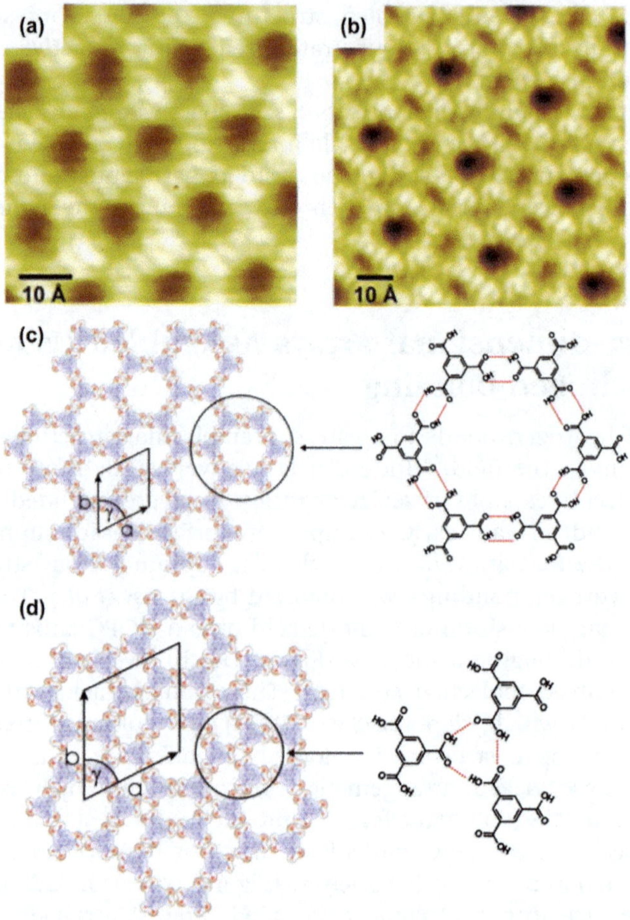

Figure 1.1 Open network arrangements of trimesic acid on graphite under UHV at low temperature. (a), (c) The 'chicken-wire' structure; (b), (d) the 'flower' arrangement.
Reproduced with permission from S. Griessl, M. Lackinger, M. Edelwirth, M. Hietschold and W. M. Heckl, Self-Assembled Two-Dimensional Molecular Host-Guest Architectures From Trimesic Acid, *Single Mol.*, 2002, **3**, 25. Copyright 2002 WILEY-VCH Verlag Berlin GmbH, Fed. Rep. of Germany.

solvent molecules desorb initiating the transition to the more densely packed and thermodynamically-favoured phase.

A simple example of a unimolecular self-assembled structure that uses hydrogen bonds is that of naphthalene-1,4,5,8-tetracarboxylic diimide (NTCDI).[27] NTCDI contains imide moieties at opposing ends of the rod-shaped molecule which can adopt imide–imide hydrogen bonds, again adopting an $R^2{}_2(8)$ intermolecular interaction, similar to that observed for carboxylic acid dimers. As a result of the divergent arrangement of the imide moieties linear chains are observed when the molecule is deposited onto a Ag/Si(111) surface (Figure 1.2a). Imaging using DFM of this molecule and the self-assembled array on a variety of surfaces[28–30] reveals sub-molecular details of the molecular arrangement. It is interesting to note that features in the DFM images appear to coincide with where hydrogen bonds would be expected to be observed (Figure 1.2b), however calculations indicate that these features do not arise as a result of the hydrogen bonds but due to some other, as yet undefined, phenomenon; the origin of such intermolecular features remains a highly debated topic.[31]

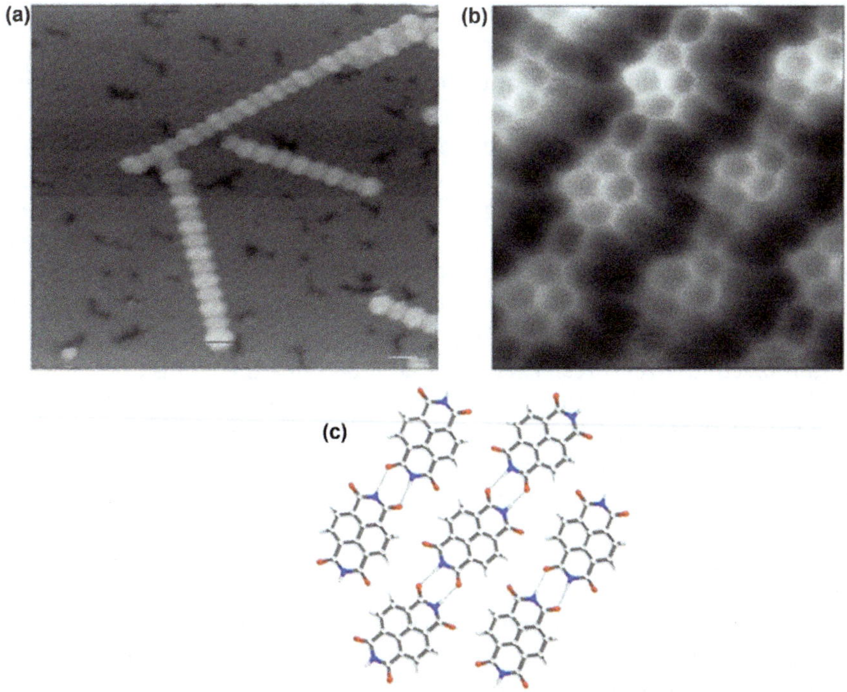

Figure 1.2 (a) STM image of three molecular chains of NTCDI adsorbed on Ag/Si(111); (b) constant height DFM image of NTCDI adsorbed on Ag/Si(111) at 77 K (2.1×2.0 nm); (c) view of hydrogen-bonded chains in the single crystal X-ray structure of NTCDI.
(a) and (c) Reproduced with permission from D. L. Keeling, N. S. Oxtoby, C. Wilson, M. J. Humphry, N. R. Champness and P. H. Beton, Assembly and Processing of Hydrogen Bond Induced Supramolecular Nanostructures, *Nano Lett.*, 2003, **3**(1), 9–12. Copyright 2003 American Chemical Society. (b) Reproduced with permission from A. M. Sweetman, S. Jarvis, H. Sang, I. Lekkas, P. Rahe, Y. Wang, J. Wang, N. R. Champness, L. Kantorovich and P. J. Moriarty, Mapping the force field of a hydrogen-bonded assembly, *Nat. Commun.*, 2014, **5**, 3931. Copyright 2014 Nature Publishing Group. Published under a creative commons attribution 3.0 unported license (http://creativecommons.org/licenses/by/3.0/).

Moving beyond simple unimolecular systems it is possible to use the full force of supramolecular design concepts to bear upon the construction of surface-based arrays. Learning from nature, a particularly attractive approach is to use the specific hydrogen-bonding capability of DNA nucleobases. The use of DNA nucleobases in supramolecular chemistry has been a persistent theme since some of the seminal work of Seeman,[32] which has been expanded by a number of groups to great effect.[33] Indeed, Seeman and Winfree demonstrated the specific preparation of 2D crystals on surfaces using his 'sticky-end' approach.[34] Examples of such surface-based arrays include the self-assembly of striped lattices, amongst other structures, successfully produced on mica surfaces[34] and hexagonal arrays constructed from three-point star motifs, in turn built from seven DNA single strands (Figure 1.3).[35]

In addition to the studies of self-assembly on surfaces of large DNA fragments, a particular focus for hydrogen-bonding studies has been the use of the individual DNA bases to create self-assembled structures on surfaces. The assembly of guanine and its derivatives on surfaces has received particular attention,[36–39] perhaps as a reflection of the importance of guanine quartets in anti-cancer drug design.[40] For example the adsorption of guanine onto a Au(111) surface under UHV conditions leads to the formation of guanine quartets as imaged by STM.[36,39] Interestingly in this example the quartets form through Hoogsteen-style hydrogen-bonding and are associated through further N–H–N hydrogen bonds to give rise to two-dimensional supramolecular structures. This example illustrates the complexity of using DNA bases due to the variety of potential hydrogen-bonding motifs that can be adopted, from classic Watson–Crick pair formation, Hoogsteen interactions, and reverse Watson–Crick and reverse Hoogsteen arrangements (Figure 1.4). The area has been recently reviewed[41] and thus only very recent examples are highlighted here.

It is of course possible to exploit the hydrogen-bonding capability of the DNA nucleobases as appendages to more complex molecules leading to enhanced control over supramolecular structure. An example of such a strategy is given by the study of a porphyrin molecule (tetra-TP) in which the porphyrin core is functionalised in each *meso*-position by a phenylthymine moiety such that each thymine presents a hydrogen-bonding face *exo* to the porphyrin core.[42] The molecule self-assembles on a HOPG substrate to give rise to a two-dimensional grid structure wherein the molecules interact through $R^2_2(8)$ intermolecular thymine–thymine hydrogen bonds (Figure 1.5a). The asymmetric arrangement of the thymine groups appended to the tetra-TP molecules suggests the potential for the molecules to adopt a chiral arrangement when adsorbed onto a surface and this is observed. The almost perfectly square 2D unit cell observed for the tetra-TP structure suggests that all of the thymine groups within an individual tetra-TP molecule adopt the same orientation with respect to the porphyrin, and that individual domains contain only molecules of the same handedness. Overall the array of tetra-TP remains globally achiral by forming an equal

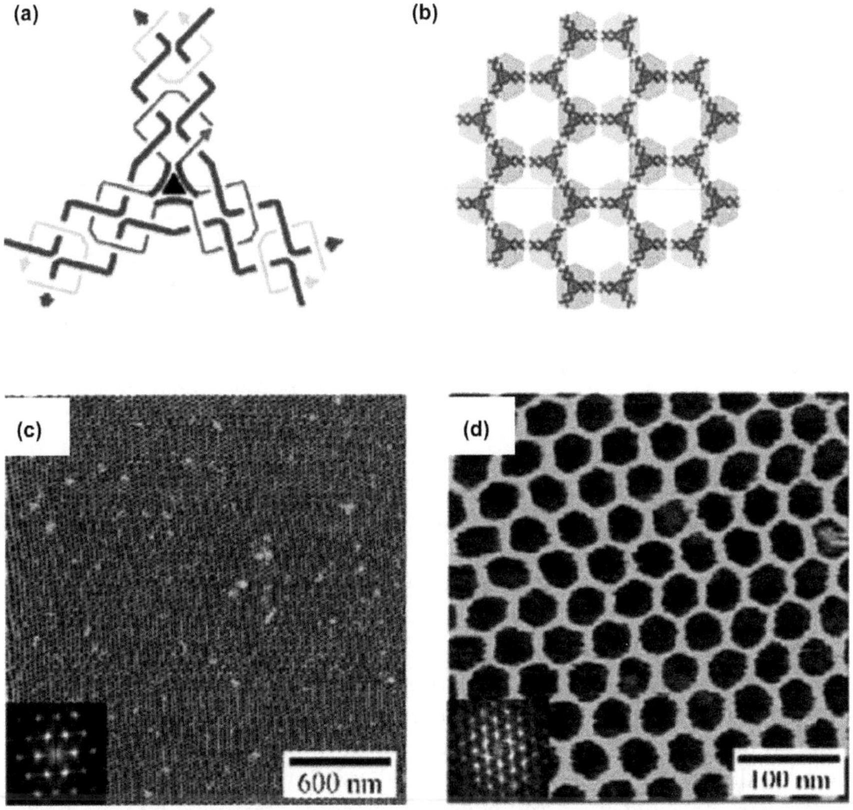

Figure 1.3 (a) The three-point star motif made of seven DNA single strands exploited by He *et al.*; (b) a model of the assembly of several motifs arranged to form a hexagonal network; (c) and (d) AFM images of the assembly on a mica surface.
Reproduced with permission from Y. He, Y. Chen, H. Liu, A. E. Ribbe and C. Mao, Self-Assembly of Hexagonal DNA Two-Dimensional (2D) Arrays, *J. Am. Chem. Soc.*, 2005, **127**, 12202. Copyright 2005 American Chemical Society.

area of mirror domains containing either right, or left-handed molecules. It is worth noting that many previous studies have shown that prochiral molecules tend to assemble into homochiral domains on surfaces containing molecules of only a single handedness.[43] Addition of a substituted adenine component, in this case 9-propyladenine, to the assembly of tetra-TP produced a molecular network with a combination of disordered regions interspersed with small domains of an ordered co-crystal structure containing both tetra-TP and 9-propyladenine. As anticipated Watson–Crick thymine–adenine interactions are observed but interestingly dimers of adenine are also formed, presumably in order to achieve maximum surface coverage, and forming an additional N–H–N hydrogen bond to a further

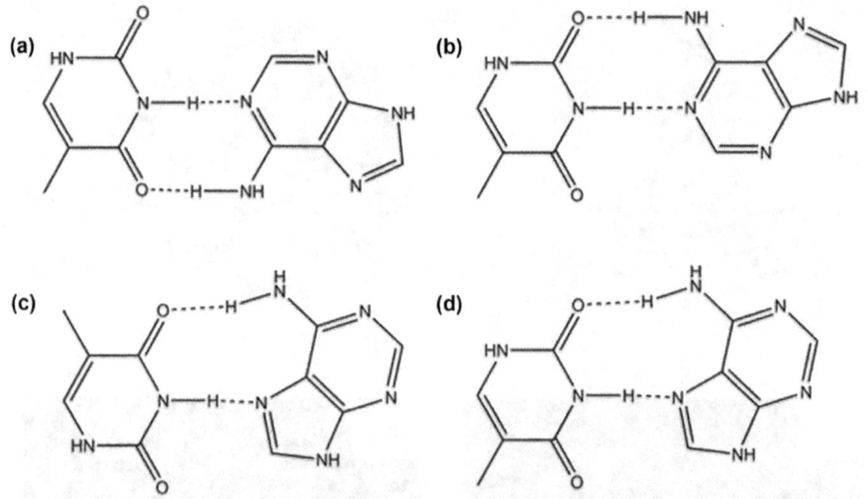

Figure 1.4 Potential hydrogen-bonding interactions between thymine and adenine; (a) standard Watson–Crick; (b) reverse Watson–Crick; (c) Hoogsteen; (d) reverse Hoogsteen arrangements.

tetra-TP molecule (Figure 1.5b). The complexity of the structure indicates the complexity in designing such structures, even when relatively predictable hydrogen-bonding synthons are present.

A related approach has been described by González-Rodríguez and De Feyter[44] who have designed rod-shaped molecules appended by DNA bases, either with guanine (G) and cytosine (C) or adenine (A) and uracil (U) (Figure 1.6a). The rods are designed such that opposing ends will form complementary hydrogen-bonding groups. Additionally, the authors have functionalised the rods with alkyl tails such that any residual space in the structure is occupied and no additional molecules are required to fill space. Self-assembly on an HOPG substrate results in the formation of cyclic structures, as imaged by STM (Figure 1.6b), and indicates the successful employment of the complementary hydrogen-bonding moieties in forming heteromolecular hydrogen-bonding interactions (G–C or A–U). Despite the alkyl chain appendages filling the space within the cyclic structures the G–C system is capable of adsorbing coronene with the alkyl chains assumed to vacate the space due to preferential physisorption of the polyaromatic guest onto the HOPG surface.

Moving beyond DNA systems it is possible to create related, non-natural, hydrogen bonding synthons to create extended structures. For example the triple hydrogen bonding interaction between 2,6-di(acetylamino)pyridines and imide groups has received extensive study in supramolecular chemistry[23,45] and can also be exploited to form bimolecular structures on surfaces.[46] An early example of a bimolecular network was prepared from perylene-3,4,9,10-tetracarboxylicdiimide (PTCDI), which bears two imide

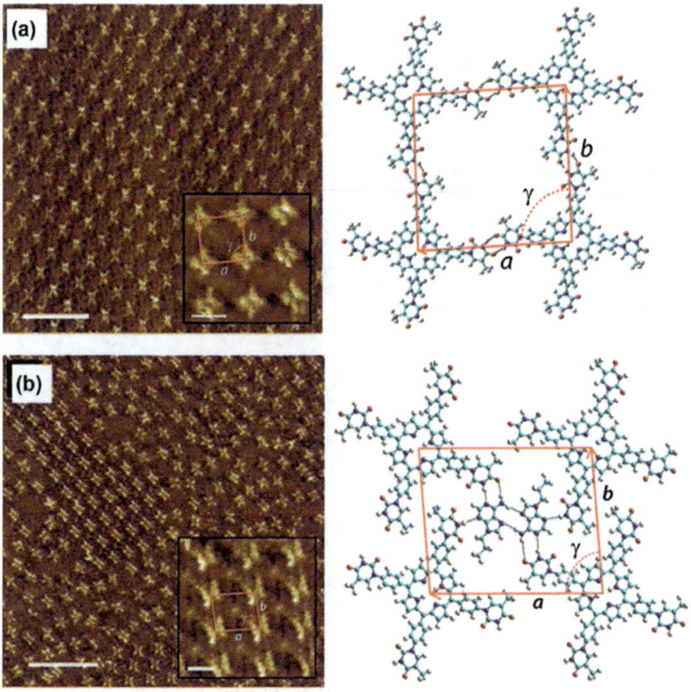

Figure 1.5 (a) 2D self-assembled network of tetra-TP adsorbed on HOPG liquid–solid interface, left: STM image. The insert shows a high resolution, drift corrected STM image of the network with an individual 2D unit cell marked in red: Scale bar = 20 nm (insert = 2 nm). Right: Molecular model of the tetra-T-porphyrin network from MM simulations; (b) 2D self-assembled network of tetra-TP and 9-propyladenine adsorbed on HOPG liquid–solid interface, left: STM image. The insert shows a high resolution, drift corrected STM image of the network with an individual 2D unit cell marked in red: Scale bar = 20 nm (insert = 1.6 nm). Right: Molecular model of the tetra-T-porphyrin-9-propyladenine network from MM simulations.
Reproduced from ref. 42 with permission from the Royal Society of Chemistry.

appendages at opposing ends of the rod, in a similar fashion to NTCDI, and melamine (Figure 1.7).[47] Co-deposition of the two building blocks onto a Ag/Si(111) surface under UHV conditions results in the formation of a honeycomb network with triple hydrogen bonds formed between the PTCDI imide moieties and each side of the melamine triangle. Network formation was achieved following annealing at ~100 °C which provides sufficient thermal energy for molecules to detach from PTCDI islands and diffuse across the surface, resulting in the desired network topology. The network was found to be commensurate with the underlying Ag/Si(111) surface, suggesting an influence from the substrate in the formation of the network.

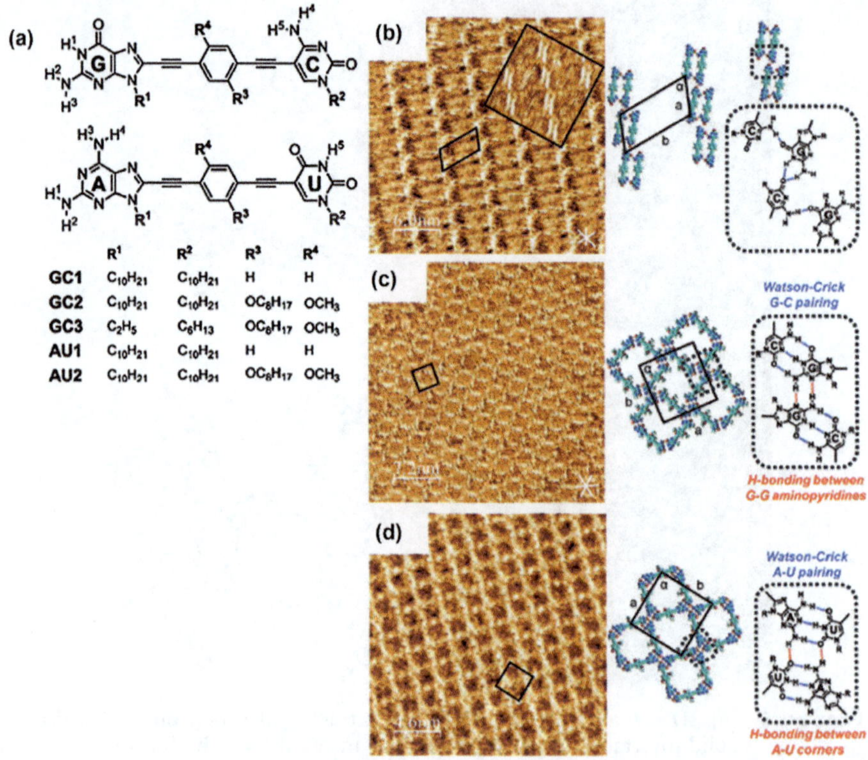

Figure 1.6 (a) Mixed guanine-cytosine and adenine-uracil hydrogen-bonding rods employed for self-assembly; (b) STM image of GC1 on HOPG showing pairs of monomers and model of observed structure; (c) STM image of GC2 on HOPG and model of observed cyclic structure; (d) STM image of AU2 on HOPG and model of observed cyclic structure.
Reproduced with permission from N. Bilbao, I. Destoop, S. De Feyter and D. González-Rodríguez, Two-Dimensional Nanoporous Networks Formed by Liquid-to-Solid Transfer of Hydrogen-Bonded Macrocycles Built from DNA Bases, *Angew. Chem., Int. Ed.*, 2016, **55**, 659–663. Copyright 2016 Wiley-VCH Verlag GmbH and Co. KGaA.

The large pores formed by the network, with a cross section of 2.4 nm, are capable of housing several large molecules and this was initially demonstrated by the introduction of C_{60}[47] into the framework pores (Figure 1.7). The deposition of C_{60} by sublimation onto the hexagonal network leads to the formation of heptameric C_{60} clusters within the pores of the network, with a compact hexagonal arrangement of fullerenes aligned parallel to the principle axes of the Ag/Si(111) surface. The arrangement and properties of the C_{60} clusters were found to differ from close-packed fullerene on the same surface, which does not align with the principle axes of the underlying Ag/Si(111) substrate or form heptamers, thus demonstrating the templating and stabilising effects of the network.[47] Indeed the ability of the

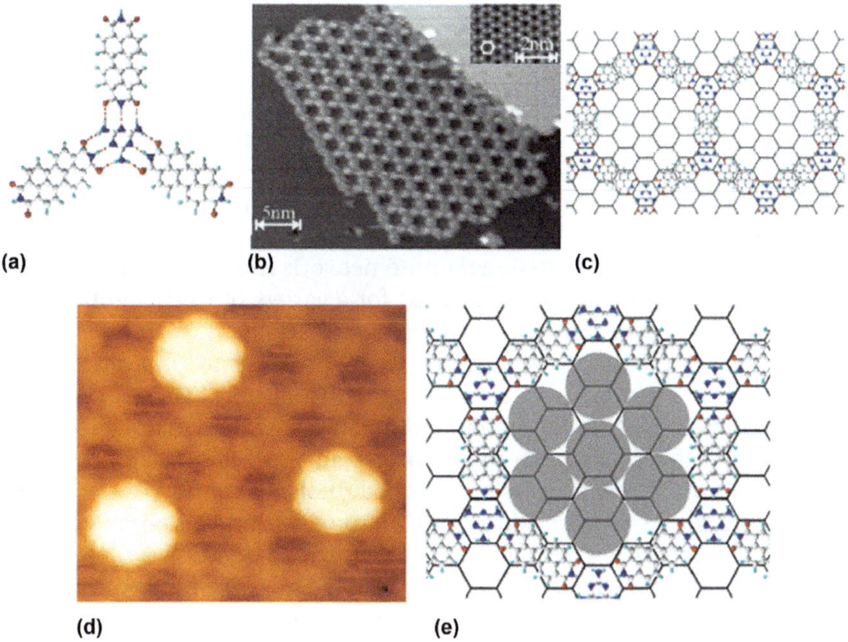

Figure 1.7 (a) A schematic of the PTCDI-melamine junction, showing the 9 hydrogen bonds that make up the structural node as dashed red lines; (b) STM image of the PTCDI-melamine network on Ag/Si(111); inset, high resolution view of the Ag/Si(111) surface; (c) schematic of the network showing the registry with the hexagonal substrate; (d) STM image of fullerenes trapped within the pores of the hexagonal network, seen as bright white features; (e) a schematic diagram of a C_{60} heptamer sitting within a pore.
Reproduced with permission from J. A. Theobald, N. S. Oxtoby, M. A. Phillips, N. R. Champness and P. H. Beton, Controlling molecular deposition and layer structure with supramolecular surface assemblies, *Nature*, 2003, **424**, 1029–1031, Nature Publishing Group.

PTCDI-melamine family of arrays has been shown to host other fullerenes, including C_{84}[48] and Lu@C_{82},[49] and a range of molecules, as discussed below.

The PTCDI-melamine network[47] can similarly be prepared on a Au(111) surface[50] leading to an analogous honeycomb arrangement to that observed on a Ag/Si(111) substrate. However, after annealing at higher temperatures, a parallelogram phase was also observed on Au(111), which has the same stoichiometric ratio as the honeycomb structure, but is more dense.[51] The parallelogram phase is also able to trap guest molecules such as C_{60}, but due to the restricted cavity size C_{60} dimers are observed, the size of the cluster being defined by the physical size of the network pores. Similar studies have demonstrated entrapment of Lu@C_{82} by the parallelogram phase but in this case only single fullerenes are accommodated within each pore.[49] The parallelogram PTCDI-melamine phase also traps two decanethiol molecules which sit parallel to the surface under the UHV conditions used.

The entrapment of a significantly different and potentially more reactive molecule confirmed that such networks have the potential for trapping a range of different species. It is interesting to note that adsorption of decanethiol onto the PTCDI-melamine arrays lead to the destruction of the honeycomb phase, but not the parallelogram phase, suggesting that the latter is preferentially stabilised by the thiol guest.[52]

Indeed the adsorption of thiol guests within the pores of PTCDI-melamine networks has been advanced significantly by Buck *et al.*[53,54] It has been demonstrated that the PTCDI-melamine network can be assembled onto a Au(111) surface from solution and used for a variety of further applications and particularly as a versatile patterning tool (Figure 1.8).[53,54] Due to the solution-based preparative procedure it is possible to apply traditional Au-thiolate SAM-synthesis in the presence of the array. The adsorption of adamantanethiol within the pores of the network is demonstrated with the thiolates sitting perpendicular to the surface, in contrast to related studies in UHV conditions.[52] The clusters of adsorbed thiols, or confined SAMs, can be exploited to guide the deposition of Cu atoms within the pores using underpotential deposition (UPD). It was demonstrated that Cu was successfully deposited only on areas of the surface that were not covered by the network, *i.e.* those areas covered by thiols,[53] and a subsequent study demonstrated that the PTCDI-melamine network acts as a diffusion barrier for Cu adatoms.[55]

Thus, molecules can be adsorbed onto hydrogen-bonded arrays by sublimation or solution-phase deposition but these techniques may not always be suitable for sensitive molecules and thus a further technique has

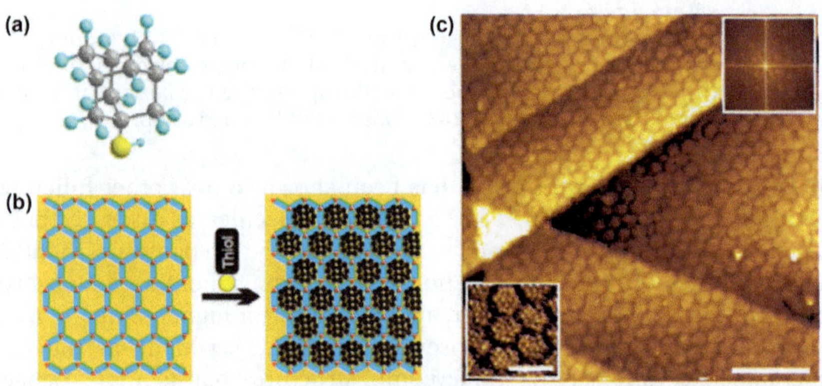

Figure 1.8 (a) The adamantanethiol molecule; (b) schematic of network filling process; (c) STM image of the PTCDI-melamine network on Au(111)/mica filled with adamantanethiol, scale bar 20 nm. Insets: higher resolution STM image and Fourier transforms, scale bar 5 nm. Reproduced with permission from R. Madueno, M. T. Räisänen, C. Silien and M. Buck, Functionalizing hydrogen-bonded surface networks with self-assembled monolayers, *Nature*, 2008, **454**, 618. Copyright 2008 Nature Publishing Group.

been developed using electrospray deposition.[56,57] Electrospray techniques have been used to deposit $Mn_{12}O_{12}(O_2CCH_3)_{16}(H_2O)_4$ clusters, a molecule of interest due to its ability to act as a single molecule magnet, onto the PTCDI-melamine array on a Au(111) substrate.[57] The softer deposition technique is required as sublimation is not possible for $Mn_{12}O_{12}(O_2CCH_3)_{16}(H_2O)_4$ due to its decomposition at elevated temperatures. In the absence of the hydrogen-bonded array the $Mn_{12}O_{12}(O_2CCH_3)_{16}(H_2O)_4$ clusters assemble generating filamentary structures on the surface (Figure 1.9). In contrast adsorption of $Mn_{12}O_{12}(O_2CCH_3)_{16}(H_2O)_4$ clusters onto the PTCDI-melamine array results in a low degree of ordering with some molecules accommodated within the pores of the network but with others positioned on top of the array (Figure 1.9). The entrapment of the $Mn_{12}O_{12}(O_2CCH_3)_{16}(H_2O)_4$ clusters in the array is not as efficient as that of other species, including fullerenes, most probably due to a mismatch of dimensions between the cluster (1.6 nm) and pore (2.5 nm). This study indicates that maximum organisation of guest molecules by porous frameworks can be achieved by idealising size match between host and guest and by maximising interactions between host and guest.

An advantage of the bimolecular approach to array formation is that each component can be altered allowing modification of the resulting structure. Thus, for the PTCDI-melamine arrays the PTCDI molecule[58,59] can be readily functionalised leading to manipulation of the framework and cavity dimensions within the array. Introducing steric bulk to the PTCDI molecules can control the available space for guest entrapment leading to networks that can trap individual molecules potentially in a regular array. Self-assembled arrays formed between melamine and PTCDI derivatives, and in some examples networks formed by the functionalised PTCDI derivatives alone, act as hosts to guest molecules. Whereas the honeycomb array formed by 1,7-dibromo-PTCDI and melamine on Ag/Si(111) accommodates C_{60} heptamers, similar to the parent unfunctionalized PTCDI-melamine array, a thiopropyl-functionalised PTCDI—$(SPr)_2$-PTCDI—network formed in combination with melamine does not trap heptamers or hexamers.[58] Although some fullerenes can be observed in STM images following deposition, the C_{60} molecules are positioned in an irregular fashion both within pores and also on top of the honeycomb array, similar to $Mn_{12}O_{12}(O_2CCH_3)_{16}(H_2O)_4$ clusters (see above) indicating that the thiopropyl groups act to inhibit fullerene adsorption by this network.

Self-assembly of thioadmantyl functionalised PTCDI, $(SAdam)_2$-PTCDI, and melamine leads to the anticipated honeycomb array.[59] It should be noted that the thioadmantyl groups are both considerably bulkier and more rigid than the thiopropyl appendages of $(SPr)_2$-PTCDI and therefore would be expected to provide an effective mechanism for inhibiting the space within the network pores. However, upon formation of a $(SAdam)_2$-PTCDI-melamine network some of the thioadamantyl groups are cleaved from the PTCDI moieties.[59] Consequently a variety of distinct pore sizes and configurations are formed which can be visualised following C_{60} adsorption

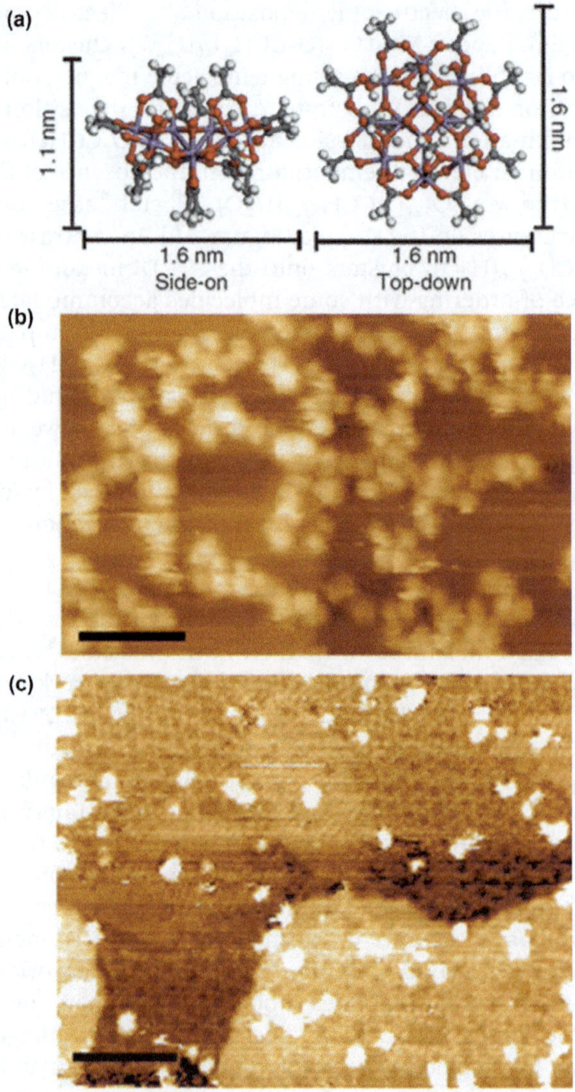

Figure 1.9 (a) Schematic representation of $Mn_{12}O_{12}(O_2CCH_3)_{16}(H_2O)_4$; STM images showing (b) the filamentary structures formed by molecular aggregates comprised of individual $Mn_{12}O_{12}(O_2CCH_3)_{16}(H_2O)_4$ molecules. Scale bar, 10 nm; (c) $Mn_{12}O_{12}(O_2CCH_3)_{16}(H_2O)_4$ molecules deposited onto a PTCDI-melamine network on a Au(111) surface. Scale bar, 20 nm. Reproduced with permission from A. Saywell, G. Magnano, C. J. Satterley, L. M. A. Perdigão, A. J. Britton, N. Taleb, M. C. Giménez-López, N. R. Champness, J. N. O'Shea and P. H. Beton, Self-assembled aggregates formed by single-molecule magnets on a gold surface, *Nat. Commun.*, 2010, **1**, 75. Nature Publishing Group.

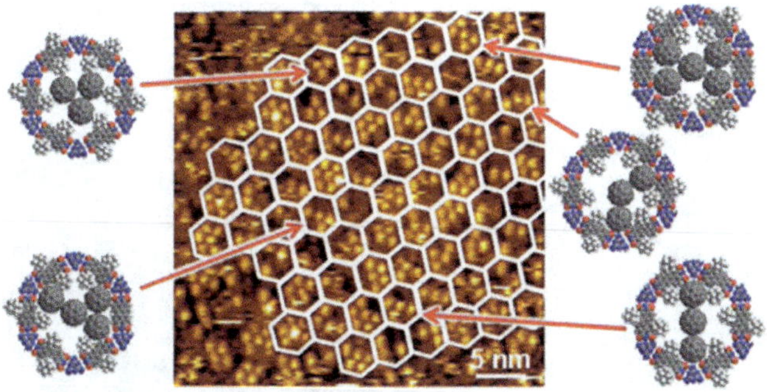

Figure 1.10 C_{60} entrapment in a $(SAdam)_2$-PTCDI-melamine network on Au(111), the honeycomb network is indicated as a guide to the eye. Schematic figures illustrate and identify the different arrangements of C_{60} within the pores of the structure which arise due to missing adamantyl units. Reproduced from ref. 59 with permission from the Royal Society of Chemistry.

onto the network and subsequent STM imaging. Individual C_{60} molecules are more readily imaged by STM than the underlying thioadamantyl groups. Using this approach, it is possible to identify the different orientations of the molecular clusters within each pore (Figure 1.10). The number of C_{60} molecules adsorbed within a given pore is determined by the degree of thioadamantyl cleavage and clusters ranging from dimers to heptamers are observed. Considering the dimensions of the honeycomb pores and the size of the thioadamantyl appendages, clusters of greater than five fullerenes can only be explained by cleavage of the thioadamantyl groups, the heptamers corresponding to a pore where no thioadamantyl appendages remain.

Further constriction of pore size can be used to control the number of guest molecules, a principal demonstrated by the array formed by 1,7-bis(4-benzoic acid)-3,4,9,10-perylenetetracarboxylic diimide (Bz_2-PTCDI).[60] As with many disubstituted PTCDI species,[46] Bz_2-PTCDI assembles into a unimolecular honeycomb array, without the need to add melamine, through the adoption of a tri-molecular hydrogen-bonded junction (see below). The rigid protrusion of the phenylcarboxylate moieties into the framework cavities leads to the encapsulation of single C_{60} molecules within each pore, which are in turn spaced in a regular fashion across the surface (Figure 1.11).

The material employed as substrate for the self-assembly process has a clear influence on the resulting framework structure and thus cannot be considered an innocent agent in the synthetic strategy. An interesting example that demonstrates the influence of the substrate on supramolecular assembly is the use of graphene or boron nitride (BN) 'nanomesh' monolayers grown on Rh(111) crystals[61] that exhibit moiré patterns introducing distinct adsorption sites on the surface. The self-assembly of PTCDI and related di-functionalised

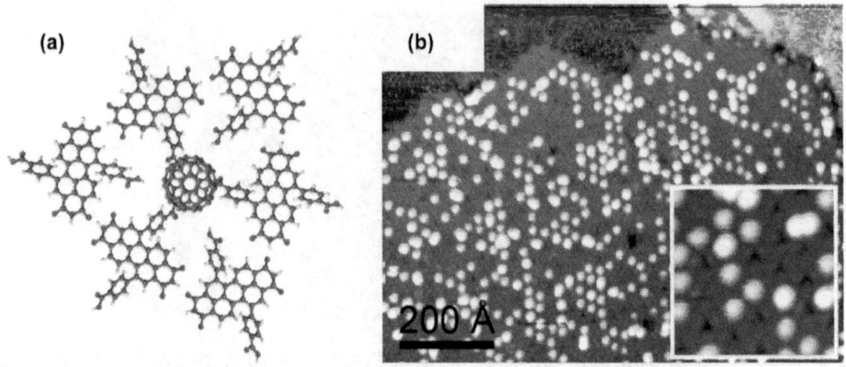

Figure 1.11 (a) Model and (b) STM image of a self-assembled network of 1,7-bis(4-benzoic acid)-3,4,9,10-perylenetetracarboxylic diimide encapsulating regularly spaced individual C_{60} molecules.
Reproduced from ref. 60 with permission from the Royal Society of Chemistry.

derivatives, 1,7-dipropylthio-perylene-3,4,9,10-tetracarboxydiimide (DP-PTCDI) and 1,7-di(butyl)-coronene-3,4,9,10-tetracarboxylic acid bisimide (DB-CTCDI), have been studied on graphene and the structure of the resulting structures determined using STM (Figure 1.12).[61] Two distinct junctions, formed by either two or three molecules are observed leading to different assemblies (Figure 1.12). For example, unfunctionalised PTCDI forms rows up to 25 nm in length running parallel to the principal directions of the substrate and adopting only simple dimeric hydrogen-bonding arrangements. The molecular arrangement on the graphene superstructure differs significantly from that observed for a graphite substrate, on which three-dimensional islands are formed.[62] The importance of a commensurate match between molecular dimensions and the moiré periodicity is demonstrated by a comparison with adsorption of PTCDI on a BN "nanomesh" monolayer (Figure 1.12). The BN monolayer is isoelectronic with graphene, and also displays a moiré pattern on a Rh(111) crystal, but with a slightly larger periodicity of 3.2 nm, which is non-commensurate with the dimensions of PTCDI. On BN [on Rh(111)] individual isolated PTCDI molecules trapped in energy minima associated with the moiré pattern are observed demonstrating the influence of the underlying substrate on supramolecular assembly. Both DP-PTCDI and DB-CTCDI form both linear rows, built using dimeric hydrogen bond junctions, and more complex arrays which include junctions where three molecules meet in a hydrogen-bonded trimeric arrangement. In the case of DP-PTCDI the ratio of dimer : trimer junctions is 75 : 25 in comparison to less than 1% of junctions being trimers for PTCDI. In the case of DB-CTCDI the three molecule junctions dominate the self-assembly process, with no linear dimers unambiguously identified. The array constructed from trimeric junctions results in a honeycomb framework aligned with the graphene monolayer superstructure and encloses the areas of bright contrast arising from the moiré pattern.

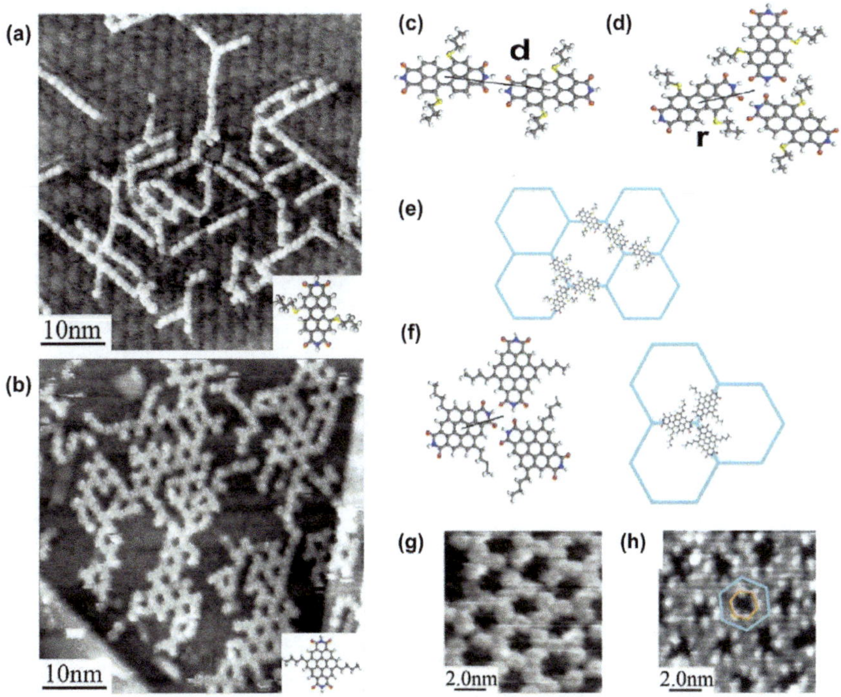

Figure 1.12 STM images acquired following deposition of: (a) DP-PTCDI, (b) DB-CTCDI on a graphene monolayer formed on Rh(111); diagram of junctions between DP-PTCDI dimers (c) and trimers (d) stabilized, respectively, by two and three C=O···NH hydrogen bonds between neighbouring molecules with dimer centre–centre spacing of d and trimer vertex to molecule centre spacing r; (e) placement of DP-PTCDI trimers and dimers; (f) DB-CTCDI trimer junction analogous to (d) with vertex to molecule centre spacing r and placement of DB-CTCDI trimer on the graphene superstructure; (g), (h) STM images of DB-CTCDI showing chirality of junctions and intramolecular detail of molecules. The hexagons in (h) highlight the chirality of the molecular arrangement.
Reproduced with permission from A. J. Pollard, E. W. Perkins, N. A. Smith, A. Saywell, G. Goretzki, A. G. Phillips, S. P. Argent, H. Sachdev, F. Müller, S. Hüfner, S. Gsell, M. Fischer, M. Schreck, J. Osterwalder, T. Greber, S. Berner, N. R. Champness and P. H. Beton, Supramolecular Assemblies Formed on an Epitaxial Graphene Superstructure, *Angew. Chem., Int. Ed.*, 2010, **49**, 1794–1799. Copyright 2010 Wiley-VCH Verlag GmbH and Co. KGaA.

The hydrogen-bonding synthon exploited in the PTCDI-melamine arrays can be readily incorporated into other building blocks (*e.g.* thymine includes the same imide moiety seen in PTCDI) and by so doing many possible arrays can be targeted. Perhaps the simplest example is the self-assembled array formed between cyanuric acid and melamine which has been extensively studied.[63,64] Samorì *et al.*[65] have elegantly demonstrated the

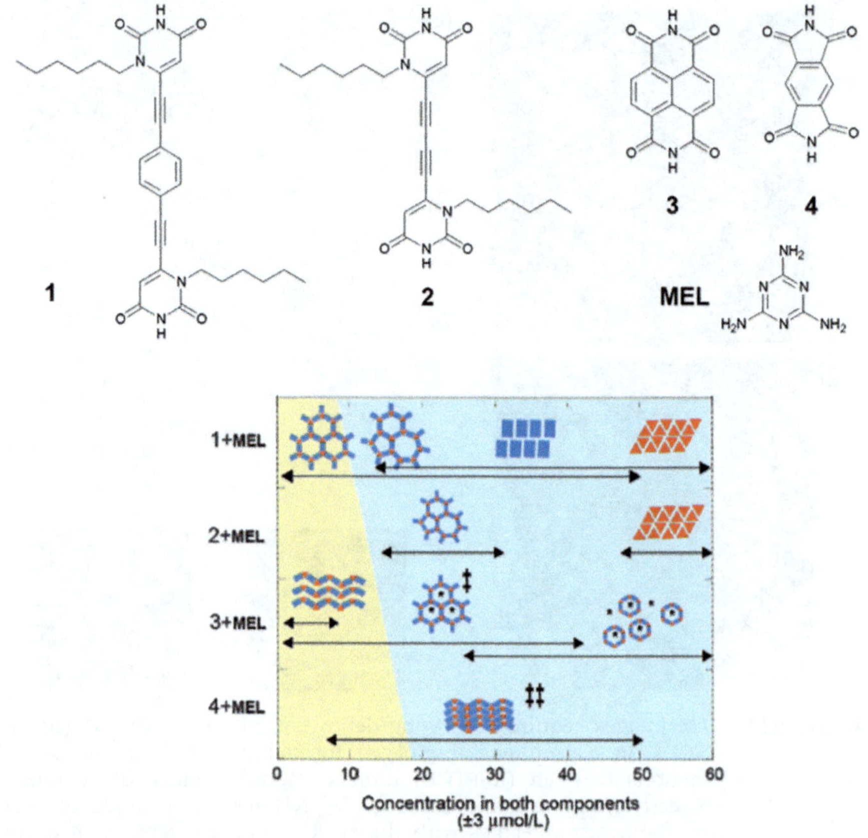

Figure 1.13 Diimide molecules used by Samorì *et al.* in combination with melamine (MEL) to afford self-assembled structures. The nature of the structure is concentration dependent giving rise to a range of polymorphic products (bottom).
Reproduced with permission from C.-A. Palma, J. Bjork, M. Bonini, M. S. Dyer, A. Llanes-Pallas, D. Bonifazi, M. Persson and P. Samorì, Tailoring Bicomponent Supramolecular Nanoporous Networks: Phase Segregation, Polymorphism, and Glasses at the Solid-Liquid Interface, *J. Am. Chem. Soc.*, 2009, **131**, 13062. Copyright 2009 American Chemical Society.

self-assembly of a range of diimide molecules and melamine (Figure 1.13) at the solution/solid interface on a HOPG surface following deposition from a 1,2,4-trichlorobenzene–dimethylsulfoxide solution. The formation of hexagonal, or honeycomb networks is anticipated for such systems in an analogous fashion to the PTCDI-melamine systems, and such arrays are observed, but detailed studies of concentration variation leads to the adoption of a range of structures. The nature of the structures formed by combinations of the diimide molecules (labelled 1–4 in Figure 1.13) with melamine ranges from open hexagonal networks to close-packed structures and significantly, varies with component concentration. Ultimately a phase diagram of polymorphs is generated for these systems which gives perspective to the complexity of the self-assembly processes involved (Figure 1.13).

1.3 Two-dimensional Arrays Assembled Using van der Waals Interactions

The role of van der Waals interactions in the creation of self-assembled surface frameworks cannot be understated. Naturally such interactions are important in all such systems and play a highly significant role in the interaction between molecules and substrates. A large number of studies have reported using van der Waals interactions to create supramolecular structures and these have been reviewed on a number of occasions.[7–9,66,67] A particularly notable interaction that has been widely exploited is the strong adsorption between alkyl chains and HOPG. The role of the alkyl chain is clearly significant in the resulting assembly process, and interdigitation of alkyl chains from adjacent molecules is commonly observed.

A representative example of a self-assembled framework controlled by van der Waals interactions was investigated by Schull *et al.*[68] 1,3,5-tris[(*E*)-2-(3,5-didecyloxyphenyl)-ethenyl]-benzene (TSB35), a trimeric compound with six pendent decyl chains was deposited onto an HOPG substrate. Self-assembly through interdigitation of the alkyl chains leads to formation of a porous network within which coronene and hexabenzocoronene could be accommodated (Figure 1.14). It can be clearly seen that the co-adsorption of coronene or hexabenzocoronene modifies the nature of the resulting structure, in a similar fashion to solution-phase host–guest chemistry. Indeed the interdigitation of alkyl chains has been extensively used by de Feyter and co-workers in a number of elegant studies.[7–9,66,67,69–74] One avenue that has been a particular focus of research has been the study of alkoxylated dehydrobenzo[12]annulenes.[69] A representative study of rhombic-shaped dehydrobenzo[12]annulenes assembled at a 1,2,4-trichlorobenzene/HOPG interface[69] demonstrates that variation in the chain length of alkyl substituents results in five distinct network structures, three porous and two nonporous structures.[69] Shorter alkyl chain substituents on the dehydrobenzo[12]annulene core favours formation of porous frameworks, whereas those with longer alkyl chains tend towards nonporous arrangements. Concentration of the building blocks in solution also plays a key role in the nature of the structures formed, thus, dilution of the solutions of dehydrobenzo[12]annulene leads to the transformation of nonporous structures into porous networks. Favouring of porous structures at lower concentrations is a result of factors related to overall surface coverage and, in solution phase studies, it is also possible that porous structures can be stabilized by the co-adsorption of solvent molecules.

In order to establish control over the self-assembly process researchers have designed building blocks that combine more than one type of intermolecular interaction, for example hydrogen-bonding and van der Waals interactions.

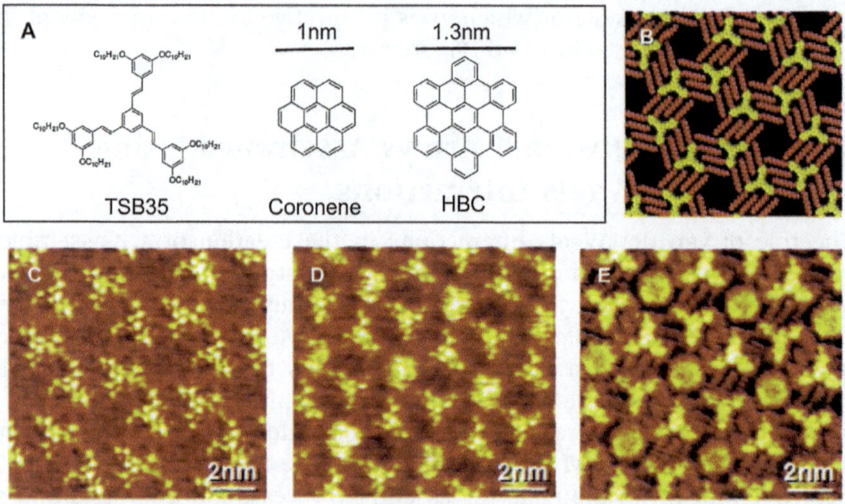

Figure 1.14 Networks formed from TSB35, 1,3,5-tris[(*E*)-2-(3,5-didecyloxyphenyl)-ethenyl]-benzene. (a) Molecular structures of the three compounds; (b) graphical representation of the resulting network; (c) STM image of the monolayer matrix; (d) after addition of coronene; and (e) after addition of hexabenzocoronene.
Reproduced with permission from G. Schull, L. Douillard, C. Fiorini-Debuisschert and F. Charra, Single-Molecule Dynamics in a Self-Assembled 2D Molecular Sieve, *Nano Lett.*, 2006, **6**, 1360. Copyright 2006 American Chemical Society.

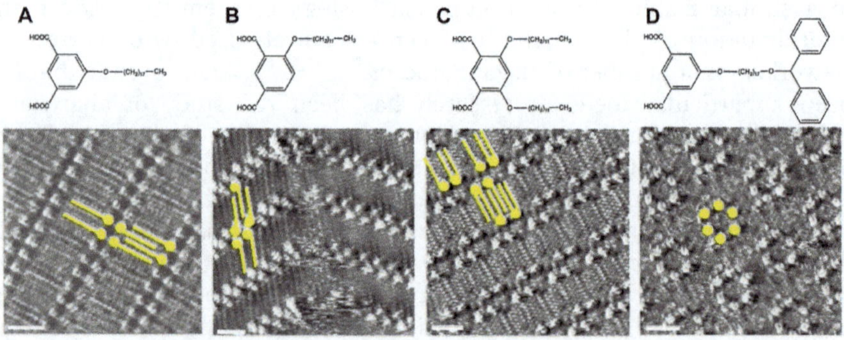

Figure 1.15 STM images of isophthalic acid derivatives A, B, C and D on graphite. The yellow discs represent isophthalic acid groups and lines, alkyl chains.
Reproduced from ref. 8 with permission from the Royal Society of Chemistry.

This approach has been widely used and examples include N^9-alkylated guanine derivatives,[37,75] and 5-alkoxyisophthalic acid derivatives on HOPG.[7,8,72] For the latter example, 5-alkoxyisophthalic acids, monolayers of close-packed arrays of interdigitating hydrogen-bonding ribbons are formed (Figure 1.15). Thus, the isophthalic acid groups form hydrogen-bonded ribbons whilst the alkyl groups interdigitate, acting as spacers between adjacent ribbons. If larger groups are appended to the alkyl chain, such as diphenylmethyl moieties, then the hydrogen-bonding observed in the ribbons is disrupted so that hexameric rings are formed (Figure 1.15).

A remarkable recent study by de Feyter[76] describes an approach to establish the range of polymorphs for a given system using building blocks that exploit van der Waals interactions or hydrogen-bonding interactions. The formation of polymorphs in two-dimensional supramolecular arrays is one of the biggest challenges in the research field as polymorphic arrangements always provide a potential barrier to the successful preparation of well-defined, ordered arrays. De Feyter *et al.* approached the question of polymorph evaluation by designing a mechanism to induce a solution flow across a substrate surface generating a lateral density gradient. This *in situ* generation of a gradient allows both discovery and separation of multiple polymorphs in a single experiment. The authors describe three separate systems: hexadecyloxy substituted dehydrobenzo[12]-annulene (DBA-OC16), hexadecyl substituted bis(dehydrobenzo[12] annulene) (bisDBA-C16) and 1,3,5-tris(4-carboxyphenyl)benzene (BTB) allowing demonstration of their approach with systems that exploit either van der Waals or hydrogen-bonding interactions. In each instance, more than one polymorph was found for each system by STM imaging of the respective samples: DBA-OC16 – 2 polymorphs; bisDBA-C16 – 4 polymorphs; BTB – 2 polymorphs (Figure 1.16). The approach allows not only identification and structural characterisation of the polymorphs observed but also quantification of surface coverage of each arrangement.

1.4 Covalently Coupled Two-dimensional Arrays: Planar Covalent Organic Frameworks

Both hydrogen-bonding and van der Waals interactions are excellent for preparing well-ordered structures due to the ability of these relatively weak interactions to be reversibly formed, allowing structural correction. However, to form robust two-dimensional structures an approach that slows the formation of covalently coupled arrays is the most desirable target. A number of approaches have been used to target such systems, including aryl–aryl coupling[77–80] and alkyne–alkyne coupling[81] processes. Other approaches have been taken in recent studies[82] and the field has clear parallels to studies of covalent organic frameworks or COFs.[15]

One of the earliest approaches taken to the development of covalently coupled arrays was the use of the coupling between halo-aryl species on Au surfaces. Grill *et al.*[77] elegantly demonstrated the use of porphyrin building blocks with bromophenyl 'arms' protruding from the *meso*-positions of the porphyrin.[77] The number of bromophenyl arms can be varied such that a range of building blocks with one, two or three reactive appendages can be utilised. The reaction of mono-bromophenyl substituted porphyrin, by heating the building blocks on a Au(111) substrate, leads to dimer formation

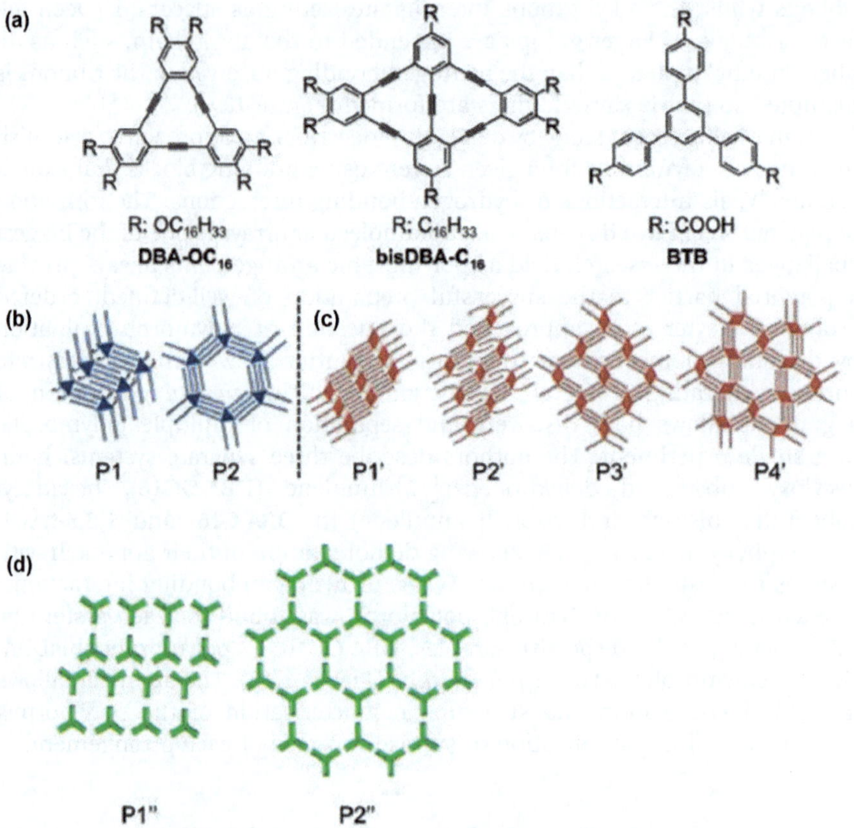

Figure 1.16 (a) Molecular structures of DBA-OC16, bisDBA-C16 and BTB and representations of the polymorphs observed for each compound (b) DBA-OC16; (c) bisDBA-C16; (d) BTB.
Reproduced from ref. 76 with permission from the Royal Society of Chemistry.

(Figure 1.17a) and *trans*-di-bromophenyl substituted porphyrin reacts to give chains (Figure 1.17b). Lastly tetra-bromophenyl substituted porphyrin can react in all four positions to give well-ordered two-dimensional grids of porphyrins (Figure 1.17c).

The approach was developed further by Grill, Hecht and co-workers through the construction of covalently coupled arrays built from two different molecular components. The strategy exploits the different reactivity of bromophenyl and iodo-phenyl groups attached to a porphyrin core in a *trans* arrangement leading to different reactivity between the different faces of the porphyrin molecules.[78,79] Deposition of the dibromo-diiodo-porphyrin (Br_2I_2TPP) building-block onto a Au(111) substrate followed by heating the sample leads to the covalent coupling between adjacent porphyrin species. The lower bond dissociation energy of the C–I bonds, in comparison to C–Br bonds, allows preferential coupling of the iodo-functionalised arms at a lower

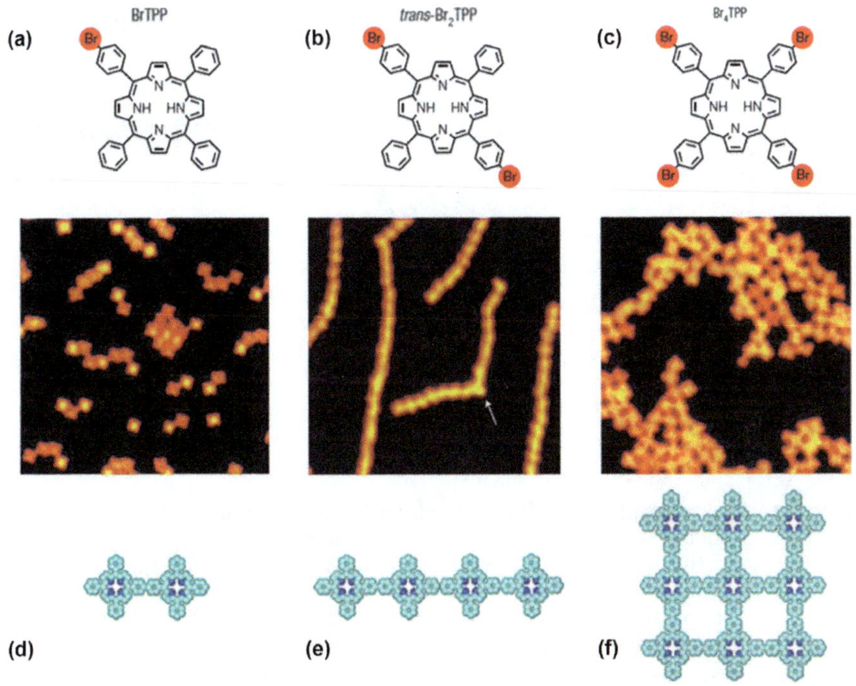

Figure 1.17 (a–c) Bromophenyl functionalised porphyrin building blocks with increasing numbers of functional moieties for subsequent polymerisation. STM images and representative models of (d) dimers prepared from mono-bromophenyl functionalised building-block BrTPP; (e) one-dimensional chains prepared from *trans*-Br_2TPP. The arrow indicates where two chains are held together by weak non-covalent interactions; (f) two-dimensional sheets prepared from covalently couple Br_4TPP. Reproduced with permission from L. Grill, M. Dyer, L. Lafferentz, M. Persson, M. V. Peters and S. Hecht, Nano-architectures by covalent assembly of molecular building blocks, *Nat. Nanotechnol.*, 2007, **2**, 687. Copyright 2007 Nature Publishing Group.

temperature, leading to the formation of one-dimensional chains of coupled porphyrin building blocks (Figure 1.18). Below 200 °C only the iodo-functionalised arms react leaving the bromo substituents intact, allowing the possibility of further reaction. Above 200 °C the bromo-functionalised arms can react leading to coupling of the one-dimensional chains forming two-dimensional porphyrin islands, mirroring the previously reported observations (see above).[77] However, following formation of the one-dimensional chains of coupled porphyrin building blocks, which are functionalised with bromophenyl appendages, reaction with a secondary species can be successfully achieved. The authors demonstrated the success of their approach by co-deposition of dibromoterfluorene which can cross-couple with the bromo-functionalised one-dimensional porphyrin array. Thus, coupling of the bromo-appendages on the porphyrin chain and the terfluorene molecules

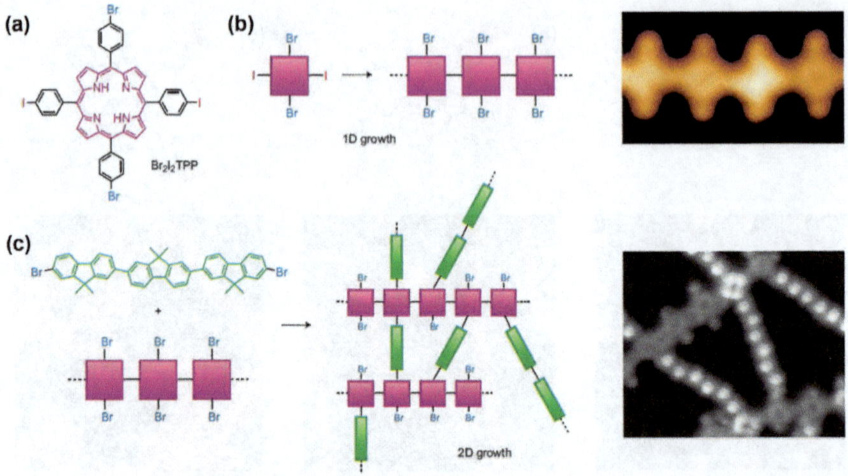

Figure 1.18 (a) A porphyrin building-block with two bromophenyl and two iodophenyl appendages in a *trans*-arrangement; (b) aryl–aryl coupling leads to the formation of one-dimensional chains; (c) subsequent reaction with dibromoterfluorene leads to the formation of a mixed, covalently coupled framework structure. The STM images confirm the molecular arrangement within the covalently coupled structure.
Reproduced with permission from N. R. Champness, Surface Chemistry: Making the right connections, *Nat. Chem.*, 2012, **4**, 149–150. Copyright 2012 Nature Publishing Group.

leads to decoration of the porphyrin chains with terfluorene groups and in some instances cross-linking of adjacent porphyrin chains is observed, leading to a complex two-dimensional array of coupled porphyrin chains and terfluorene moieties (Figure 1.18). The simple strategy of hierarchical coupling, exploiting dissimilar chemical reactivity and thermal activation barriers for the reactive components, leads to the formation of a bimolecular array and opens up the possibility for the creation of more complex covalently coupled molecular arrays.

It has also been demonstrated that it is possible to transfer such porphyrin-based covalently coupled arrays from a gold thin film deposited on mica to other surfaces using a C_{60} adhesion layer.[83] It was demonstrated that a thin film of C_{60}, with a thickness ranging from 10 to 100 nm, can promote adhesion between a thin film of Au and a solution-deposited layer of polymethyldisolaxane (PDMS). The adhesion facilitated by the C_{60} thin film allows removal of the gold film from the mica support by simple peeling. Following removal from the mica support the monolayers may be isolated and transferred to other substrates, *e.g.* SiO_2, by etching of the thin film of gold, mechanical transfer and removal of the fullerene layer by annealing and then dissolution.

The strength of the strategy of using covalent coupling to form surface-based arrays is demonstrated by a recent study by Amabilino and Raval.[84]

The study reports the coupling of polyaromatic molecules, (pentacene, tetramesitylporphyrin, perylene) and porphyrins (H_2-porphyrin and Zn(II)-diphenylporphyrin), through C–H bond activation reactions on a copper surface. It is demonstrated that all of the molecules are capable of undergoing C–H activation and can be coupled on the surface chosen. The study not only demonstrated the ability to homo-couple the molecules into extended arrays, *e.g.* Zn(II)diphenylporphyrin can be cross-coupled through the unsubstituted edges (*i.e.* not supporting phenyl groups) to afford covalently coupled one-dimensional chains of porphyrins, but also hetero-coupling between the different building blocks. A large variety of coupled structures are described (Figure 1.19) with up to three different building blocks

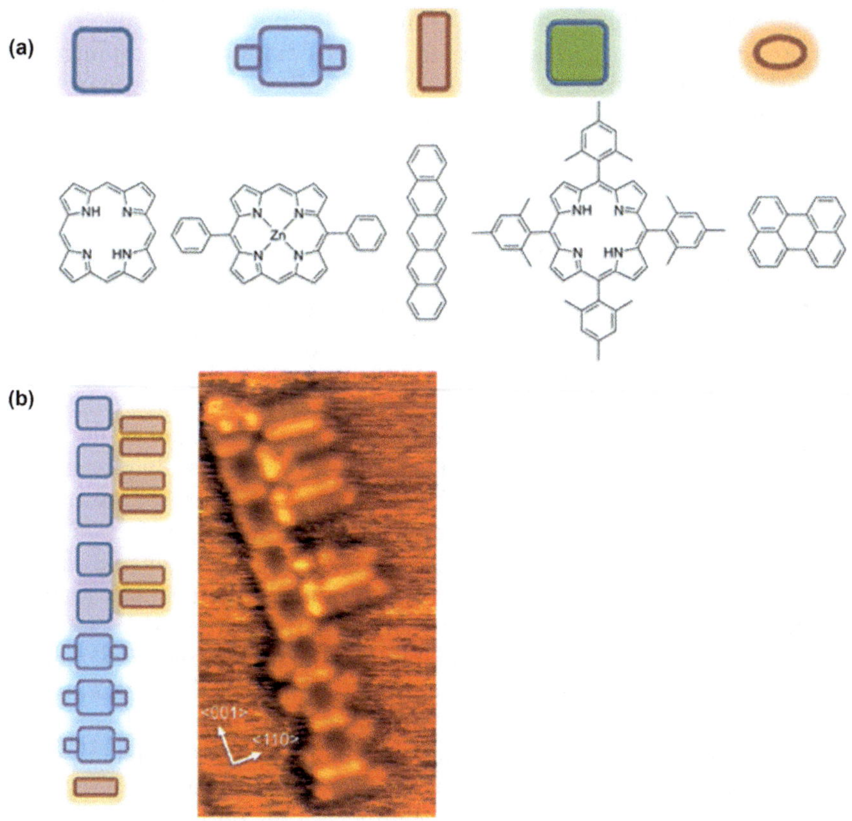

Figure 1.19 (a) Building blocks capable of undergoing C–H activation on a Cu(110) surface and (b) STM image demonstrating the covalent coupling of three distinct components.
Reproduced with permission from S. Haq, F. Hanke, J. Sharp, M. Persson, D. B. Amabilino and R. Raval, Versatile Bottom-Up Construction of Diverse Macromolecules on a Surface Observed by Scanning Tunneling Microscopy, *ACS Nano*, 2014, **8**, 8856–8870. Copyright 2014 American Chemical Society.

cross-coupled. STM studies provide sufficient insight to allow characterisation of each individual component of the covalently coupled structure. In summary, this study demonstrates the ability to prepare highly complex covalently coupled structures from multiple building blocks and indicates the potential for creating bespoke covalent structures on surfaces.

1.5 Complex Structures with Low Degrees of Symmetry

One of the most exciting features of creating surface-based supramolecular frameworks is the ability to image the resulting structures at the molecular and sub-molecular[22,28] level. This allows a level of detailed understanding that is extremely challenging, if not impossible, in solid-state materials due to the nature of the characterisation that is employed – scanning probe microscopies for surface-based structures *vs.* diffraction techniques in the solid-state. The additional level of detail that can be achieved using STM, or AFM, has allowed characterisation of structures that would prove extremely challenging by any other technique.

One of the first examples of such a structure is that formed by terphenyl-3,3″,5,5″-tetracarboxylic acid.[85] Terphenyl-3,3″,5,5″-tetracarboxylic acid, adsorbed onto HOPG, leads to the formation of a two-dimensional hydrogen-bonded structure that utilizes $R^2_2(8)$ intermolecular carboxylic acid–carboxylic acid interactions. STM imaging allows identification of the position of each molecule in the extended structure and direct visualisation of a non-ordered structure. Indeed, the relative position of molecules within the array is random and reminiscent of dynamically-arrested systems, such as glasses (Figure 1.20). The intermolecular hydrogen-bonding leads to the formation of hexagonal junctions which are formed from three, four, five, or six molecules as a result of the dimensions of the molecule. As such the structure forms an extremely rare example of a random, entropically stabilised, rhombus tiling. It is important to note that the detailed structure of this framework, and other related frameworks, can only be appreciated with a molecular level understanding and that this can only be readily achieved by scanning probe microscopy. The degree of randomness of the rhombus tiling can be evaluated by detailed analysis of each image and subsequent studies demonstrated that both solvent used for deposition and temperature of experiment can affect how random the tiling structure is.[86] It is important to note that the rhombus tiling arrangement is only observed for molecules which have the appropriate dimensions, *i.e.* those which are rhombus shaped, and analogous molecules such as quaterphenyl-3,3‴,5,5‴-tetracarboxylic acid form regular two-dimensional structures.[87]

The random rhombus tiling structure formed by terphenyl-3,3″,5,5″-tetracarboxylic acid on HOPG is able to trap guest molecules.[88] C_{60} can be deposited onto the random array with preferential adsorption into one of the five possible pores, each pore is defined by different numbers and

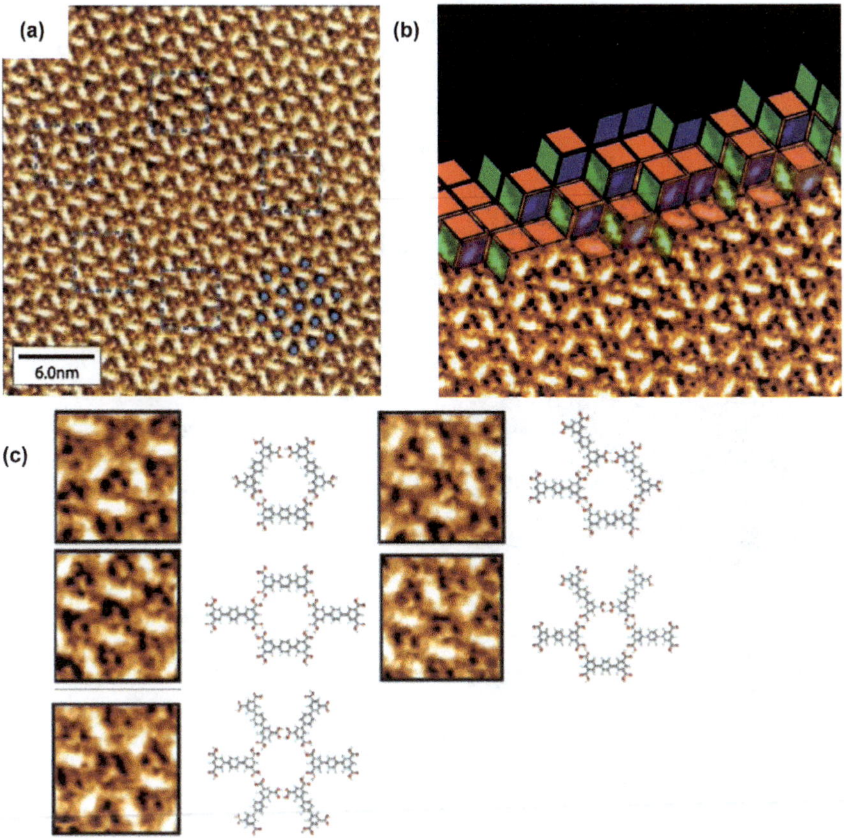

Figure 1.20 (a) STM image of a typical area of terphenyl-3,3″,5,5″-tetracarboxylic acid network at the nonanoic acid/HOPG interface. The group of three phenyl rings of the molecule backbone appear as bright features in the image. The hexagonal orientational order of the structure is indicated by the group of blue dots in the lower right-hand corner of the image, marking the location of pores in the network; (b) illustration of how the molecular arrangement, and each molecule, maps onto a rhombus tiling; (c) diagrams representing the five possible arrangements of terphenyl-3,3″,5,5″-tetracarboxylic acid molecules around a network pore, accompanied by magnified STM image examples of each pore type. The locations of the magnified regions are marked in (a) by blue dashed squares.
Reproduced from M. O. Blunt, J. Russell, M. C. Giménez-López, J. P. Garrahan, X. Lin, M. Schröder, N. R. Champness and P. H. Beton, Random Tiling and Topological Defects in a Two-Dimensional Molecular Network, *Science*, 2008, **322**, 1077–1081 with permission from the AAAS.

arrangements of molecules that define the periphery of the pore. The images reveal a strong preference for adsorption in pores of type A (Figure 1.21), which make up approximately 40% of pores in the initial framework but *ca.* 76% of pores that trap a C_{60} molecule. Calculations confirm that adsorption

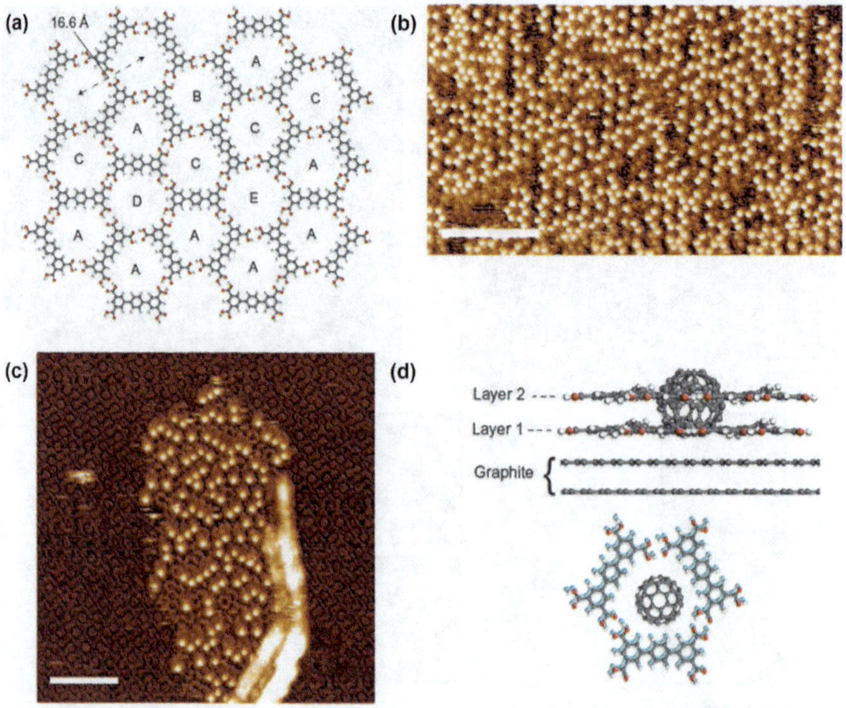

Figure 1.21 (a) Schematic representation of a section of the random rhombus tiling adopted by the terphenyl-3,3″,5,5″-tetracarboxylic acid framework that highlights the hexagonally ordered network of pores, A–E; (b) STM image of an area of terphenyl-3,3″,5,5″-tetracarboxylic acid framework *ca.* 24 hours after deposition of C_{60}. The locations of C_{60} are clearly visible as the bright spots in the image; the underlying terphenyl-3,3″,5,5″-tetracarboxylic acid framework structure is not visible. Scale bar = 160 Å; (c) STM image of terphenyl-3,3″,5,5″-tetracarboxylic acid framework immediately after C_{60} deposition. An island of C_{60} and bilayer terphenyl-3,3″,5,5″-tetracarboxylic acid framework grows away from a surface defect. The initial layer of terphenyl-3,3″,5,5″-tetracarboxylic acid framework is visible with an altered contrast and the terphenyl-3,3″,5,5″-tetracarboxylic acid molecules in the second layer appear with the long axis of the molecules as bright, rod-like features. Scale bar = 110 Å; (d) side-view of the C_{60}–bilayer network that consists of two overlying pores of type A and a view perpendicular to the surface plane with the C_{60} placed at its minimum-energy position for both the first layer (light blue) and second layer (grey) pores. The two layers are displaced slightly with respect to each other, which aids clarity, but in addition is expected on the basis of calculations.
Reproduced with permission from M. O. Blunt, J. C. Russell, M. C. Giménez-López, N. Taleb, X. Lin, M. Schröder, N. R. Champness and P. H. Beton, Guest-induced growth of a surface-based supramolecular bilayer, *Nat. Chem.*, 2011, **3**, 74–78. Nature Publishing Group.

in pore type A is favoured because of the higher proportion of phenyl edges in this configuration. Intriguingly images show that in regions of the network where C_{60} molecules are adsorbed a second framework layer is assembled sitting over the initial layer, thus forming a bilayer structure (Figure 1.21). This was the first time that growth of a bilayer has been observed in such assemblies. The spherical C_{60} molecule acts as a template for the formation of the second layer, acting as an anchor that allows the hydrogen-bonded structure to assemble. The importance of the trapped C_{60} molecules to the formation of the second layer is demonstrated by the addition of coronene to bilayer samples which results in displacement of the entrapped C_{60} by the flat, and more strongly adsorbed, coronene molecule and concomitant removal of the second supramolecular layer.

The rhombus tiling observed with terphenyl-3,3″,5,5″-tetracarboxylic acid is related to a Penrose tiling.[89] The first observation of a structure more directly related to such a tiling was reported for the self-assembled structure formed by ferrocenecarboxylic acid [Fc(COOH)] on a Au(111) substrate.[90] Penrose tilings are related to quasicrystal structures in that they exhibit long range, non-periodic order, and unusual rotational symmetry. Fc(COOH) assembles through intermolecular hydrogen-bonding to form pentagonal arrangements of molecules (Figure 1.22) with carboxylic acid – carboxylic acid O–H···O interactions being additionally stabilised by C–H···O hydrogen bonds between adjacent molecules. Importantly the cyclic pentamer observed for the surface-based array was found to be more stable than other potential hydrogen-bonding arrangements, such as dimeric systems, by DFT

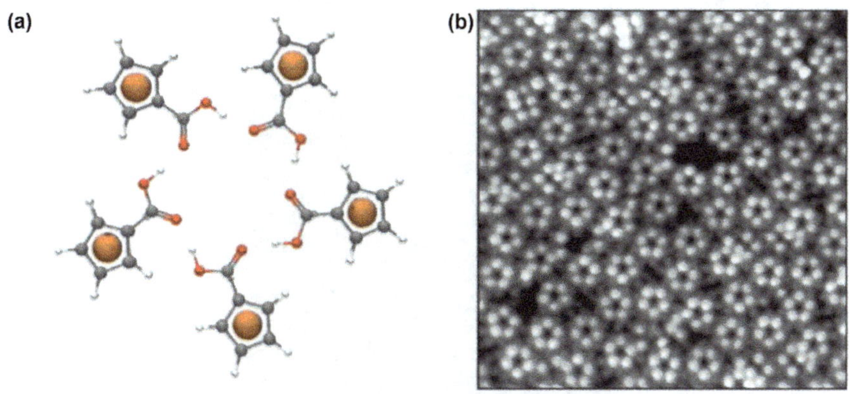

Figure 1.22 (a) Schematic representation of the pentagonal arrangement formed by Fc(COOH) through intermolecular hydrogen-bonding; (b) pentagonal arrangements can be clearly seen in STM images of Fc(COOH) adsorbed on a Au(111) substrate.
Reproduced with permission from N. A. Wasio, R. C. Quardokus, R. P. Forrest, C. S. Lent, S. A. Corcelli, J. A. Christie, K. W. Henderson and S. A. Kandel, Self-assembly of hydrogen-bonded two-dimensional quasicrystals, *Nature*, 2014, **507**, 86–89. Copyright 2014 Nature Publishing Group.

calculations. It was found that a related compound—ferroceneacetic acid [Fc(CH_2COOH)]—does not form the pentagonal arrangements, adopting a more conventional dimeric arrangement. This can be accounted for by the observation that the length of the additional methylene group, in comparison to Fc(COOH), prevents interactions between the carboxyl oxygen of one carboxylic acid and the hydrogens on an adjacent cyclopentadienyl fragment. The pentagonal arrangements observed are related to subunits of the Penrose P1 tiling.[89]

The last example of a highly unusual surface-based structure was reported for the angular molecules 4,4″-dibromo-1,1′:3′,1″-terphenyl (B3PB) and 4,4‴-dibromo-1,1′:3′,1″:4″,1‴-quaterphenyl (B4PB).[91] Both B3PB and B4PB are functionalised on their termini with bromo-aryl moieties which are instrumental in forming intermolecular halogen bonding interactions (Figure 1.23). These interactions result in the formation of extended

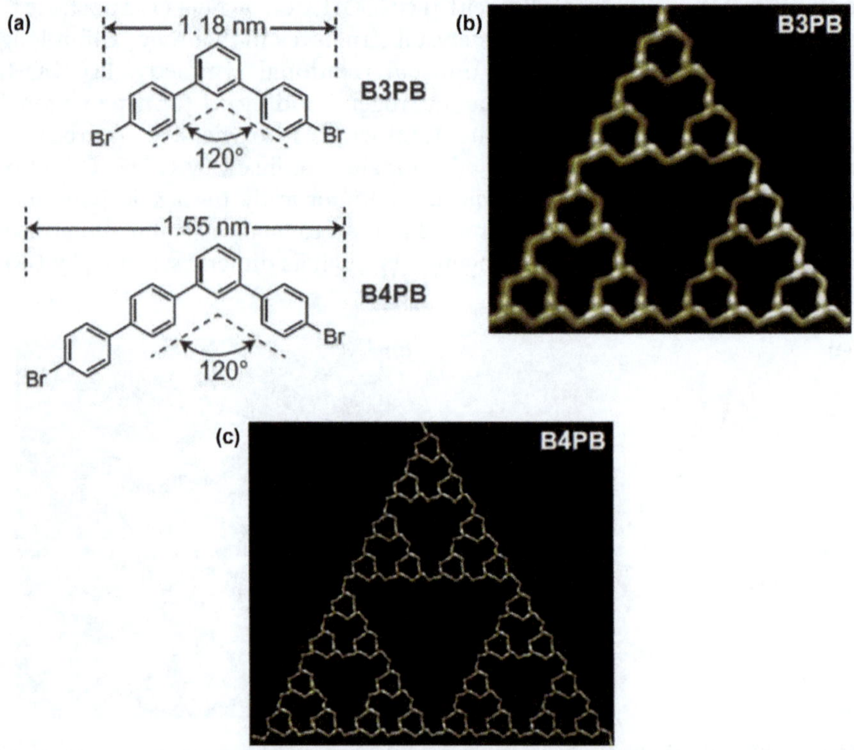

Figure 1.23 (a) Molecular structures of B3PB and B4PB with dimensions; (b) high resolution STM images of B3PB (13×11 nm); and (c) B4PB (33×29 nm) illustrating the formation of Serpinski triangles by self-assembly of the molecules through halogen bonding.
Reproduced with permission from J. Shang, Y. Wang, M. Chen, J. Dai, X. Zhou, J. Kuttner, G. Hilt, X. Shao, J. M. Gottfried and K. Wu, Assembling molecular Sierpinski triangle fractals, *Nat. Chem.*, 2015, 7, 389–393. Copyright 2015 Nature Publishing Group.

structures which share the same topological arrangement as Serpinski triangles, which are fractal structures. The formation of the fractal arrangements is truly remarkable and it is clear that the energy balance between such unusual structures and other potential arrangements is very finely balanced. As with the rhombus and Penrose tilings it is important to note that a complete appreciation of the structures formed is only possible due to a molecular level understanding of the self-assembled arrangement. This can only be readily achieved using scanning probe microscopies and it is perhaps for this reason that such structures are found in surface supramolecular framework structures.

1.6 Conclusions

It can be clearly seen from the systems described above that surface-based supramolecular chemistry has a rich future. The ability to create well-ordered two-dimensional frameworks shows great promise for the development of bespoke materials. Thus far hydrogen-bonded frameworks have received a great deal of attention, as have those structures that rely on van der Waals interactions to control framework formation, but increasingly the principles of two-dimensional framework formation are being applied to covalently bonded structures. Traditional host–guest chemistry is also feasible using surface-based frameworks, opening up avenues of research for the preparation of nanoscale devices.

Although many of the approaches that are employed by supramolecular chemists in either the solution phase or solid state are generally applicable to surface-based processes, significant differences are also evident. Notably the surface does not play a passive role in the two-dimensional self-assembly process; for molecules to adsorb on the surface there is inherently an interaction between substrate and the molecule and this can lead to subtle differences, even affecting the conformations of molecules[92] and importantly the way in which they self-assemble. As the area of surface-based self-assembled frameworks continues to develop it is important that the role of the substrate becomes increasingly understood and ultimately exploited to control self-assembly. In no area is this more important than in the formation of covalently bonded surface frameworks for which the substrate often plays an integral role in the formation of the covalent bond.

This chapter demonstrates that a number of successful strategies have already been developed for the synthesis of surface-based framework structures and some of these approaches have led to the discovery of highly unusual structures based upon unusual tiling processes, frameworks which are unlikely to be discovered using other strategies or environments. Although it is clear that further reliable pathways to robust frameworks still need to be developed it is also clear that significant progress has already been made in the field. Now that significant success in synthetic strategies has been achieved the first few steps towards functional materials are already underway.

References

1. J. M. Lehn, *Proc. Natl. Acad. Sci. U. S. A.*, 2002, **99**, 4763.
2. D. N. Reinhoudt and M. Crego-Calama, *Science*, 2002, **295**, 2403.
3. C. Aakeröy, N. R. Champness and C. Janiak, *CrystEngComm*, 2010, **12**, 22.
4. L. Bartels, *Nat. Chem.*, 2010, **2**, 87.
5. D. Bonifazi, S. Mohnani and A. Llanes-Pallas, *Chem. – Eur. J.*, 2009, **15**, 7004.
6. K. E. Plass, A. L. Grzesiak and A. J. Matzger, *Acc. Chem. Res.*, 2007, **40**, 287.
7. S. De Feyter, A. Gesquière, M. M. Abdel-Mottaleb, P. C. M. Grim and F. C. De Schryver, *Acc. Chem. Res.*, 2000, **33**, 520.
8. S. De Feyter and F. C. De Schryver, *Chem. Soc. Rev.*, 2003, **32**, 139.
9. T. Kudernac, S. Lei, J. A. A. W. Elemans and S. De Feyter, *Chem. Soc. Rev.*, 2009, **38**, 402.
10. A. G. Slater (née Phillips), P. H. Beton and N. R. Champness, *Chem. Sci.*, 2011, **2**, 1440–1448.
11. C. Vericat, M. E. Vela, G. Benitez, P. Carro and R. C. Salvarezza, *Chem. Soc. Rev.*, 2010, **39**, 1805.
12. Y. Liu, A. H. Flood, P. A. Bonvallet, S. A. Vignon, B. Northrop, H.-R. Tseng, J. Jeppesen, T. J. Huang, B. Brough, M. Baller, S. Magonov, S. Solares, W. A. Goddard III, C.-M. Ho and J. F. Stoddart, *J. Am. Chem. Soc.*, 2005, **127**, 9745.
13. G. Pace, A. Petitjean, M.-N. Lalloz-Vogel, J. Harrowfield, J.-M. Lehn and P. Samorì, *Angew. Chem., Int. Ed.*, 2008, **47**, 2484; N. R. Champness, *Nat. Nanotechnol.*, 2008, **3**, 324.
14. R. Ricco, C. Pfeiffer, K. Sumida, C. J. Sumby, P. Falcaro, S. Furukawa, N. R. Champness and C. J. Doonan, *CrystEngComm*, 2016, **18**, 6532–6542.
15. S.-Y. Ding and W. Wang, *Chem. Soc. Rev.*, 2013, **42**, 548–568.
16. S. Narasimhan and D. Vanderbilt, *Phys. Rev. Lett.*, 1992, **69**, 1564.
17. M. D. Upward, P. Moriarty and P. H. Beton, *Phys. Rev. B: Condens. Matter Mater. Phys.*, 1997, **56**, R1704.
18. M. Ruben, J. M. Lehn and P. Muller, *Chem. Soc. Rev.*, 2006, **35**, 1056.
19. J. Coraux, A. T. N'Diaye, C. Busse and T. Michely, *Nano Lett.*, 2008, **8**, 565; P. W. Sutter, J.-I. Flege and E. A. Sutter, *Nat. Mater.*, 2008, **7**, 406.
20. S. Berner, M. Corso, R. Widmer, O. Groening, R. Laskowski, P. Blaha, K. Schwarz, A. Goriachko, H. Over, S. Gsell, M. Schreck, H. Sachdev, T. Greber and J. Osterwalder, *Angew. Chem., Int. Ed.*, 2007, **46**, 5115.
21. V. V. Korolkov, S. A. Svatek, S. Allen, C. J. Roberts, S. J. B. Tendler, T. Taniguchi, K. Watanabe, N. R. Champness and P. H. Beton, *Chem. Commun.*, 2014, **50**, 8882–8885.
22. L. Gross, F. Mohn, N. Moll, P. Liljeroth and G. Meyer, *Science*, 2009, **325**, 1110–1114.
23. G. M. Whitesides, E. E. Simanek, J. P. Mathias, C. T. Seto, D. N. Chin, M. Mammen and D. M. Gordon, *Acc. Chem. Res.*, 1995, **28**, 37.

24. S. Griessl, M. Lackinger, M. Edelwirth, M. Hietschold and W. M. Heckl, *Single Mol.*, 2002, **3**, 25.
25. M. C. Etter, *Acc. Chem. Res.*, 1990, **23**, 120–126.
26. R. Gutzler, T. Sirtl, J. F. Dienstmaier, K. Mahata, W. M. Heckl, M. Schmittel and M. Lackinger, *J. Am. Chem. Soc.*, 2010, **132**, 5084.
27. D. L. Keeling, N. S. Oxtoby, C. Wilson, M. J. Humphry, N. R. Champness and P. H. Beton, *Nano Lett.*, 2003, **3**(1), 9–12.
28. A. M. Sweetman, S. Jarvis, H. Sang, I. Lekkas, P. Rahe, Y. Wang, J. Wang, N. R. Champness, L. Kantorovich and P. J. Moriarty, *Nat. Commun.*, 2014, **5**, 3931.
29. A. M. Sweetman, S. P. Jarvis, N. R. Champness, P. Rahe, L. Kantorovich and P. Moriarty, *Phys. Rev. B: Condens. Matter Mater. Phys.*, 2014, **90**, 165425.
30. S. P. Jarvis, A. M. Sweetman, I. Lekkas, N. R. Champness, L. Kantorovich and P. Moriarty, *J. Phys.: Condens. Matter*, 2015, **27**, 054004.
31. S. P. Jarvis, M. A. Rashid, A. Sweetman, J. Leaf, S. Taylor, P. Moriarty and J. Dunn, *Phys. Rev. B: Condens. Matter Mater. Phys.*, 2015, **92**, 241405.
32. N. C. Seeman, *Nature*, 2003, **421**, 427.
33. T. G. W. Edwardson, K. L. Lau, D. Bousmail and H. Sleiman, *Nat. Chem.*, 2016, **8**, 162–170; C. A. Mirkin, R. L. Letsinger, R. C. Mucic and J. J. Storhoff, *Nature*, 1996, **382**, 607–609.
34. E. Winfree, F. Liu, L. A. Wenzler and N. C. Seeman, *Nature*, 1998, **394**, 539.
35. Y. He, Y. Chen, H. Liu, A. E. Ribbe and C. Mao, *J. Am. Chem. Soc.*, 2005, **127**, 12202.
36. R. Otero, M. Schock, L. M. Molina, E. Laegsgaard, I. Stensgaard, B. Hammer and F. Besenbacher, *Angew. Chem., Int. Ed.*, 2005, **44**, 2270.
37. A. Ciesielski, R. Perone, S. Pieraccini, G. Piero Spada and P. Samorì, *Chem. Commun.*, 2010, **46**, 4493.
38. A. Ciesielski, S. Haar, M. El Garah, M. Surin, S. Masiero and P. Samori, *Actual. Chim.*, 2015, **399**, 31–36.
39. M. El Garah, R. C. Perone, A. Santana Bonilla, S. Haar, M. Campitiello, R. Gutierrez, G. Cuniberti, S. Masiero, A. Ciesielski and P. Samorì, *Chem. Commun.*, 2015, **51**, 11677–11680.
40. S. Neidle and S. Balasubramanian, *Quadruplex Nucleic Acids*, Royal Society of Chemistry, Cambridge, 2006.
41. A. Ciesielski, M. El Garah, S. Masiero and P. Samorì, *Small*, 2016, **12**, 83–95.
42. A. G. Slater, Y. Hu, L. Yang, S. P. Argent, W. Lewis, M. O. Blunt and N. R. Champness, *Chem. Sci.*, 2015, 6, 1562–1569.
43. J. A. A. W. Elemans, I. De Cat, H. Xu and S. De Feyter, *Chem. Soc. Rev.*, 2009, **38**, 722–736.
44. N. Bilbao, I. Destoop, S. De Feyter and D. González-Rodríguez, *Angew. Chem., Int. Ed.*, 2016, **55**, 659–663.
45. J. A. Zerkowski, J. P. Mathias and G. M. Whitesides, *J. Am. Chem. Soc.*, 1994, **116**, 4305.

46. A. G. Slater, L. M. A. Perdigao, P. H. Beton and N. R. Champness, *Acc. Chem. Res.*, 2014, **47**, 3417–3427.
47. J. A. Theobald, N. S. Oxtoby, M. A. Phillips, N. R. Champness and P. H. Beton, *Nature*, 2003, **424**, 1029–1031.
48. J. A. Theobald, N. S. Oxtoby, N. R. Champness, P. H. Beton and T. J. S. Dennis, *Langmuir*, 2005, **21**, 2038.
49. F. Silly, A. Q. Shaw, K. Porfyrakis, J. H. Warner, A. A. R. Watt, M. R. Castell, H. Umemoto, T. Akachi, H. Shinohara and G. A. D. Briggs, *Chem. Commun.*, 2008, 4616.
50. L. M. A. Perdigão, E. W. Perkins, J. Ma, P. A. Staniec, B. L. Rogers, N. R. Champness and P. H. Beton, *J. Phys. Chem. B*, 2006, **110**, 12539.
51. P. A. Staniec, L. M. A. Perdigão, A. Saywell, N. R. Champness and P. H. Beton, *ChemPhysChem*, 2007, **8**, 2177.
52. L. M. A. Perdigão, P. A. Staniec, N. R. Champness and P. H. Beton, *Langmuir*, 2009, **25**, 2278.
53. R. Madueno, M. T. Räisänen, C. Silien and M. Buck, *Nature*, 2008, **454**, 618.
54. C. Silien, M. T. Räisänen and M. Buck, *Small*, 2010, **6**, 391.
55. C. Silien, M. T. Raisanen and M. Buck, *Angew. Chem., Int. Ed.*, 2009, **48**, 3349.
56. A. Saywell, G. Magnano, C. J. Satterley, L. M. A. Perdigão, N. R. Champness, P. H. Beton and J. N. O'Shea, *J. Phys. Chem. C*, 2008, **112**, 7706–7709.
57. A. Saywell, G. Magnano, C. J. Satterley, L. M. A. Perdigão, A. J. Britton, N. Taleb, M. C. Giménez-López, N. R. Champness, J. N. O'Shea and P. H. Beton, *Nat. Commun.*, 2010, **1**, 75.
58. L. M. A. Perdigão, A. Saywell, G. N. Fontes, P. A. Staniec, G. Goretzki, A. G. Phillips, N. R. Champness and P. H. Beton, *Chem. – Eur. J.*, 2008, **14**, 7600.
59. M. T. Räisänen, A. G. Slater (née Phillips), N. R. Champness and M. Buck, *Chem. Sci.*, 2012, **3**, 84–92.
60. A. G. Phillips, L. M. A. Perdigão, P. H. Beton and N. R. Champness, *Chem. Commun.*, 2010, **46**, 2775.
61. A. J. Pollard, E. W. Perkins, N. A. Smith, A. Saywell, G. Goretzki, A. G. Phillips, S. P. Argent, H. Sachdev, F. Müller, S. Hüfner, S. Gsell, M. Fischer, M. Schreck, J. Osterwalder, T. Greber, S. Berner, N. R. Champness and P. H. Beton, *Angew. Chem., Int. Ed.*, 2010, **49**, 1794–1799.
62. N. Berdunov, A. J. Pollard and P. H. Beton, *Appl. Phys. Lett.*, 2009, **94**, 043110.
63. L. M. A. Perdigão, N. R. Champness and P. H. Beton, *Chem. Commun.*, 2006, 538.
64. P. A. Staniec, L. M. A. Perdigão, B. L. Rogers, N. R. Champness and P. H. Beton, *J. Phys. Chem. C*, 2007, **111**, 886–893.
65. C.-A. Palma, J. Bjork, M. Bonini, M. S. Dyer, A. Llanes-Pallas, D. Bonifazi, M. Persson and P. Samorì, *J. Am. Chem. Soc.*, 2009, **131**, 13062.

66. K. S. Mali, J. Adisoejoso, E. Ghijsens, I. De Cat and S. De Feyter, *Acc. Chem. Res.*, 2012, **45**, 1309–1320.
67. K. S. Mali and S. De Feyter, *Philos. Trans. R. Soc., A*, 2013, **371**, 20120304.
68. G. Schull, L. Douillard, C. Fiorini-Debuisschert and F. Charra, *Nano Lett.*, 2006, **6**, 1360.
69. K. Tahara, S. Okuhata, J. Adisoejoso, S. Lei, T. Fujita, S. De Feyter and Y. Tobe, *J. Am. Chem. Soc.*, 2009, **131**, 17583.
70. J. A. A. W. Elemans, S. B. Lei and S. De Feyter, *Angew. Chem., Int. Ed.*, 2009, **48**, 7298.
71. S. Lei, M. Surin, K. Tahara, J. Adisoejoso, R. Lazzaroni, Y. Tobe and S. De Feyter, *Nano Lett.*, 2008, **8**, 2541.
72. S. Lei, K. Tahara, F. C. De Schryver, M. Van der Auweraer, Y. Tobe and S. De Feyter, *Angew. Chem., Int. Ed.*, 2008, **47**, 2964.
73. S. Furukawa, H. Uji-i, K. Tahara, T. Ichikawa, M. Sonoda, F. C. De Schryver, Y. Tobe and S. De Feyter, *J. Am. Chem. Soc.*, 2006, **128**, 3502.
74. K. Tahara, S. Furukawa, H. Uji-i, T. Uchino, T. Ichikawa, J. Zhang, M. Sonoda, F. C. De Schryver, S. De Feyter and Y. Tobe, *J. Am. Chem. Soc.*, 2006, **128**, 16613.
75. A. Ciesielski, S. Lena, S. Masiero, G. Piero Spada and P. Samorì, *Angew. Chem., Int. Ed.*, 2010, **49**, 1963.
76. S.-L. Lee, J. Adisoejoso, Y. Fang, K. Tahara, Y. Tobe, K. S. Mali and S. De Feyter, *Nanoscale*, 2015, **7**, 5344–5349.
77. L. Grill, M. Dyer, L. Lafferentz, M. Persson, M. V. Peters and S. Hecht, *Nat. Nanotechnol.*, 2007, **2**, 687.
78. L. Lafferentz, V. Eberhardt, C. Dri, C. Africh, G. Comelli, F. Esch, S. Hecht and L. Grill, *Nat. Chem.*, 2012, **4**, 215–220.
79. N. R. Champness, *Nat. Chem.*, 2012, **4**, 149–150.
80. M. O. Blunt, J. C. Russell, N. R. Champness and P. H. Beton, *Chem. Commun.*, 2010, **46**, 7157–7159.
81. F. Klappenberger, Y.-Q. Zhang, J. Björk, S. Klyatskaya, M. Ruben and J. V. Barth, *Acc. Chem. Res.*, 2015, **48**, 2140–2150.
82. S. Clair, M. Abel and L. Porte, *Chem. Commun.*, 2014, **50**, 9627–9635.
83. M. B. Wieland, A. G. Slater, B. Mangham, N. R. Champness and P. H. Beton, *Beilstein J. Nanotechnol.*, 2014, **5**, 394–401.
84. S. Haq, F. Hanke, J. Sharp, M. Persson, D. B. Amabilino and R. Raval, *ACS Nano*, 2014, **8**, 8856–8870.
85. M. O. Blunt, J. Russell, M. C. Giménez-López, J. P. Garrahan, X. Lin, M. Schröder, N. R. Champness and P. H. Beton, *Science*, 2008, **322**, 1077–1081.
86. A. Stannard, J. C. Russell, M. O. Blunt, C. Sallesiotis, M. C. Gimenez-Lopez, N. Taleb, M. Schroder, N. R. Champness, J. P. Garrahan and P. H. Beton, *Nat. Chem.*, 2012, **4**, 112–117.
87. M. Blunt, X. Lin, M. C. Gimenez-Lopez, M. Schröder, N. R. Champness and P. H. Beton, *Chem. Commun.*, 2008, 2304.

88. M. O. Blunt, J. C. Russell, M. C. Gimenez-Lopez, N. Taleb, X. Lin, M. Schröder, N. R. Champness and P. H. Beton, *Nat. Chem.*, 2011, **3**, 74–78.
89. R. Penrose, *Eureka*, 1978, **39**, 16–32.
90. N. A. Wasio, R. C. Quardokus, R. P. Forrest, C. S. Lent, S. A. Corcelli, J. A. Christie, K. W. Henderson and S. A. Kandel, *Nature*, 2014, **507**, 86–89.
91. J. Shang, Y. Wang, M. Chen, J. Dai, X. Zhou, J. Kuttner, G. Hilt, X. Shao, J. M. Gottfried and K. Wu, *Nat. Chem.*, 2015, **7**, 389–393.
92. S. P. Jarvis, S. Taylor, J. D. Baran, N. R. Champness, J. A. Larsson and P. Moriarty, *Nat. Commun.*, 2015, **6**, 8338.

CHAPTER 2

Designing Soft Supramolecular Materials Using Intermolecular Interactions

PARTHASARATHI DASTIDAR,* RAJDIP ROY, RUMANA PARVEEN, SUMI GANGULY, JOYDEB MAJUMDER AND MITHUN PAUL

Department of Organic Chemistry, Indian Association for the Cultivation of Science, 2A and 2B Raja S C Mullick Road, Jadavpur, Kolkata 700032, India
*Email: ocpd@iacs.res.in

2.1 Introduction

Soft supramolecular materials such as gels encompass human life – cytoplasm, hair gel, jam/jelly at breakfast, after dinner dessert are all gels. A large volume of liquid (solvent) and a small amount of solid (gelator) are the main ingredients of gels. The flow characteristic of a gel is best defined as viscoelastic, displaying both solid and liquid like behaviour under mechanical stress. Gel is formed when a small amount of gelator is dissolved in a suitable solvent and subjected to various stimuli such heat-cool cycles, pH variation, light exposure, sound *etc.* The resulting solution is then converted to a solid-like material which is capable of withstanding its own weight against gravity.[1–12] The most popular and well-accepted hypothesis behind gelation is that the gelator molecules form a 3D network within which the solvent molecules are immobilized due to surface tension (capillary force

Monographs in Supramolecular Chemistry No. 22
Functional Supramolecular Materials: From Surfaces to MOFs
Edited by Rahul Banerjee

Published by the Royal Society of Chemistry, www.rsc.org

action). Two different categories of gels (polymeric and supramolecular) are known. If covalent bond formation is responsible for the 3D network, the resulting gel is known as chemical or polymeric gel which is usually thermo-irreversible *i.e.* the gel–sol transition is irreversible upon heating and cooling. Supramolecular gel, on the other hand, is formed when the 3D network is generated *via* supramolecular (non-covalent) interactions such as hydrogen bonding, halogen bonding, π–π stacking, van der Waals interactions *etc.* These gels are thermo-reversible and sometimes display thixotropic behaviour *i.e.* reversible gel–sol transition under severe mechanical stress (vortex) followed by rest.[13–22] While hydrogels are made from pure water or aqueous solvents, the gelling solvent in organogels, as the name suggests, are organic solvents. Gelator molecules having molecular weights of less than 3000 are popularly known as low molecular weight gelators (LMWGs).[23,24] The gelation phenomenon was reported for the first time as early as the mid-nineteenth century.[25] However, it took more than 100 years, *i.e.* the early 90s of the last century, to initiate systematic research on LMWGs. The recent upsurge in developing various types of LMWGs is understandable as these soft materials offer various potential applications that include sensors,[26–29] structure directing agents,[30–33] electro/optoelectronic materials,[34–37] cosmetics,[38–41] materials for conservation of arts,[42–44] biomedical applications[45–55] *etc.*

The other class of supramolecular gels that has attracted wide attention is metallogels;[56–61] as the name suggests, in these gels, the gel network contains metal atoms. Discrete metal coordination complexes,[62–68] well-defined coordination polymers,[69–72] and cross-linked coordination polymers[73] are reported to have imparted gelation. The term 'metallogel' emphasizes the essential role of metal–ligand bonds in gel network formation. Metallogels are technologically important as they offer potential applications in sensing,[74] catalysis,[75–80] magnetic materials,[81] photophysics[82–85] *etc.*

2.2 Designing LMWGs

One of the major problems associated with the designing of gelators is the lack of molecular level information about the self-assembly process through which the gelator molecule can form SAFiNs.[1] More importantly, how exactly the solvent molecules interact with the surface of the SAFiNs to become immobilized is virtually unknown. Therefore, designing a gelator molecule *a priori* is a monumental task. Moreover, the wide variety of molecules displaying gelation ability also adds to the difficulty of designing gelators. Therefore, many of the gelators reported thus far are serendipitously obtained and the next generation gelators are often derived by modifying the known gelators. 1D fibrous objects highly entangled and/or hyperbranched are the common morphological characteristics of gels as revealed by various high resolution microscopic studies. It is, therefore, highly opined that gelator molecules capable of self-assembling into a 1D network would promote the growth of 1D fibres that eventually are entangled and/or hyperbranched to form SAFiNs within which solvent molecules are immobilized resulting in gels. Thus, molecules capable of forming 1D supramolecular self-assembly are potential candidates for gelation. Various attempts have been made to synthesize new gelator molecules. Overall, the designing of LMWGs may be classified into two broad strategies: (i) molecular engineering approach wherein known gelators are being modified or gel-inducing functionalities (such as amide, urea, long alkyl chain *etc.*) are used to decorate a molecular scaffold to produce potential gelators; and (ii) structure–property correlation approach based on supramolecular synthon in the context of crystal engineering. In the following sections, some seminal examples pertaining to the design of gelators belonging to each category are discussed. A few examples of designing metallogelators are also discussed in a separate section.

2.2.1 Molecular Engineering Approach

One of the early examples of serendipitous discoveries of gelators was reported by Lin and Weiss in late 1980s.[86–89] While studying the photochemistry of cholesteryl 4-(2-anthryloxy) butyrate (CAB), they discovered the gelation ability of this molecule. Systematic investigation based on the CAB scaffold that contains three distinctly different structural units, namely aromatic, linker, and steroid (ALS) units led to the development of second generation ALS gelators. Studies showed that large π surface, subtle changes in the S moiety, even and odd methylene group in the L part *etc.* had profound effect on gelation. It was also observed that ester, amide, carbamate and urea moieties in the L part promoted gelation. The shape of the ALS molecule was also found to be important in gelation; rod-shaped ALS molecules turned out to be better gelators than the bent analogues.[1]

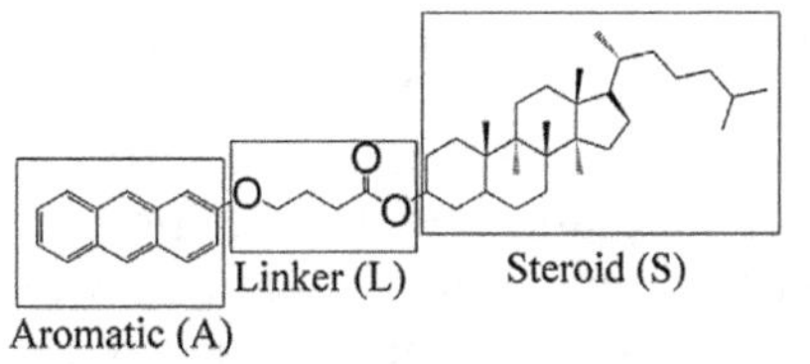

Cholesteryl 4-(2-anthryloxy) butyrate (**CAB**)

It was subsequently recognized that the following factors were important for inducing gelation: (a) strong and directional supramolecular interactions that promote aggregation of the gelator molecules to form fibres; (b) ability to form entangled or intertwined fibres (SAFiNs); and (c) factors preventing neat crystallization of the gelling agents. Among the various non-covalent interactions important in supramolecular chemistry, hydrogen bonding[90,91] is one of the most reliable supramolecular interactions as it is reasonably strong as well as directional. Thus, it became a general strategy to install hydrogen bonding functionalities for promoting the features (a), (b), and long alkyl chain for encouraging the factor (c) as stated above for designing gelators.

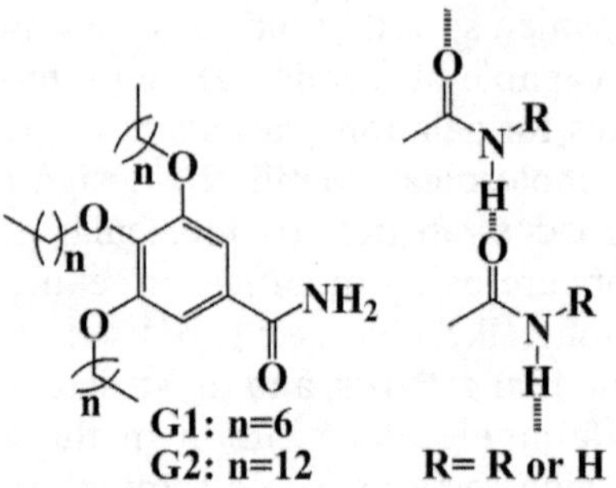

Figure 2.1 Tris(alkoxy) benzamides gelators; usually observed 1D HBN in primary or secondary amides.

Primary and secondary amide moieties are known to form extended 1D hydrogen bonding networks (HBN) and a large number of LMWGs containing amide functionality have been reported thus far. Early examples of LMWGs containing primary amides are tris(alkoxy)benzamides (**G1** and **G2**) which were capable of gelling both polar and non-polar organic solvents;[92] while amide functionality provided the required hydrogen-bonded network to form fibrils, the long alkyl chains understandably discouraged neat crystallization resulting in gelation (Figure 2.1).

Secondary amide functionality was elegantly exploited to design an alkanoic acid amide of alanine, *N*-lauroyl-L-Ala (**G3**), which was capable of gelling various lipophilic solvents including commercial fuels like petrol. **G3** was also shown to be selective towards the gelation of petrol from a biphasic mixture of petrol and water, indicating plausible applications in oil transport or containing oil spills (Figure 2.2).[93]

In an elegant design, Esch *et al.* installed alternating hydrophobic and hydrophilic regions on a cyclohexane scaffold to synthesize a series of LMWGs (**G4–G7**) containing multiple amide functionality. The inner hydrophilic core containing three amide moieties were buried between hydrophobic regions (cyclohexane ring and peripheral hydrophobic amino acid moieties) in order to promote amide···amide hydrogen bonding leading to 1D HBN. The crystal structure of one such tyrosine based on a non-gelator model compound displayed 1D HBN (Figure 2.3).[94]

Multiple amide functionality was employed to decorate a well-known non-steroidal anti-inflammatory drug (NSAID), namely naproxen (Np), that resulted in a new series of hydrogelators (**G8–G11**) also capable of gelling aqueous NaCl (0.9 wt% w/v) and phosphate-buffered saline (PBS) (pH 7.4) solution. The β peptide derivatives of Np were also shown to be biocompatible (MTT assay) and their anti-inflammatory response (PGE_2 assay) was as potent as the parent drug Np. Single crystal structure of one of the gelators (**G11**) was shown to possess β-sheet HBN involving the multiple amide moieties (Figure 2.4).[95]

Ulijn *et al.* demonstrated that the self-assembly of bolaamphiphile-type conjugates of dipeptides and perylene bisimide (PBI) in water and other

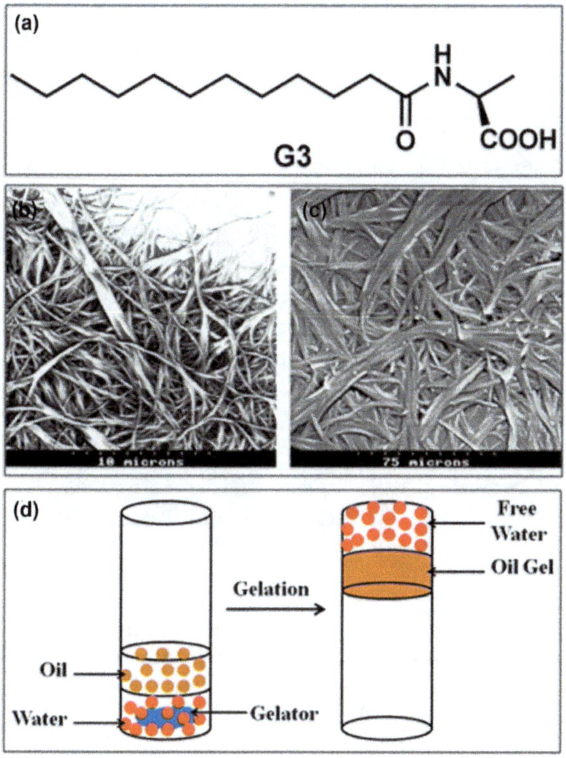

Figure 2.2 (a) Alkanoic acid amide **G3** studied for gelation; SEM images of xerogel of **G3** in (b) *n*-heptane and (c) toluene; (d) schematic representation of gelation of a two-phase solvent mixture of water and oil (either commercial fuel or pure organic solvent).
Partly reproduced from ref. 93 with permission from the Royal Society of Chemistry.

polar solvents had a profound effect on the morphology of the nanostructures and their fluorescence property. Thus, the glycine–tyrosine derivative (**G12**) produced hydrogel displaying fibrous morphology whereas the glycine–aspartic acid derivative (**G13**) produced spherical aggregates. The resulting nanostructures showed highly tuneable fluorescence (Figure 2.5).[96]

Tetrathiafulvalene (TTF)—a well-known organic conductor—was conjugated with a 3,4,5-trialkoxybenzoic acid backbone through bis-amide linkage to produce a gelator (**G14**) capable of gelling various hydrocarbon solvents, such as hexane, decane, and cyclohexane. The nanostructures present in the thin film prepared from the hexane gel of **G14** under TEM were found to be highly aligned 1D fibres leading to the possibility of developing thin films having anisotropic conductivity (Figure 2.6).[97]

Spatially resolved multicomponent gels were developed by Adams *et al.* In this work, two bis-amide gelators (**G15, G16**) were mixed to form gel. While **G15**—having a *trans*-stilbene backbone—was photo-responsive, **G16** was

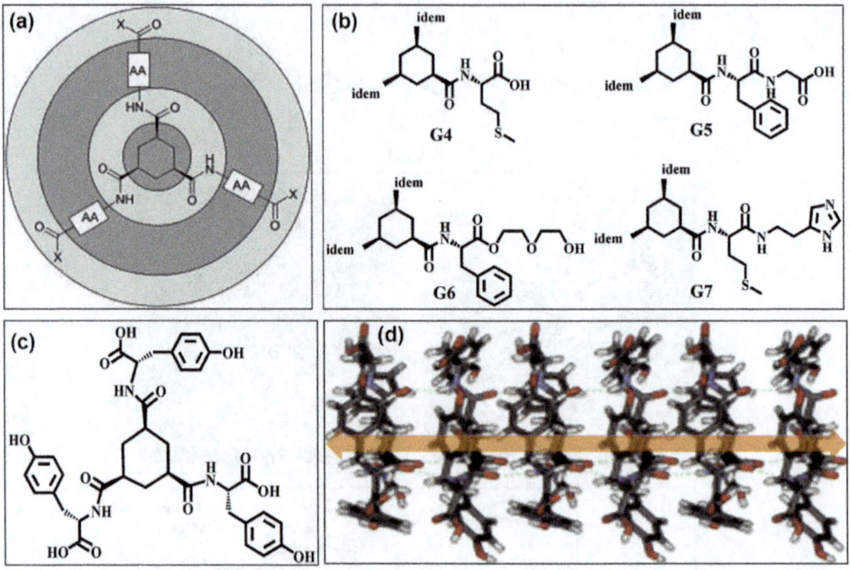

Figure 2.3 (a) Gelling scaffold (dark grey – hydrophobic, light grey – hydrophilic); (b) gelators **G4–G7**; (c) non-gelator model compound; (d) 1D self-assembly of the non-gelator model compound in its single crystal structure. idem: all the compounds have three identical side chains, for simplicity only one is shown for each compound.
Partly reproduced with permission from K. J. C. van Bommel, C. van der Pol, I. Muizebelt, A. Friggeri, A. Heeres, A. Meetsma, B. L. Feringa and J. van Esch, Responsive Cyclohexane-Based Low-Molecular-Weight Hydrogelators with Modular Architecture, *Angew. Chem., Int. Ed.*, 2004, **43**, 1663. Copyright 2004, Wiley-VCH Verlag GmbH & Co. KGaA.

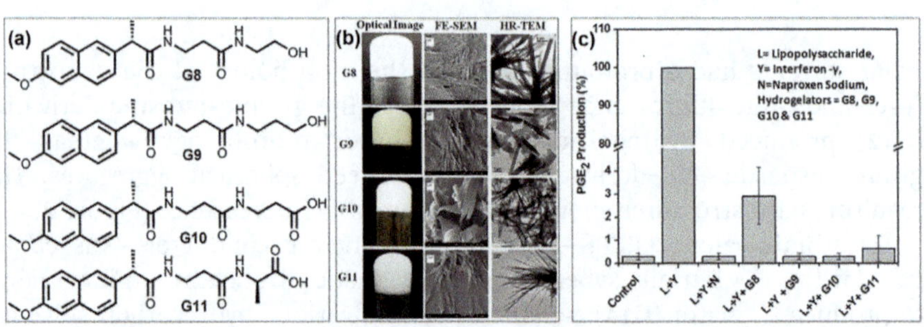

Figure 2.4 (a) Naproxen (Np) derivatives **G8–G11** studied for hydrogelation; (b) optical images, FESEM and HRTEM images of the opaque xerogels; (c) PGE_2 assay of the hydrogelators; adapted with permission from J. Majumder, M. R. Das, J. Deb, S. S. Jana and P. Dastidar, β-Amino Acid and Amino-Alcohol Conjugation of a Nonsteroidal Anti-Inflammatory Drug (NSAID) Imparts Hydrogelation Displaying Remarkable Biostability, Biocompatibility, and Anti-Inflammatory Properties, *Langmuir*, 2013, **29**, 10254. Copyright 2013, American Chemical Society.

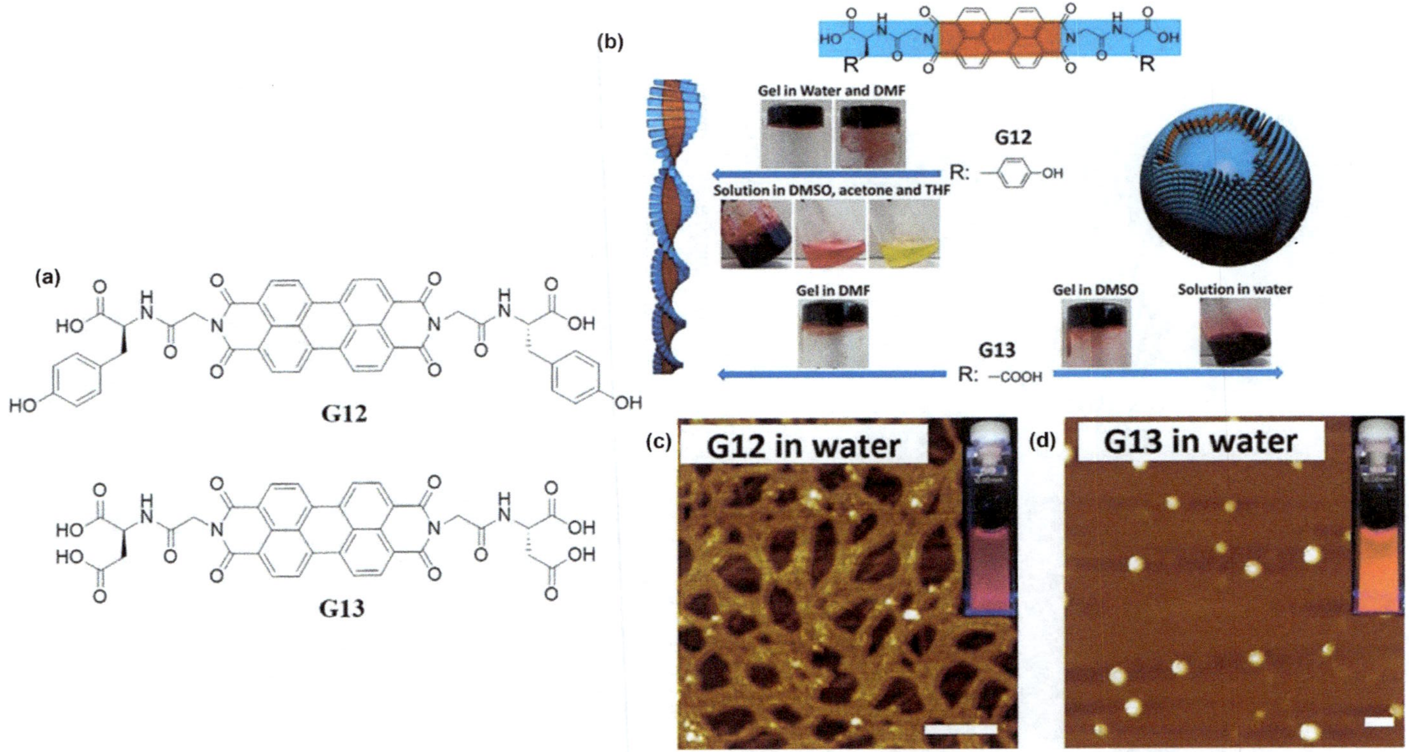

Figure 2.5 (a) Bolaamphiphile-type conjugates of dipeptides and perylene bisimide (PBI) **G12** and **G13** studied for self-assembly; (b) cartoon representation of the self-assembly of **G12** and **G13** to chiral fibrous and spherical structures in different polar solvents; AFM images of the structures of **G12** (c) and **G13** (d) assembled in buffer solution (pH 10.8), concentration 1×10^{-4} M. The scale bar is 500 nm. The insets are optical photographs of **G12** and **G13** at the same concentration under a UV lamp (365 nm).
Adapted with permission from S. Bai, S. Debnath, N. Javid, P. W. J. M. Frederix, S. Fleming, C. Pappas and R. V. Ulijn, Differential Self-Assembly and Tunable Emission of Aromatic Peptide Bola-Amphiphiles Containing Perylene Bisimide in Polar Solvents Including Water, *Langmuir*, 2014, **30**, 7576. Copyright 2014, American Chemical Society.

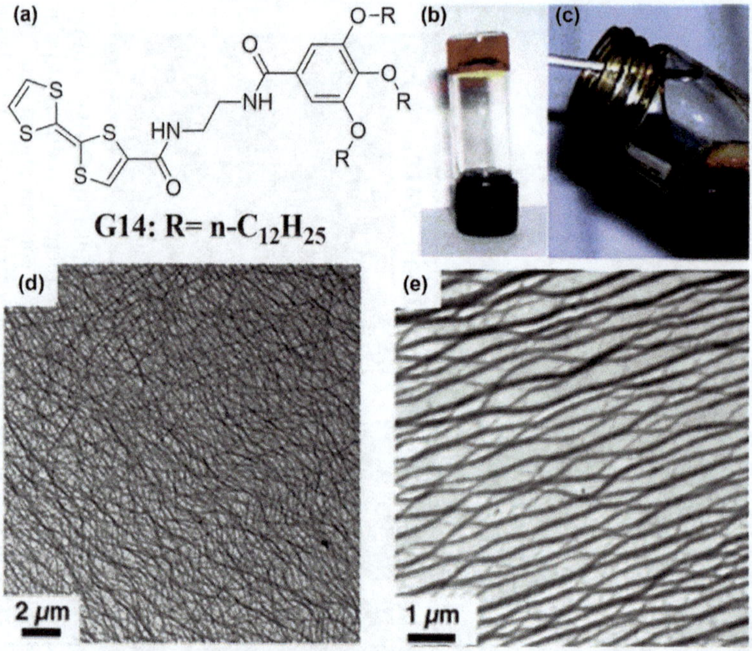

Figure 2.6 (a) Chemical structure of **G14**; photographs showing (b) transparent and (c) viscous nature of the gel prepared from **G14** with hexane (100 g dm^{-3}); TEM images of the hexane gel of **G14**; the samples were prepared at (d) 50 g dm^{-3} and (e) at 100 g dm^{-3}.
Adapted with permission from T. Kitahara, M. Shirakawa, S. Kawano, U. Beginn, N. Fujita and S. Shinkai, Creation of a Mixed-Valence State from One-Dimensionally Aligned TTF Utilizing the Self-Assembling Nature of a Low Molecular-Weight Gel, *J. Am. Chem. Soc.*, 2005, **127**, 14980. Copyright 2005, American Chemical Society.

photo-inert. Interestingly, the multicomponent gel derived from **G15/G16** formed a 'self-sorted network' structure as evident from the photoirradiation of the gel that selectively induced the gel–sol transition of the part of the gel produced by **G15** (Figure 2.7).[98]

Like amides, urea is also known to form a 1D extended hydrogen bonded network. A large number of gelators containing urea functionality have been reported.[99–101] The mono and bis-urea derivatives (**G17** and **G18**, respectively) developed by Hamilton and co-workers displayed gelation ability with supercritical CO_2. The resultant aerogels obtained by removal of CO_2 were microcellular materials having highly porous structures (Figure 2.8).[102]

A strong polymer gel was obtained *via* the supramolecular route when the bis-urea gelator **G19** was decorated with a photopolymerizable moiety, such as methacrylate was irradiated with 200 W high-pressure Hg lamp in presence of 2-phenylacetophenone. The resultant dried gel (xerogel) displayed remarkable porosity and shape-sustaining ability. Interestingly, the xerogel

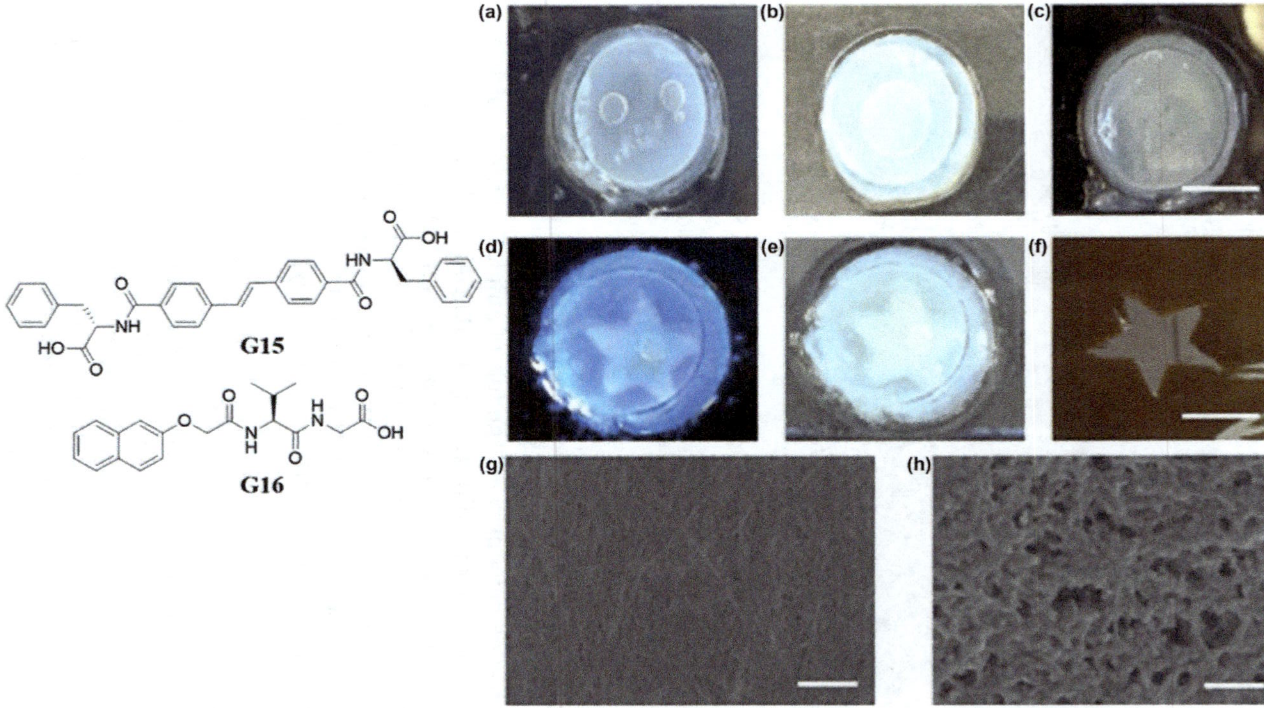

Figure 2.7 Spatially resolved removal of one network: chemical structure of **G15** and **G16**; (a)–(c) **G15 : G16** hydrogel (1 : 1 **G15/G16B**, w/w, conc. 10 mg mL^{-1} gel produced by adjusting the pH from 10 to 4 by the slow hydrolysis of glucono-δ-lactone (GdL) to gluconic acid) before and after irradiation with a handheld 365 nm LED (a) before irradiation, (b) after 30 minutes of irradiation and (c) after two hours of irradiation; (d)–(f) spatial patterning by combining the irradiation step with a mask; photographs of the gel (d) irradiated with a handheld 365 nm ultraviolet light, (e) under daylight, (f) photograph of the start shaped mask; (g)–(h) SEM images of the (g) non-irradiated and (h) irradiated part of the gel. Scale bars, a–f, 1 cm; g–h, 1 μm.
Adapted with permission from E. R. Draper, E. G. B. Eden, T. O. McDonald and D. J. Adams, Spatially resolved multicomponent gels, *Nat. Chem.*, 2015, 7, 848. Copyright 2015, Nature Publishing Group.

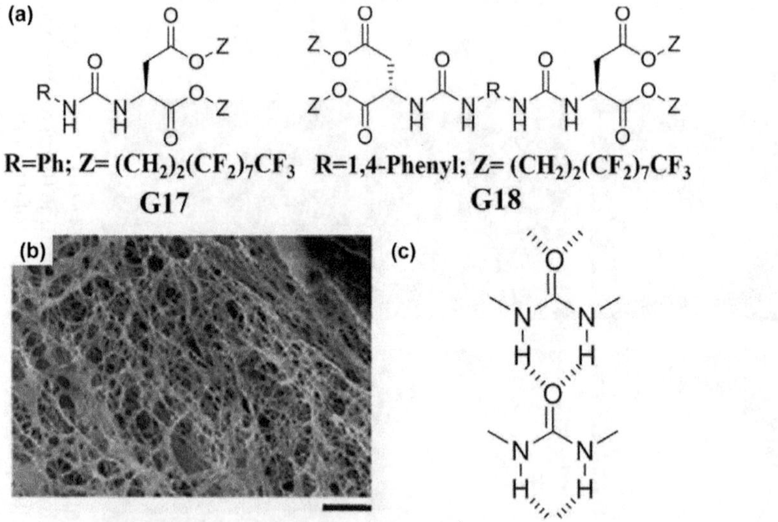

Figure 2.8 (a) Urea derivatives **G17** and **G18** capable of gelling supercritical CO_2; (b) SEM image of the foam produced by **G17** (bar 5 μm); (c) urea···urea hydrogen bonding.
Reproduced with permission from C. Shi, Z. Huang, S. Kilic, J. Xu, R. M. Enick, E. J. Beckman, A. J. Carr, R. E. Melendez and A. D. Hamilton, The Gelation of CO_2: A Sustainable Route to the Creation of Microcellular Materials, *Science*, 1999, **286**, 1540. Copyright 1999, AAAS.

of the polymer gel derived from **G19** was able to produce a transparent gel when resolvated with MeOH or $CHCl_3$ (Figure 2.9).[103]

Recently, Ghosh *et al.* demonstrated the role of amide and urea moiety as a part of the external structure directing agent (ESDA) used to generate a three-component charge-transfer hydrogen-bonded gel (**G20**). Interestingly, it was observed that the amide containing EDSA (*i.e.* EDSA-A) led to gelation whereas ESDA-U (urea containing EDSA) failed to gel. The gel and sol formations in these two cases were attributed to the contrasting morphologies (elongated fibres and ill-defined particles, respectively) (Figure 2.10).[104]

2.2.2 Crystal Engineering Approach

In this approach, the question one asks is more fundamental – does the molecular self-assembly as observed in the crystal structure of a molecule have anything to do with the gelation ability of that molecule? In other words, is there any structure–property (gelling–nongelling) correlation? The success of this approach lies in the determination of the crystal structure of a gel network. The information one gets out of such an exercise is understandably precious as it provides molecular level insights into the various self-assembly modes of the gelator molecules within the gel network which is obviously fundamental to designing gelators. It is clear from the forgoing

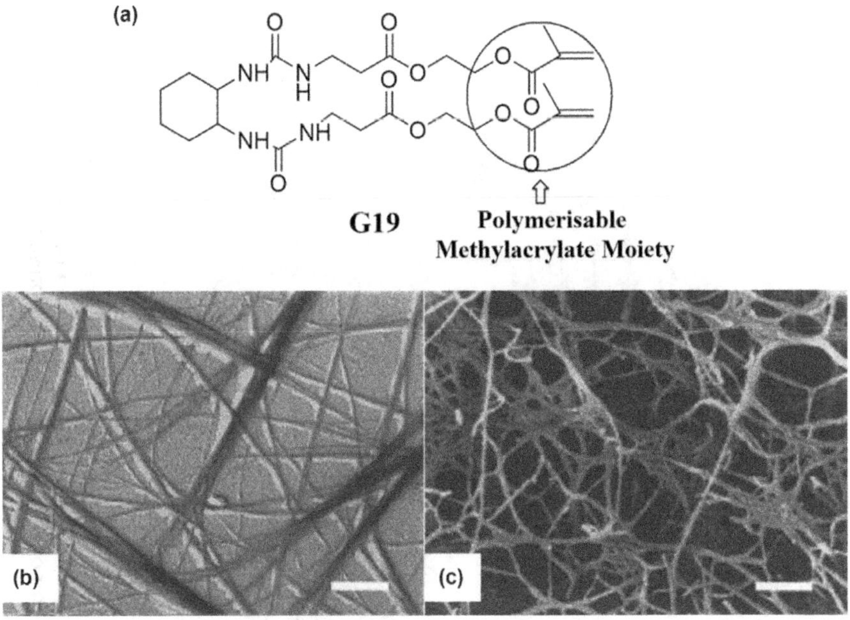

Figure 2.9 (a) Polymerizable bis-urea gelator **G19** capable of producing highly porous dried gel; (b) TEM micrographs of butyl acetate gel of **G19** before polymerization (5 mg mL^{-1}, bar 1000 nm); (c) SEM of a benzene gel of **G19** after irradiation (5 mg mL^{-1}, bar 400 nm).
Reproduced with permission from M. de Loos, J. van Esch, I. Stokroos, R. M. Kellogg and B. L. Feringa, Remarkable Stabilization of Self-Assembled Organogels by Polymerization, *J. Am. Chem. Soc.*, 1997, **119**, 12675. Copyright 1997, American Chemical Society.

discussions delineated in the previous sections that the molecules capable of forming 1D supramolecular self-assembly are potential candidates for gelation. Is it really so? If it is, then one can install 1D self-assembly forming functionality on a molecular scaffold with the hope of producing a potential gelator. The concept of supramolecular synthon in the context of crystal engineering comes handy in this endeavour. Detailed discussion on supramolecular synthons and crystal engineering is beyond the scope of this chapter. However, for a fundamental understanding of the subject, the pioneering work of Prof. G. R. Desiraju may be consulted.[105,106]

In crystal engineering "one attempts to understand crystal structure in terms of intermolecular interactions; one attempts to define a reliable design strategy using these interactions; and finally one attempts to direct such a crystal design exercise toward a property that may be needed."[107] Supramolecular synthons may be considered the tools in the crystal engineering kit, which are kinetically defined spatial arrangements of intermolecular interactions robust enough to ensure generality and predictability of the resulting crystalline solid and convey the essential features of a crystal structure.[105,108]

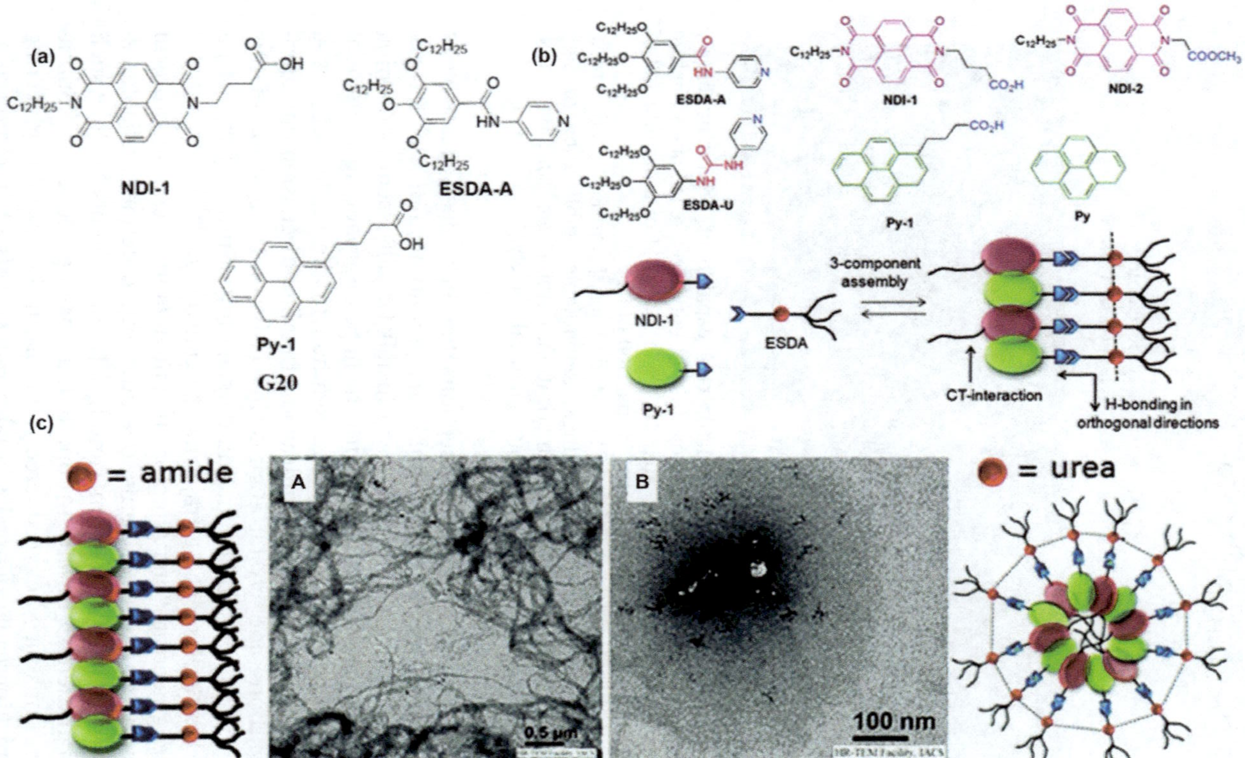

Figure 2.10 (a) Three component (NDI-1, ESDA-A and Py-1) charge-transfer hydrogen-bonded gel **G20**; (b) (top) structure of donor–acceptor building blocks and ESDA; (bottom) schematic showing the three component assembly; (c) possible mode of assembly that leads to contrasting TEM morphology for the three-component organization in the presence of (A) ESDA-A (left) and (B) ESDA-U (right).
Adapted from ref. 104 with permission from the Royal Society of Chemistry.

Is crystal engineering relevant in designing soft material (gel) producing agent like LMWGs? Studies showed that gel networks were quite crystalline, as xerogels or even gels were shown to produce reasonably ordered powder X-ray diffraction (PXRD) patterns. Because of the nano or micro dimensions of the gel network, single crystal X-ray diffraction (SXRD) technique cannot be exploited to determine the crystal structure of a gel network. An alternative strategy, introduced by Weiss *et al.* wherein the PXRD patterns of the xerogel/gel were compared with that of the simulated PXRD patterns obtained from the SXRD data of the gelator molecules,[109] turned out to be quite useful; an excellent match of these patterns would then establish the gel network structure. Two major drawbacks of this method, however, are (a) difficulty in getting a SXRD quality single crystal of the gelator molecule and (b) relatively weak diffraction of the gel sample attributed to the scattering contribution of the solvent and low gelator content. Perhaps the only option to get around these drawbacks is to solve the structure of the gel network by *ab intio* structure determination using PXRD data of the gel sample. However, such approach would require PXRD data collection in synchrotron beamline which is not easily accessible. Moreover, structure determination *ab inito* from PXRD data is still not a routine procedure. These impediments compel one to compare simulated PXRD patterns with that of xerogel—which shows much stronger diffraction. One has to keep in mind that the network structure in xerogel might be quite different from that of the gel state as a new nucleation event or polymorphism of the gel network may take place (during xerogel formation by evaporating the solvents), leading to a different crystalline phase. However, there is no guaranty that such a crystalline phase change would always occur. Under these given constraints, comparing simulated PXRD with that of the xerogel appears to be a reasonable compromise.

Thus, once the structure of the gel network is so determined, supramolecular synthons responsible for the formation of such self-assembled gel network are identified. Installing functionalities leading to the identified supramolecular synthons therefore is the first step in the design of LMWGs. Thus, supramolecular synthon in the context of crystal engineering is very much apt and attractive in designing LMWGs.

Shinkai *et al.* attempted the structure–property correlation in a series of sugar based molecules (**G21**, **NG22**, **NG23** and **NG24**). They demonstrated, based on the crystal structures of these sugar derivatives, that 1D HBN promoted gelation whereas 2D and 3D HBN either produced weak gel or no gelation at all. It was argued that that 1D HBN directed the molecules to self-assemble into 1D thereby triggering the growth of 1D fibres resulting in SAFiNs that ultimately lead to the gel formation (Figure 2.11).[110]

Fully fledged structure-property correlation and consequent application of supramolecular synthon in the context of crystal engineering in designing LMWGs started with the serendipitous discovery of the gelator imidazolium hydrogen cyclobutane-1,1-dicarboxylate (**G25**).[111] It was observed during a crystallization attempt that **G25** was an efficient gelator of nitrobenzene (minimum gelator concentration or MGC = 0.137 wt% at RT). Later the gelator

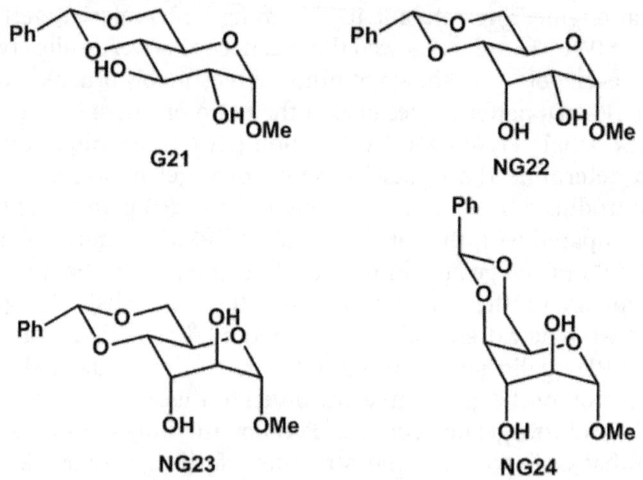

Figure 2.11 Various sugar derivatives studied for structure–property correlation.

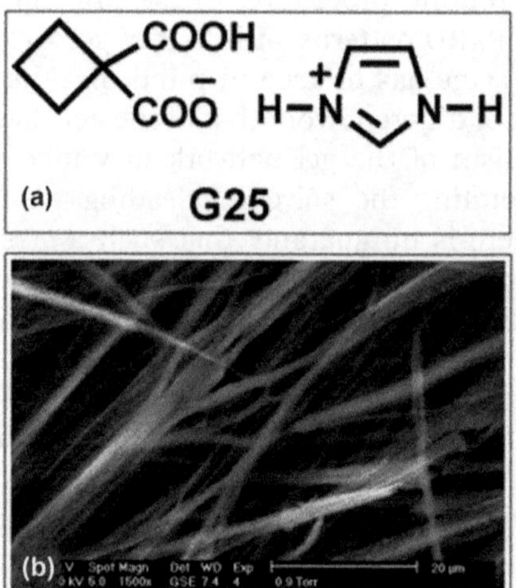

Figure 2.12 (a) Imidazolium salt **G25** studied for gelation; (b) SEM image of the xerogel obtained from **G25** in nitrobenzene (1 wt%).
Adapted from A. Ballabh, D. R. Trivedi and P. Dastidar, Structural Studies of a New Low Molecular Mass Organic Liquids Based on Simple Salt, *Chem. Mater.*, 2003, **15**, 2136. Copyright 2003, American Chemical Society.

salt was crystallized from MeOH and the structure revealed that it contained 3D HBN containing 1D HBN involving the imidazolium cation and hydrogen carboxylate anion. It was argued that the primary 1D HBN observed in the crystal structure of **G25** was somehow responsible for the growth of 1D fibres (observed in the SEM image of the nitrobenzene xerogel of **G25**) that eventually entangled to form SAFiNs resulting in gelation (Figure 2.12).

After this serendipitous entry of **G25** in the literature, developing organic-salt-based gelators became an attractive target as it was realized that carrying out salt formation reactions was much easier than multi-step organic synthesis for making molecular scaffolds as potential gelators. Salt formation is, perhaps, the easiest reaction to carry out that often provides a near quantitative yield without going through tedious and expensive purification processes, allowing one to generate a large library of organic salts for quick scanning for gelation. Moreover, organic salts are governed by charge-assisted, relatively stronger, highly directional hydrogen bonding that is understandably advantageous for fabricating these materials that are useful for real-life applications.

Furthermore, a virtually infinite number of combinations of commercially available acids and amines offers an opportunity to discover new gelators by scanning such a combinatorial library. Extensive structure–property correlation based studies towards the goal of designing organic salt based LMWGs led to the identification of a number of gel-inducing supramolecular synthons which are discussed in the following sections.

2.2.2.1 Secondary Ammonium Monocarboxylate (SAM) Synthon

Studies revealed that secondary ammonium monocarboxylate (SAM) salts usually display two different types of supramolecular synthons – one, an extended 1D polymer chain involving alternating cations and anions sustained by N–H···O hydrogen bonding and the other, a discrete or 0D HBN wherein the ion pairs formed a discrete cyclic structure sustained by N–H···O interactions.[112] Early examples of SAM salts as LMWGs variously substituted cinnamate/benzoate salts of dicyclohexyl amine (**G26**$_i$).[113–115] While many of the cinnamate salts were gelators, none of the benzoate salts were able to gel the set of solvents studied. The gelation ability of these salts was reported to be dependent on the position of the substitution in the phenyl ring of the cinnamate moiety; in most of the cases, the 3- and 4- substitution led to gelation whereas the 2-substitution appeared to be ineffective in inducing gelation. Selective gelation of commercial fuel (like petrol) from a biphasic mixture of petrol and water was possible by dicyclohexyl 4-chlorocinnamate, demonstrating the proof of the concept for developing technology for containing oil spill and safe transport.[93,116] The number of crystal structures of gelator and non-gelator SAM salts in this series revealed that the 1D SAM synthon played a crucial role in gelation as six gelator salts displayed 1D SAM synthon, whereas none of the SAM salts displaying synthon B (0D HBN) were gelators. These data clearly supported Shinkai's hypothesis as discussed earlier (Figure 2.13).[110]

The view that 1D SAM synthons might be playing an important role in promoting the growth of 1D fibres leading to gelation in these salts was further reinforced when it was observed that by merely changing the cation from dicyclohexyl amine (DCA) to dibenzyl amine (DBA), the gelation

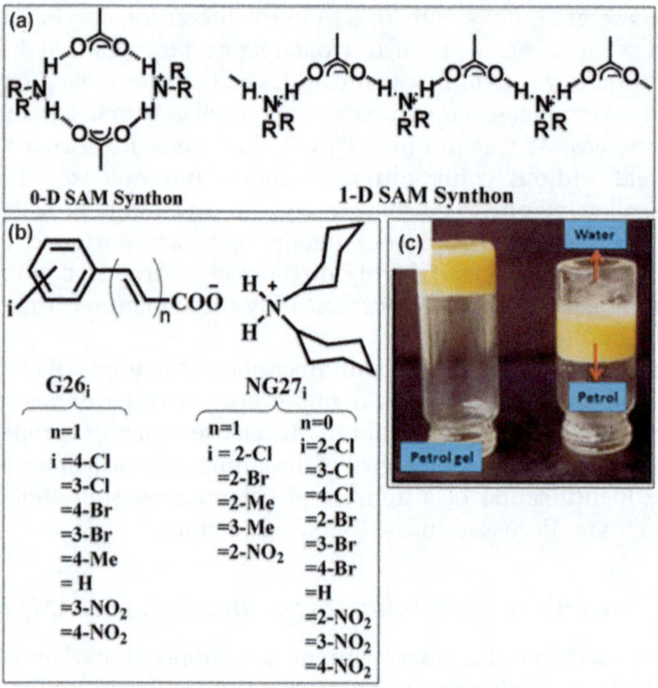

Figure 2.13 (a) Schematic representation of SAM synthons and; (b) various SAM salts studied for gelation; (c) selective gelation of petrol by **G26$_{4\text{-Cl}}$**.

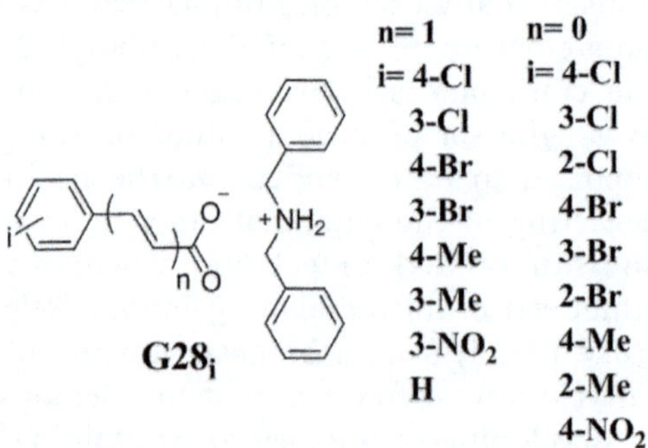

Figure 2.14 Various DBA SAM salts as gelators.

behaviour of the DBA salts (**G28$_i$**) was greatly changed; now the majority of the SAM salts were gelators, including the benzoates which were non-gelators when the cation of DCA. SXRD studies revealed that none of the SAM salts displayed 0D SAM synthon B; all the cinnamates (six salts) and benzoates (12 salts) displayed 1D SAM synthon A emphasizing the role of cationic moiety on the HBN isomerization and the importance of 1D HBN in gelation (Figure 2.14).[117]

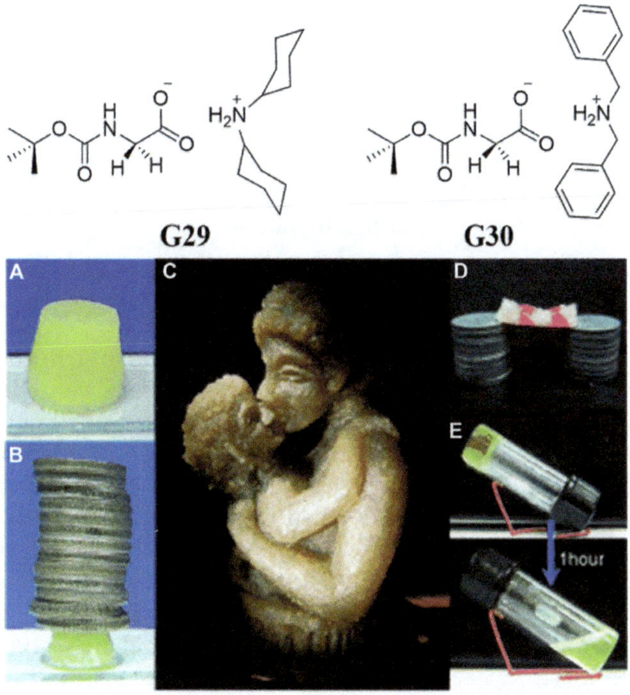

Figure 2.15 Top: chemical structure of **G29** and **G30**; bottom: nitrobenzene gel of **G29** – (A) a piece of free standing gel; (B) load-bearing gel; (C) gel sculpture of 'mother and child'; (D) self-healing; (E) the nitrobenzene gel **G30** and its transition of sol at room temperature over a period of 1 h. Adapted with permission from P. Sahoo, R. Sankolli, H.-Y. Lee, S. R. Raghavan and P. Dastidar, Gel Sculpture: Moldable, Load-Bearing and Self-Healing Non-Polymeric Supramolecular Gel-Derived from a Simple Organic Salt, *Chem. – Eur. J.*, 2012, **18**, 8057. Copyright 2012, Wiley-VCH Verlag GmbH & Co. KGaA.

The effect of cations on gelation was most remarkably observed in a series of SAM salts derived from N-*tert*-butoxycarbonyl(Boc)-protected L-amino acids and secondary amines (DCA and DBA). Among these SAM salts, a DCA salt of *N*-Boc-glycine (**G29**) displayed amazing gel properties; nitrobenzene gel of **G29** displayed free-standing, load-bearing and self-healing properties. The gel was so strong that a sculpture 'mother and child' was moulded and the sculpture remained stable for many years! However, the corresponding DBA salt (**G30**) could not produce a stable gel with nitrobenzene. SXRD studies revealed that both these salts possessed 1D HBN similar to SAM synthon A with additional hydrogen bonding (Figure 2.15).[118]

2.2.2.2 Secondary Ammonium Dicarboxylate (SAD) Synthon

Studies on SAM synthon revealed that not a single SAM salt displaying 0D HBN (synthon B) was reported to be a gelator and the probability of

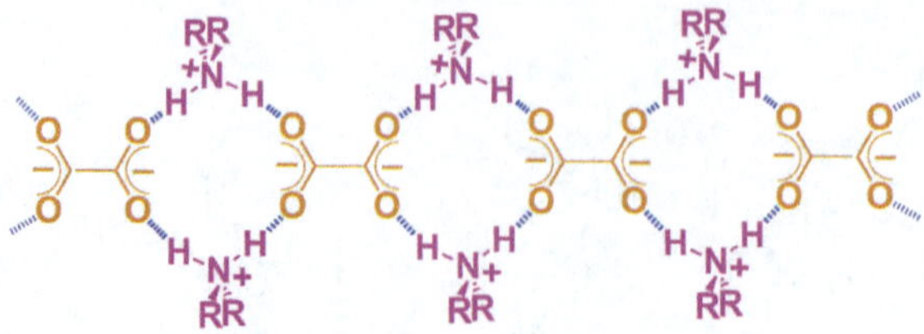

Figure 2.16 Schematic representation of 1D SAD synthon.

formation of both 1D and 0D SAM synthons is equal. To overcome this problem, secondary ammonium dicarboxylate (SAD) salts were considered in order to ascertain 1D HBN in these salts. The first such deliberate attempt was a systematic study wherein various aromatic dicarboxylic acids were reacted with DCA in both 1 : 1 and 1 : 2 molar ratios. SXRD data on some of the 1 : 2 SAD salts in this series revealed the existence of 1D HBN, wherein SAM synthon B propagated along 1D by virtue of the dicarboxylate functionality of the acid part (Figure 2.16). Unfortunately, none of these salts were capable of gelling the set of solvents under the study, which emphasized the role of surface compatibility of the gel network and target solvent molecules.[119]

A few years after this unsuccessful attempt to produce gelator salts based on the SAD synthon the first successful crystal engineering-based design of organometallic gelator was reported. In this study, the SAD synthon was exploited to generate an organometallic gelator, namely dicyclohexylammonium ferrocene-1,1′-dicarboxylate (**G31**) which was capable of gelling a wide range of solvents that included both polar and non-polar solvents.[120] Crystal structure determination of **G31** revealed the presence of the SAD synthon. Interestingly, the morphology of the gel network as observed under SEM was comprised of twisted fibres (showing both right- and left-handed twists) (Figure 2.17).

Unfortunately, the SAD synthon appeared to be not so promising in regard to designing LMWGs.[119,121,122]

2.2.2.3 Primary Ammonium Monocarboxylate (PAM) Synthon

Sada, Miyata and co-workers reported that primary ammonium monocarboxylate (PAM) salts were prone to display 1D columnar HBN with the occasional occurrence of 2D HBN.[123] Various attempts to exploit such 1D PAM synthons led to the discovery of a number of LMWGs useful for various applications that ranged from containing oil spills to drug delivery. In the majority of the PAM salts, two types of 1D PAM synthons, namely synthon W and X are present; synthon W is comprised of ten-membered hydrogen-bonded rings, whereas synthon X is made of alternating 12- and 8-membered rings (Figure 2.18).

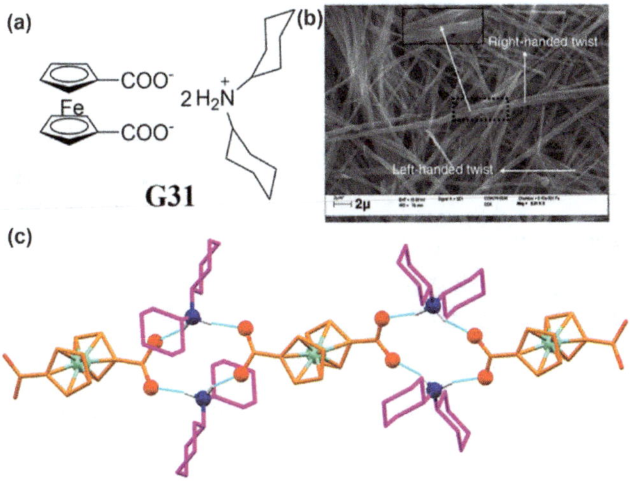

Figure 2.17 (a) SAD salt based organometallic gelator **G31**; (b) twisted helical fibres with both right- and left-handedness seen in the SEM of DMF-xerogel of **G31**; (c) typical 1D SAD synthon present in the crystal structure of **G31**. Adapted from P. Sahoo, D. K. Kumar, D. R. Trivedi and P. Dastidar, An easy access to an organometallic low molecular weight gelator: a crystal engineering approach, *Tetrahedron Lett.*, 2008, **49**, 3052. Copyright 2008, with permission from Elsevier.

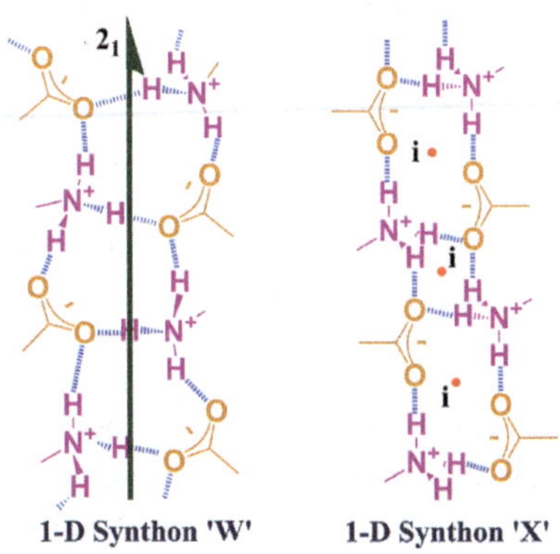

Figure 2.18 Schematic representation of 1D PAM synthon W and X. Synthon W is intrinsically asymmetric possessing a screw axis (2_1) whereas synthon X is centrosymmetric containing centres of inversion (i).

One of the early attempts to exploit PAM synthon in designing LMWGs was a combinatorial library generated by reacting variously-substituted cinnamic acids and four selected *n*-alkyl primary amines ($Me\text{-}(CH_2)_n\text{-}NH_2$; $n=3$, 4, 5 and 15). Gelation experiments with a set of selected solvents revealed that long alkyl chains played an important role in gelation as only three such salts (**G32**, **G33** and **G34**) capable of gelation contained hexadecyl

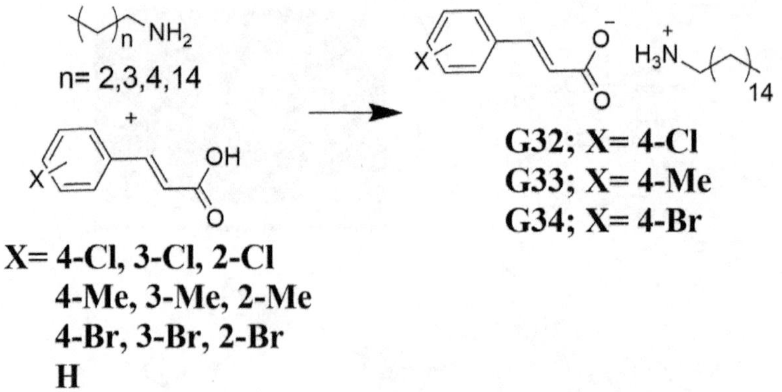

Figure 2.19 Combinatorial library of PAM salts studied for gelation.

alkyl chain in the cationic part. As many as 12 crystal structures of the non-gelator salts in the series were determined and the majority (10 salts) displayed synthon X whereas the rest were characterized by synthon W; none of these non-gelator salts contained cationic parts having alkyl chain more than 14, thereby emphasizing the role of long alkyl chains in gelation. Interestingly, **G34** was found to be capable of gelling selectively commercial fuels (petrol, diesel and kerosene) from the corresponding oil–water mixture (Figure 2.19).[124]

It may be noted that synthon W is intrinsically asymmetric possessing a 2_1 screw axis along the propagation of 1D HBN while synthon X is intrinsically achiral, having an inversion centre in each of the propagating hydrogen bonded rings (Figure 2.18). A detailed CSD analysis of various PAM salts (considering all possible combination of aromatic/aliphatic acids and amines) revealed that in majority of the cases, asymmetric synthon W was observed and it was also evident that the chirality of either the ammonium cation or carboxylate anion was playing a crucial role in the formation of synthon W in these salts. Assisted by these data, a systematic study of PAM salts derived from variously substituted phenyl acetic acid (PA) and chiral amines – (*R*)- and (*S*)-1-phenylethyl amines (PEA) (**G35***ᵢ* and **G36***ᵢ*) was launched. As predicted, all the 32 PAM salts for which crystal structures were obtained possessed 2_1 screw axes displaying exclusively synthon W and ~28% salts were gelators of various solvents. Interestingly left-handed twisted fibres were also observed in the AFM image of one of the gels (Figure 2.20).[125]

PAM synthon was successfully employed in a series of LMWGs derived from the free amine of L-phenylalanine-3-pyridyl amide and variously substituted benzoic acids (**G37**$_i$ and **NG38**$_i$). the majority of the PAM salts thus derived showed gelation ability with various solvents including pure water (hydrogelation). Synthon W was observed in the crystal structures of four gelator salts in the series. The fact that synthon W was formed despite

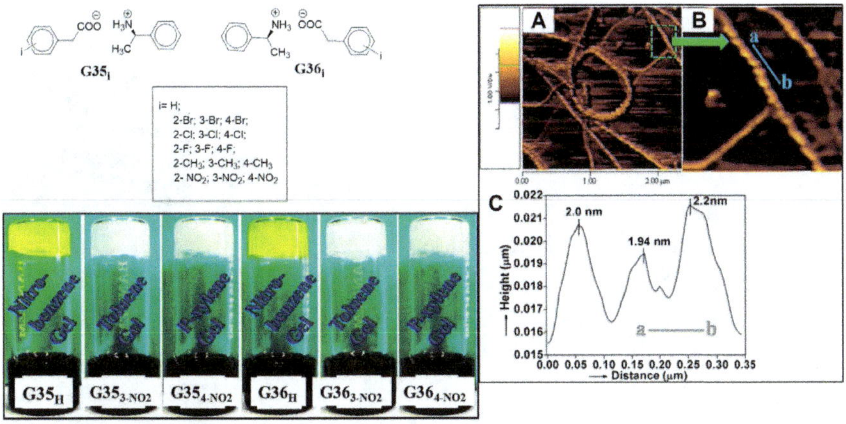

Figure 2.20 Left panel – various PAM salts and the optical images of four wt% gels derived from these PAM salts; AFM images showing left-handed twisted fibres.
Adapted with permission from T. K. Adalder and P. Dastidar, Crystal Engineering Approach toward Selective Formation of an Asymmetric Supramolecular Synthon in Primary Ammonium Monocarboxylate (PAM) Salts and Their Gelation Studies, *Cryst. Growth Des.*, 2014, **14**, 2254. Copyright 2014, American Chemical Society.

having additional hydrogen bonding functionality like amide and pyridyl N, indicates the robustness of the PAM synthon. Most interestingly, Au nanoparticles could be formed over the hydrogel bed of **G37***$_i$* (i = 4-Me) without using any exogenous reducing agent (Figure 2.21).[126]

Very recently PAM salts have found applications in the area of drug delivery. In conventional drug delivery system, a drug molecule is loaded in a vehicle (carrier molecule) and then delivered at the target site. The major concerns associated with this approach are biocompatibility, biodegradability, cytotoxicity, synthetic access of the carrier molecule, drug loading and its timely release to the target site *etc.* An alternative approach namely 'self-delivery' wherein the carrier molecule is not needed is being explored.[45–55] In this approach, the drug molecule is modified either supramolecularly or covalently to a supramolecular gelator; the resulting gel then can be used directly without the help of an external carrier molecule either by topical or subcutaneous route. One such recent example is the PAM salts derived from a NSAID diflunisal and various primary amines (**G39–G42**). These salts were capable of gelling various solvents that included methylsalicylate which is an important ingredient in the commercial formulation of topical gel. Cytotoxicity (MTT) and anti-inflammatory activity assay (PGE_2 assay) on some the select gelator salts revealed that the modified diflunisal drug was biocompatible as well as anti-inflammatory and therefore, were suited for drug delivery *via* topical route (Figure 2.22).[127]

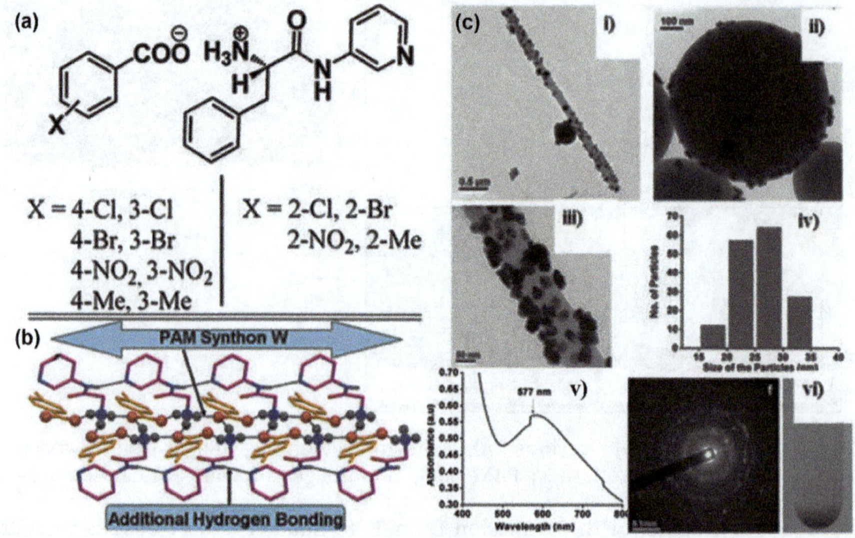

Figure 2.21 (a) variously substituted benzoate salts of 1-phenylalanine-3-pyridyl amide studied for gelation; (b) 1D HBN displaying PAM synthon W with additional hydrogen bonding in the crystal structures of the gelators **G37**$_i$ (i = 4-Cl, 4-Br, 4-NO_2, 4-Me); (c) – (i) Au NPs adhered to the rod and spherical shaped gel network, (ii), (iii) enlarged view of the Au NPs adhered to spherical and rod gel network, respectively, (iv) size distribution of the Au NPs, (v) surface plasmon peak of the Au NPs, selected area electron diffraction of the Au NPs, (vi) aggregated reduced Au particles produced in the aqueous solution of **G37**$_i$ (i = 4-Me). Adapted with permission from U. K. Das, S. Banerjee and P. Dastidar, Primary Ammonium Monocarboxylate Synthon in Designing Supramolecular Gels: A New Series of Chiral Low-Molecular-Weight Gelators Derived from Simple Organic Salts that are Capable of Generating and Stabilizing Gold Nanoparticles, *Chem. – Asian J.*, 2013, **8**, 3022. Copyright 2013, Wiley-VCH Verlag GmbH & Co. KGaA.

2.2.2.4 Primary Ammonium Dicarboxylate (PAD) Synthon

The foregoing discussions emphasized on the role of 1D supramolecular synthons on gelation. What about 2D synthon? Does it have any role in gelation? To probe that PAD salts were considered. It was argued that if the 1D PAM synthons (W or X) prevailed at each end of a dicarboxylate, the resulting assembly would be a 2D HBN. A series of PAD salts derived from 1,1-cyclobutane dicarboxylic acid and *n*-alkyl amines (CH_3-$(CH_2)_n$-NH_2; n = 3–15) showed interesting results. It was observed that the salts having *n*-alkylammonium cation with $n > 10$ were gelators (**G43**$_i$) whereas the salts having $n < 10$ were nongelators (**NG44**$_i$). Crystal structure of the non-gelator salt (n = 5) revealed the existence of 2D PAD synthon. Remarkably two gelators salts (n = 11, 12) for which crystal structures could be determined displayed supramolecular hydrogen bond isomerism producing a 1D nanotubular HBN. These results once again emphasized the role of 1D network in gelation (Figure 2.23).[128]

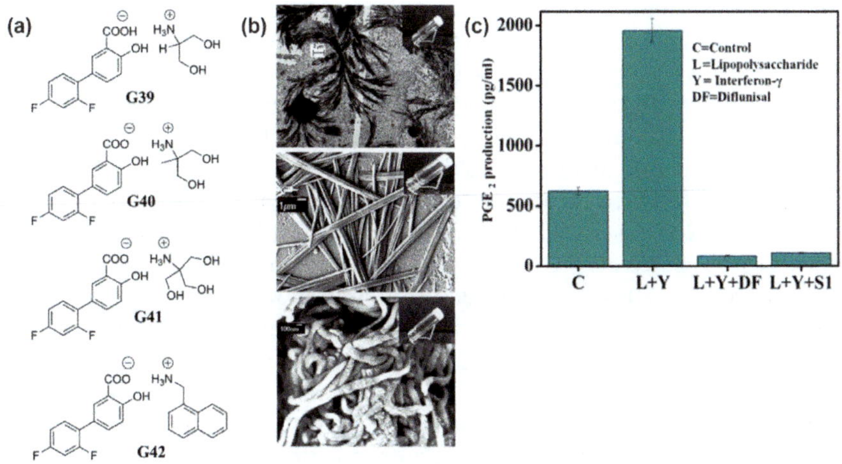

Figure 2.22 (a) PAM salts derived from diflunisal and various primary amines; (b) physical appearance of the gels (insets) and SEM micrographs of the gel networks; (c) PGE_2 assay of diflunisal and gelator **G39** in RAW 264.7 cells.
Adapted with permission from R. Parveen and P. Dastidar, Easy Access to Supramolecular Gels of the Nonsteroidal Anti-Inflammatory Drug Diflunisal: Synthesis, Characterization, and Plausible Biomedical Applications, *Chem. – Asian J.*, 2015, **10**, 2427. Copyright 2015, Wiley-VCH Verlag GmbH & Co. KGaA.

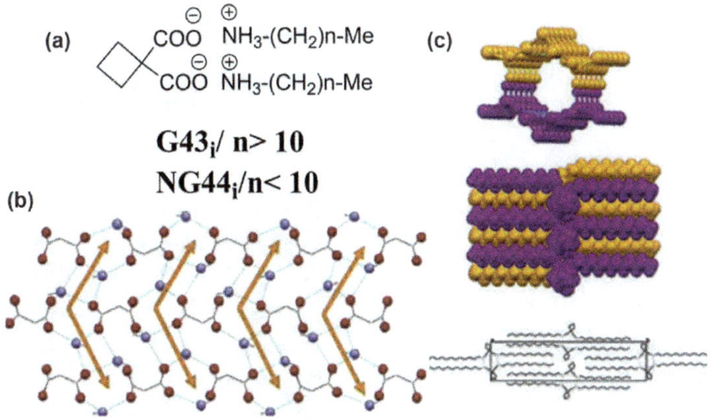

Figure 2.23 (a) Various PAD salts studied for gelation; (b) 2D PAD seen in the crystal structure of **NG44**$_i$ ($n=5$); (c) crystal structure illustration of **G43**$_i$ ($n=11$): *top* – nanotubular network, *middle* – side view of the nanotubular network, *bottom* – interdigital packing of **G43**$_i$ ($n=11$) displaying long alkyl chain interactions.
Adapted with permission from A. Ballabh, D. R. Trivedi and P. Dastidar, From Nonfunctional Lemellae to Functional Nanotubes, *Org. Lett.*, 2006, **8**, 1271. Copyright 2006, American Chemical Society.

In another attempt wherein PAD synthon was employed, a combinatorial library of PAD salts derived from various dicarboxylic acids and *n*-alkyl amines ($n = 11$–15) was prepared and scanned for gelation. Majority of the salts (~66.0%) of the salts were found be moderate to good gelator of various polar and non-polar solvents including commercial fuels such as petrol. Single crystal structures of two gelator salts (**G45, G46**) and two non-gelator salts (**NG47, NG48**) revealed intriguing features. While **G45** displayed the same 1D columnar HBN as observed in the previous case just discussed (see above), **G46** showed the expected 2D PAD synthon. On the other hand, both the non-gelator salts showed 2D HBN. These structural features indicated that 1D HBN is important but may not be the only criterion for gelation (Figure 2.24).[129]

PAD synthon was successfully employed to develop ferrocene based metallogels. In this study, a series of PAD salts were prepared by reacting ferrocene dicarboxylic acid and n-alkyl amines (**G49**$_i$; $i = 3$–15). Nearly all the salts in the series turned out to be gelator of various solvents. Single crystals structures of six gelator salts (**G49**$_i$; $i = 5$, 9, 11–14) could be determined and all of them turned out to be isostructural having same space group *P*-1 with varying *c*-axis presumably to accommodate the increasing chain length of

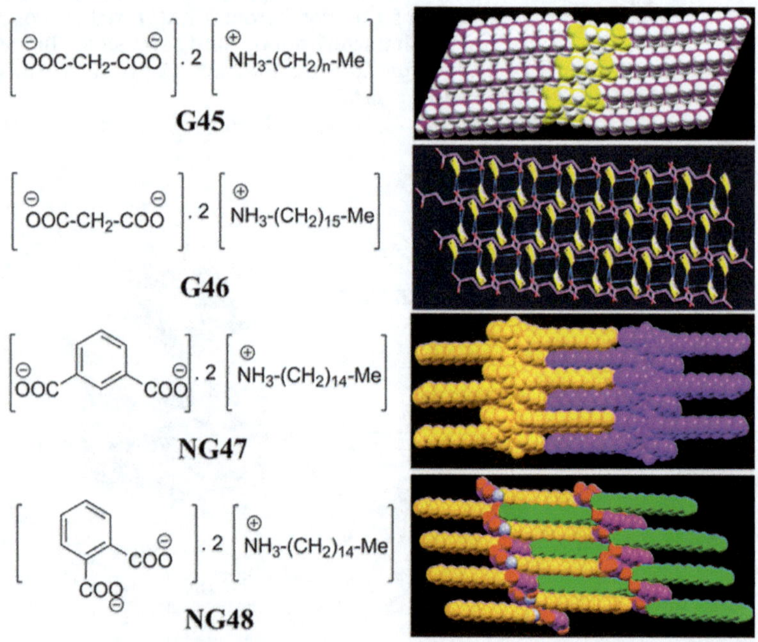

Figure 2.24 Various PAD salts studied for gelation and their corresponding single crystal structures.
Adapted with permission from P. Sahoo, N. N. Adarsh, G. E. Chacko, S. R. Raghavan, V. G. Puranik and P. Dastidar, Combinatorial Library of Primaryalkylammoium Dicarboxylate Gelators: A Supramolecular Synthon Approach, *Langmuir*, 2009, **25**, 8742. Copyright 2009, American Chemical Society.

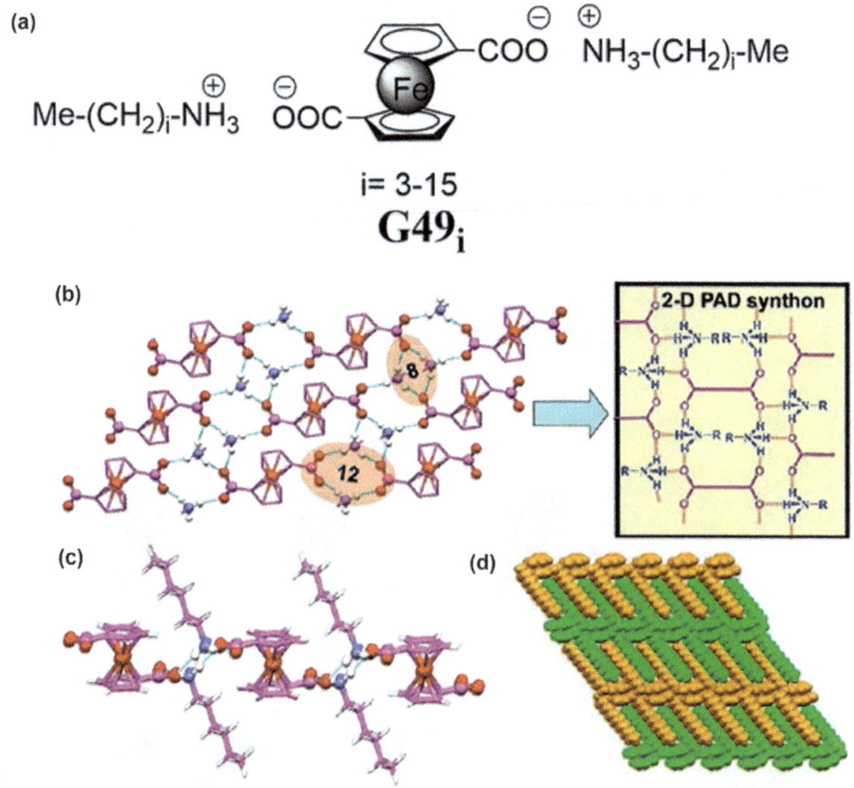

Figure 2.25 (a) Various PAD salts studied for gelation and Crystal structure illustration; (b) 2D PAD synthon present in the crystal structures reported herein; (c) side view of the 2D hydrogen bonded sheet; (d) interdigited packing of the 2D sheets maximizing the hydrophobic interactions of the long alkyl chain.
Reproduced in part from ref. 130 with permission from the Royal Society of Chemistry.

the alkyl group. Analyses of the structures, revealed that it was 2D PAD synthon having PAM synthon X in the two terminal carboxylate positions. These results clearly demonstrate that 2D HBN is also important in gelation in favorable cases (Figure 2.25).[130]

In an attempt to produce and stabilize metal nanoparticles (MNPs), PAD derived gel bed was exploited. Metal nanoparticles (MNPs) are known for their various applications in catalysis,[131,132] bioimaging,[133] and sensing.[134,135] Controlling and stabilizing the size of MNPs are a big challenge. Slow nucleation and growth of MNPs[136] appeared to be important. Slow and sustained release of weak reducing agent was found to be effective for this purpose. Since both acid and amines are known to act as reducing agents in synthesizing MNPs, salt based gel bed was considered as the fertile ground for slow and sustained release of the components of salts as possible reducing agent.

For this purpose, a series of PAD gelators derived from (1*R*,3*S*)-(+)-camphoric acid (CA) and *n*-alkyl amines having varying alkyl chain length ($n = 4$–15) was synthesized and scanned for gelation with a number of solvents. It was observed that 100% of the salts (**G50***i*; $n = 4$–15) were gelators. DMF gel of one such gelator salt (**G50***i*; $n = 14$) was able to produce gold nanoparticles (AuNPs) without the use of any exogenous reducing agent. Control experiments revealed that the amine component was the reducing agent (Figure 2.26).[137]

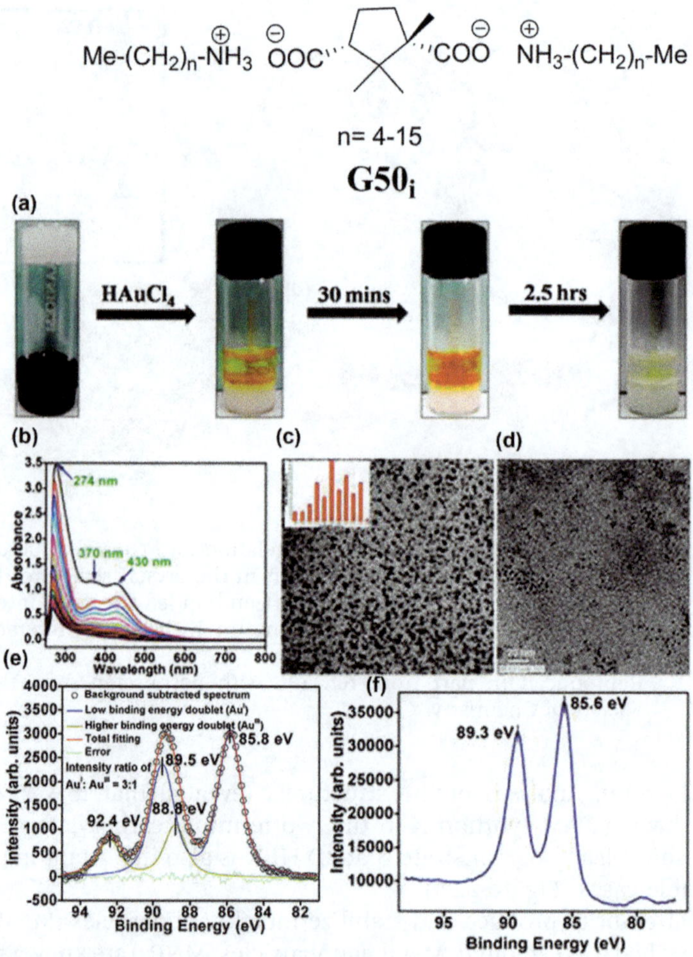

Figure 2.26 Camphoric acid derived PAD salts studied for gelation and AuCs formation; (a) synthesis of AuCs over a gel bed 5.0 wt% DMF gel of **G50***i* ($n = 14$); (b) time-resolved UV–vis spectra of the solution above the gel bed; (c and d) FEG-TEM of wine-red and colorless solution, respectively; (e and f) XPS of wine-red and colorless solution, respectively. Adapted with permission from T. K. Adalder, D. P. Kumar and P. Dastidar. High-Throughput Crystal Engineering Based Synthesis of Supramolecular Gels: Blue-Emitting Fluorescent Gold Clusters Synthesized and Stabilized on the Gel-Bed, *Cryst. Growth Des.*, 2014, **14**, 11. Copyright 2014, American Chemical Society.

2.3 Metallogels

Metallogels are special types of gels derived from molecules containing metal atoms. These classes of gels are becoming increasingly important because of their various potential applications (*see introduction*). Akin to LMWGs, designing metallogelators is also a difficult task as the molecular level understanding of gelation mechanism is unclear. However, there are attempts to design metallogelators that include appending an ALS (a well-known gel forming fragment) to an organometallic moiety,[138] aborting crystallization by increasing the randomness of mixed ligand carboxylates of AgI coordination compounds,[139] introducing a long alkyl chain on co-ordination complex,[140] flexible ligand coordination and slow formation of coordination polymers,[56] and also the supramolecular synthon approach.[120]

One of the most systematic attempts to design metallogelators was inspired by the structural analogy that exists between a gel and lattice inclusion compounds; in both the cases, a large number of solvent molecules are entrapped within the corresponding network formed due to the supramolecular self-assembly. It has also been shown in a few cases that a considerably large number of solvent molecules are entrapped with the lattice in the crystal structure of some LMWGs. It was proposed therefore that if structural features conducive to lattice occlusion were installed in a co-ordination polymeric system, metallogels may be obtained under suitable conditions. The proof of the concept was demonstrated in a mixed ligand based coordination polymeric system wherein hydrogen bond equipped bis-pyridyl-bis-amide ligands (**L1** and **L2**) were reacted with various di-carboxylates in presence of metal salts (Cu^{II}/Co^{II}) in 1 : 1 : 1 molar ratio. It was envisaged that if the carboxylates coordinated to the metal center as a bi-dentate ligand, a 2D grid like architecture would be formed; the pores of the grid would then be occupied by hydrogen bond capable guest molecules driven by hydrogen bonding backbone of the ligands. If the other possibilities (carboxylates acting as a bridge or chelate) prevailed, 1D polymeric chain would be the result; even then, solvent occlusion was possible because of the hydrogen bonding backbone of the ligand. Remarkably both **L1** and **L2** were found to gel DMF/water when reacted with the metal salts (Cu^{II}/Co^{II}) in presence of various dicarboxylates (oxalate, malonate, terephthalate, and naphthalate) in 1 : 1 : 1 molar ratio. The gels were not thermo-reversible indicating their coordination polymeric nature of the network and were visco-elastic as revealed by the rheological data (frequency sweep). Single crystals of the corresponding coordination polymers (CPs) were obtained by slow evaporation of the much diluted solution of the reaction mixtures. SXRD analyses revealed that most of the CPs were indeed lattice occluded solids. Remarkably PXRD patterns of the xerogels matched pretty well with the simulated and experimental patterns of the corresponding CPs in most the cases supporting the proof of the concept based on which these metallogels were designed (Figure 2.27).[141]

A bis-pyridyl-bis-amide ligand derived from L-tartaric acid (**L3**) was shown to produce metallogels as well as single crystals of the corresponding CPs when reacted with various metal salts of Cu^{II} under different conditions. Single

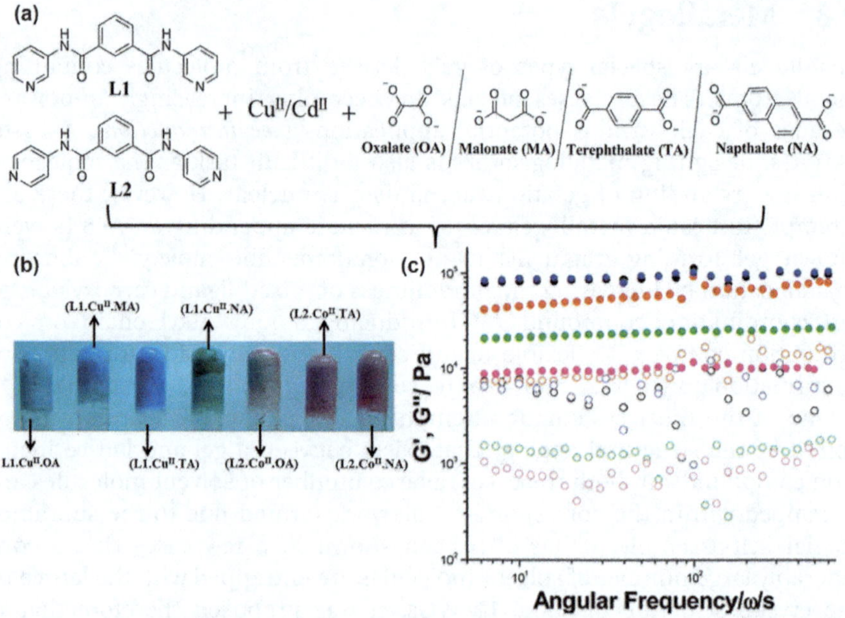

Figure 2.27 (a) Different ligands, metal salts and linker used for gelation; (b) optical images of various metallogels; (c) rheological responses (frequency sweeps) of the metallogels.
Adapted with permission from N. N. Adarsh, P. Sahoo and P. Dastidar, Is a Crystal Engineering Approach Useful in Desgining Metallogels? A Case Study, *Cryst. Growth Des.*, 2010, **10**, 4976. Copyright 2010, American Chemical Society.

crystal structure of one such CP namely $[\{(H_2O)Cu_2(\mu\text{-L3})_2\mu\text{-Cl}\}\cdot(SO_4)(Cl)]_\infty$ and PXRD of the corresponding xerogel established that a SnS type topological network present in the CP was the gel network (Figure 2.28).[142]

Recently Cu^{II} based metallogels have been developed from the lowest molecular weight ligand (oxalate) that displayed unprecedented proton conductivity and formation of solvent free Cu-oxalate polymorph from the metallogel as well as xerogel (Figure 2.29).[143]

Luminescent metallogels have been developed from a bis-amide-dicarboxylate ligand (**H_2L**) derived from pyridine-2,6-dicarboxylic acid. Highly luminescent gels were obtained when **H_2L** was reacted with lanthanides (Ln(III) = Eu(III), Tb(III)). Variety of different colors could be produced by mixing Eu and Tb gel in appropriate amount. Studies showed that these gels prepared using only Eu(III) or only Tb(III) had different morphological and rheological properties that were different from those obtained from the gels prepared by mixing of Eu(III) and Tb(III) gels. The gels were also found to have possessed self-healing properties. It was concluded that the host–guest interactions expected between the lanthanide metal and the ligand **H_2L** were responsible for gelation as well as self-healing behaviour (Figure 2.30).[144]

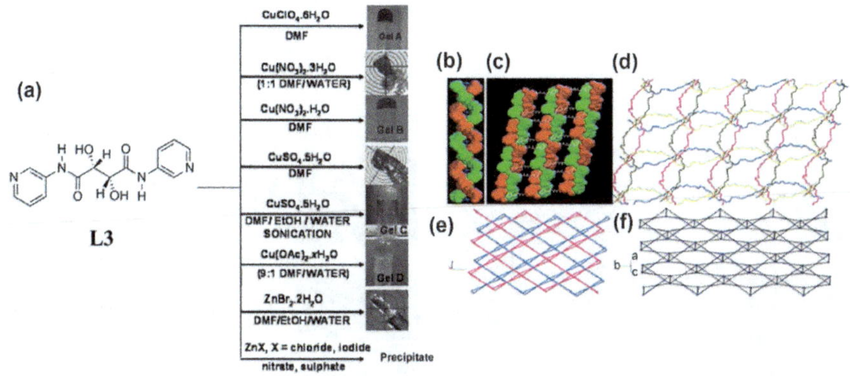

Figure 2.28 (a) Metallogels and crystals obtained from **L3**; single crystal structure illustration of a CP: (b) double-helical strand bridged by chloride; (c) 2D grid composed of all double-helical sides (the horizontal sides are shown in caped stick (white)); (d) another representation of a 5-c 2D network (part of the aromatic rings and hydroxyl moieties and hydrogen not shown for better clarity); (e) TOPOS diagram of the 2D grid (uncoordinated chloride and sulfate not shown for better clarity); (f) TOPOS diagram of SnS.
Reproduced with permission from U. K. Das and P. Dastidar, Supramolecular Chirality in Organo-, Hydro-, and Metallogels Derived from Bis-amides of L-(+)-Tartaric Acid: Formation of Highly Aligned 1D Silica Fibres and Evidence of 5-c Net SnS Topology in a Metallogel Network, *Chem. – Eur. J.*, 2012, **18**, 13079. Copyright 2012, Wiley-VCH Verlag GmbH & Co. KGaA.

Metallogels derived from pyridine-3,5-bis(benzimidazole-2-yl) (**L**) and Cu^{II}/Cd^{II} displayed remarkable mechanical properties. Studies revealed that the gelation behaviour was dependent on the nature of the counter anions; halides (chlorides or bromides) and sulfates were found to be more efficient than other univalent ions such as nitrate, aceteate, perchlorate, and tetrafluoro borate. Rheological data supported the observed mechanical properties (self sustainability, load-bearing, self-healing) of the metallogels. The xerogels were porous and displayed nice N_2-gas sorption isotherms (Figure 2.31).[145]

2.4 Summing Up and Future Direction

Soft supramolecular materials such as gels are indeed interesting and research attempts towards the design of such materials are challenging and rewarding. In this chapter, attempt has been made to delineate the progress of such research activities towards the designing of supramolecular gels. Designing gelators *a priori* is indeed daunting task as the molecular level understanding of gelation mechanism is still in its infancy. There are innumerous parameters that control gelation. Starting from the nucleation of gel fibre, entanglement of fibres to form SAFiNs and finally immobilization

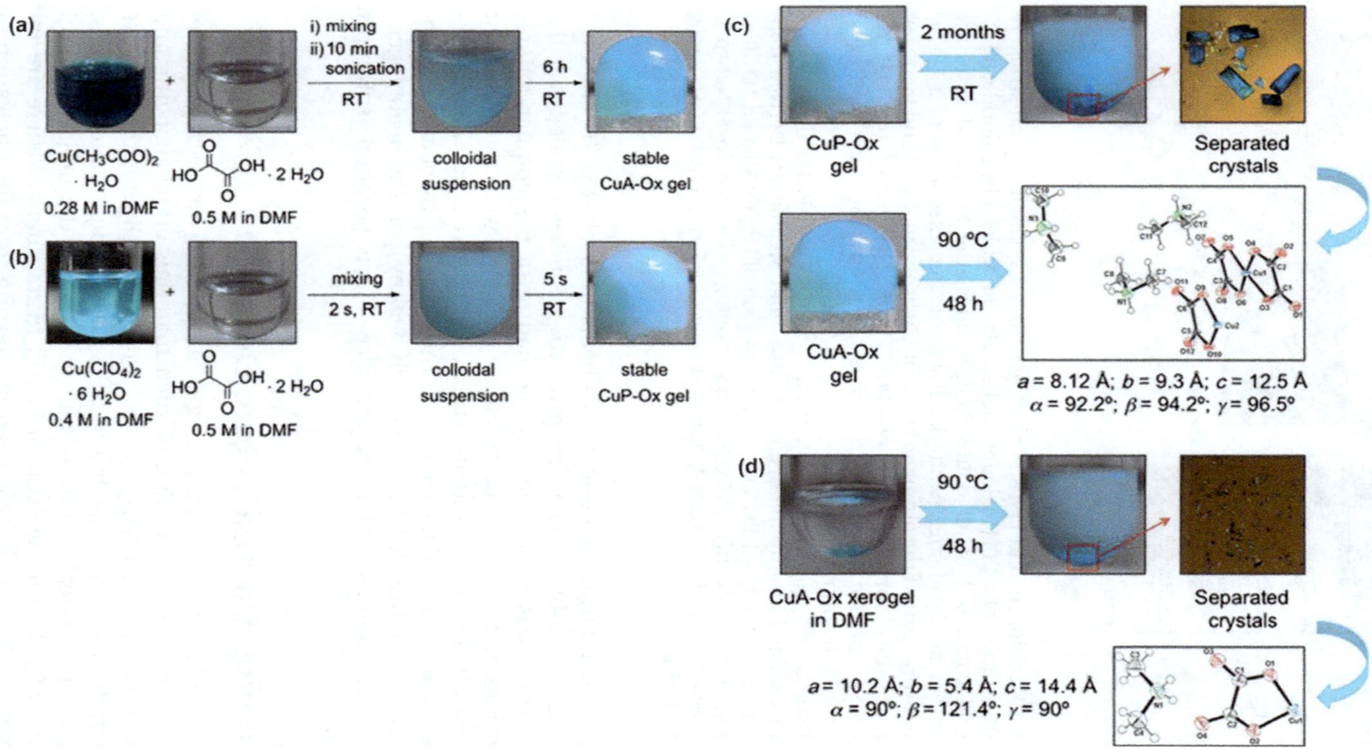

Figure 2.29 Synthesis of (a) CuA-Ox gel; (b) CuP-Ox gel upon mixing stock solutions of the corresponding Cu^{II} salt and oxalic acid in DMF at optimized concentrations; a transient light-blue colloidal solution precedes in each case the formation of the stable and homogeneous metallogels; formation of blue block-shaped Cu-Ox polymorphs from (c) CuP-Ox or CuA-Ox gels; (d) CuA-Ox xerogel. Reproduced with permission from S. Saha, E-M. Schön, C. Cativiela, D. Díaz Díaz and R. Banerjee, Proton-Conducting Supramolecular Metallogels from the Lowest Molecular Weight Assembler Ligand: A Quote for Simplicity, *Chem. – Eur. J.*, 2013, **19**, 9562. Copyright 2013, Wiley-VCH Verlag GmbH & Co. KGaA.

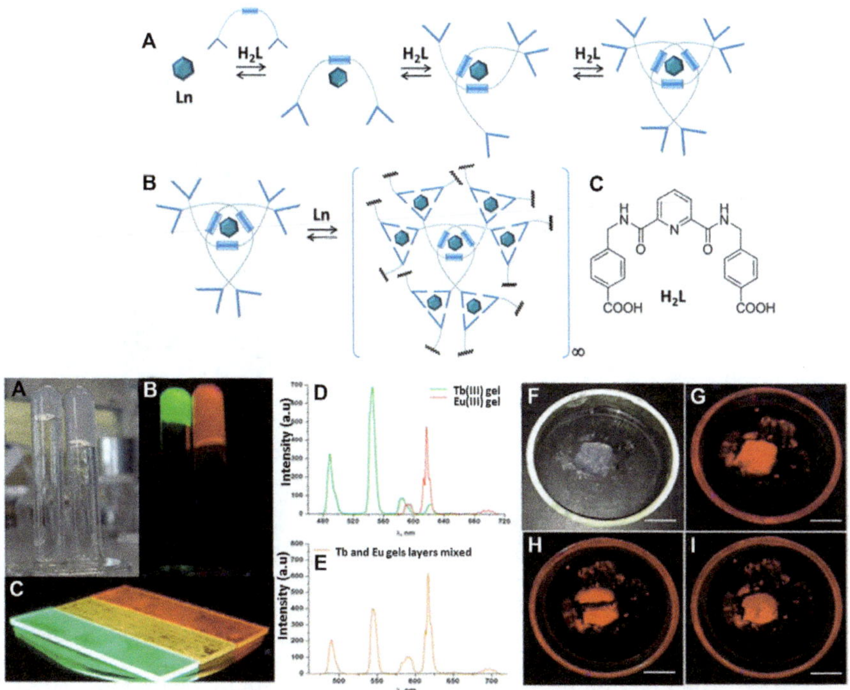

Figure 2.30 Self-assembly formation of **Ln:H_2L** with the following formation of higher order metallo-supramolecular polymers with structural formula of **H_2L**; Eu(III) and Tb(III) gels (A) in day light; (B) their luminescence under UV light; (C) Luminescence of Eu(III), Tb(III), and Eu(III)/Tb(III) gels on quartz plates; (D) Eu(III), Tb(III); (E) Eu(III)/Tb(III) mixed gel luminescence spectra at 25 °C ($\lambda_{ex} = 275$ nm); (F–I) Healing experiment of Eu(III) gel with (F) Eu(III) gel in the day light; (G) the same gel under UV light; (H) gel after being cut in half; (I) self-healing properties of the gel (scale bars, 1 cm).
Reproduced with permission from M. Martínez-Calvo, O. Kotova, M. E. Möbius, A. P. Bell, T. McCabe, J. J. Boland and T. Gunnlaugsson, Healable Luminescent Self-Assembly Supramolecular Metallogels Possessing Lanthanide (Eu/Tb) Dependent Rheological and Morphological Properties, *J. Am. Chem. Soc.*, 2015, **137**, 1983. Copyright 2015, American Chemical Society.

of the target solvent molecules resulting in gelation are all governed by various parameters on which the researchers have hardly any control. However, based on the current understanding of gelation, two major approaches (molecular and crystal engineering) have been discussed with seminal and contemporary examples. Hydrogen bonding being one of the most reliable and relatively strong non-covalent interactions in supramolecular chemistry, most of the design of LMWGs was dependent on the clever use of such interactions. In crystal engineering approach, supramolecular synthons exploited in designing organic salt based LMWGs turned out to be

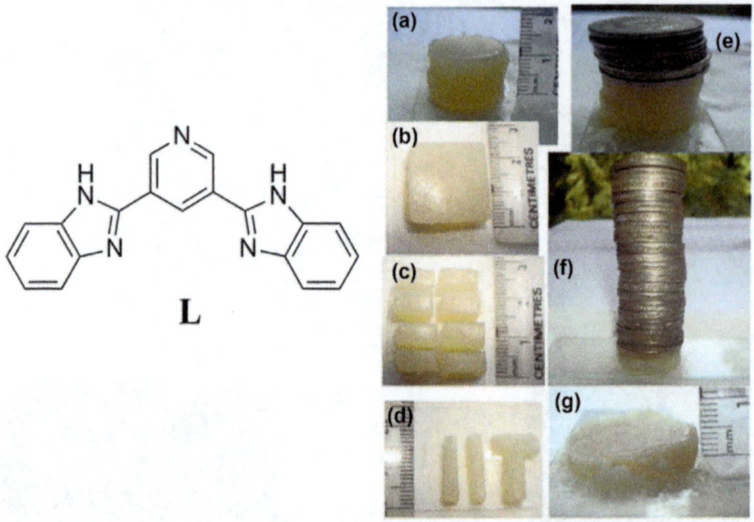

Figure 2.31 Molecular structure of pyridine-3,5-bis(benzimidazole-2-yl) (**L**); self-sustaining behaviour of MOG-3; (a) cylindrical; (b) rectangular box shapes of gel material; (c) cutting the gel into pieces; (d) tailoring the gel to make the shape of IIT; the sustainability of the gel material after carrying the (e) three coins; (f) twenty coins; (g) the compression in the height of the gel (12 mm to 6 mm) after carrying twenty coins.
Adapted with permission from S. Samai and K. Biradha, Chemical and Mechano Responsive Metal-Organic Gels of Bis(benzimidazole)-Based Ligands with Cd(II) and Cu(II) Halide Salts: Self Sustainability and Gas and Dye Sorptions, *Chem. Mater.*, 2012, **24**, 1165. Copyright 2012, American Chemical Society.

quite attractive and successful. Large body of data pertaining to the design of LMWGs based on structure–property correlation approach (crystal engineering approach) clearly indicated that 1D HBN indeed played a crucial role in gelation. However, that is not the necessary and sufficient conditions as many organic salts displaying 1D HBN failed to show gelation. Interactions of the SAFiNs with the target solvent molecules must be understood at molecular level. Another class of supramolecular gelators wherein metal ion is involved in the gel network formation (metallogel) is also discussed with some seminal and recent examples. Crystal engineering approach was also found to be quite useful in designing such metallo-supramolecular soft materials.

Acknowledgements

PD thanks DST, DBT and CSIR for financial support. Central Salt and Marine Chemicals Research Institute (CSMCRI), Bhavnagar and Indian Association for the Cultivation of Science (IACS) for providing infrastructure and financial support are also thankfully acknowledged. PD also thanks the former group members for their contributions. RR, RP, SG, JM, MP thank CSIR, New Delhi and IACS for their research fellowships.

References

1. M. George and R. G. Weiss, *Acc. Chem. Res.*, 2006, **39**, 489.
2. D. K. Smith, *Chem. Soc. Rev.*, 2009, **38**, 684.
3. P. Dastidar, *Chem. Soc. Rev.*, 2008, **37**, 2699.
4. M. Suzuki and K. Hanabusa, *Chem. Soc. Rev.*, 2009, **38**, 967.
5. L. E. Buerkle and S. J. Rowan, *Chem. Soc. Rev.*, 2012, **41**, 6089.
6. A. Ajayaghosh, V. K. Praveen and C. Vijayakumar, *Chem. Soc. Rev.*, 2008, **37**, 109.
7. D. K. Kumar and J. W. Steed, *Chem. Soc. Rev.*, 2014, **43**, 2080.
8. G. Yu, X. Yan, C. Han and F. Huang, *Chem. Soc. Rev.*, 2013, **42**, 6697.
9. M. D. Segarra-Maset, V. J. Nebot, J. F. Miravet and B. Escuder, *Chem. Soc. Rev.*, 2013, **42**, 7086.
10. A. Dawn, T. Shiraki, S. Haraguchi, S. Tamaru and S. Shinkai, *Chem. – Asian J.*, 2011, **6**, 266.
11. Z. Yang, G. Liang and B. Xu, *Acc. Chem. Res.*, 2008, **41**, 315.
12. Z. Qi and C. A. Schalley, *Acc. Chem. Res.*, 2014, **47**, 2222.
13. M. Lescanne, P. Grondin, A. d'Aleo, F. Fages, J.-L. Pozzo, O. M. Monval, P. Reinheimer and A. Colin, *Langmuir*, 2004, **20**, 3032.
14. Y. Zhao, J. B. Beck, S. J. Rowan and A. M. Jamieson, *Macromolecules*, 2004, **37**, 3529.
15. J. Brinksma, B. L. Feringa, R. M. Kellog, R. Vreeker and J. van Esch, *Langmuir*, 2000, **16**, 9249.
16. L. Garlaschelli, F. Ramaccini and S. D. Sala, *Nature*, 1991, **353**, 507.
17. H. A. J. Barnes, *J. Non-Newtonian Fluid Mech.*, 1997, **70**, 1.
18. F. M. Menger and A. V. Peresypkin, *J. Am. Chem. Soc.*, 2003, **125**, 5340.
19. K. Lalitha, Y. S. Prasad, V. Sridharan, C. U. Maheswari, George Johnb and S. Nagarajan, *RSC Adv.*, 2015, **5**, 77589.
20. M. Suzuki, Y. Hayakawa and K. Hanabusa, *Gels*, 2015, **1**, 81.
21. W. Fang, Z. Sun and T. Tu, *J. Phys. Chem. C*, 2013, **117**, 25185.
22. D. Higashi, M. Yoshida and M. Yamanaka, *Chem. – Asian J.*, 2013, **8**, 2584.
23. R. G. Weiss and P. Terech, *Molecular Gels: Materials with Self-assembled Fibrillar Networks*, Springer, Dordrecht, 2005.
24. P. Terech and R. G. Weiss, *Chem. Rev.*, 1997, **97**, 3133.
25. V. Lipowitz and A. Liebigs, *Ann. Chem. Pharm.*, 1841, **38**, 348.
26. J. J. D. de Jong, L. N. Lucas, R. M. Kellogg, J. H. van Esch and B. L. Feringa, *Science*, 2004, **304**, 278.
27. K. Murata, M. Aoki, T. Nishi, A. Ikeda and S. Shinkai, *J. Chem. Soc., Chem. Commun.*, 1991, 1715.
28. M. Yamauchi, S. Kubota, T. Karatsu, A. Kitamura, A. Ajayaghosh and S. Yagai, *Chem. Commun.*, 2013, **49**, 4941.

29. Q. Lin, T.-T. Lu, X. Zhu, B. Sun, Q.-P. Yang, T.-B. Weia and Y.-M. Zhang, *Chem. Commun.*, 2015, **51**, 1635.
30. K. J. C. van Bommel, A. Friggeri and S. Shinkai, *Angew. Chem., Int. Ed.*, 2003, **42**, 980.
31. H. Basit, A. Pal, S. Sen and S. Bhattacharya, *Chem. – Eur. J.*, 2008, **14**, 6534.
32. T. M. Babu and E. Prasad, *Chem. – Eur. J.*, 2015, **21**, 11972.
33. A. Corma, F. Rey, J. Rius, M. J. Sabater and S. Valencia, *Nature*, 2004, **431**, 287.
34. A. Ajayaghosh, V. K. Praveen, C. Vijayakumar and S. George, *Angew. Chem., Int. Ed.*, 2007, **46**, 6260.
35. T. Kato, *Science*, 2000, **295**, 2414.
36. A. Ajayaghosh, C. Vijayakumar, V. K. Praveen, S. S. Babu and R. Varghese, *J. Am. Chem. Soc.*, 2006, **128**, 7174.
37. S. S. Babu, V. K. Praveen and A. Ajayaghosh, *Chem. Rev.*, 2014, **114**, 1973.
38. V. Jenning, A. Gysler, M. Schafer-Korting and S. H. Gohla, *Eur. J. Pharm. Biopharm.*, 2000, **49**, 211.
39. A. Wynne, M. Whitefield, A. J. Dixon and S. Anderson, *J. Dermatol. Treat.*, 2002, **13**, 61.
40. W. Weng, J. B. Beck, A. M. Jamieson and S. J. Rowan, *J. Am. Chem. Soc.*, 2006, **128**, 11663.
41. H. Svobodová, V. Noponen, E. Kolehmainen and E. Sievänen, *RSC Adv.*, 2012, **2**, 4985.
42. E. Carretti, M. Bonini, L. Dei, B. H. Berrie, L. V. Angelova, P. Baglioni and R. G. Weiss, *Acc. Chem. Res.*, 2010, **43**, 751.
43. E. Carretti and L. Dei, in *Molecular Gels: Materials with Self-assembled Fibrillar Networks*, ed. R. G. Weiss and P. Terech, Springer, Dordrecht, 2005, p. 929.
44. E. Carretti, L. Dei and R. G. Weiss, *Soft Matter*, 2005, **1**, 17.
45. R. Langer, *Nature*, 1998, **392**, 5.
46. X. Du, J. Zhou, J. Shi and B. Xu, *Chem. Rev.*, 2015, **115**, 13165.
47. K. E. Uhrich, S. M. Cannizzaro, R. S. Langer and K. M. Shakesheff, *Chem. Rev.*, 1999, **99**, 3181.
48. M. J. Webber, J. B. Matson, V. K. Tamboli and S. I. Stupp, *Biomaterials*, 2012, **33**, 6823.
49. J. Li, Y. Gao, Y. Kuang, J. Shi, X. Du, J. Zhou, H. Wang, Z. Yang and B. Xu, *J. Am. Chem. Soc.*, 2013, **135**, 9907.
50. K. Y. Lee and D. J. Mooney, *Chem. Rev.*, 2001, **101**, 1869.
51. H. Wang, J. Wei, C. Yang, H. Zhao, D. Li, Z. Yin and Z. Yang, *Biomaterials*, 2012, **33**, 5848.
52. W. T. Truong, Y. Su, J. T. Meijer, P. Thordarson and F. Braet, *Chem. – Asian J.*, 2011, **6**, 30.
53. R. Lin, A. G. Cheetham, P. Zhang, Y. Lin and H. Cui, *Chem. Commun.*, 2013, **49**, 4968.
54. F. Zhao, M. L. Ma and B. Xu, *Chem. Soc. Rev.*, 2009, **38**, 883.

55. Y. Gao, F. Zhao, Q. Wang, Y. Zhang and B. Xu, *Chem. Soc. Rev.*, 2010, **39**, 3425.
56. B. Xing, M. Choi and B. Xu, *Chem. Commun.*, 2002, **38**, 362.
57. Q. Wang, L. Li and B. Xu, *Chem. - Eur. J.*, 2009, **15**, 3168.
58. J. W. Steed, *Chem. Soc. Rev.*, 2010, **39**, 3686.
59. B. Xing, M. Choi and B. Xu, *Chem. - Eur. J.*, 2002, **8**, 5028.
60. A. Y. Tam and V. W. Yam, *Chem. Soc. Rev.*, 2013, **42**, 1540.
61. Y. Liu, L. He, J. Zhang, X. Wang and C. Su, *Chem. Mater.*, 2009, **21**, 557.
62. D. Lopez and J.-M. Guenet, *Macromolecules*, 2001, **34**, 1076.
63. M. Kimura, T. Muto, H. Takimoto, K. Wada, K. Ohta, K. Hanabusa, H. Shirai and N. Kobayashi, *Langmuir*, 2000, **16**, 2078.
64. P. Terech, G. Gebel and R. Ramasseul, *Langmuir*, 1996, **12**, 4321.
65. T. Ishi-I, R. Iguchi, E. Snip, M. Ikeda and S. Shinkai, *Langmuir*, 2001, **17**, 5825.
66. O. Kotova, R. Daly, C. M. G. dos Santos, M. Boese, P. E. Kruger, J. J. Boland and T. Gunnlaugsson, *Angew. Chem., Int. Ed.*, 2012, **51**, 7208.
67. W. J. Gee and S. R. Batten, *Chem. Commun.*, 2012, **48**, 4830.
68. Z. Džolić, M. Cametti, D. Milić and M. Žinic, *Chem. - Eur. J.*, 2013, **19**, 5411.
69. P. Terech, V. Schaffhauser, P. Maldivi and J. M. Guenet, *Langmuir*, 1992, **8**, 2104.
70. S. Bhattacharjee and S. Bhattacharya, *Chem. Commun.*, 2014, **50**, 11690.
71. W. L. Leong, A. Y. Tam, S. K. Batabyal, L. W. Koh, S. Kasapis, V. W. Yam and J. J. Vittal, *Chem. Commun.*, 2008, **44**, 3628.
72. M. M. Piepenbrock, N. Clarke and J. W. Steed, *Soft Matter*, 2011, **7**, 2412.
73. K. Hanabusa, Y. Maesaka, M. Suzuki, M. Kimura and H. Shirai, *Chem. Lett.*, 2000, **29**, 1168.
74. S. R. Bull, M. O. Guler, R. E. Bras, T. J. Meade and S. I. Stupp, *Nano Lett.*, 2005, **5**, 1.
75. M. D. Segarra-Maset, V. J. Nebot, J. F. Miravet and B. Escuder, *Chem. Soc. Rev.*, 2013, **42**, 7086.
76. B. Escuder, F. Rodríguez-Llansola and J. F. Miravet, *New J. Chem.*, 2010, **34**, 1044.
77. B. Escuder and J. F. Miravet, *Chem. Commun.*, 2005, **41**, 5796.
78. T. Tu, X. Bao, W. Assenmacher, H. Peterlik, J. Daniels and K. H. Dötz, *Chem. - Eur. J.*, 2009, **15**, 1853.
79. J. Huang, L. He, J. Zhang, L. Chen and C.-Y. Su, *J. Mol. Catal. A: Chem.*, 2010, **317**, 97.
80. Y.-R. Liu, L. He, J. Zhang, X. Wang and C.-Y. Su, *Chem. Mater.*, 2009, **21**, 557.
81. O. Roubeau, A. Colin, V. Schmitt and R. Clerac, *Angew. Chem., Int. Ed.*, 2004, **43**, 3283.
82. A. Kishimura, T. Yamashita, K. Yamaguchi and T. Aida, *Nat. Mater.*, 2005, **4**, 546.
83. A. Y.-Y. Tam, K. M.-C. Wong, G. Wang and V. W.-W. Yam, *Chem. Commun.*, 2007, **43**, 2028.

84. M. Shirakawa, N. Fujita, T. Tani, K. Kaneko and S. Shinkai, *Chem. Commun.*, 2005, **41**, 4149.
85. S. Kume, K. Kuroiwa and N. Kimizuka, *Chem. Commun.*, 2006, **42**, 2442.
86. Y. C. Lin and R. G. Weiss, *Liq. Cryst.*, 1989, **4**, 367.
87. Y. C. Lin, B. Kachar and R. G. Weiss, *J. Am. Chem. Soc.*, 1989, **111**, 5542.
88. Y. C. Lin and R. G. Weiss, *U.S. Pat. 4*, 1988, **790**, 961.
89. Y. C. Lin and R. G. Weiss, *Macromolecules*, 1987, **20**, 414.
90. J. C. MacDonald and G. M. Whitesides, *Chem. Rev.*, 1994, **94**, 2383.
91. C. B. Aakeröy and K. R. Seddon, *Chem. Soc. Rev.*, 1993, **22**, 397.
92. U. Beginn, S. Sheiko and M. Moeller, *Macromol. Chem. Phys.*, 2000, **201**, 1008.
93. S. Bhattacharya and Y. K. Ghosh, *Chem. Commun.*, 2001, 85.
94. K. J. C. van Bommel, C. van der Pol, I. Muizebelt, A. Friggeri, A. Heeres, A. Meetsma, B. L. Feringa and J. van Esch, *Angew. Chem., Int. Ed.*, 2004, **43**, 1663.
95. J. Majumder, M. R. Das, J. Deb, S. S. Jana and P. Dastidar, *Langmuir*, 2013, **29**, 10254.
96. S. Bai, S. Debnath, N. Javid, P. W. J. M. Frederix, S. Fleming, C. Pappas and R. V. Ulijn, *Langmuir*, 2014, **30**, 7576.
97. T. Kitahara, M. Shirakawa, S. Kawano, U. Beginn, N. Fujita and S. Shinkai, *J. Am. Chem. Soc.*, 2005, **127**, 14980.
98. E. R. Draper, E. G. B. Eden, T. O. McDonald and D. J. Adams, *Nat. Chem.*, 2015, 7, 848.
99. G. O. Lloyd and J. W. Steed, *Nat. Chem.*, 2009, **1**, 437.
100. H. Maeda, *Chem. – Eur. J.*, 2008, **14**, 11274.
101. E.-M. Schön, E. Marqués-López, R. P. Herrera, C. Alemán and D. Díaz Díaz, *Chem. – Eur. J.*, 2014, **20**, 10720.
102. C. Shi, Z. Huang, S. Kilic, J. Xu, R. M. Enick, E. J. Beckman, A. J. Carr, R. E. Melendez and A. D. Hamilton, *Science*, 1999, **286**, 1540.
103. M. de Loos, J. van Esch, I. Stokroos, R. M. Kellogg and B. L. Feringa, *J. Am. Chem. Soc.*, 1997, **119**, 12675.
104. H. Kar and S. Ghosh, *Chem. Commun.*, 2014, **50**, 1064.
105. G. R. Desiraju, *Angew. Chem., Int. Ed. Engl.*, 1995, **34**, 2311.
106. G. R. Desiraju, J. J. Vittal and A. Ramanan, *Crystal Engineering – A Textbook*, World Scientific, Singapore, 2011.
107. G. R. Desiraju, *J. Am. Chem. Soc.*, 2013, **135**, 9952.
108. V. R. Thalladi, B. S. Goud, V. J. Hoy, F. H. Allen, J. A. K. Howard and G. R. Desiraju, *Chem. Commun.*, 1996, 401.
109. E. Ostuni, P. Kamaras and R. G. Weiss, *Angew. Chem., Int. Ed. Engl.*, 1996, **35**, 1324.
110. R. Luboradzki, O. Gronwald, M. Ikeda, S. Shinkai and D. N. Reinhoudt, *Tetrahedron*, 2000, **56**, 9595.
111. A. Ballabh, D. R. Trivedi and P. Dastidar, *Chem. Mater.*, 2003, **15**, 2136.
112. K. Biradha, D. Dennis, V. A. MacKinnon, C. V. K. Sharma and M. J. Zaworotko, *J. Am. Chem. Soc.*, 1998, **120**, 11894.

113. D. R. Trivedi, A. Ballabh and P. Dastidar, *Chem. Mater.*, 2003, **15**, 3971.
114. D. R. Trivedi, A. Ballabh, P. Dastidar and B. Ganguly, *Chem. – Eur. J.*, 2004, **10**, 5311.
115. D. R. Trivedi, A. Ballabh and P. Dastidar, *J. Mater. Chem.*, 2005, **15**, 2606.
116. S. R. Jadhav, P. K. Vemula, R. Kumar, S. R. Raghavan and G. John, *Angew. Chem., Int. Ed.*, 2010, **49**, 7695.
117. D. R. Trivedi and P. Dastidar, *Cryst. Growth Des.*, 2006, **6**, 2114.
118. P. Sahoo, R. Sankolli, H.-Y. Lee, S. R. Raghavan and P. Dastidar, *Chem. – Eur. J.*, 2012, **18**, 8057.
119. A. Ballabh, D. R. Trivedi and P. Dastidar, *Cryst. Growth Des.*, 2005, **5**, 1545.
120. P. Sahoo, D. K. Kumar, D. R. Trivedi and P. Dastidar, *Tetrahedron Lett.*, 2008, **49**, 3052.
121. P. Sahoo and P. Dastidar, *Cryst. Growth Des.*, 2012, **12**, 5917.
122. D. R. Trivedi, A. Ballabh and P. Dastidar, *Cryst. Growth Des.*, 2006, **6**, 763.
123. K. Sada, K. Inoue, T. Tanaka, A. Tanaka, A. Epergyes, S. Nagahama, A. Matsumoto and M. Miyata, *J. Am. Chem. Soc.*, 2004, **126**, 1764.
124. A. Ballabh, D. R. Trivedi and P. Dastidar, *Chem. Mater.*, 2006, **18**, 3795.
125. T. K. Adalder and P. Dastidar, *Cryst. Growth Des.*, 2014, **14**, 2254.
126. U. K. Das, S. Banerjee and P. Dastidar, *Chem. – Asian J.*, 2013, **8**, 3022.
127. R. Parveen and P. Dastidar, *Chem. – Asian J.*, 2015, **10**, 2427.
128. A. Ballabh, D. R. Trivedi and P. Dastidar, *Org. Lett.*, 2006, **8**, 1271.
129. P. Sahoo, N. N. Adarsh, G. E. Chacko, S. R. Raghavan, V. G. Puranik and P. Dastidar, *Langmuir*, 2009, **25**, 8742.
130. P. Sahoo, V. G. Puranik, A. K. Patra, P. U. Sastry and P. Dastidar, *Soft Matter*, 2011, **7**, 3634.
131. A. Shivhare, S. J. Ambrose, H. Zhang, R. W. Purves and R. W. J. Scott, *Chem. Commun.*, 2013, **49**, 276.
132. H. Yamamoto, H. Yano, H. Kouchi, Y. Obora, R. Arakawa and H. Kawasaki, *Nanoscale*, 2012, **4**, 4148.
133. R. Jin, *Nanoscale*, 2010, **2**, 343.
134. J. Sun, Y. Yue, P. Wang, H. He and Y. Jin, *J. Mater. Chem. C*, 2013, **1**, 908.
135. H. Liu, X. Zhang, X. Wu, L. Jiang, C. Burda and J.-J. Zhu, *Chem. Commun.*, 2011, **47**, 4237.
136. S. Palmal, S. K. Basiruddin, A. R. Maity, S. C. Ray and N. R. Jana, *Chem. – Eur. J.*, 2013, **19**, 943.
137. T. K. Adalder, D. P. Kumar and P. Dastidar, *Cryst. Growth Des.*, 2014, **14**, 11.
138. T. Klawonn, A. Gansäuer, I. Winkler, T. Lauterbach, D. Franke, R. J. M. Nolte, M. C. Feiters, H. Börner, J. Hentschel and K. H. Dötz, *Chem. Commun.*, 2007, 1894.
139. S. Yoon, W. J. Kwon, L. Piao and S.-H. Kim, *Langmuir*, 2007, **23**, 8295.

140. A. Y.-Y. Tam, K. M.-C. Wong and V. W.-W. Yam, *J. Am. Chem. Soc.*, 2009, **131**, 6253.
141. N. N. Adarsh, P. Sahoo and P. Dastidar, *Cryst. Growth Des.*, 2010, **10**, 4976.
142. U. K. Das and P. Dastidar, *Chem. – Eur. J.*, 2012, **18**, 13079.
143. S. Saha, E.-M. Schön, C. Cativiela, D. Díaz Díaz and R. Banerjee, *Chem. – Eur. J.*, 2013, **19**, 9562.
144. M. Martínez-Calvo, O. Kotova, M. E. Möbius, A. P. Bell, T. McCabe, J. J. Boland and T. Gunnlaugsson, *J. Am. Chem. Soc.*, 2015, **137**, 1983.
145. S. Samai and K. Biradha, *Chem. Mater.*, 2012, **24**, 1165.

CHAPTER 3

Anion Binding in Tripodal Receptors and Supramolecular Capsular Dimeric Assemblies

SOURAV CHAKRABORTY, TAMAL KANTI GHOSH, RANJAN DUTTA AND PRADYUT GHOSH*

Department of Inorganic Chemistry, Indian Association for the Cultivation of Science, India, E-mail: stsc@iacs.res.in; ictkg@iacs.res.in; icrd@iacs.res.in
*Email: icpg@iacs.res.in

3.1 Introduction

Supramolecular chemistry of anions has grown over the last few decades due to research into receptor dimensionalities, functionalities, platforms and different non-covalent interactions. Anion recognition by catapinands by Park and Simmons in 1968 is the beginning of this area of research.[1–3] For the next twenty five years or so mostly macrobicyclic and macropolycyclic receptors were utilized for anion recognition by Lehn *et al.*[4–7] and Schmidtchen *et al.*[8–11] through ammonium···anion hydrogen bonding and electrostatic interactions respectively. In the early 90's anion binding by tripodal amide receptors started with work by Beer *et al.*[12–14] and Reinhoudt *et al.*[15] and these receptors were selective towards $H_2PO_4^-$. Then urea and thiourea functionalities were introduced into tripodal receptors for the recognition of oxyanions by Morán *et al.*[16] Until then a wide variety of tripodal hosts containing different recognition elements, such as ammonium, amide, urea, thiourea, guanidinium, pyrrole, pyridinium *etc.*, on various

Monographs in Supramolecular Chemistry No. 22
Functional Supramolecular Materials: From Surfaces to MOFs
Edited by Rahul Banerjee

Published by the Royal Society of Chemistry, www.rsc.org

platforms—TREN (tris-(2-aminoethyl)amine), arene, triazine-trione and cholic acid *etc.*—have been developed for anion recognition and encapsulation studies. Numerous weak interactions, such as hydrogen bonding, coulombic anion···π, and halogen bonding, are found to be responsible for such recognition processes. Importantly, tripodal receptors have shown potentiality towards CO_2 sequestration, SO_4^{2-} and F^- extraction and separation, anion transportation, recognition of anionic carbohydrates and hydrated anion clusters *etc.* This book chapter synchronizes the chronological development of different tripodal receptors for anion recognition based on their functionalities. Potential applications of such anion receptors are also discussed in corresponding sections. For further details on anion recognition by tripodal receptors interested readers are encouraged to go through previous review articles.[17,18]

3.2 Anion Recognition with Tripodal Amide Receptors

The first report on a tripodal (acyclic) receptor, which turned out be an amide-based receptor, came in 1992 when Beer *et al.* reported arene and TREN-based tripodal cobalticinium [Co(Cp)$_2$] attached to neutral amide receptors, **L1** & **L2** (Chart 3.1).[12] However, the first amide-based anion receptors reported were macrobicyclic (Pascal *et al.*).[19] **L1** showed significant cathodic shifts

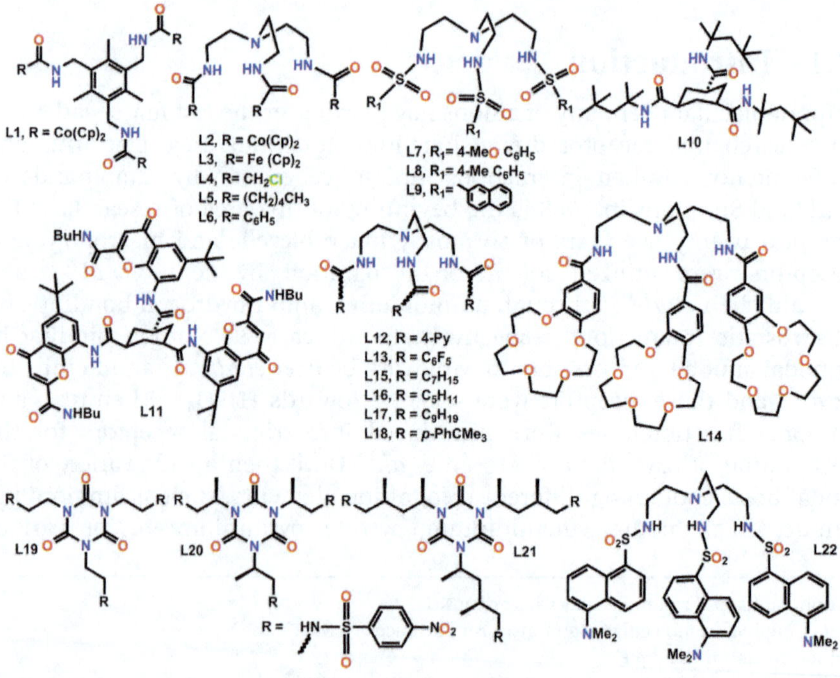

Chart 3.1 Chemical structures of receptors **L1–L22**.

($\Delta E = 30$ mv for Cl^- and $\Delta E = 15$ mv for Br^-) when anion interacted with the amide –NH units.[14] Whereas the ^{1}H-NMR titration **L2** with Cl^- in *ACN-d_3* produced a change in chemical shift ($\Delta\delta = 1.28$ ppm) of amide –NH protons suggesting binding of Cl^-. Here $Co(Cp)_2$ moiety along with pendant amide –NH protons created a favourable microenvironment for successful complexation. Anion binding to **L2** showed significant cathodic shifts ($\Delta E = 30$ mv for Cl^- and $\Delta E = 15$ mv for Br^-). The next year, Beer *et al.* first reported a ferrocenium $[Fe(Cp)_2]$ appended receptor **L3** for the electrochemical detection of $H_2PO_4^-$ (of cathodic shift of 180 mV) in the presence of excess HSO_4^- and Cl^-. Interestingly, with increasing guest concentration, they observed a change in shape of the oxidation wave from a reversible redox process to an electrode-chemical mechanism. The cathodic shift of the ferrocene–ferricinium redox couple in the presence of $H_2PO_4^-$ alone was comparable to that observed in presence of tenfold excess HSO_4^- and Cl^-.[13,14]

During this period, Reinhoudt *et al.* reported another set of neutral tri-s-amide receptors **L4–L9** (Chart 3.1) and manifested anion binding through ^{1}H-NMR and conductometric experiments.[15] Downfield shift of amide –NH of **L4–L9** ($\Delta\delta = 1.5$–2.0 ppm) in ^{1}H-NMR titrations with $(n\text{-Bu})_4N^+X^-(X^- = H_2PO_4^-, HSO_4^-, Cl^-)$ in *$CDCl_3$* in 1 : 1 (host : guest) stoichiometry established anion binding. On the contrary, $H_2PO_4^-$ showed 1 : 2 (host : guest) binding stoichiometry which was confirmed by the characteristic ^{31}P values for all host–guest complexes in *ACN-d_3*. A 1 : 2 (host : guest) ratio was also reported by others and was attributed to dimerization of $H_2PO_4^-$ through intermolecular hydrogen bonding interaction.[20] The association constant values of **L4–L9** revealed selectivity towards $H_2PO_4^-$ ($H_2PO_4^- > Cl^- > HSO_4^-$) and the highest binding affinity was observed for **L9** ($K_a = 14\,200$ M^{-1}).

Morán *et al.* reported two cyclohexane-platform-based tripodal triamide (**L10**) and hexamide (**L11**) receptors for the preferential binding of phosphates. The association constant obtained for the binding of PO_4^{3-} with receptor **L10** was 1.0×10^2 M^{-1} in *DMSO-d_6*. Receptor **L11**, having six amide binding sites, showed binding constant values of 1.5×10^4 M^{-1} and $>10^5$ M^{-1} with phenyl phosphonate and phosphate respectively in *DMSO-d_6*.[21] In 1997, Stibor *et al.* investigated anion binding properties with receptors **L12** and **L13** (Chart 3.1).[22] ^{1}H-NMR titration studies showed 1 : 1 (host : guest) binding stoichiometry. Studies revealed that **L12** was selective towards $H_2PO_4^-$ ($K_a = 7550 \pm 310$ M^{-1} in *ACN-d_3*) whereas **L13** was selective towards HSO_4^- ($K_a = 5120 \pm 740$ M^{-1}) in *$CDCl_3$*. Later on, our group further utilized **L13** for removal of KF and KCl *via* the liquid–liquid extraction technique, utilizing a dual host approach with 18-Crown-6 and a tripodal amide, **L13.**[23] The extracted complexes were analysed by single crystal X-ray structural studies and powder XRD studies. ^{1}H-NMR spectra indicated a 1 : 1 involvement of **L13** and 18-Crown-6 in the course of extraction. Single crystal X-ray studies of [**L13**(F)(H_2O)][N(Bu)$_4$] showed recognition of monohydrated F^- inside the C_{3v} symmetric cleft of **L13** and Cl^- was encapsulated inside the tripodal cavity. ^{1}H-NMR titrations revealed **L13** was selective towards F^- with 8.9×10^3 M^{-1} in $CDCl_3$. Beer *et al.* designed the crown-amide functionalized hetero-di-topic receptor **L14** for binding with Cl^-, I^- and ReO_4^- having binding constants in the range of 40–60 M^{-1} in $CDCl_3$ (Chart 3.1). However, in presence of Na^+, a ten-fold increase in binding constants were observed due to the cooperative effect. **L14** was also utilized for efficient extraction and transportation of the toxic anions of nuclear waste.[24] Bowman-James *et al.* designed *N*-octan-amide, **L15** (Chart 3.1) which on protonation with HNO_3 showed first bilayer ladder-like cascade of the receptor.[25] As this bilayer was formed due to protonation with HNO_3, the channels were filled by nitrate guests. They reported that **L15** was not a selective receptor for nitrate. Single crystals of lipophilic amide bilayer were also obtained. The same group employed a dual host approach for efficient extraction of $CsNO_3$ into the organic phase using several tripodal amide receptors, **L15–L18** (Chart 3.1) and tetrabenzo-24-crown-8 which is selective for binding the Cs^+ ion.[26] Binding constants of the nitrate anion with **L15–L18** were estimated as 56 M^{-1}, 52 M^{-1}, 42 M^{-1} and 33 M^{-1} respectively with 1 : 1 binding stoichiometry in CD_2Cl_2. Liquid–liquid extraction of aqueous solutions of $CsNO_3$ and HNO_3 were carried out with organic solutions of crown ether and **L15, L16** and **L18.** The synergistic effect increased the extraction efficiency by factors of 2.4, 1.7, and 4.4 for **L15, L16** and **L18** respectively.

Hoffmann *et al.* reported triazine–trione based neutral tripodal sulphonamide based receptors, **L19–L21** (Chart 3.1) for the preferential binding of the Cl^- anion.[27] Binding constant values for anions obtained *via* isothermal titration calorimetry (ITC) revealed that all the three receptors bind with Cl^- [$K_a = 4870$ M^{-1} (**L19**), $K_a = 12\,630$ M^{-1} (**L20**), $K_a = 4170$ M^{-1} (**L21**)] more strongly compared to other anions such as Br^- and NO_3^-. Among conformationally preorganized receptors, **L20** and **L21, L20** showed the highest binding constants with Cl^- as compared to unorganized **L19.** This preorganization led to an increase in binding constants (2.5 times) with Cl^- and a decrease in binding constant (0.4 times) with NO_3^-. $(K_a)_{Cl^-}/(K_a)_{NO_3^-} = 105$ for **L20** as compared to **L19.** The investigation of the anion binding properties of tripodal-amide-based receptors using the atmospheric pressure chemical ionization mass spectrometric (APCI-MS) method was carried out by Kavallieratos *et al.*[28] The dansylamide derivative of TREN, **L22** (Chart 3.1) was used as the receptor for anion binding. They showed formation of anionic supramolecular complexes $[\mathbf{L22} + X]^-$, ($X = Cl^-$, NO_3^-, Br^- and I^-) in negative mode APCI-MS with deprotonated receptors $[\mathbf{L22} + X - H]^-$. Hey *et al.* showed the effect of substituents on aryl rings during complexation with Cl^- and NO_3^- with ligands, **L18, L23** and **L24** (Charts 3.1 and 3.2). Extraction studies suggested that **L18** was a selective NO_3^- receptor and it could form cavities suitable for the encapsulation of NO_3^- and ClO_3^-. They calculated interaction energies and intermolecular distances between host and guests too. Interesting results were obtained in case of **L24** which showed the formation of a more stable complex. Molecular modelling showed that in addition to the amide protons the *ortho* C–H

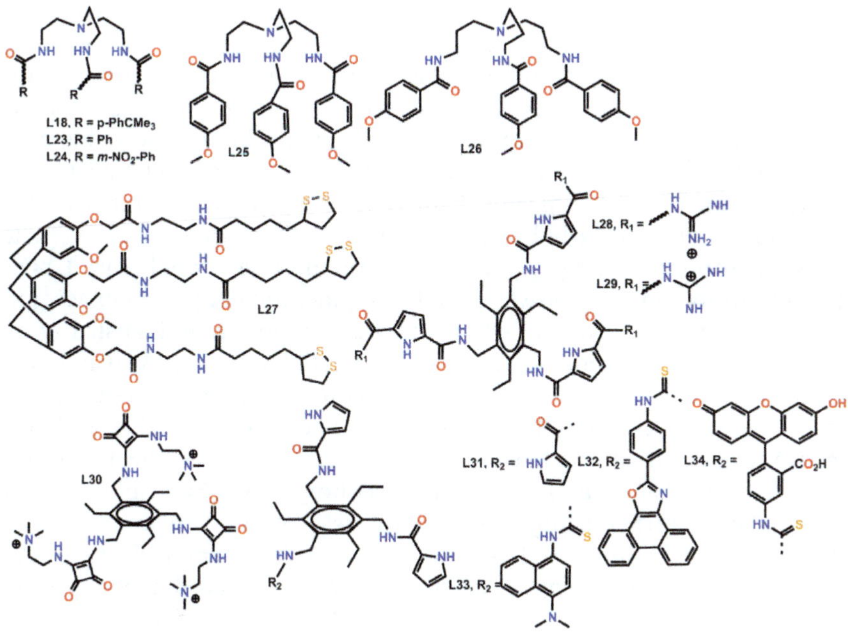

Chart 3.2 Chemical structures of receptors **L23–L34.**

protons of the aryl ring were also involved in H-bonding interaction with NO_3^-. Calculations showed that $-NO_2$ of **L24** enhanced the binding affinity by 33% as compared to the un-substituted one, **L23**.[29] Parallel in time, Tucker *et al.* reported TREN and tris(3-aminopropyl)amine-based two anion receptors **L25** and **L26** (Chart 3.2) that mimicked the active sites of vanadium haloperoxidase and acid phosphatase respectively. The binding constants of HPO_4^{2-} and HVO_4^{2-} with **L25** were calculated as 496 M^{-1} and 540 M^{-1} respectively with 1 : 1 binding stoichiometry in CD_3CN. On the other hand, the comparatively larger receptor **L26** exhibited recognition of $H_2PO_4^-$ with comparable association constant values in a 2 : 1 CD_3CN–CD_2Cl_2 mixture. V–N bond formation from the apical N atom of **L26** was confirmed by the net up-field shift of the amide -NH protons upon addition of the first two equivalents of $(n\text{-Bu})_4N^+H_2VO_4$.[30]

Zhang *et al.* prepared a cyclotriveratrylene-based tris-amide receptor, **L27** (Chart 3.2) and immobilized it on a gold surface *via* the formation of a self-assembled monolayer.[31] UV-Vis and ^{1}H-NMR titrations showed the selective binding of **L27** with the acetate anion among the investigated anions (*e.g.* Cl^-, Br^-, NO_3^-, HSO_4^- and $H_2PO_4^-$) in $CDCl_3$. Impedance spectroscopy also showed the selective binding of the acetate anion in an aqueous medium.

In 2005, Schmuck *et al.* synthesized a tri-cationic guanidiniocarbonyl pyrrole receptor **L28** (Chart 3.2) for the selective recognition of citrate and other tri-carboxylates in water.[32] Absorption and emission studies revealed high association constant (~10^5 M^{-1}) with citrate in water with 1 : 1

(host : guest) binding stoichiometry. They exhibited that **L28** recognized citrate in tris buffer medium with $K_a = 8.6 \times 10^4$ M^{-1} even in presence of a large excess of competing anions such as tartrate and maleate. Subsequently, the same group have developed another tris-cationic receptor, **L29**, for the naked eye detection of citrate in water which solely relies on reversible non-covalent interactions.[33] They showed selectivity of citrate over malate and tartrate in water *via* the indicator displacement method. They reported the recognition of anionic carbohydrate **L28** in water.[34] Among the carboxylate salts, galacturonate anion ($K_a = 1500$ M^{-1}) bound more strongly with **L28** as compared to its mono-carboxylate analogue glucuronate, ($K_a = 450$ M^{-1}) in 7 : 3 water : DMSO (v/v) binary solvent mixture. Frontera and Anslyn *et al.* designed a tris-sqauramide-ammonium substituted receptor, **L30** (Chart 3.2) for the recognition of carboxylates in aqueous medium.[35] Extensive ITC studies in 1 : 3 H_2O : EtOH binary solvent mixture revealed that **L30** was selective towards tricarboxylates (citrate and trimesate) and dicarboxylates (glutarate and succinate) over monocarboxylates. The binding of tricarboxylates were entropy-driven with association constants in the order of 10^5 M^{-1}. They used the indicator displacement technique and showed that the progressive addition of citrate displaced the dye from **L30**-fluorescein complex and thus the restoration of fluorescence of fluorescein was observed. This method was used for the quantitative detection of zinc citrate in commercial toothpaste.

Later on, we reported that monoprotonated **L24** (Chart 3.2) preferred cleft binding of anions (Cl^-, Br, NO_3^-, ClO_4^-, SiF_6^-), in all the cases, irrespective of their shapes and sizes due to preorganization of the receptor in its protonate state through N–H and C–H interactions.[36] In this study the sheet-like structure of **L24** was modified to a bilayer structure upon pronation as shown previously by Bowman-James *et al.*[25] Anzenbacher *et al.* showed the binding of phosphate anion by tripodal amide receptors **L31–L34** (Chart 3.2) in solution by ^{1}H-NMR and fluorescence titrations. These receptors showed 1 : 1 (host : guest) binding stoichiometry with the studied anions having the following order of binding constant $H_2PO_4^- > HPPi^{3-} > AcO^- \gg Cl^- > Br^-$ in DMSO. To recognize these biologically relevant anions in water, **L32–L34** was embedded in hydrophilic polyurethane matrices and were found to be highly efficient sensors for anions in buffered solution at pH = 7.4. They performed principal component analysis (PCA) which revealed that films of **L31–L34** were capable of distinguishing inorganic phosphates from adenosine monophosphate (AMP) and adenosine diphosphate (ADP) in water which mimicked the blood serum.[37]

Schmuck *et al.* developed the biomimetic receptor **L35** (Chart 3.3) *via* solid phase synthesis, which was able to bind anionic carbohydrates (*e.g.* glucose-1-phosphate and cAMP). Based on the same principle of binding of proteins in carbohydrate-binding proteins, **L35** was isolated as a hydrochloride salt, wherein two dipeptide moieties, Ser-Phe, were attached with an aromatic scaffold together with an arginine residue. This design allowed both apolar (Phe) and polar (Ser and arginine analogue) units to interact with sugars.

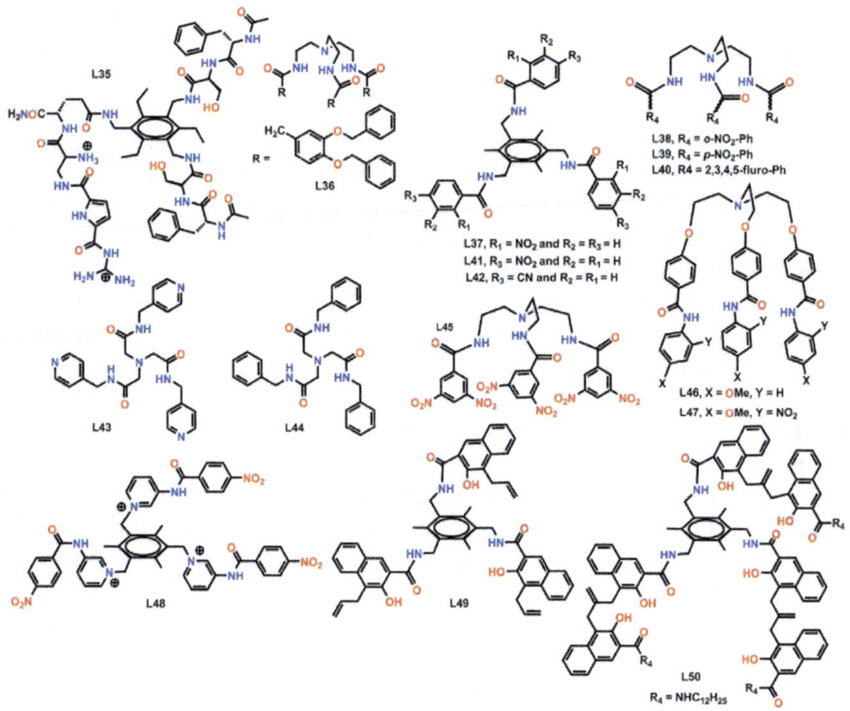

Chart 3.3 Chemical structures of receptors **L35–L50**.

Binding studies showed that **L35** bound anionic sugars such as galactose-1-phosphate ($K_a = 3400\ M^{-1}$) or glucose 1-phosphate ($K_a = 3000\ M^{-1}$) in 20% aqueous DMSO.[38] Smith *et al.* reported a TREN-based receptor, **L36** (Chart 3.3), for the transportation of HCl through an apolar medium. **L36** had an interior binding site decorated with an aromatic ether shell and had the highest affinity for Cl^- with a binding constant value of $K_a = 550\ M^{-1}$. U-Tube experiments using DCM and apolar medium was used to assess the HCl transport ability of **L36.**[39]

In 2009, our group first demonstrated a unique example of hydrated fluoride encapsulation *via* the formation of a fluoride–water cluster $[F_2(H_2O)_6]^{2-}$ in the cavity of a dimeric capsular assembly. On the other hand, other anions like Cl^-, NO_3^- and AcO^-, showed formation of non-capsular type molecular assemblies of **L37** (Chart 3.3). The fluoride–water cluster was further obtained upon charging $(n\text{-Bu})_4N^+F^-$ into the solution of **L37**$\cdot(n\text{-Bu})_4N^+NO_3^-$ in dioxane. In the presence of equal equivalents of tetrabutylammonium acetate and fluoride with **L37** in dioxane, we obtained the $[F_2(H_2O)_6]^{2-}$ cluster.[40] Later on, we reported receptors **L24, L38** and **L39** (Charts 3.2 and 3.3).[41] These receptors were selective towards halides and did not bind oxyanions. **L38** showed the highest association constant value ($4.3\times10^5\ M^{-1}$) with F^- in 1 : 1 host : guest binding stoichiometry and **L39** was selective towards F^- whereas **L24** failed to show any selectivity. Later on we explored **L13** and **L40** for anion recognition studies.[23,42] Further we have generalized our findings by a comparative anion binding studies between **L37** and its positional isomers **L41** (Chart 3.3).[40] In the case of **L41**, anion–water cluster encapsulation-driven formation of dimeric capsular assemblies were observed for F^-, Cl^-, AcO^- and NO_3^-. In general, **L41** was found to encapsulate $[F_2(H_2O)_6]_2^-$, $[Cl_2(H_2O)_4]_2^-$, $[(AcO)_2(H_2O)_4]^{2-}$ and two NO_3^- in its dimeric capsular assembly. Interestingly, a competitive crystallization experiment showed the encapsulation of the $[F_2(H_2O)_6]_2^-$ cluster by **L41** from a mixture of F^-, Cl^-, AcO^- and NO_3^-, like that in the case of **L37**.[43] Later on, our report on the *para*-CN substituted tripodal amide receptor **L42** showed recognition of nitrate in the form of a capsule and recognition of acetate in the form of a half capsule.[44]

Ozturk *et al.* reported amide-based neutral tripodal anion receptors, **L43** and **L44** (Chart 3.3), for the recognition of $H_2PO_4^-$ and $C_6H_5CO_2^-$ anions in the presence of other anions such as PF_6^-, ClO_4^-, HSO_4^- and Br^-.[45] The pyridyl-appended tripodal receptor **L43** showed a higher binding affinity towards anions as compared to its benzene analogue, **L44**. Binding constants K_a of the $H_2PO_4^-$ and $C_6H_5CO_2^-$ anions with **L44** were 241 M^{-1} and 110 M^{-1} respectively in *DMSO-d*$_6$, whereas **L43** gave binding constants of 810 M^{-1} and 286 M^{-1} in *DMSO-d*$_6$ respectively, in 1 : 1 binding stoichiometry.

In 2011, Das *et al.* developed **L45** (Chart 3.3) for selective recognition and sensing of F^- in the presence of other competing anions.[46] Intense orange/purple coloration in the presence of F^- in polar aprotic solvents was attributed to charge transfer interaction between the anion and the di-nitrophenyl receptor unit. Solid state structural analysis revealed that F^- encapsulation was within the tripodal cleft (Figure 3.1). The high degree

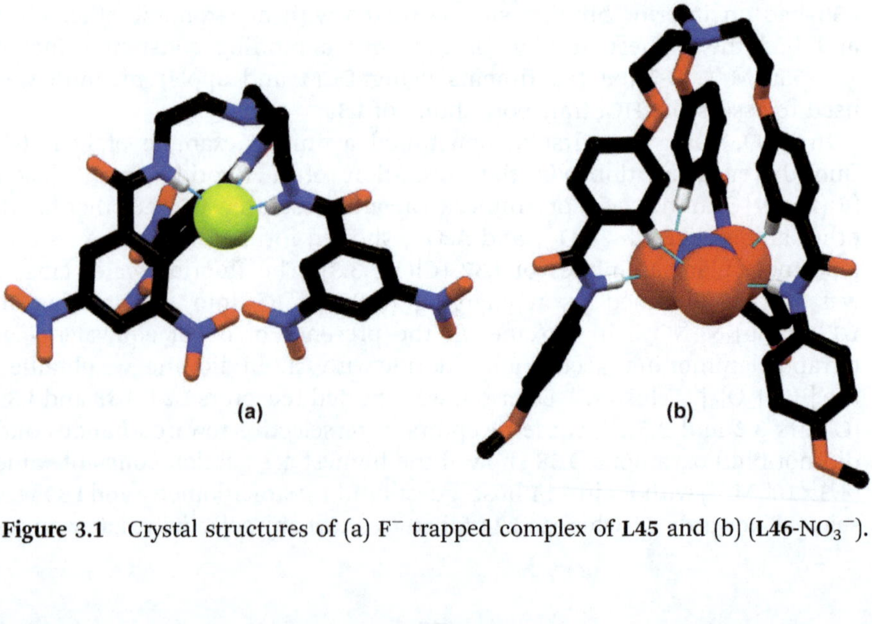

Figure 3.1 Crystal structures of (a) F^- trapped complex of **L45** and (b) (**L46**-NO_3^-).

of fluoride selectivity of **L45** ($K_a > 10^7$ M^{-1}) was utilized for the transformation of charged anion complexes [(**L45**) + A^-] ($A^- = Cl^-$, Br^-, ClO_4^-, HSO_4^-) into [$(n\text{-}Bu)_4N^+$(**L45** · F)].[47] Further, they reported that **L45**, upon protonation of the bridgehead proton, was oriented towards the ring and formed hydrogen bonding interactions with one amide centre. Thus the guest (Cl^-, Br^-, ClO_4^-, HSO_4^-) bound with the ligand outside the cavity *via* N–H···anion and C–H···anion interactions. Significant downfield chemical shift of amide N–H ($\Delta\delta = 3.55$ ppm) and *ortho*-C–H ($\Delta\delta = 3.55$ ppm) protons respectively upon the addition of $(n\text{-}Bu)_4N^+F^-$ in *DMSO-d_6* confirmed F^- binding in 1 : 1 stoichiometry. They further showed differential complexation behaviour of **L45** towards F^-.[48] X-Ray crystallographic analysis revealed that **L45** could encapsulate F^- inside the tripodal cavity, depending upon the F^- source.

Sun *et al.* showed encapsulation of NO_3^- in the cavity of tripodal receptor **L46** (Chart 3.3) upon protonation with HNO_3 in $CHCl_3$ (Figure 3.1).[49] The ^{1}H-NMR spectrum suggested formation of two distinctly different species in solution upon protonation with HNO_3. One was the NO_3^- encapsulated complex of **L46** and other was the cleft bound NO_3^- complex. **L46** and **L47** were utilized for [2 + 2] cyclo-addition reactions. In 2010, Bao *et al.* reported an amide-pyridinium functionalized tripodal anion receptor, **L48** (Chart 3.3) with nitrobenzene as a signalling unit.[50] ^{1}H-NMR and UV-Vis studies showed high selectivity of **L48** towards AcO^- and F^- binding over other anions in CH_3CN with 1 : 1 binding stoichiometry. **L48** was also capable of naked eye detection of F^- and AcO^- due to clear colour changes. In 2010, Hiratani *et al.* reported two tripodal receptors, **L49** and **L50** (Chart 3.3), on a benzene platform for recognition of F^-, AcO^- and $H_2PO_4^-$ in *$CDCl_3$*.[51] Fluorescence studies showed that **L49** sensed F^-, AcO^- and $H_2PO_4^-$ anions as their $(n\text{-}Bu)_4N^+$ salts in the presence of other anions and the selectivity was in the order $F^- > AcO^- > H_2PO_4^-$ in $CHCl_3$. On the other hand, **L50** showed a highly selective response only in presence of F^- among all the anions both in polar and non-polar solvents.

In 2012, we reported **L51–L55** (Chart 3.4) for anion recognition.[52] ITC studies showed that **L51, L52, L54, L55** formed complexes with $(n\text{-}Bu)_4N^+$ salts of Cl^-, AcO^-, $C_6H_5CO_2^-$ and HSO_4^- in 1 : 1 stoichiometry, having the highest binding affinity towards $(n\text{-}Bu)_4N^+AcO^-$. Solid state structure revealed recognition of Cl^- by **L51** in its cavity and the cavity of the receptor was capped by $(n\text{-}Bu)_4N^+$ counteraction. Binding of Cl^- was largely influenced by the counter cation involved ($Bu_4N^+ > Et_4N^+ > Me_4N^+$ in terms of binding constants). We further developed an electron deficient triazine–trione based tripodal amide receptor **L56** (Chart 3.4) for encapsulation of fluoride/chloride anions *via* the formation of C_{3v}-symmetric cleft.[53] Interestingly, upon complexation of **L56** with $(n\text{-}Bu)_4N^+F^-$ and $(n\text{-}Bu)_4N^+Cl^-$, it showed encapsulation of monomeric F^- and Cl^- *via* formation of N–H···X ($X = F^-$, Cl^-) hydrogen bonding interactions inside the C_{3v} symmetric cleft in both the complexes. Solution state ITC studies of **L56** with $(n\text{-}Bu)_4N^+$ salts of different halides and

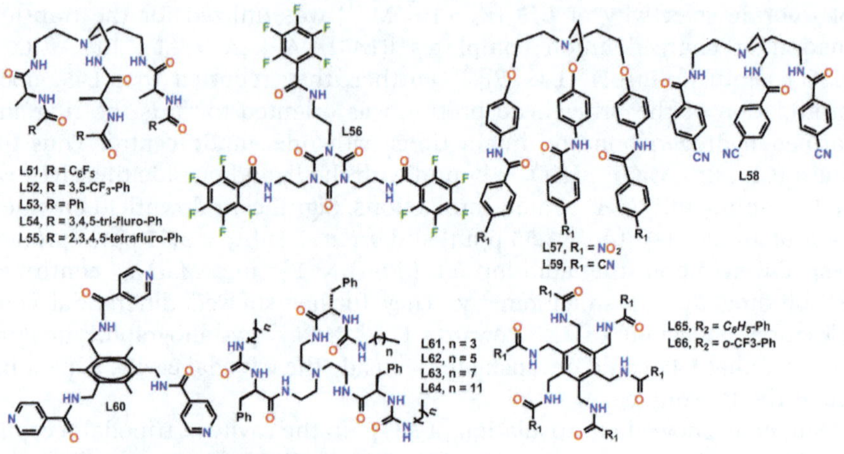

Chart 3.4 Chemical structures of receptors **L51–L66**.

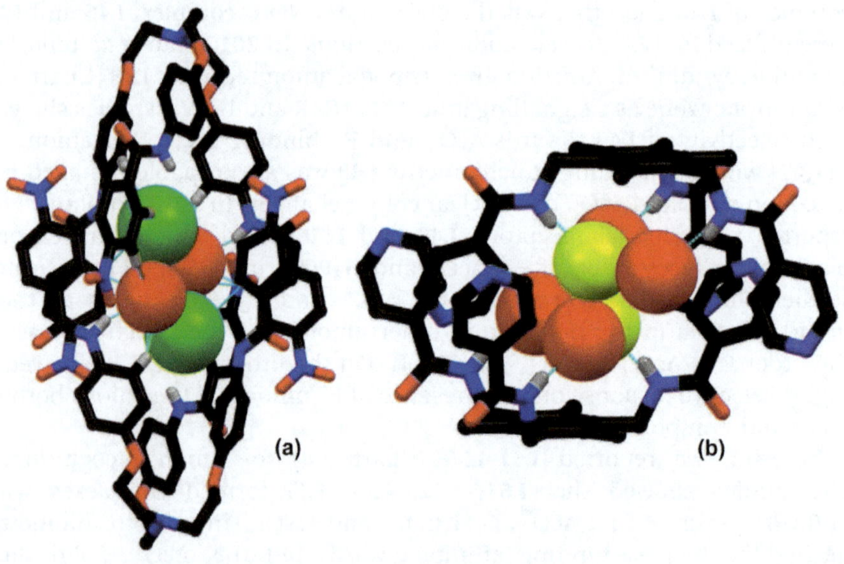

Figure 3.2 Crystal structures of (a) trapping of $[(Cl^-)_2(H_2O)_2]$ inside **L57** and (b) trapping of $[(F^-)_2(H_2O)_4]$ inside **L60**.

oxyanions in acetonitrile showed 1 : 1 (host : guest) stoichiometry for fluoride ($K_a = 7.2 \times 10^4\ M^{-1}$) and chloride ($K_a = 6.8 \times 10^3\ M^{-1}$).

Later, in 2013, Das *et al.* reported a $[X_2(H_2O)_2]^-$ ($X = Cl^-$ or Br^-) cluster inside the dimeric capsular assembly of **L57** (Chart 3.4, Figure 3.2).[54] Solid-state FT-IR analysis confirmed the existence of halide–water clusters at a distinct band at around ~3250 cm^{-1} suggesting O–H stretching vibration of water. Upon protonation with HI, a non-capsular 1D polymeric assembly was found to form. Further in 2015, Das *et al.* reported **L58** (Chart 3.4) and ^{1}H-NMR analysis suggested that **L58** was reluctant to bind oxyanions strongly but recognized only F^- and $H_2PO_4^-$ as compared to other anions.[55] As TREN-based anion receptors were reluctant towards the encapsulation of anions upon protonation, people used more flexible oxy-TREN-based receptors for anion recognition upon protonation. **L59** encapsulated F^- and SiF_6^{2-} inside a protonated cavity and ^{1}H-NMR suggested that **L59** was reluctant to bind oxyanions.

In 2014, we designed a 4-pyridyl substituted tripodal amide receptor **L60** (Chart 3.4) that was able to isolate $[F_2(H_2O)_4]^{2-}$ and $[Cl_2(H_2O)_4]^{2-}$ clusters from a acetone–water (1:1, v/v) binary solvent mixture (Figure 3.2). Solution state ^{1}H-NMR studies with **L60** in *D$_2$O*-acetone-*d$_6$* (1:19, v/v) supported 1:4 (host:guest) binding stoichiometry with F^-, Cl^-, Br^-, NO_3^-, HSO_4^- and $H_2PO_4^-$. A 1:1 binding model was utilized to calculate the binding constant values which revealed the selectivity order $NO_3^- \approx HSO_4^- > F^- \approx Cl^- \approx Br^- > H_2PO_4^-$. Further, solution state ^{19}F-NMR studies were carried out to establish the F^- binding to **L60** in *DMSO-d$_6$* in solution.[56]

Gale *et al.* reported TREN-platform-based receptors having amino acid residues and amide-urea functionalities, **L61–L64** (Chart 3.4) which were able to interact with biologically important anions, *e.g.* L-lactate, L-maleate and L-aspartate *etc.*[57] The ligands were found to bind Cl^- selectively as compared to other organic anions. Transportation of anions was affected by the nature of the aliphatic central spacer and the side chains. The receptor having the shortest aliphatic side chain and longest spacer was found to be the most active (**L61**).

Bis-tripodal amide hosts are solely utilized by our group for anion recognition. We reported a benzene platform based hexapodal amide receptor, **L65** (Chart 3.4), decorated with pentafluro-phenyl unit which showed compartmental recognition of $[(NO_3^-)_2H_2O]$ in the tripodal cavity of the receptor in *ababab* fashion where alternate arms are directed in opposite directions to create a bowl for anionic guests (Figure 3.3). A Job's plot analysis from ^{1}H-NMR titration of **L65** with $(n\text{-Bu})_4N^+NO_3^-$ in acetone-d$_6$ showed the formation of a 1:2 (host:guest) complex which matched well with solid state structures.[58] **L65** also exhibited compartmental recognition of CH_3COO^- in *aaabbb* fashion. Later on, we synthesized a benzene platform based hexapodal (bis-tripodal) amide receptor, **L66** (Chart 3.4), with an attached *o*-trifluoromethyl unit which recognized anions *via* different conformations and showed compartmental recognition of AcO^- in *ababab* fashion (Figure 3.3).[59] Further, we also showed recognition of $[(F^-)_4(H_2O)_{10}]$[60] and $[(F^-)_4(H_2O)_6]$[59] clusters in *aaaaaa* fashion where all the six arms of receptors were oriented in one direction. ^{1}H-NMR and ITC titrations were carried out with **L66** and anions as $(n\text{-Bu})_4^+$ salts, which showed highest binding constant for anion complex formation, having 1:2 (host:guest) stoichiometry. Apart from these two reports, we have carried out extensive structural and solution state studies on bis-tripodal-amide host molecules.[60,61]

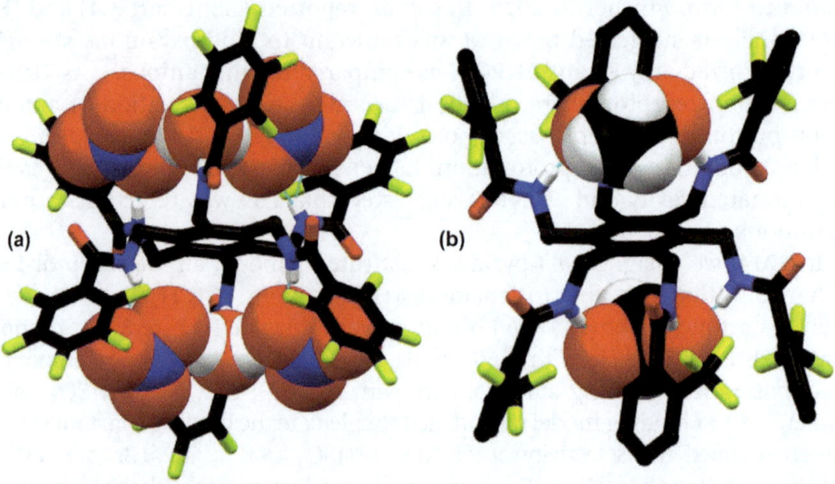

Figure 3.3 Crystal structures of (a) compartmental recognition of $[(NO_3^-)_4(H_2O)_2]$ **L65** and (b) compartmental recognition of acetate by **L66.**

3.3 Anion Recognition with Tripodal Urea Receptors

The first report on tris-urea receptor was published by Morán *et al.* in 1995. They showed that oxyanions like phosphate and sulfate preferred to bind with urea over thiourea. They have estimated the binding constant by ^{1}H-NMR titration which revealed $K_a = 1.4 \times 10^4$ M^{-1} for **L67** (Chart 3.5) and $K_a = 1.6 \times 10^3$ M^{-1} for **L68** with PO_4^{3-} in *DMSO-d*$_6$.[16] Wu *et al.* developed a TREN-based naphthyl-urea receptor **L69** (Chart 3.5) and its protonated form (**L69**···**Cl**$^-$) and showed that **L69** was found to be selective towards $H_2PO_4^-$ with 1 : 1 (host : guest) stoichiometry ($K_a = 1.1 \times 10^4$ M^{-1}) in presence of other interfering anions. The preference of interaction of $H_2PO_4^-$ with **L69**···**Cl**$^-$ was interpreted as anion induced reduction of efficiency of Photoinduced Electron Transfer (**PET**).[62] Electronic calculations by Hay *et al.* demonstrated that SO_4^{2-} preferred twelve hydrogen bonds in each of the six O–S–O planes.[63] This idea was manifested by Custelcean *et al.* to show recognition of sulfate in the metal-organic framework functionalized with urea binding motif where the $[Ag_2$(**L70**)$_2](SO_4)]$ (Figure 3.4) unit formed a 1D-polymeric assembly that binds sulfate with twelve hydrogen bonding interactions which resembles nature's sulfate-binding protein. They investigated the thermal stability of the metal-organic framework, which was up to 192 °C and Job's plot analysis showed 2 : 1 ligand to sulfate-binding stoichiometry. This represented the very first report on the formation of dimeric capsular assembly upon sulfate recognition by a 1D-polymeric assembly of the synthetic anion receptor **L70** (Chart 3.5).[64] Due to unavailability of the single crystals, they carried out extensive theoretical modelling studies (MMFF9) with **L71** (Chart 3.5) for sulfate recognition which revealed that C_3 symmetric receptors have shapes complementarity towards SO_4^{2-}.

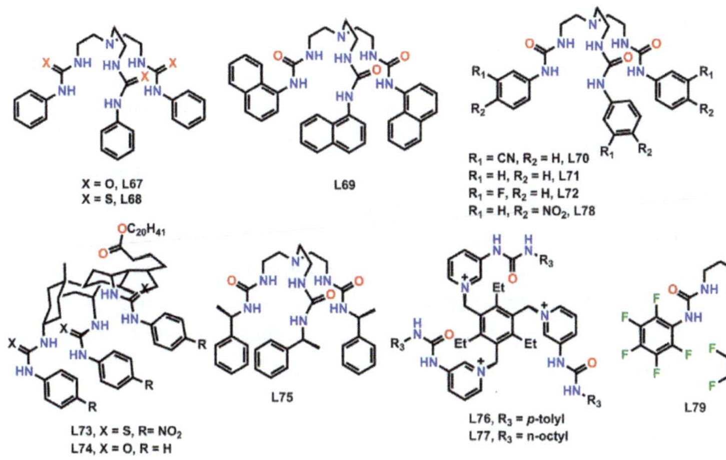

Chart 3.5 Chemical structures of receptors **L67–L79.**

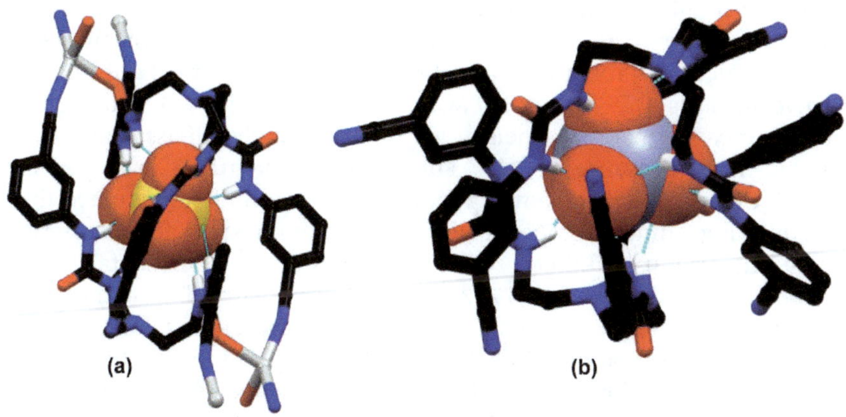

Figure 3.4 (a) Encapsulation of SO_4^{2-} inside $[Ag_2(\mathbf{L70})_2]$ and (b) encapsulation of CrO_4^{2-} by **L70**.

Our group utilized **L70** and **L72** (Chart 3.5) for isolating two different conformers of oxalate, *i.e.* staggered and planar. Among these two receptors, **L70** isolated the staggered conformer of oxalate as a dimeric capsule (capsular dimension 9.815 Å) where the O–C–C–O dihedral angle was found to be 68.8° and **L72** formed a dimeric assembly (capsular dimension 10.823 Å) with a planar conformer of oxalate where the O–C–C–O dihedral angle was found to be 0.12°. Further, we carried out ^{1}H-NMR titration studies in D_2O-$DMSO$-d_6 solvent mixture which revealed $C_2O_4^{2-}$ was slightly selective towards **L70** ($K_a = 6.6\times10^4$ M^{-1}) compared to **L72** ($K_a = 1.9\times10^4$ M^{-1}). This was the first report of its kind to trap unusual staggered conformers into the capsular assembly.[65] The receptor **L70** was also capable of forming a carbonate capsule $[(\mathbf{L70})_2CO_3^{2-}]$ from aerial CO_2 which was further utilized for efficient extraction of CrO_4^{2-}, SO_4^{2-} and $S_2O_3^{2-}$ (Figure 3.4). ^{1}H-NMR spectroscopic analysis of the bulk extract confirmed 2 : 1 (host : guest) formation. ^{53}Cr-NMR revealed a characteristic peak at −99.98 ppm which corresponded to the presence of CrO_4^{2-}. Single crystals were also obtained for each case with almost comparable capsular dimensions of ~9.7 Å. The carbonate capsule $[(\mathbf{L70})_2CO_3^{2-}]$ can efficiently extract 90% of the above-mentioned anions through anion exchange metathesis from aqueous solution. The extraction of SO_4^{2-} was also demonstrated at a highly alkaline pH of ~12.5 and for CrO_4^{2-}, UV/Vis spectroscopic measurement revealed around 50% removal of it in the presence of an equimolar amount of SO_4^{2-}.[66]

Davis and his co-workers developed tripodal receptors on steroid platform which showed high affinity towards chloride and were utilized as transporting agents. They termed this 'cholapod' as it was derived from cholic acid. **L73** (Chart 3.5) showed stability constant 1.8×10^{11} M^{-1} with Cl^- and 4.3×10^{10} M^{-1} with Br^-.[67] Further, they prepared **L74** (Chart 3.5) which was the first example of an 'allo-cholapod' receptor where the steroid has a '5-α' skeleton. They have measured the anion binding affinity of **L74** in water saturated $CHCl_3$ by salt extraction and compared it with corresponding '5-β' analogue. Both the analogues exhibited binding constants (K_a) in the order of ~10^8 with halides and oxyanions.[68]

Steed *et al.* developed the tris-urea receptor, **L75** (Chart 3.5) on the TREN platform which was able to act as a low molecular weight gelator and was selective towards Cl^- ($K_a = 1154$ M^{-1}, in *DMSO-d*$_6$) in 1 : 1 (host : guest) stoichiometry. Formation of helical gel fibres through self-association was inhibited in the presence of Cl^-.[69] Further, the same group developed **L76** and **L77** (Chart 3.5) with tris-urea pyridinium substitution and studied their anion binding affinity through solid state, solution state and theoretical studies. Studies revealed that for **L76**, binding to Cl^- in its most stable form was half-capsule whereas for **L77**, 2-up arms bound with Cl^- was the stable conformer.[70] Das *et al.*, in 2007, demonstrated the encapsulation of the $SO_4^{2-}\cdots H_2O\cdots SO_4^{2-}$ adduct in the shape of a rugby ball inside the cavity of the dimeric capsule formed with **L78** (Chart 3.5), using single crystal X-ray structural analysis (Figure 3.5). They utilized UV-vis titration techniques to find out anion binding selectivity where SO_4^{2-} ($K_a = 9.3\times10^4$ M^{-1}) and $H_2PO_4^-$ ($K = 1.8\times10^4$ M^{-1}) were found to be selective compared to the other oxyanions. The broad intense absorption band of **L78** at 345 nm was found to shift up to 368 nm in the presence of the *n*-Bu_4N^+ salt of SO_4^{2-} and $H_2PO_4^-$ in an acetonitrile–water (19 : 1, v/v) binary solvent mixture.[71]

Our group trapped an inorganic phosphate dimer inside the capsular assembly of a pentafluoro-substituted tris-urea receptor **L79** (Chart 3.5). This assembly was stabilized through several hydrogen bonding and anion···π interactions. The $[H_2PO_4]^-$ dimer was involved in eighteen coordination sites with the dimer of **L79** and DMF molecule. ^{1}H-NMR titration was carried out with **L79**, which showed binding of CH_3COO^- and $H_2PO_4^-$ with association constant (K_a) values of 2.8×10^4 M^{-1} and 3.3×10^5 M^{-1} respectively in *DMSO-d*$_6$.[72] Later on this receptor was utilized for the recognition of F^- and SO_4^{2-}. We compared the anion selectivity of **L79** with **L71** which was previously reported by Morán *et al.* which exhibited the binding of phosphate and sulfate

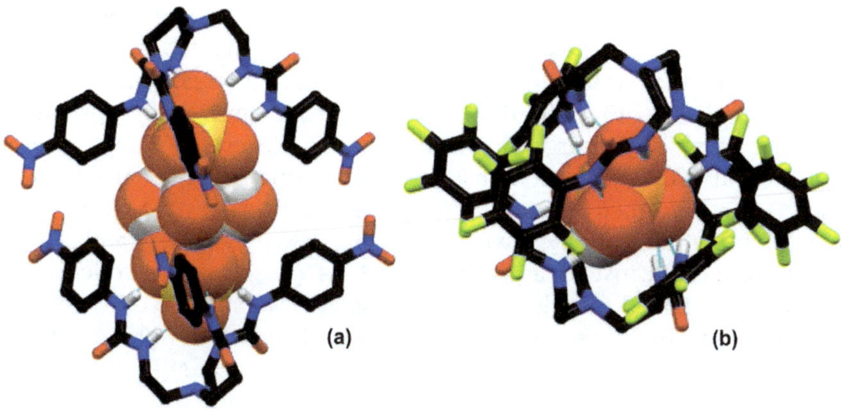

Figure 3.5 Crystal structures of (a) $[(SO_4^{2-})_2(H_2O)_3]$ trapping by **L78** and (b) $[(\mathbf{L79})_2(HPO_4)]^{2-}$.

in solution.[16] Extensive ^{1}H-NMR studies carried out with **L79** revealed the selectivity order as, $H_2PO_4^- > SO_4^{2-} > CH_3COO^- > F^- > Cl^- \gg Br^-$ whereas I^-, NO_3^- and ClO_4^- did not show any significant binding. Comparative binding studies of SO_4^{2-} and $H_2PO_4^-$ with **L79** and **L71** exhibited higher association constant (K_a) values with **L79.** The association constant (K_a) value found with **L79** for recognition of SO_4^{2-} was 5.4×10^4 M^{-1}. We illustrated that in the solution state bisulfate was converted to sulfate which supported formation of a capsular 1 : 2 (SO_4^{2-}: **L79**) assembly which matched well with the solid state structure. The SO_4^{2-} was encapsulated inside the dimeric capsule of **L79** through fourteen H-bonding interactions. Further, we utilized ^{31}P-NMR to probe the binding of **L79** to $H_2PO_4^-$ where we observed significant down field shift ($\Delta\delta = 8.33$ ppm) of $H_2PO_4^-$ compared to that of free $H_2PO_4^-$.[73] Apart from $H_2PO_4^-$ and SO_4^{2-}, **L79** was also capable in efficient fixation of aerial CO_2 in the form of CO_3^{2-} in a capsular form as $[(\mathbf{L79})_2(CO_3^{2-})]$ $(n\text{-}Bu_4N^+)_2$ in alkaline conditions. **L79** was dissolved in DMSO in the presence of n-$Bu_4N^+OH^-$ which resulted in the formation of a carbonate capsule in an almost quantitative yield. Single crystal X-ray structural studies showed that CO_3^{2-} was bound with two **L79** units *via* sixteen hydrogen bonding interactions. An ^{1}H-NMR titration study was also carried out with **L79** with n-$Bu_4N^+HCO_3^-$ which showed down field shift of urea –NH protons and the Job's plot analysis revealed 1 : 2 (**L79:** HCO_3^-) binding stoichiometry. The ^{13}C-NMR peak for free n-$Bu_4N^+HCO_3^-$ appeared at 159 ppm, which was shifted to 170 ppm ($\Delta\delta = 11$ ppm) confirming the existence of encapsulated CO_3^{2-}. FT-IR analysis revealed the existence of two new peaks in $[(\mathbf{L79})_2(CO_3^{2-})]$ $(n\text{-}Bu_4N^+)_2$ at 1377 cm^{-1} and 1688 cm^{-1} which matched with that of the asymmetric and symmetric stretching frequencies of carbonate, respectively. In order to recover **L79**, the carbonate capsule ($[(\mathbf{L5})_2(CO_3^{2-})]$ $(n\text{-}Bu_4N^+)_2$) was dissolved in methanol–water (1 : 4, v/v) binary solvent mixture and it immediately precipitated out **L79** with an

85% yield which was further characterized by IR and powder X-ray diffraction studies. Thus this work established that **L79** could be an effective potential system for reversible binding and release of carbonate.[74] Further, in 2012, we utilized **L79** for acid–base controlled capsular size modulation to trap inorganic phosphates. Single crystal X-ray structural studies and ITC studies demonstrated that mono-negative guests like F^- and $H_2PO_4^-$ formed pseudo-capsular assemblies in 1:1 (host:guest) stoichiometry, and di-negative guests like CO_3^{2-}, SO_4^{2-} and HPO_4^{2-} formed dimeric capsular assemblies in 2 : 1 (host : guest) stoichiometry (Figure 3.5). The dimension of the capsular assembly of **L79** varied depending upon the guest from $H_2PO_4^-$ (13.8 Å) to HPO_4^{2-} (9.9 Å). We employed the previously reported ([[(**L79**)$_2$(CO_3^{2-})] (n-Bu_4N^+)$_2$) crystals for liquid–liquid extraction of SO_4^{2-} from water in quantitative yield, and unambiguously demonstrated that extraction was also feasible in the presence of NO_3^- and $H_2PO_4^-$ *via* anion exchange strategy and characterized the extracted mass through NMR, FT-IR, EDX, XRD and PXRD studies. Instead of the carbonate capsule we showed that a mixture of **L79** and n-$Bu_4N^+Cl^-$ failed to extract SO_4^{2-} purely and efficiently. ITC studies with **L79** in DMSO revealed that the receptor was found to be selective towards $H_2PO_4^-$ ($K_a = 3.3\times10^5$ M^{-1}) compared to SO_4^{2-} ($K_a = 5.4\times10^4$ M^{-1}) and $H_2PO_4^-$ binding was enthalpy-driven (-8.37 kcal mol^{-1}).[75] This receptor (**L79**) was further manifested for the recognition of the environmentally important and toxic arsenate anion ($HAsO_4^{2-}$) in the cavity of its dimeric capsular assembly. It is noteworthy to mention that it was the first structural evidence of $HAsO_4^{2-}$ recognition by a neutral receptor. We have calculated the binding constant of arsenate to **L79** in *DMSO-d$_6$-D$_2$O* buffered [(borax (10 mM)/boric acid (200 mM)] media which showed the association constant (k_a) value 2.2×10^4 M^{-1}. In this solvent system $H_2PO_4^-$ and SO_4^{2-} showed association constant values of 3.62 and 3.48. The arsenate complex [(**L79**)$_2$(n-Bu_4)$_2$(HA-sO$_4$)(DMSO)] was isolated with an 80% yield and in crystal, $HAsO_4^{2-}$ was found to be trapped by two units of **L79** *via* eleven N–H···O and one O–H···O hydrogen bonding interactions which referred to 2:1 (**L79**:$HAsO_4^{2-}$) stoichiometry of binding. The capsular dimension of [(**L79**)$_2$($HAsO_4$)]$^{2-}$ was found to be 10.184 Å, which was comparable to that of [(**L79**)$_2$(HPO_4)]$^{2-}$ (9.916 Å), [(**L79**)$_2$(SO_4)]$^{2-}$ (9.18 Å) and [(**L79**)$_2$(CO_3)]$^{2-}$ (9.17 Å).[76] We have also utilized **L79** further for the encapsulation of $C_2O_4^{2-}$. Encapsulation of the planar conformer of oxalate (torsion angle 3.06°) was observed with the capsular dimension of 10.75 Å where the guest was stabilized through twelve hydrogen bonding interactions. The existence of an orthogonal C–F···C=O interaction was also observed in solid state structure of [(n-Bu_4)$_2$(**L79**)$_2$(C_2O_4)] which had a capsular dimension of 10.75 Å. ^{1}H-NMR titration studies in *DMSO-d$_6$–D$_2$O* solvent mixture showed the highest association constant (K_a) of $C_2O_4^{2-}$ with **L70** (6.6×10^4 M^{-1}) which was much higher than that with **L79** (1.8×10^4 M^{-1}) and **L72** (1.9×10^4 M^{-1}).[77]

Wu *et al.* decorated the TREN scaffold with the 3-pyridyl urea unit **L80** (Chart 3.6) which encapsulated SO_4^{2-} in the second sphere in the

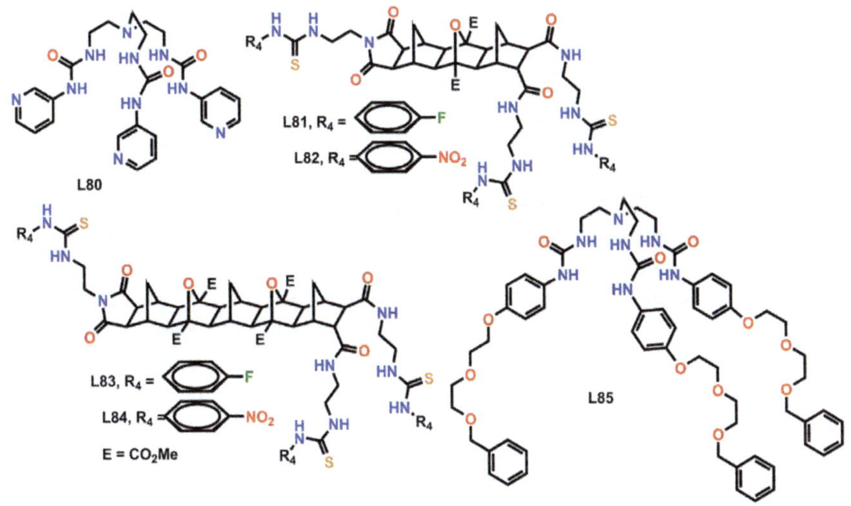

Chart 3.6 Chemical structures of receptors **L80–L85**.

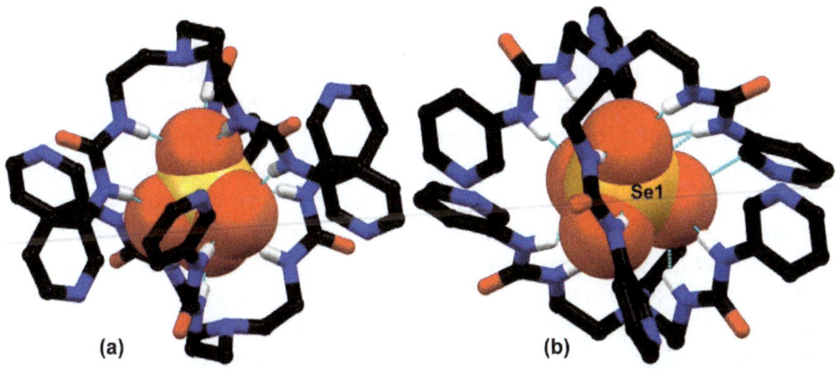

Figure 3.6 Crystal structures of (a) SO_4^{2-} encapsulation by **L80** and (b) SeO_4^{2-} encapsulation by **L80**.

presence of Mn^{+2} and Zn^{+2} to form a supramolecular cage structure $[M(H_2O)_6(\mathbf{L80}_2SO_4)][M=Mn^{+2}, Zn^{+2}]$ (Figure 3.6) which was confirmed by single crystal X-ray crystallography. The Zn^{2+} complex of **L80** showed significant down field shift of urea protons by 1.01 ppm and 0.74 ppm compared to that of **L80** in *DMSO-d_6* which confirmed the binding of SO_4^{2-}. ^{1}H-NMR titration studies were also carried out with **L80** and $ZnSO_4 \cdot 7H_2O$ in *DMSO-d_6* which showed a significant high binding constant value (K_a) of 2.6×10^6 M^{-1}. This binding constant value with sulfate was comparatively higher than those found in contemporary literature.[71,78] Negative ion mode ESI-MS spectra of $[M(H_2O)_6(\mathbf{L80}_2SO_4)]$ $[M=Mn^{+2}, Zn^{+2}]$ exhibited anionic peaks at *m*/*z* 1109.8 for the monovalent $[(\mathbf{L80})_2+SO_4+H]^-$ and *m*/*z* 554.6 for divalent $[(\mathbf{L80})_2+SO_4]^{2-}$ which proved the tight binding of sulfate in solution.[79] Further in 2009, they have utilized **L80** for cleft binding of SO_4^{2-} in the presence of the self-assembly of $[Fe(DABP)_3]SO_4$ (DABP = 5,5′-diamino-2,2′-bipyridine) or $[Fe(bipy)_3]SO_4$ (bipy = 2,2′-bipyridine).[80] Single crystal X-ray structural analysis revealed formation of similar sheets by anionic units $[(\mathbf{L80})SO_4^{2-}]$ and cationic units $[([Fe(DABP)_3]^{2+}, [Fe(bipy)_3]^{2+})]$. The only difference lay in the water cluster that connected the adjacent layers. In $[Fe(DABP)_3]SO_4 \cdot \mathbf{L80} \cdot 10H_2O$, the water cluster formed parallelograms and quasi-water cubes, whereas in $[Fe(bipy)_3]SO_4 \cdot \mathbf{L80} \cdot 9H_2O$ water dimers or hexamers were present. The secondary coordination sphere in $[M(H_2O)_6(\mathbf{L80}_2SO_4)]$ $[M = Mn^{+2}, Zn^{+2}]$ helped the formation of a cation–anion octahedral complex, but in $[Fe(DABP)_3]SO_4 \cdot \mathbf{L80} \cdot 10H_2O$ and $[Fe(bipy)_3]SO_4 \cdot \mathbf{L80} \cdot 9H_2O$ the water molecules satisfied the H-bonding demand of **L80** which resulted in the formation of the non-spherical and water-fenced half-capsule of $[(\mathbf{L80})SO_4^{2-}]$. FT-IR, PXRD and elemental analysis showed the sulfate selectivity of these two supramolecular architectures *via* competitive experiments in aqueous medium carried out in presence of NO_3^-, AcO^- and ClO_4^-.[80]

The same receptor **L80** has also been utilized by Custelcean *et al.* to trap sulfate inside the capsule formed as $MSO_4(\mathbf{L80})_2(H_2O)_6$ [M = Zn, Cd, Co, Mg] *via* second sphere coordination in a methanol–water (1 : 1, v/v) binary solvent mixture which was confirmed by single crystal X-ray crystallography and elemental analysis. The crystals obtained were all isostructural with the space group $P2_1/n$ regardless of the metal ion. This metal-organic framework was composed of a centrosymmetric SO_4^{2-} which was bound to six urea centres through twelve hydrogen bonding interactions. The quasi-octahedral arrangement of the water molecules around the hydrogen-bonded capsules matched the coordinating geometry of $Mg^{2+}(H_2O)_6$ hydrates, which formed a 3D framework with distorted NaCl topology. ^1H-NMR Job's plot analysis confirmed 1 : 1 SO_4^{2-} : **L80** complex formation in $CD_3OD : D_2O$ 1 : 1 binary solvent mixture. Moreover, **L80** was capable of isolating SO_4^{2-} in the presence of a 100-fold excess of $NaNO_3$ with almost 88% yield through competitive crystallization. To probe the sulfate selectivity, they crystallized isostructural complexes of $MgX(\mathbf{L80})_2(H_2O)_6$ $(X = SO_3^{2-}, CO_3^{2-}, SeO_4^{2-})$ (Figure 3.6) with the same space group but containing different cell parameters. Competitive crystallization was also performed in the presence of two equivalents of **L80** and 1 : 1 molar mixtures of $MgSO_4$ and Na_2X salts $(X = SO_3^{2-}, CO_3^{2-}, SeO_4^{2-})$ and precipitated solids were characterized by FT-IR spectroscopy. The anion selectivity found in this case was $SO_4^{2-} > SeO_4^{2-} > CO_3^{2-} > SO_3^{2-}$ which differed from the Hofmeister series.[81] They continued their study with **L80** further and utilized it for selective separation of SO_4^{2-}. In the context of their work they mentioned that sulfate-binding proteins can selectively separate out SO_4^{2-}, compared to that of SeO_4^{2-}, by a factor of 40 and till that time, there were no synthetic anion receptors known which showed such competitive selectivity. Allosteric modification was regulated by the external bridging cation, *i.e.* $Mg(H_2O)^{2+}$ to $Li(H_2O)^+$, and showed formation of a new capsule with reduced cavity size to

Table 3.1 Thermodynamic parameters of crystallization in borax-buffered water.

Complexes	K_{sp}	ΔG°_{cryst} (kJ mol^{-1})	ΔH°_{cryst} (kJ mol^{-1})	$T\Delta S^{\circ}_{cryst}$ (kJ mol^{-1})
$MgSO_4$(**L80**)$_2$(H_2O)$_6$	$2.0 \pm 0.3 \times 10^{-17}$	-95.3 ± 0.4	-99.1 ± 1.9	-3.8 ± 2.3
$MgSeO_4$(**L80**)$_2$(H_2O)$_6$	$5.5 \pm 0.6 \times 10^{-16}$	-87.1 ± 0.3	-108.5 ± 2.3	-21.4 ± 2.6
$MgSO_3$(**L80**)$_2$(H_2O)$_6$	$6.6 \pm 0.3 \times 10^{-16}$	-86.6 ± 0.1	-64.6 ± 0.8	22.0 ± 0.9

bind SO_4^{2-} extensively with selectivity, exceeding that of sulfate-binding protein.[82] Further, in order to justify the entropic–enthalpic contribution to capsule formation and selectivity order in respect of thermodynamics *versus* kinetics, they continued their research on **L80**. The thermodynamic parameters associated with the crystallization process measured in borax-buffered aqueous solution revealed that crystallization was enthalpy-driven (Table 3.1) which was in contrast with anion binding nature of TREN-based protonated receptors in water. The major factor that governed the recognition process was the size of the urea functionalized cavity that preferred SO_4^{2-} and SeO_4^{2-}. Among these two guests, SeO_4^{2-} was preferred due to its favourable enthalpy of crystallization for size and shape complementarity whereas SO_4^{2-} was favoured for its dehydration entropy and less tight binding in the crystalline capsules of **L80**. Thus, entropy of crystallization played a simultaneous role with shape complementarity to attain selectivity. Thermodynamic measurements indicated the non-existence of hydrogen-bonded guest bound capsules in aqueous medium and the K_{sp} disclosed that $MgSO_4$(**L80**)$_2$ was the least aqueous soluble followed by $MgSeO_4$(**L80**)$_2$ and $MgSO_3$(**L80**)$_2$.[83]

The same group further utilized **L80** to illustrate a viable approach for SO_4^{2-} separation in a highly alkaline solution (pH = 14) which mimicked that of the conditions of the nuclear waste. The previously reported $MgSO_4$(**L80**)$_2$(H_2O)$_6$ was of limited utility for SO_4^{2-} separation in alkaline conditions (pH = 14) as it turned out to be precipitated out in the form of $Mg(OH)_2$. Thus they utilized Na^+ and K^+ as metals for the 3-pyridyl centre as they were present in abundance in the nuclear waste and no external ionic components need to be added to the waste. Moreover, due to the common ion effect the solubility of the capsule was found to be decreased. They grew suitable crystals for Na_2SO_4(**L80**)$_2$(H_2O)$_4$ and K_2SO_4(**L80**)$_2$(H_2O)$_2$ from methanol–water (1 : 1, v/v) binary solvent mixture. Twelve hydrogen bonding interactions through six chelating urea groups in Na_2SO_4(**L80**)$_2$(H_2O)$_4$ and K_2SO_4(**L80**)$_2$(H_2O)$_2$ resulted to form a SO_4^{2-} encapsulated 3D network of NaCl topology. Even in presence of 114 times more nitrate compared to that of sulfate, such selective and competitive crystallization was found to occur which was further characterized by PXRD and Single crystal X-ray structural studies.[84]

Pfeffer *et al.* reported tris thio urea functionalized fused [3] and [5] polynorboranane-based receptors **L81–L84** (Chart 3.6). All these receptors were found to bind terephthalate^{2-} in 1 : 1 (host : guest) stoichiometry. Among these two frameworks the [5] polynorboranane-based receptors **L83** and **L84** showed higher binding constant (K_a) values of 2.0×10^4 M^{-1} and 7.9×10^4 M^{-1} with terephthalate^{2-} respectively.[85] Beer *et al.* synthesized receptor **L85** (Chart 3.6) and studied its anion binding propensity in *CD$_3$CN–DMSO-d$_6$* 5 : 1 solvent mixture. The urea –NH protons were found to shift up to $\Delta\delta$ 1.09 and 2.09 ppm upon the addition of one equivalent of $(n\text{-}Bu_4)_2SO_4$. The hydroquinone protons were also found to shift upon the addition of 0.5 equivalent of $(n\text{-}Bu_4)_2SO_4$ to form a 2 : 1 (**L85** : SO_4^{2-}) adduct which dissociated to a 1 : 1 (**L85** : SO_4^{2-}) adduct upon the addition of higher equivalents of guest. Negative mode ESI-MS experiment showed a molecular ion peak at *m/z* 1133.7 which corresponded to the [(**L85**)$_2$SO$_4$]$^{2-}$ adduct. Further they modified the **L85** moiety to utilize its SO_4^{2-} selectivity for the preparation of sulfate-templated tris-urea based triply-interlocked molecule[86] and we exclude discussion on this aspect as it falls beyond the scope of this chapter.

Gale *et al.* developed urea- and thiourea-substituted receptors on a TREN platform for studying transmembrane transporters for bicarbonate. The receptors **L86–L89** (Chart 3.7) were also utilized for anion binding studies to assess the stability of the anion complex receptor. ^{1}H-NMR studies conducted in *DMSO-d$_6$* –0.5% water solvent revealed that all these receptors (**L86–L89**) showed association constant (K_a) values above 10^4 M^{-1} with sulfate. HCO_3^- bound the receptor with significant down field shift but not in 1 : 1 (host : guest) stoichiometry, except in the case of **L88**. The association constants (K_a) measured in case of Cl^- were found to be 658 M^{-1}, 830 M^{-1}, 447 M^{-1} and 191 M^{-1} in cases of **L86–L89** respectively. Single crystal X-ray structural analysis showed CO_3^{2-} encapsulated in between two units of **L89**. The encapsulated CO_3^{2-} was bound through twelve hydrogen bonding interactions which ranged within 2.824 Å (6)–3.070 Å (7). Further, they utilized the receptors for bicarbonate transport which is not discussed here as it falls beyond the scope of the content.[87] In the following year, they reported a series of TREN-based urea- and thiourea-based receptors (**L90–L93**) (Chart 3.7) and investigated their anion selectivity and membrane

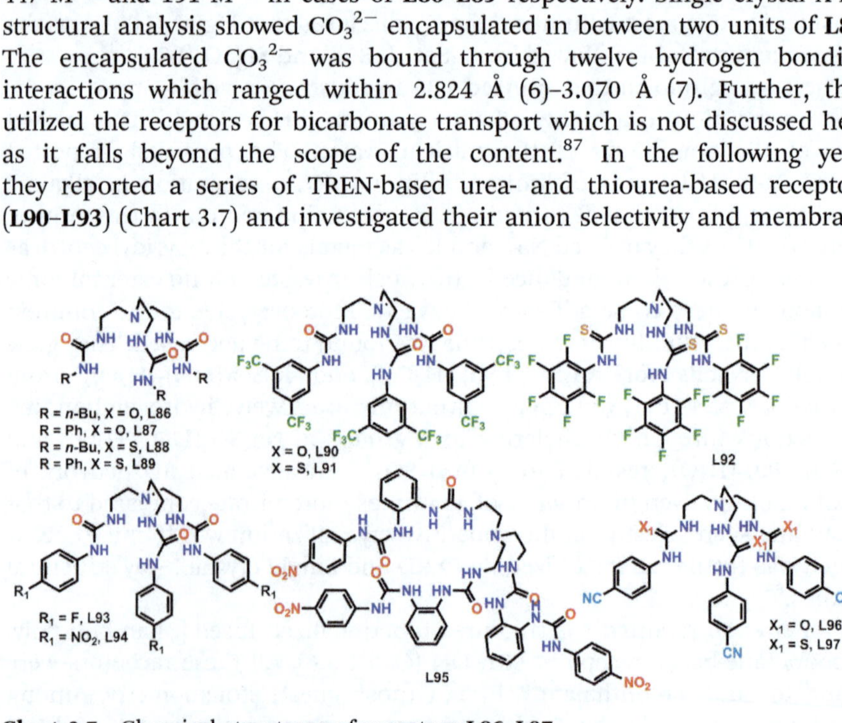

Chart 3.7 Chemical structures of receptors **L86–L97**.

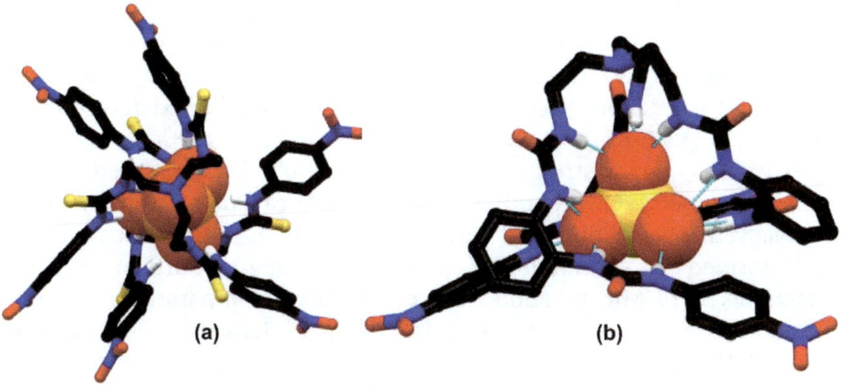

Figure 3.7 Crystal structures of (a) PO_4^{3-} trapping by **L94** and (b) SO_4^{2-} recognition by **L95**.

transportation properties. Among these receptors, **L90** and **L91** were two such receptors that encapsulated CO_3^{2-} and HPO_4^{2-} to form respective capsules. Both these receptors bound SO_4^{2-} with association constant (K_a) values higher than 10^4 M^{-1}. Titration with n-$Bu_4N^+H_2PO_4^-$ resulted in the generation of new peaks in the aromatic region upon addition of guest beyond one equivalent. This phenomenon was explained by the deprotonation of bound dihydrogen phosphate and the subsequent formation of a mono-hydrogen phosphate and thus prevented from association constant estimation. In the solid state, **L91** was also found to encapsulate HPO_4^{2-} (Figure 3.7) inside the dimeric capsular assembly (**L91** : HPO_4^{2-} = 2 : 1) in the presence of excess n-$Bu_4N^+H_2PO_4$.[88]

Das *et al.* decorated the TREN scaffold with p-NO_2-thiourea functionality, **L94** (Chart 3.7) which eventually formed PO_4^{3-} encapsulated dimeric capsules in the presence of an excess amount of n-$Bu_4N^+H_2PO_4^-$. ^{1}H-NMR titration with **L94** resulted in the generation of new peaks upon the addition of guest beyond one equivalent which resembled that of observed in **L91**. The dimeric capsule, $[(Bu_4N)_3(\mathbf{L94})_2(PO_4)]\cdot 2MeCN$ (Figure 3.7) was obtained from acetonitrile (MeCN) by slow evaporation of solution containing **L94** and excess n-$Bu_4N^+H_2PO_4^-$. The axial and facial mode position of two **L94** units formed a caged supramolecular structure where twelve hydrogen bonding interactions stabilized PO_4^{3-}. The shift of urea –NH protons ($\Delta\delta$) in $[(Bu_4N)_3(\mathbf{L94})_2(PO_4)]\cdot 2MeCN$ was found to be 3.76 and 2.80 ppm compared to that observed in **L94** which confirmed strong binding. Further, the binding constants (K_a) estimated with F^- and Cl^- with **L94** were found to be 3.0×10^4 M^{-1} and 3.5×10^3 M^{-1} respectively.[89]

Wu and co-workers reported the hexa-urea substituted, TREN-platform-based receptor, **L95** (Chart 3.7) which was the first organic receptor to show saturated coordination of SO_4^{2-} by a single host. The $SO_4^{2-}\cdots$**L95** formed a tetrahedral cage in which the guest was encapsulated *via* twelve N–H$\cdots$O interactions (Figure 3.7). ^{1}H-NMR titrations showed an association constant > 10^4 M^{-1} for SO_4^{2-} in 3 : 1 (v/v) *DMSO-d*$_6$ and H_2O binary solvent mixture. The extraction of sulfate in the presence of $NaNO_3$ was also shown.[90] **L96** and **L97** (Chart 3.7) were synthesized by us to study anion binding and extraction phenomena. The tris-thiourea-based receptor **L97** exhibited 1 : 1 (**L97** : F^-) complex formation in single crystal X-ray structural studies with F^- *via* six strong N–H···F^- interactions in a distorted trigonal prismatic geometry. On the other hand, dimeric capsular assembly (9.5 Å) was observed in case of SO_4^{2-} with fifteen N–H···O interactions whereas CO_3^{2-} formed a sandwich-like assembly through fourteen N–H···O interactions. ^{1}H-NMR titration studies revealed comparatively stronger binding with F^- ($K_a = 3.7 \times 10^3$ M^{-1}) and Cl^- ($K_a = 1.1 \times 10^3$ M^{-1}) compared to that of HSO_4^- ($K_a = 6.7 \times 10^2$ M^{-1}) and $H_2PO_4^-$ ($K_a = 1.6 \times 10^2$ M^{-1}) in *DMSO*-d$_6$. ITC studies with **L97** in acetonitrile evoked the same selectivity pattern as was observed in ^{1}H-NMR titration, *i.e.* F^- ($K_a = 1.0 \times 10^7$ M^{-1}) ≈ Cl^- ($K_a = 5.6 \times 10^6$ M^{-1}) > HSO_4^- ($K_a = 2.0 \times 10^5$ M^{-1}) ≈ $H_2PO_4^-$ ($K_a = 8.5 \times 10^4$ M^{-1}). Further, **L97** was utilized for liquid–liquid extraction of F^- (70% extraction) and SO_4^{2-} (40% extraction) in presence of $Bu_4N^+I^-$.[91] **L96** was also capable of forming a capsular dimeric assembly with the planar form of $C_2O_4^{2-}$ with a capsular dimension of around 10.08 Å and ^{1}H-NMR titration studies in *DMSO-d*$_6$*–D*$_2$*O* with $C_2O_4^{2-}$ revealed association constant (K_a) value 2.3×10^5 M^{-1}.

Another tris-*m*-chloro substituted thiourea receptor, **L98** (Chart 3.8) was developed by Das *et al.* almost parallel in time. They reported the formation of a dimeric capsular assembly in the presence of both SO_4^{2-} [(**L98**)$_2$$SO_4^{2-}$] and $S_2O_3^{2-}$ [(**L98**)$_2$$S_2O_3^{2-}$] guests in a DMSO–water binary solvent mixture. This was the first example of $S_2O_3^{2-}$ encapsulation by a neutral receptor **L98** (Figure 3.8). The structural orientations of these complexes were mainly governed by halogen bonding interactions, *i.e.* Cl···Cl in case of [(**L98**)$_2$$S_2O_3^{2-}$] and Cl···S in case of [(**L98**)$_2$$SO_4^{2-}$]. Solution state ^{1}H-NMR and 2D-NOESY experiments confirmed binding of these guests in the pseudo-cavity of **L98** in 1 : 1 (host : guest) stoichiometry which contradicted the solid state stoichiometry. Association constant values (K_a) estimated for SO_4^{2-} and $S_2O_3^{2-}$ with **L98** were found to be 3.5×10^4 M^{-1} and 2.2×10^3 M^{-1} respectively in *DMSO-d*$_6$.[92] Electron deficient cyanuric acid platform based pentafluorophenyl group attached tripodal urea receptors **L99** (Chart 3.8) was synthesized and explored for sulfate recognition studies by us. Pentafluorophenyl substituted receptor **L99** showed selective recognition of SO_4^{2-} over all other anions as evident from a ^{1}H-NMR titration study ($K_a = 5.6 \times 10^5$ M^{-1}). Interestingly, structural analysis of the SO_4^{2-} complex of **L99** revealed formation of a SO_4^{2-} encapsulated unimolecular capsule (Figure 3.8) where two $(n\text{-Bu})_4N^+$ countercations act as a stopper of the capsule. SO_4^{2-} encapsulation was facilitated by strong H-bonding interactions with the urea –NH protons and weak H-bonding interactions with $(n\text{-Bu})_4N^+$ C–H protons. This unimolecular capsular assembly and disassembly process was nicely demonstrated by temperature as an external stimulus. A ^{1}H-DOSY experiment showed the formation of a single species in

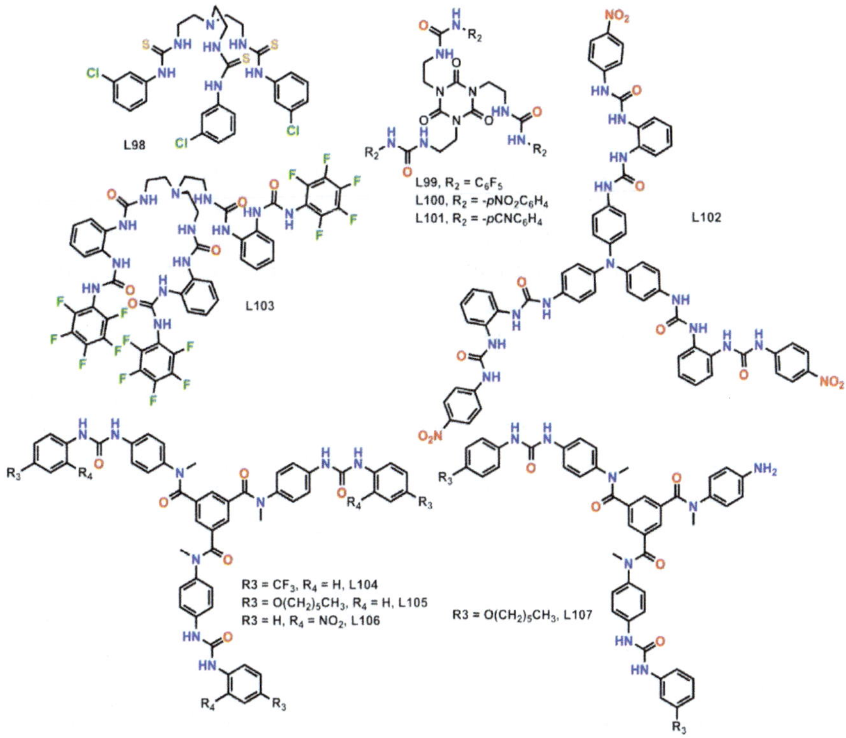

Chart 3.8 Chemical structures of receptors **L98–L107**.

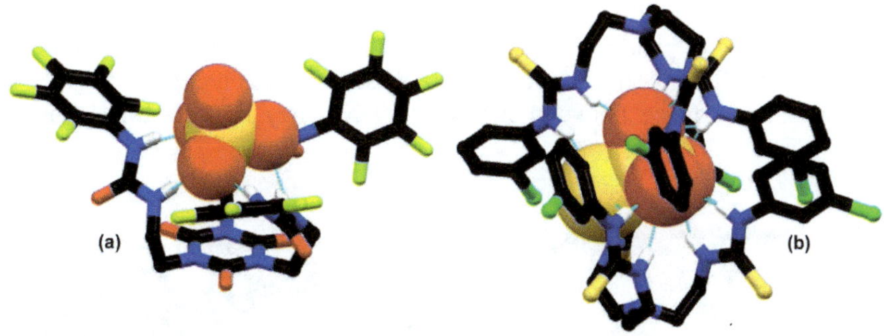

Figure 3.8 Crystal structures of (a) SO_4^{2-} encapsulation by **L99** and (b) $S_2O_3^{2-}$ encapsulation by **L98**.

solution at room temperature (298 K) composed of one **L99** unit, one SO_4^{2-} and two $(n\text{-Bu})_4N^+$ cations. The calculated size of the capsule from the experimental diffusion coefficient was perfectly matched with the observed capsular size from the X-ray structure. This assembly was ruptured when the temperature was raised to 323 K by the breaking of weak C–H···O interactions with the $(n\text{-Bu})_4N^+$ group.[93]

Electron deficient cyanuric acid platform based tripodal urea receptors **L100–L101** (Chart 3.8) were synthesized and explored for SO_4^{2-} recognition studies by us. We reported structural variation of SO_4^{2-} encapsulation by two other tripodal urea receptors **L100** and **L101** having *p*-nitrophenyl and *p*-cyanophenyl groups respectively. **L100** showed encapsulation of SO_4^{2-} by the aid of one $(n\text{-Bu})_4N^+$ counter cation whereas SO_4^{2-} assisted dimeric capsular assembly was observed for **L101**. Thus we concluded that either participation of $(n\text{-Bu})_4N^+$ counteraction(s) or dimeric capsular assembly was required to furnish the higher coordination number of SO_4^{2-}.[94]

Wu *et al.* synthesized a tris(bis-urea) ligand **L102** (Chart 3.8) appended to a central C_3-symmetric triphenylamine platform. This was the first report on hexa-urea receptors utilized for anion recognition. Complexation of **L102** with $(NMe_4)_3PO_4$ resulted in the formation of a tetrahedral anion cage of composition $[(\mathbf{L102})_4(PO_4)_4]^{12-}$ (Figure 3.9) which was unprecedented in anion-coordination chemistry. The four C_3-symmetric **L102** units were positioned on the four triangular faces of the tetrahedron, and each vertex was occupied by a phosphate ion which was coordinated by three bis-urea arms. The $PO_4^{3-}\cdots PO_4^{3-}$ distance inside the cage was about 15 Å and the internal volume of the cage was found to be 181 Å^3 which was calculated by the VOIDOO program. They had also shown that the ^{31}P-NMR of the $[(\mathbf{L102})_4(PO_4)_4]^{12-}$ showed substantial downfield shift by 8.03 ppm compared to that of free PO_4^{3-} in *DMSO-d*$_6$. In contrast to phosphate, SO_4^{2-} formed a pinwheel complex with a composition $[(\mathbf{L102})_2(SO_4)_3]^{6-}$, where each sulfate was bound by eight N–H···O hydrogen bonds.[95]

In 2014, Hossain *et al.* reported another pentafluro substituted second generation tripodal hexa-urea receptor **L103** (Chart 3.8) which absorbed aerial CO_2 to produce a CO_3^{2-} capsule. The carbonate complex $[(\mathbf{L103})_2CO_3^{2-}]$ was found to be stabilized by twelve strong N–H···O interactions ranging between 2.703–2.989 Å. They dissolved **L103** in the

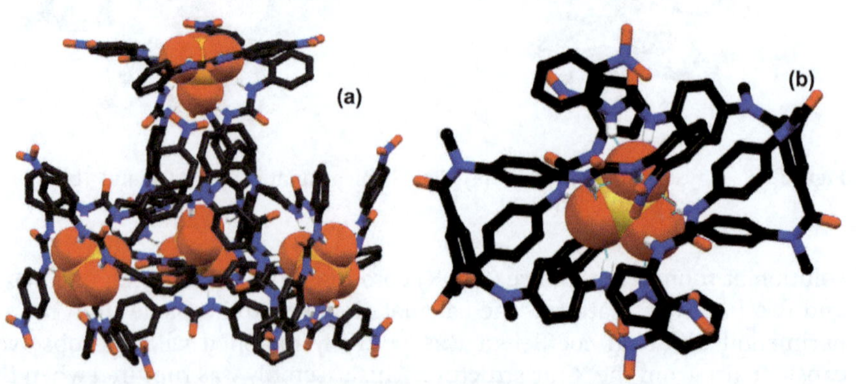

Figure 3.9 Crystal structures of (a) $[(\mathbf{L102})_4(PO_4)_4]^{12-}$ and (b) $H_2PO_4^-$ assisted 2:2 cage of **L104**.

presence of $Bu_4N^+F^-$ in DMSO and obtained needle-like crystals within two days. ^{1}H-NMR titration in *DMSO-d*$_6$ revealed 1:1 (host:guest) binding stoichiometry with an association constant (K_a) value of 1780 M^{-1}. Time dependent NMR of **L103** in the presence of $Et_4N^+HCO_3^-$ showed no change in NMR signals which suggested that CO_3^{2-} was not formed by the deprotonation of HCO_3^-. The association constant (K_a) value for $Et_4N^+HCO_3^-$ was found to be 564 M^{-1} which was much smaller than that of CO_3^{2-}. A ^{13}C-NMR peak of [(**L103**)$_2$$CO_3^{2-}$] appeared at 176.7 ppm which was different from that of $Et_4N^+HCO_3^-$. Further, they recovered **L103** from [(**L103**)$_2$$CO_3^{2-}$] from a methanol–water solvent mixture which was previously demonstrated by us.[74]

Gunnalaugsson *et al.* reported synthesis and tetrahedral oxy-anion binding studies on *N*-methyl-1,3,5-benzenetricarboxamide platform based tripodal urea receptors **L104–L107** (Chart 3.8). SO_4^{2-} template dimeric capsular assemblies were observed in the cases of **L104–L106.** The capsular assemblies clearly showed that SO_4^{2-} was fully engulfed within a cavity formed from two interdigitated ligands through the hydrogen bonding interactions of all twelve urea –NH groups. On the other hand, **L104** was found to encapsulate $H_2PO_4^-$ in a 2:2 (host:guest) stoichiometry, where the two $H_2PO_4^-$ ions were bridged by a water molecule within the cage (Figure 3.9). Interestingly receptor **L107**, a side product of **L105**, formed self-assembled capsule with a 4:4 (host:guest) complex with SO_4^{2-}.[96]

3.4 Anion Recognition with Tripodal Amine Receptors

In 2004, Bowman-James *et al.* reported anion binding affinity of a phenyl group attached TREN-based tripodal amine receptor **L108** (Chart 3.9) in its

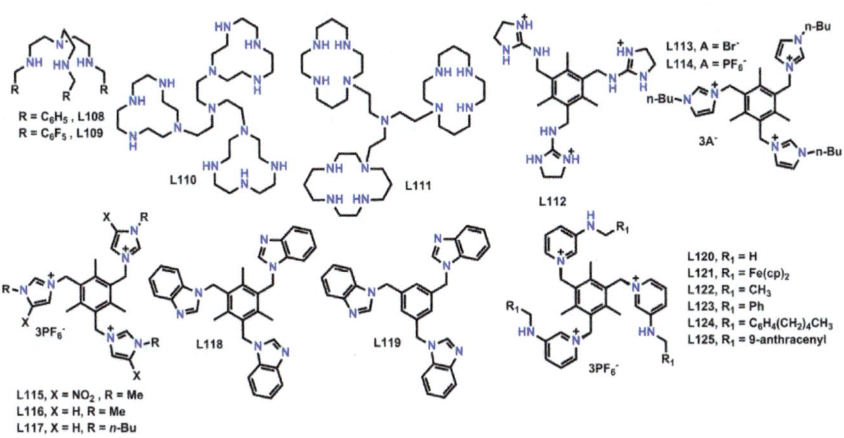

Chart 3.9 Chemical structures of receptors **L108–L125.**

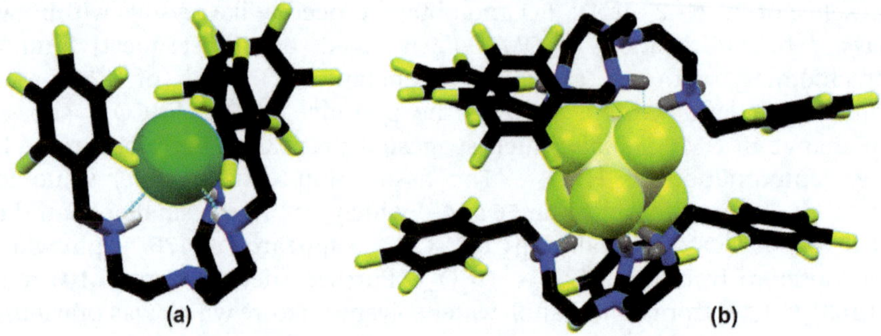

Figure 3.10 Crystal structures of (a) Cl^- trapping by **L109** and (b) Capsular recognition of SiF_6^{2-} by **L109**.

protonated form. ^1H-NMR titration showed that tris-protonated ligand possessed a higher affinity for $H_2PO_4^-$ and HSO_4^- over nitrate and halides. The crystal structure of the bromide complex revealed the receptor adopted C_{2v}-symmetric conformation.[97] Later on, we structurally demonstrated that the pentafluorophenyl group attached tripodal amine receptor **L109** (Chart 3.9) adopted C_{3v} symmetric conformation upon binding with bromide and chloride anions (Figure 3.10). Both the Cl^- and Br^- binding was facilitated by hydrogen bonding and anion–Π interactions with the receptor functionalities.[98] Bencini *et al.* reported two TREN-based macrocyclic receptors having three [12]aneN_4, **L110** and [14]aneN_4, **L111** (Chart 3.9) units which bound three positional isomers of the benzenetricarboxylic acid (BTC). Calorimetric experiments in an aqueous medium showed the formation of 1 : 1 complexes with the three guests having highest binding affinity towards the 1,3,5-BTC isomer. Calorimetric studies pointed out that the complexation with trianionic forms were entropy-driven whereas with the protonated forms, the complexation processes were mainly enthalpy-driven.[99] Further, we reported a TREN-based pentaflurophenyl amine ligand, **L109** for the encapsulation of F^- inside the triprotonated cavity of **L109**. Octahedral hexaflurosilicate was also encapsulated *via* the formation of a dimeric capsular assembly (Figure 3.10). Crystal structural analysis showed that the SiF_6^- anion was encapsulated in the capsular cavity *via* twelve NH···F^- interactions. Solution state potentiometric studies in a 1 : 1 methanol–water binary solvent mixture revealed increased affinity of **L109** towards highly basic anions such as fluoride and acetate.[100]

3.5 Anion Recognition with Tripodal Guanidinium Receptor

Anslyn and co-workers developed **L112** (Chart 3.9) with three guanidinium groups which were complementary towards guests with three carboxylate

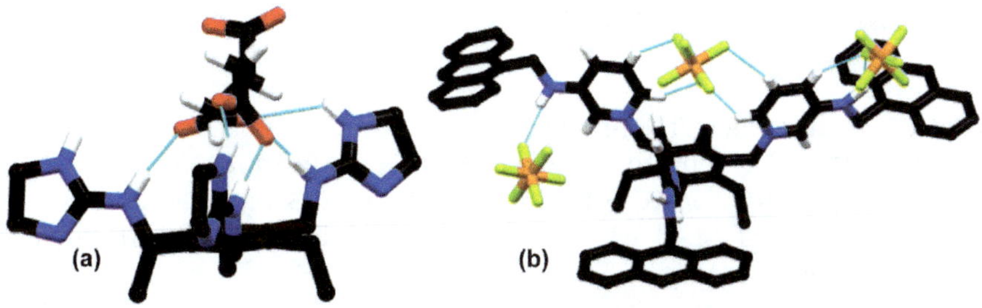

Figure 3.11 Crystal structures of (a) citrate recognition by **L112** and (b) recognition of SiF_6^{2-} by **L125.**

groups. They estimated association constants from ^{1}H-NMR titrations in D_2O which revealed selectivity towards citrate (6.9×10^3 M^{-1}) compared to guests with fewer carboxylate groups. Single crystal X-ray structural studies revealed the binding of citrate to **L112** (Figure 3.11).[101] Further, they utilized **L112** for estimation of citrate in beverages using the dye (5-Carboxyfluorescein) displacement method.[102]

3.6 Anion Recognition with Tripodal Imidazole and Imidazolium Receptors

Sato *et al.* developed **L113** and **L114** (Chart 3.9) for recognition through C–H···anion interactions. They measured the stability constants for these receptors by ^{1}H-NMR titrations with halides in acetonitrile-d_3 which revealed preference of halides towards **L113** (Cl^-; 7.5×10^4 M^{-1}, Br^-; 4.6×10^4 M^{-1} and I^-; 7.2×10^3 M^{-1}) compared to **L114** (Cl^-; 1.5×10^3 M^{-1}).[103] These results were in contrast to that of receptors reported by Anslyn *et al.*[101,104] Kim *et al.* developed imidazolium-based receptors (**L115–L117**) and manifested their anion binding through ^{1}H-NMR titrations. **L115** showed an association constant of 1.1×10^6 M^{-1} with Cl^- in a 9:1 *DMSO-d_6–MeCN-d_3* mixture. F^- exhibited nucleophilic attack on the receptor **L115** and thus they developed **L116** and **L117** which showed association constants 1.2×10^3 M^{-1} and 2.4×10^3 M^{-1} with F^- respectively.[105] We developed the tris-benzimidazole substituted receptor **L118** which upon protonation in water produced capsular dimeric assembly $[H_3(\mathbf{L118})(NO_3)_3(H_2O)]_2$. The capsule was formed by six nitrates and two water molecules. When $(n\text{-Bu})_4N^+Cl^-$ was charged into $[H_3(\mathbf{L118})(NO_3)_3(H_2O)]_2$ in water, the dimeric assembly broke to form $[H_3(\mathbf{L118})Cl(NO_3)_2(H_2O)_4]$ which was confirmed by single crystal X-ray structural studies.[106] Later on we reported that the same receptor was capable of forming a dimeric capsular assembly upon protonation with $HClO_4$ in water as $[H_3(\mathbf{L118})(ClO_4)_3(HClO_4)(H_2O)_n]$. We have also investigated the effect of substitution on the arene platform. In order to do so, we prepared **L119** which was unable to form capsular dimeric assembly in presence of NO_3^-, I^- and ClO_4^-.[107]

3.7 Anion Recognition with Tripodal Pyridinium Receptors

Steed *et al.* developed 3-amino-pyridinium substituted tripodal receptors, **L120–L125** (Chart 3.9) on the tri-alkyl benzene platform. **L123** was found to be selective towards Cl^- ($K_a > 10^6$ M^{-1}) whereas **L125** was selective towards acetate ($K_a = 49\,000$ M^{-1}) which was estimated by ^{1}H-NMR titrations in *MeCN-d*$_3$ (Figure 3.11).[108,109]

3.8 Steroid-based Tripodal Anion Receptors with Diversified Binding Motifs

Davis *et al.* developed tripodal receptors on a cholic acid platform with carbamate, **L126** and amide groups, **L127** (Chart 3.10). The cholic acid platform acted to orient the hydrogen bonding donating groups in space to bind the anion. Among the studied anions (halides and tosylate), **L126** was found to be selective towards F^- (1.54×10^4 M^{-1}) whereas **L127** was selective towards Cl^- (9.2×10^4 M^{-1}) which was measured through ^{1}H-NMR titrations.[110] They functionalized the cholic acid platform with two urea motifs and one quaternary ammonium group to show spherical anion binding. They undertook U-tube experiments to show the ability of **L128** in anion transportation where halides were favoured over oxyanions of comparable lipophilicity.[111] Further, they functionalized the cholic acid platform with an urea motif (**L129**) for the transportation of Cl^- and translocation of phospholipids through vesicles and cell membranes. They utilized the voltammetry method to estimate binding constant of **L129** with F^- (1×10^{12} M^{-1}), Cl^- (5×10^{12} M^{-1}) and Br^- (2×10^{11} M^{-1}).[112] They have continued their work on cholic acid platform based tripodal receptors for

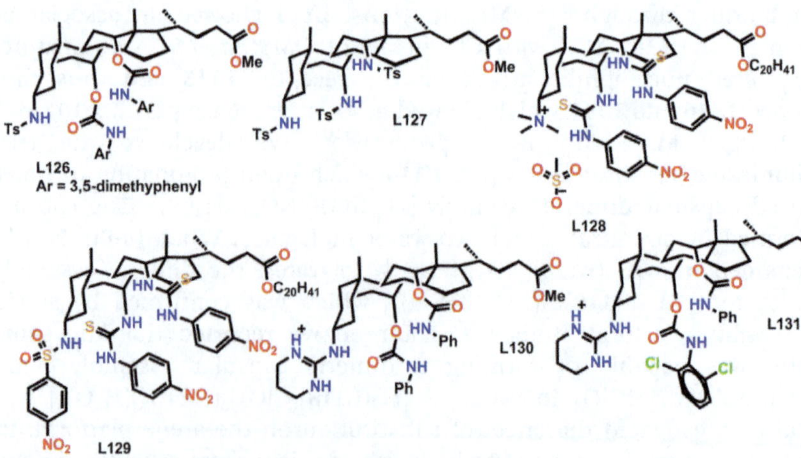

Chart 3.10 Chemical structures of receptors **L126–L131**.

extraction and transportation of anions.[113,114] Davis and Lawless explored the extraction properties of guanidinium functionalized steroids receptors with *N*-acetyl-amino acids. The $CHCl_3$ solution of **L130** · Cl^- and **L131** · Cl^- was found to be an effective extraction agent for carboxylates and *N*-acetyl-*a*-amino acids.[115]

3.9 Anion Recognition Through Tripodal Halogen Bond Donors

Halogen bonding (XB) has seen remarkable growth in last decade. Resnati *et al.* was the first to report tripodal neutral ditopic XB donor receptors **L132**–**L133** (Chart 3.11).[116] Among these two, **L132** was an efficient XB donor, binding I^- and Na^+ simultaneously which was observed in the single crystal X-ray structure of [NaI···**L132**]. ESI-MS experiments confirmed the higher affinity of **L132** towards I^- than Cl^- and Br^- (Figure 3.12). In 2010 Taylor *et al.* monitored the effect of the number and position of fluorine atoms on the strength of halogen bonding with the receptors **L134–L144** (Chart 3.11). ^{19}F-NMR titrations with Cl^- confirmed 1 : 1 (host : guest) binding stoichiometry with **L134–L144.** Among **L134–L137**, highest binding was observed in the case of **L34** with Cl^- (1.9×10^4 M^{-1}). Correlation between complex stability and attached number and position of fluorine atom was also demonstrated. For instance, **L138** exhibited the association constant value of 18 M^{-1}, much smaller than that of **L141** (38 M^{-1}) and **L144** (2.8×10^2 M^{-1}).[117] Huber *et al.* reported tridentate tri-cationic XB donor receptors **L145** and **L146** (Chart 3.11) on a 5-iodo-1,2,3 triazolium scaffold and utilized them as activators for Ritter-type solvolysis (carbon–bromine bond activation).

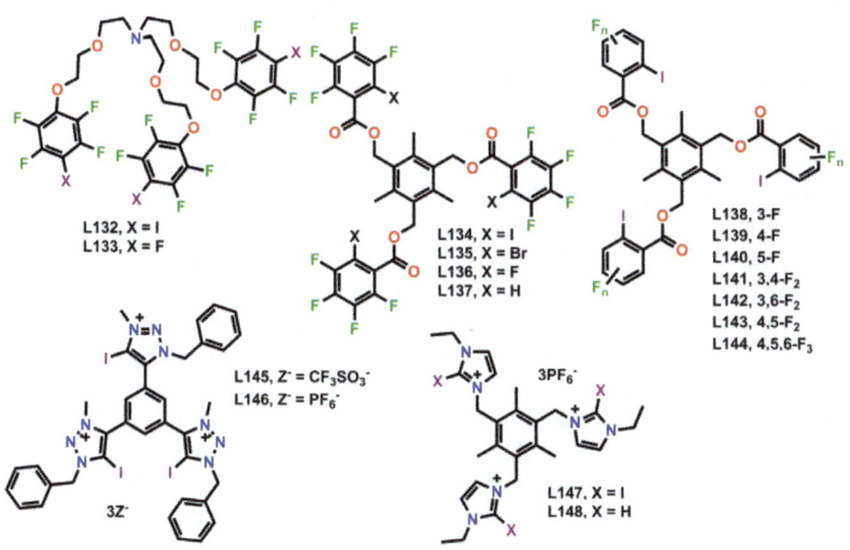

Chart 3.11 Chemical structures of receptors **L132–L148**.

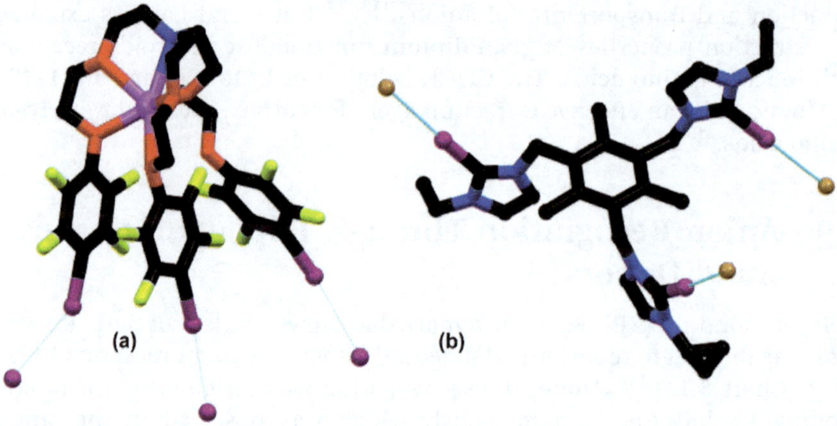

Figure 3.12 Crystal structures of (a) [I$^-$···**L132**], and (b) [Br$^-$···**L132**].

They found that **L145** acted as a more efficient activator, with a product yield of 95% within 48 hours. **L145** showed a 73% product yield after 48 hours, although it increased to 92% after 96 hours which was estimated by ^{1}H-NMR studies.[118]

In 2015, our group developed the arene-platform-based 2-iodo-imidazole substituted receptor **L147** (Chart 11). ^{13}C-NMR titration studies on **L147** with $(n\text{-Bu})_4N^+Cl^-$ in *DMSO-d*$_6$ suggested 1 : 3 (host : guest) binding and ITC studies showed highest binding $(n\text{-Bu})_4N^+Cl^-$ to **L147** (1.11×10^{13} M^{-3}). We also demonstrated selectivity role reversal of **L147** with its hydrogen bond donor analogue **L148**. Such selectivity of **L147** was utilized for selective removal of Br^- from the mixture of interfering anions. We also showed solid state evidence against XB interaction (Figure 3.12).[119]

3.10 Concluding Remarks

Early developments on the recognition of anionic guests were mainly targeted towards $H_2PO_4^-$. Further, capsular dimeric assemblies were mainly utilized for SO_4^{2-} encapsulation. Later on, capsular dimeric assemblies were found to be applicable for inorganic phosphates, CO_3^{2-} and halide–water clusters. An important achievement of capsular dimeric assembly was the trapping of aerial CO_2 in the form of CO_3^{2-} capsule, which was further utilized for extraction of SO_4^{2-}, CrO_4^{2-}, F^- *etc.* The removal of toxic anions like SeO_4^{2-} and $HAsO_4^{2-}$ in the form of capsular assembly was also an achievement worthy of mention. Zwitter ionic receptors were also capable of showing recognition of halides in aqueous medium. Recently, the recognition of anions through halogen bonding has been used in molecular recognition, crystal engineering, non-covalent organo-catalysis, anion transport and interlocked molecule formation. In short, mainly urea-based receptors were utilized for such capsule formation. In this chapter, we have also discussed how structural modifications on an arene platform could generate bis-tripodal amide receptors for anion binding in the tripodal clefts *via* conformational diversities. Finally, we have highlighted the importance of these tripodal receptors in the area of anion recognition and liquid–liquid extraction processes. We hope this comprehensive review on the aspect of anion binding to tripodal receptors will definitely provide a consolidated overview on this specified area of interest.

References

1. C. H. Park and H. E. Simmons, *J. Am. Chem. Soc.*, 1968, **90**, 2431–2432.
2. H. E. Simmons and C. H. Park, *J. Am. Chem. Soc.*, 1968, **90**, 2428–2429.
3. C. H. Park and H. E. Simmons, *J. Am. Chem. Soc.*, 1968, **90**, 2429–2431.
4. D. Heyer and J. M. Lehn, *Tetrahedron Lett.*, 1986, **27**, 5869–5872.
5. B. Dietrich, J. M. Lehn, J. Guilhem and C. Pascard, *Tetrahedron Lett.*, 1989, **30**, 4125–4128.
6. M. W. Hosseini, A. J. Blacker and J. M. Lehn, *J. Am. Chem. Soc.*, 1990, **112**, 3896–3904.
7. J. M. Lehn, R. Meric, J. P. Vigneron, I. Bkouche-Waksman and C. Pascard, *J. Chem. Soc., Chem. Commun.*, 1991, 62–64.
8. F. P. Schmidtchen, *Angew. Chem.*, 1977, **89**, 751–752.
9. F. P. Schmidtchen, *J. Am. Chem. Soc.*, 1986, **108**, 8249–8255.
10. F. P. Schmidtchen, *J. Org. Chem.*, 1986, **51**, 5161–5168.
11. F. P. Schmidtchen, *J. Chem. Soc., Perkin Trans. 2*, 1986, 135–141.
12. P. D. Beer, D. Hesek, J. Hodacova and S. E. Stokes, *J. Chem. Soc., Chem. Commun.*, 1992, 270–272.
13. P. D. Beer, Z. Chen, A. J. Goulden, A. Graydon, S. E. Stokes and T. Wear, *J. Chem. Soc., Chem. Commun.*, 1993, 1834–1836.
14. P. D. Beer, C. Hazlewood, D. Hesek, J. Hodacova and S. E. Stokes, *J. Chem. Soc., Dalton Trans.*, 1993, 1327–1332.
15. J. F. J. E. Suresh Valiyaveettil and D. N. R. Willem Verboom, *Angew. Chem., Int. Ed. Engl.*, 1993, **32**, 900–901.
16. C. Raposo, M. Almaraz, M. Martin, V. Weinrich, M. L. Mussons, V. Alcazar, M. C. Caballero and J. R. Moran, *Chem. Lett.*, 1995, 759–760.
17. B. Kuswandi, Nuriman, W. Verboom and D. N. Reinhoudt, *Sensors*, 2006, **6**, 978–1017.
18. R. Custelcean, *Chem. Commun.*, 2013, **49**, 2173–2182.
19. R. A. Pascal, J. Spergel and D. Van Engen, *Tetrahedron Lett.*, 1986, **27**, 4099–4102.
20. L. S. Flatt, V. Lynch and E. V. Anslyn, *Tetrahedron Lett.*, 1992, **33**, 2785–2788.
21. C. Raposo, N. Perez, M. Almaraz, M. L. Mussons, M. C. Caballero and J. R. Moran, *Tetrahedron Lett.*, 1995, **36**, 3255–3258.
22. I. Stibor, D. S. M. Hafeed, P. Lhotak, J. Hodacova, J. Koca and M. Cajan, *Gazz. Chim. Ital.*, 1998, **127**, 673–685.
23. I. Ravikumar, S. Saha and P. Ghosh, *Chem. Commun.*, 2011, **47**, 4721–4723.

24. P. D. Beer, P. K. Hopkins and J. D. McKinney, *Chem. Commun.*, 1999, 1253–1254.
25. A. Danby, L. Seib, K. Bowman-James and N. W. Alcock, *Chem. Commun.*, 2000, 973–974.
26. K. Kavallieratos, A. Danby, G. J. Van Berkel, M. A. Kelly, R. A. Sachleben, B. A. Moyer and K. Bowman-James, *Anal. Chem.*, 2000, **72**, 5258–5264.
27. F. Hettche and R. W. Hoffmann, *New J. Chem.*, 2003, **27**, 172–177.
28. K. Kavallieratos, A. J. Sabucedo, A. T. Pau and J. M. Rodriguez, *J. Am. Soc. Mass Spectrom.*, 2005, **16**, 1377–1383.
29. V. S. Bryantsev and B. P. Hay, *Org. Lett.*, 2005, **7**, 5031–5034.
30. S. Tapper, J. A. Littlechild, Y. Molard, I. Prokes and J. H. R. Tucker, *Supramol. Chem.*, 2006, **18**, 55–58.
31. S. Zhang and L. Echegoyen, *J. Am. Chem. Soc.*, 2005, **127**, 2006–2011.
32. C. Schmuck and M. Schwegmann, *J. Am. Chem. Soc.*, 2005, **127**, 3373–3379.
33. C. Schmuck and M. Schwegmann, *Org. Biomol. Chem.*, 2006, **4**, 836–838.
34. C. Schmuck and M. Schwegmann, *Org. Lett.*, 2005, **7**, 3517–3520.
35. A. Frontera, J. Morey, A. Oliver, M. N. Pina, D. Quinonero, A. Costa, P. Ballester, P. M. Deya and E. V. Anslyn, *J. Org. Chem.*, 2006, **71**, 7185–7195.
36. P. S. Lakshminarayanan, E. Suresh and P. Ghosh, *Inorg. Chem.*, 2006, **45**, 4372–4380.
37. G. V. Zyryanov, M. A. Palacios and P. Anzenbacher, *Angew. Chem., Int. Ed.*, 2007, **46**, 7849–7852.
38. M.-M. Titirici, A. Thomas and M. Antonietti, *New J. Chem.*, 2007, **31**, 787–789.
39. K. J. Winstanley, S. J. Allen and D. K. Smith, *Chem. Commun.*, 2009, 4299–4301.
40. M. Arunachalam and P. Ghosh, *Chem. Commun.*, 2009, 5389–5391.
41. I. Ravikumar, P. S. Lakshminarayanan and P. Ghosh, *Inorg. Chim. Acta*, 2010, **363**, 2886–2895.
42. S. Saha, B. Akhuli, I. Ravikumar, P. S. Lakshminarayanan and P. Ghosh, *CrystEngComm*, 2014, **16**, 4796–4804.
43. M. Arunachalam and P. Ghosh, *Inorg. Chem.*, 2010, **49**, 943–951.
44. M. Arunachalam and P. Ghosh, *Indian J. Chem., Sect. A: Inorg., Bio-inorg., Phys., Theor. Anal. Chem.*, 2011, **50A**, 1343–1349.
45. G. Ozturk, M. Colak and M. Togrul, *J. Inclusion Phenom. Macrocyclic Chem.*, 2010, **68**, 49–54.
46. S. Kumar Dey and G. Das, *Chem. Commun.*, 2011, **47**, 4983–4985.
47. S. K. Dey and G. Das, *Cryst. Growth Des.*, 2011, **11**, 4463–4473.
48. S. K. Dey, B. K. Datta and G. Das, *CrystEngComm*, 2012, **14**, 5305–5314.
49. A. S. Singh and S.-S. Sun, *Chem. Commun.*, 2011, **47**, 8563–8565.
50. S. Bao, W.-T. Gong, W.-D. Chen, J.-W. Ye, Y. Lin and G.-L. Ning, *J. Inclusion Phenom. Macrocyclic Chem.*, 2011, **70**, 115–119.
51. J. Hao, K. Hiratani, N. Kameta and T. Oba, *Supramol. Chem.*, 2011, **23**, 319–328.

52. P. Bose, I. Ravikumar, B. Akhuli and P. Ghosh, *Chem. – Asian J.*, 2012, **7**, 2373–2380.
53. R. Dutta and P. Ghosh, *Eur. J. Inorg. Chem.*, 2012, 3456–3462, S3456/3451–S3456/3416.
54. A. Basu and G. Das, *Chem. Commun.*, 2013, **49**, 3997–3999.
55. M. N. Hoque, A. Gogoi and G. Das, *Dalton Trans.*, 2015, **44**, 15220–15231.
56. S. Chakraborty, R. Dutta, M. Arunachalam and P. Ghosh, *Dalton Trans.*, 2014, **43**, 2061–2068.
57. I. Marti, M. I. Burguete, P. A. Gale and S. V. Luis, *Eur. J. Org. Chem.*, 2015, 5150–5158.
58. M. Arunachalam and P. Ghosh, *Org. Lett.*, 2010, **12**, 328–331.
59. S. Chakraborty, R. Dutta, B. M. Wong and P. Ghosh, *RSC Adv.*, 2014, **4**, 62689–62693.
60. M. Arunachalam and P. Ghosh, *Chem. Commun.*, 2011, **47**, 6269–6271.
61. S. Chakraborty, M. Arunachalam, R. Dutta and P. Ghosh, *RSC Adv.*, 2015, **5**, 48060–48070.
62. H. Xie, S. Yi, X. Yang and S. Wu, *New J. Chem.*, 1999, **23**, 1105–1110.
63. B. P. Hay, T. K. Firman and B. A. Moyer, *J. Am. Chem. Soc.*, 2005, **127**, 1810–1819.
64. R. Custelcean, B. A. Moyer and B. P. Hay, *Chem. Commun.*, 2005, 5971–5973.
65. P. Bose, R. Dutta and P. Ghosh, *Org. Biomol. Chem.*, 2013, **11**, 4581–4584.
66. R. Dutta, S. Chakraborty, P. Bose and P. Ghosh, *Eur. J. Inorg. Chem.*, 2014, **2014**, 4134–4143.
67. J. P. Clare, A. J. Ayling, J.-B. Joos, A. L. Sisson, G. Magro, M. N. Perez-Payan, T. N. Lambert, R. Shukla, B. D. Smith and A. P. Davis, *J. Am. Chem. Soc.*, 2005, **127**, 10739–10746.
68. K. M. Bhattarai, V. del Amo, G. Magro, A. L. Sisson, J.-B. Joos, J. P. H. Charmant, A. Kantacha and A. P. Davis, *Chem. Commun.*, 2006, 2335–2337.
69. C. E. Stanley, N. Clarke, K. M. Anderson, J. A. Elder, J. T. Lenthall and J. W. Steed, *Chem. Commun.*, 2006, 3199–3201.
70. M. H. Filby, T. D. Humphries, D. R. Turner, R. Kataky, J. Kruusma and J. W. Steed, *Chem. Commun.*, 2006, 156–158.
71. D. A. Jose, D. K. Kumar, B. Ganguly and A. Das, *Inorg. Chem.*, 2007, **46**, 5817–5819.
72. P. S. Lakshminarayanan, I. Ravikumar, E. Suresh and P. Ghosh, *Chem. Commun.*, 2007, 5214–5216.
73. I. Ravikumar, P. S. Lakshminarayanan, M. Arunachalam, E. Suresh and P. Ghosh, *Dalton Trans.*, 2009, 4160–4168.
74. I. Ravikumar and P. Ghosh, *Chem. Commun.*, 2010, **46**, 1082–1084.
75. B. Akhuli, I. Ravikumar and P. Ghosh, *Chem. Sci.*, 2012, **3**, 1522–1530.
76. R. Dutta, P. Bose and P. Ghosh, *Dalton Trans.*, 2013, **42**, 11371–11374.

77. R. Dutta, B. Chowdhury, P. Bose and P. Ghosh, *J. Chem. Sci.*, 2014, **126**, 1303–1309.
78. T. Clifford, A. Danby, J. M. Llinares, S. Mason, N. W. Alcock, D. Powell, J. A. Aguilar, E. Garcia-Espana and K. Bowman-James, *Inorg. Chem.*, 2001, **40**, 4710–4720.
79. B. Wu, J. Liang, J. Yang, C. Jia, X.-J. Yang, H. Zhang, N. Tang and C. Janiak, *Chem. Commun.*, 2008, 1762–1764.
80. F. Zhuge, B. Wu, J. Liang, J. Yang, Y. Liu, C. Jia, C. Janiak, N. Tang and X.-J. Yang, *Inorg. Chem.*, 2009, **48**, 10249–10256.
81. R. Custelcean, P. Remy, P. V. Bonnesen, D.-E. Jiang and B. A. Moyer, *Angew. Chem., Int. Ed.*, 2008, **47**, 1866–1870.
82. R. Custelcean and P. Remy, *Cryst. Growth Des.*, 2009, **9**, 1985–1989.
83. R. Custelcean, A. Bock and B. A. Moyer, *J. Am. Chem. Soc.*, 2010, **132**, 7177–7185.
84. A. Rajbanshi, B. A. Moyer and R. Custelcean, *Cryst. Growth Des.*, 2011, **11**, 2702–2706.
85. A. J. Lowe and F. M. Pfeffer, *Org. Biomol. Chem.*, 2009, **7**, 4233–4240.
86. Y. Li, K. M. Mullen, T. D. W. Claridge, P. J. Costa, V. Felix and P. D. Beer, *Chem. Commun.*, 2009, 7134–7136.
87. N. Busschaert, P. A. Gale, C. J. E. Haynes, M. E. Light, S. J. Moore, C. C. Tong, J. T. Davis and W. A. Harrell Jr., *Chem. Commun.*, 2010, **46**, 6252–6254.
88. N. Busschaert, M. Wenzel, M. E. Light, P. Iglesias-Hernandez, R. Perez-Tomas and P. A. Gale, *J. Am. Chem. Soc.*, 2011, **133**, 14136–14148.
89. S. K. Dey and G. Das, *Dalton Trans.*, 2011, **40**, 12048–12051.
90. C. Jia, B. Wu, S. Li, X. Huang, Q. Zhao, Q.-S. Li and X.-J. Yang, *Angew. Chem., Int. Ed.*, 2011, **50**, 486–490.
91. P. Bose, R. Dutta, S. Santra, B. Chowdhury and P. Ghosh, *Eur. J. Inorg. Chem.*, 2012, 5791–5801.
92. A. Basu and G. Das, *Dalton Trans.*, 2012, **41**, 10792–10802.
93. I. Ravikumar and P. Ghosh, *Chem. Commun.*, 2010, **46**, 6741–6743.
94. R. Dutta and P. Ghosh, *Eur. J. Inorg. Chem.*, 2013, 2673–2681.
95. B. Wu, F. Cui, Y. Lei, S. Li, N. D. S. Amadeu, C. Janiak, Y.-J. Lin, L.-H. Weng, Y.-Y. Wang and X.-J. Yang, *Angew. Chem., Int. Ed.*, 2013, **52**, 5096–5100.
96. K. Pandurangan, J. A. Kitchen, S. Blasco, E. M. Boyle, B. Fitzpatrick, M. Feeney, P. E. Kruger and T. Gunnlaugsson, *Angew. Chem., Int. Ed.*, 2015, **54**, 4566–4570.
97. M. A. Hossain, J. A. Liljegren, D. Powell and K. Bowman-James, *Inorg. Chem.*, 2004, **43**, 3751–3755.
98. P. S. Lakshminarayanan, I. Ravikumar, E. Suresh and P. Ghosh, *Inorg. Chem.*, 2007, **46**, 4769–4771.
99. C. Bazzicalupi, A. Bencini, A. Bianchi, L. Borsari, C. Giorgi, B. Valtancoli, C. Anda and A. Llobet, *J. Org. Chem.*, 2005, **70**, 4257–4266.
100. P. Bose, I. Ravikumar and P. Ghosh, *Inorg. Chem.*, 2011, **50**, 10693–10702.

101. A. Metzger, V. M. Lynch and E. V. Anslyn, *Angew. Chem., Int. Ed. Engl.*, 1997, **36**, 862–865.
102. A. Metzger and E. V. Anslyn, *Angew. Chem., Int. Ed.*, 1998, **37**, 649–652.
103. K. Sato, S. Arai and T. Yamagishi, *Tetrahedron Lett.*, 1999, **40**, 5219–5222.
104. L. A. Cabell, M.-K. Monahan and E. V. Anslyn, *Tetrahedron Lett.*, 1999, **40**, 7753–7756.
105. H. Ihm, S. Yun, H. G. Kim, J. K. Kim and K. S. Kim, *Org. Lett.*, 2002, **4**, 2897–2900.
106. M. Arunachalam and P. Ghosh, *Chem. Commun.*, 2009, 3184–3186.
107. M. Arunachalam, S. Chakraborty, S. Marivel and P. Ghosh, *Cryst. Growth Des.*, 2012, **12**, 2097–2108.
108. L. O. Abouderbala, W. J. Belcher, M. G. Boutelle, P. J. Cragg, J. Dhaliwal, M. Fabre, J. W. Steed, D. R. Turner and K. J. Wallace, *Chem. Commun.*, 2002, 358–359.
109. K. J. Wallace, W. J. Belcher, D. R. Turner, K. F. Syed and J. W. Steed, *J. Am. Chem. Soc.*, 2003, **125**, 9699–9715.
110. A. P. Davis, J. J. Perry and R. P. Williams, *J. Am. Chem. Soc.*, 1997, **119**, 1793–1794.
111. A. L. Sisson, J. P. Clare, L. H. Taylor, J. P. H. Charmant and A. P. Davis, *Chem. Commun.*, 2003, 2246–2247.
112. R. A. W. Dryfe, S. S. Hill, A. P. Davis, J.-B. Joos and E. P. L. Roberts, *Org. Biomol. Chem.*, 2004, **2**, 2716–2718.
113. A. L. Sisson, J. P. Clare and A. P. Davis, *Chem. Commun.*, 2005, 5263–5265.
114. B. A. McNally, A. V. Koulov, T. N. Lambert, B. D. Smith, J.-B. Joos, A. L. Sisson, J. P. Clare, V. Sgarlata, L. W. Judd, G. Magro and A. P. Davis, *Chem. – Eur. J.*, 2008, **14**, 9599–9606.
115. A. P. Davis and L. J. Lawless, *Chem. Commun.*, 1999, 9–10.
116. A. Mele, P. Metrangolo, H. Neukirch, T. Pilati and G. Resnati, *J. Am. Chem. Soc.*, 2005, **127**, 14972–14973.
117. E. Dimitrijevic, O. Kvak and M. S. Taylor, *Chem. Commun.*, 2010, **46**, 9025–9027.
118. F. Kniep, L. Rout, S. M. Walter, H. K. V. Bensch, S. H. Jungbauer, E. Herdtweck and S. M. Huber, *Chem. Commun.*, 2012, **48**, 9299–9301.
119. P. Ghosh, S. Chakraborty and R. Dutta, *Chem. Commun.*, 2015, 14793–14796.

CHAPTER 4

Functional Molecular and Supramolecular Materials for Electron Transfer Reactions and their Applications

SUDHIR KUMAR KESHRI, SRIKANTA DANA, M. R. AJAYAKUMAR, DEEPAK ASTHANA AND PRITAM MUKHOPADHYAY*

Supramolecular and Material Chemistry Lab, School of Physical Science, Jawaharlal Nehru University, New Delhi 110067, India
*Email: m_pritam@mail.jnu.ac.in

4.1 Introduction

Electron transfer (ET) reactions have vital importance in life's fundamental processes such as photosynthesis and respiration.[1,2] In recent times, the application of ET reactions in synthesis and in materials based on organic, inorganic, organometallic or hybrid scaffolds has increased phenomenally.[3–5] The advent of innovative molecular designs, efficient synthetic methodologies coupled with the understanding of weak supramolecular interactions[6,7] and crystal engineering[8–14] principles have led to molecular and supramolecular materials with multifaceted properties, which have ET processes as the fundamental driving force.[15–22]

For a general understanding, a typical ET reaction involves the exchange of an electron between two chemical entities, an electron donor (D) and an

Monographs in Supramolecular Chemistry No. 22
Functional Supramolecular Materials: From Surfaces to MOFs
Edited by Rahul Banerjee

Published by the Royal Society of Chemistry, www.rsc.org

electron acceptor (A). Such reactions may or may not involve chemical bond breaking or bond formations. Generally, as a consequence of the ET reaction, these entities undergo changes in their oxidation and reduction states. To explain the rate of ET reactions, Rudolph A. Marcus developed a theory based on outer-sphere ET reactions.[23–27] Marcus theory gives a valuable platform for understanding ET reactions and for his immense contribution in this area of research he received Nobel Prize in 1992. As a consequence of the ET reaction an electron donating organic scaffold can lose its electrons to form a neutral radical or a radical cation. In contrast, an electron withdrawing scaffold can gain electrons to form a radical anion.

The unpaired electron with its spin and charge in an organic π-conjugated framework provides multidimensional opportunities towards realization of conductive systems, electrochromic materials, organic magnets, reversible switches, and so on. While there is a plethora of π-conjugated electron-rich and electron-deficient scaffolds studied for ET reactions and paramagnetic materials, in this book chapter, we have limited our focus to a handful of organic scaffolds which accept one or more electrons. Our choice of the π-conjugated scaffolds is based on the enormity of interest they have generated in today's contemporary research of ET-based reactions, design and stabilization of organic radical systems and electron transport based materials. Also, our choice is based on identifying scaffolds, which have detailed structural information in their different redox states. In addition, our identified π-conjugated scaffolds of tetracyanoquinodimethanes, naphthalenediimides (NDIs), perylenediimides (PDIs), corannulenes, and bipyridinium ions provide easy synthetic tailorability and several attractive opto-electronic attributes.

This chapter describes the diverse aspects of ET of the five identified π-conjugated scaffolds (Sections 4.2–4.6). In the first section on tetracyanoquinodimethanes, we will see how bioorganic semiconductor materials, porous electroactive materials and magnetic materials can be designed from the radical anion state of TCNQ. We highlight an interesting report, which underlines synthetic routes to overcome the difficulty in handling of the TCNQ dianion under ambient conditions. In the next section on NDIs, we will see how new design strategies of using noncovalent intramolecular interaction and through-space orbital interaction is shaping up towards the realization of stable organic open-shell as well as highly electron deficient systems. Here, we also consider the studies of the photophysical properties of the excited doublet states of the naphthalenediimide radical ions. Light-driven intramolecular ET reactions from the electron-rich pendants to the electron deficient NDI is also discussed, which can be an important strategy to have photo-controlled colour switchable materials. In the third section on PDIs, we see how molecular and crystal engineering principles can work in tandem towards the synthesis of n-type semiconductors with high charge mobility. Non-planarity as an intriguing design element is deliberated towards the realization of non-fullerene-based electron acceptors in organic solar cells. Also, we discuss the recent advances towards the stabilization of PDI-based zwitterionic radicals and dianions formed from one- and two-electron reduction, respectively. We see how ET processes in combination with other stimuli can reversibly switch physical states of electroactive PDI-based soft materials. In this section, we also discuss an exemplary example of a photocatalytic redox process mediated by the PDI radical anion, which allows C–C bond formation and reduction of aryl halides. In the fourth section on corannulenes, we see the recent advances of multi-electron transfer reactions of these π-scaffolds and the supramolecular structures that emanate from these ET reactions. The systematic control gained over the intercalated alkali metal ions and their stoichiometry between the corannulene tetraanions is quite remarkable. In the section on bipyridinium ions, we look at the multi-ET reactions, radical–radical interactions and their supramolecular stabilization. In the final section, we discuss the application of the radical ion systems as probes for toxic ions and detection of volatile electron-rich organic systems. It is now evident that the unique optical properties of π-conjugated open-shell systems can be utilized as new platform for sensing diverse analytes. Furthermore, we see how electroactive materials for organic Li-ion batteries can be designed to have attractive specific power and energy.

4.2 Tetracyanoquinodimethanes

4.2.1 Introduction to Tetracyanoquinodimethanes

Tetracyanoquinodimethanes are well-known as a class of strong electron acceptors.[28,29] A diverse range of these π-acceptors, for example, have been synthesized by varying the π-ring size (Figure 4.1a).[30] Among these π-acceptors, 7,7,8,8-tetracyanoquinodimethane (TCNQ) has attracted a tremendous amount of attention because of its ability to sequentially accept two electrons and form a variety of D-A charge transfer (CT) complexes. TCNQ undergoes a one-electron reduction to form a radical monoanion $TCNQ^{\bullet-}$ which exhibits good stability. Further, $TCNQ^{\bullet-}$ can accept another electron to form highly air- and moisture- sensitive dianion, $TCNQ^{2-}$ (Figure 4.1b).

$TCNQ^{-}$ can be readily synthesized by chemical reduction, photoreduction, or electrochemical methods. Importantly, investigations of CT complexes of TCNQ indicate that TCNQ is able to carry a partial negative charge in addition to the formal 0 and −1 oxidation states.[31–33] This significant feature has driven interesting physical properties such as electrical conductivity. One of the seminal CT complexes incorporating TCNQ is that of TTF-TCNQ (TTF = tetrathiafulvalene), which is hailed as the first example of a purely organic electrical conductor.[34] In this D-A crystalline complex, the TTF molecules are arranged in infinite parallel stacks while the acceptor TCNQ molecules are arranged in neighbouring infinite parallel stacks. The TCNQ molecules are partially reduced with $\rho = -0.59$ (ρ = calculated charge).

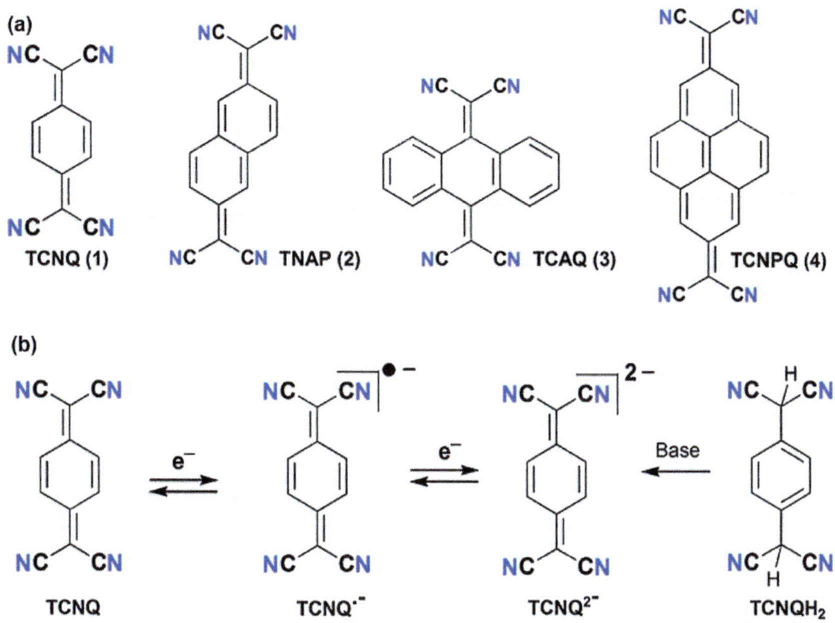

Figure 4.1 (a) Molecular structures of TCNQ, TNAP, TCAQ and TCNPQ, and (b) One- and two-electron reduction of TCNQ to form $TCNQ^{\bullet -}$ form and $TCNQ^{2-}$. An alternative route to $TCNQ^{2-}$ from $TCNQH_2$.

Crystals of this CT complex exhibited remarkably high metallic conductivity with σ values of 10^4 S cm^{-1}.

4.2.2 TCNQ-based Radical Anion and Dianion

In the quest to synthesize hybrid organic–inorganic architectures, Dunbar *et al.* reported highly conducting coordination polymers applying the TCNQ radicals, $M(NO_3)_2 \cdot 6H_2O$ (M = Mn or Zn) and 4,4′-bipyridine units (Figure 4.2a).[35] The TCNQ-Mn complex exhibited high conductivity with a σ of 3.8 S cm^{-1}, due to the high regularity of the conducting stacks formed by the direct bonding of the TCNQ radicals to the metal ions. However, an analysis of the magnetic properties of the complexes revealed that the spins on the TCNQ stacked radicals are strongly correlated and no appreciable interaction between the localized Mn(II) spins were found as initially predicted. It has now been realized that $TCNQ^{\bullet -}$ based CT complexes can even have superconductivity under certain conditions. To date, $TCNQ^{\bullet -}$ materials have been formed in combination with many cations, including metal ions (*e.g.*, Na^+, Mg^{2+}, Cu^{2+}, *etc.*), organometallic complexes (*e.g.* ferrocene, $[Ru(bipy)_3]^{2+}$) (bipy = bipyridine), as well as with organic cations (*e.g.* Me_4N^+, TTF^+, *etc.*).

Recently Martin *et al.* reported a novel bioorganic semiconductor material formed as a CT complex between L-proline (Pro) and TCNQ.[36] X-ray crystallographic analysis confirmed the formation of $(Pro_2H^+)_2(TCNQ^{\bullet -})_2 \cdot TCNQ$,

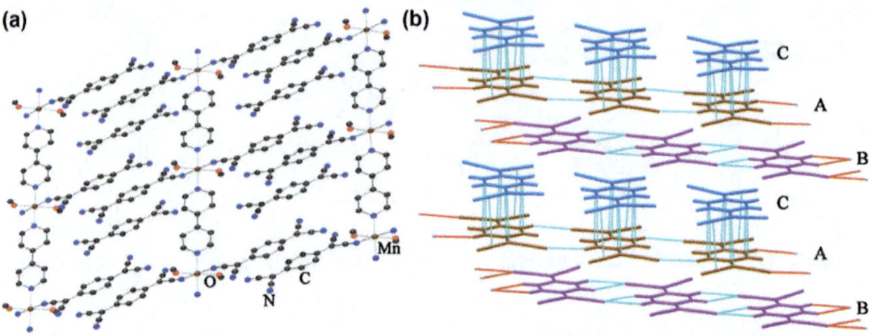

Figure 4.2 (a) Highly conducting coordination polymers applying TCNQ radicals and $Mn(NO_3)_2 \cdot 6H_2O$ and 4,4′-bipyridine units. Reproduced with permission from M. B.-Rivas, A. Ota, E. Reinheimer, A. Prosvirin, J. V. Martinez and K. R. Dunbar, Highly Conducting Coordination Polymers Based on Infinite M(4,4′-bpy) Chains Flanked by Regular Stacks of Non-Integer TCNQ Radicals, *Angew. Chem., Int. Ed.*, 2011, **50**, 9703–9707. Copyright 2011 Wiley-VCH Verlag GmbH & Co. KGaA, and (b) A bioorganic semiconductor material $(Pro_2H^+)_2(TCNQ^{\bullet -})_2 \cdot TCNQ$ having two forms of the TCNQ radical anion *i.e.*, TCNQ-A and TCNQ-C, and a neutral TCNQ as TCNQ-B. Proline (Pro) moiety and the H atoms have been removed for clarity. Colour code for Figure 4.2a: C, dark grey; N, blue; O, red; Mn, brown. Reproduced with permission from X. Qu, J. Lu, C. Zhao, J. F. Boas, B. Moubaraki, K. S. Murray, A. Siriwardana, A. M. Bond and L. L. Martin, $(Pro_2H^+)_2(TCNQ^{\bullet -})_2 \cdot TCNQ$: An Amino Acid Derived Semiconductor, *Angew. Chem., Int. Ed.*, 2011, **50**, 1589–1592. Copyright 2011 Wiley-VCH Verlag GmbH & Co. KGaA.

constituting two forms of the TCNQ radical anion *i.e.* TCNQ-A and TCNQ-C, while a neutral TCNQ molecule was also realized and named as TCNQ-B (Figure 4.2b).

The structure consists of alternating layers of proline cations and TCNQ moieties. Strong π–π interactions between the chains containing the TCNQ-A and TCNQ-C molecules with an interplanar distance of 3.237 Å was revealed, with TCNQ-B chains located between TCNQ-A and TCNQ-C chains; however no π–π interactions are formed. The solid-state conductivity of the complex was measured as 2.5 mS cm^{-1} at 295 K and the material behaved as an intrinsic semiconductor.

In an innovative approach, Miyasaka *et al.* demonstrated that the control of inter-lattice magnetic interactions becomes a critical issue when long-range ordered magnets based on low-dimensional magnetic frameworks are designed. The authors showed that a pillared layer framework (PLF) model could be an efficient system to realize this goal. For the magnetic layers, an array of CT D_2A compounds composed of a paddle-wheel-type diruthenium(II, II) complex ($Ru_2^{II,II}$) as the D and TCNQ as the A was chosen.[37] The CT system $[FeCp^*_2]TCNQ$ ($[FeCp^*_2]^+$ = decamethylferrocenium) was chosen in order for the pillars to be able to stabilize the interlayer interactions. Upon integration with the layered D_2A, the $[FeCp^*_2]^+$ unit acts as a

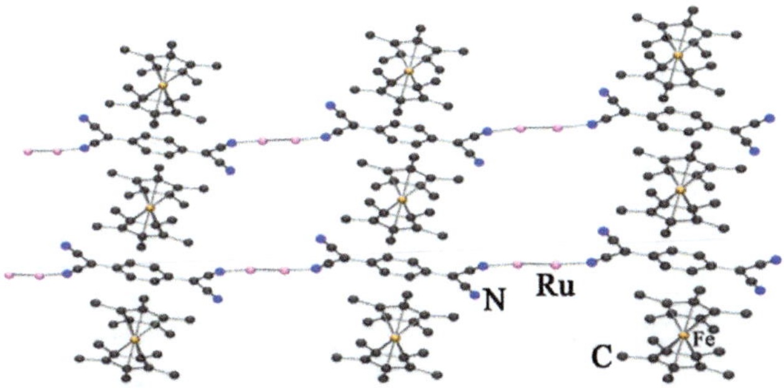

Figure 4.3 A ferrimagnet from a π-stacked pillared layer framework of $[FeCp^{*}_2][\{Ru_2\}_2TCNQ]$. Hydrogen atoms, crystallization solvents, and the 2,3,5,6-$F_4C_6HCO_2^-$ ligands around the Ru centres are omitted for clarity. Colour code: C, dark grey; N, blue; Fe, orange; Ru, pink. Reproduced with permission from H. Fukunaga and H. Miyasaka, Magnet Design by Integration of Layer and Chain Magnetic Systems in a π-Stacked Pillared Layer Framework *Angew. Chem., Int. Ed.*, 2015, **54**, 569–573. Copyright 2015 Wiley-VCH Verlag GmbH & Co. KGaA.

π-stacked pillar for the D_2A layers sandwiched between the $TCNQC^{\bullet -}$ moieties, which consequently results in the π-stacked PLF $[FeCp^{*}_2][\{Ru_2\}_2TCNQ]$ (Figure 4.3).

Thus a ferrimagnet based on a π-stacked PLF with a phase transition temperature of 82 K was rationally constructed. Furthermore, the transition temperature could be increased to 107 K by applying a pressure of 12.5 k bar. This new strategy which assimilates a combination of layers and chains holds immense promise for new magnet design, and also towards the realization of multifunctional porous materials.

The dianionic form ($TCNQ^{2-}$) is known to be highly sensitive to aerial oxidation and therefore has received significantly less attention compared to the neutral and radical forms. In 2006, Kitagawa *et al.* reported a coordination polymer $\{[Zn(TCNQ)bipy]\cdot 6MeOH\}_n$ with an open framework having CT-based guest inclusion.[38] The charge number of the TCNQ was estimated to be −2 (Zn: +2, TCNQ: −2, bipy: 0), since no counteranion was observed in the X-ray structure. The Zn ions were found to be linked by TCNQ molecules to generate a two-dimensional (2D) corrugated layer, while the bipy ligands act as pillars, with the Zn ions in the adjacent layers linked to form a 3D pillared layer structure. However, the cause of production of the resulting $TCNQ^{2-}$ was predicted to be due to a disproportionation reaction. Therefore, a controlled production of $TCNQ^{2-}$ and its stabilization under ambient conditions remained elusive.

In a pioneering discovery, Robson *et al.* recently reported that the acid form of this dianion, $(CN)_2CH\cdot C_6H_4\cdot CH(CN)_2$ referred to as $TCNQH_2$ (Figure 4.1b), is quite air-stable and provides a convenient entry for making coordination polymers avoiding the elusive $TCNQ^{2-}$ as a starting material.[39] A diverse range of $TCNQ^{2-}$ based ionic complexes were obtained. The synthetic methodology was found to be simple and involved the reaction

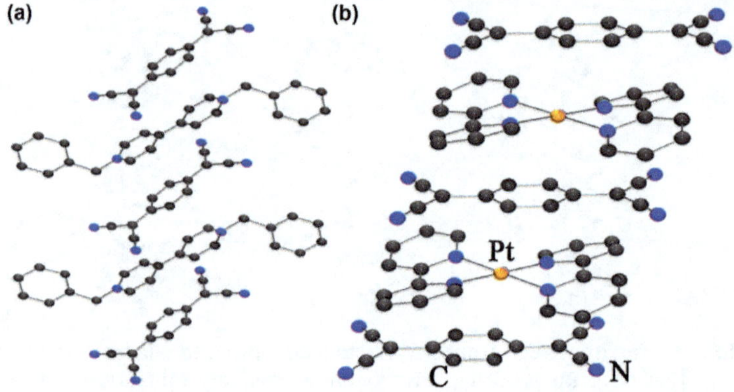

Figure 4.4 (a) Stack of alternating $TCNQ^{2-}$ anions and methyl viologen (MV^{2+}) cations in ($TCNQ^{2-}$)(MV^{2+}), (b) Stack of alternating $TCNQ^{2-}$ and $Pt(bipy)_2{}^{2+}$. Colour code: C, dark grey; N, blue; Pt, orange.
Reprinted with permission from T. A. Hudson and R. Robson. A New Class of TCNQ Derivatives Easily Generated from $TCNQH_2$ Containing Discrete $TCNQ^{2-}$ Anions and Noncoordinating Cations. *Cryst. Growth Des.*, 2009, **9**, 1658–1662. Copyright 2009 American Chemical Society.

of $TCNQH_2$ with various cations at room temperature and was successful in a variety of media such as methanol, ethanol, DMF and dioxane. As representative examples, Figure 4.4a shows a stack of alternating $TCNQ^{2-}$ anions and methyl viologen (MV^{2+}) cations in ($TCNQ^{2-}$)(MV^{2+}), while Figure 4.4b depicts a stack of alternating $TCNQ^{2-}$ and $Pt(bipy)_2{}^{2+}$.

This novel chemistry of bypassing the unstable $TCNQ^{2-}$ and in turn applying the air-stable $TCNQH_2$ could be replicated for the tetra-fluorinated derivative *i.e.* F_4TCNQ^{2-}. Robson *et al.*, from structures of 14 ionic complexes, demonstrated that relative arrangements of the ions could be categorized into Type I–IV (Figure 4.5a).[40]

Type I involves face-to-face interactions between oppositely charged ions to form infinite ···DADADA··· stacks. For example, Figure 4.5b shows the type I interaction with (F_4TCNQ)(MV). In type II, segregated parallel stacks of cations and anions could be realized, while type III and IV structures are similar to type I structures except that pairs of cationic units are located between the F_4TCNQ^{2-} anions. Interestingly, the optical band gaps could be nicely tuned depending on the electronic and structural aspects of the dications.

4.3 Arylenediimides

Arylenediimides have attracted a tremendous amount of attention as electron acceptors and as scaffolds for ET reactions in studies related to models for photosynthesis, solar energy conversion, n-type semiconductors, electrochromic devices and photorefractive materials. Among these, pyrromelliticdiimides (PMDIs), naphthalenediimides (NDIs) and perylenediimides (PDIs) have drawn the maximum attention.[41,42] The larger analogues like

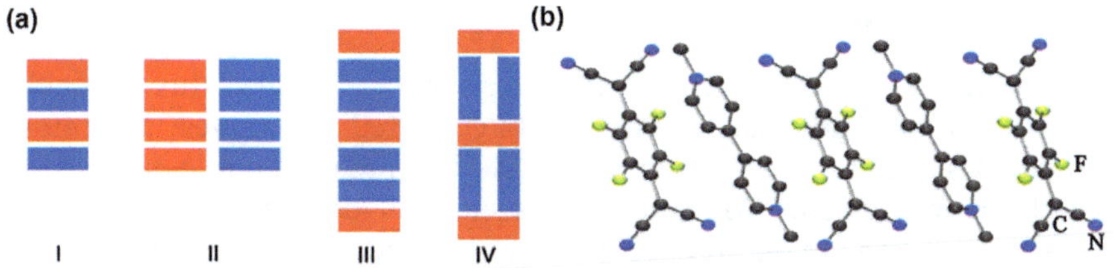

Figure 4.5 (a) Arrangements of the donor (red blocks) and acceptor (blue blocks) ions categorized into Type I–IV, and (b) An example of a type I interaction depicting alternative stacking of F_4TCNQ and MV ions. Colour code: C, dark grey; N, blue; F, yellow-green.
Reproduced with permission from ref. 40. Copyright 2014 Royal Society of Chemistry.

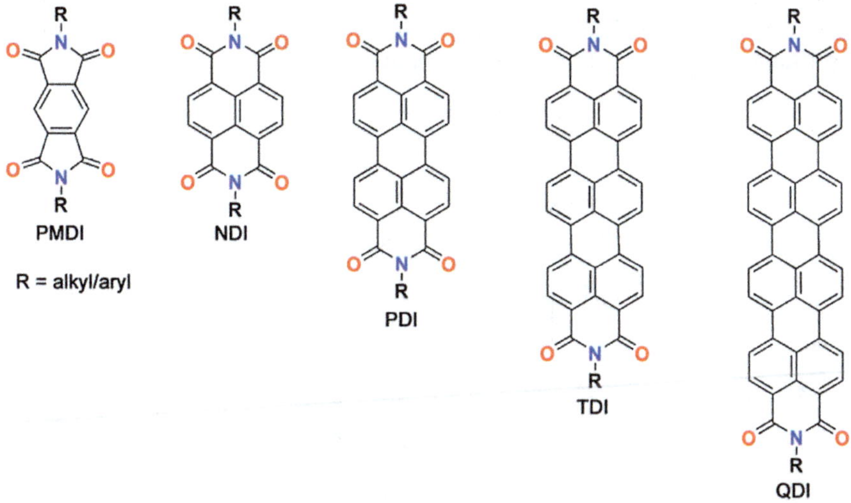

Figure 4.6 Molecular structures of various arylenediimides, PMDI, NDI, PDI, TDI and QDI.

terrylenediimides (TDIs) and quaterrylenediimides (QDIs) have also been synthesized and their opto-electronic properties delineated (Figure 4.6).

A large volume of research has been done on the CT driven self-assembly of these scaffolds, mostly NDI[43–46] and PMDI[47] with different organic donors. These scaffolds find major applications related to ET studies since they undergo sequential two-electron reversible reduction at modest potentials to form radical anions and dianions. The radical anions formed from the arylenediimides are excellent chromophores having signature absorption bands in the visible and near-infrared (NIR) region, which aid in their easy and straightforward identification. The dianions formed from the arylenediimides have significantly blue-shifted absorption characteristics compared to the corresponding radical anions. Herein, we discuss the recent studies associated to ET reactions with NDIs and PDIs and the emerging new directions and applications.

4.3.1 Introduction to NDIs

NDIs bestow favourable π-acidic characteristics and due to this particular aspect remarkable advances in the areas of CT-based assemblies,[43–47] anion–π interactions, anion transport and catalysis[48–53] and organic electronics[54–57] have been formulated. To modulate its electron deficient nature a large number of synthetic efforts have been directed, utilizing axial- and/or core-functionalization with electron deficient groups. As axial substituents, pyridine rings (NDI-1), aryl-sulfonate (NDI-2), aryl-nitro (NDI-3), aryl-hydrazines (NDI-4), *etc.* have been functionalized to generate the corresponding electron-deficient NDIs (Figure 4.7).

After the successful core-halogenation and diverse substitution reactions initiated by the group of Würthner and others,[58–63] a gateway to diverse highly electron-deficient NDIs with the cyano-(NDI-5), sulfonyl-(NDI-6), trifluoromethyl-(NDI-7) and phosphonium-(NDI-8) groups at the core of the naphthalene moiety has opened up (Figure 4.8).

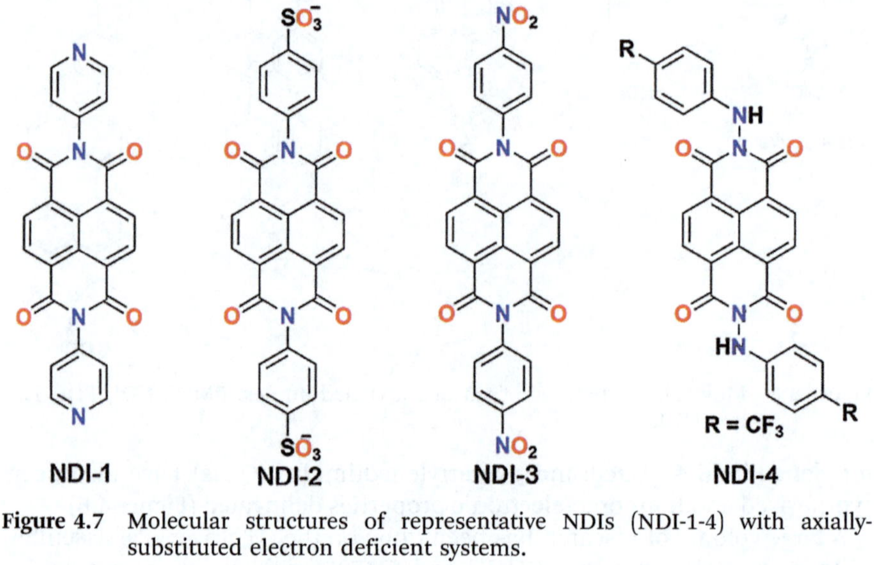

Figure 4.7 Molecular structures of representative NDIs (NDI-1-4) with axially-substituted electron deficient systems.

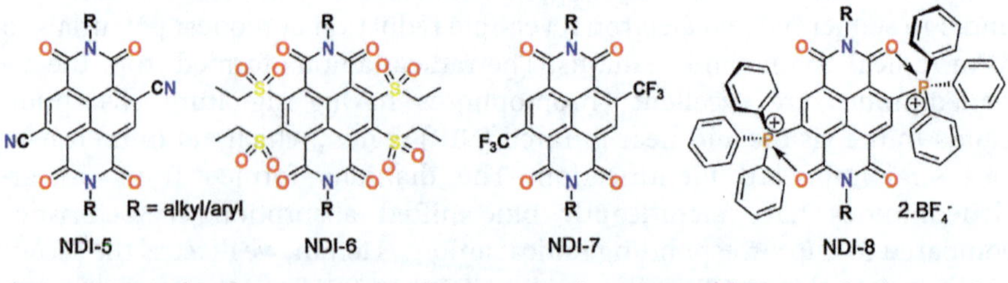

Figure 4.8 Molecular structures of representative NDIs with core-substituted electron deficient systems (NDI-5-8).

In addition to the core-substitution, core extension at the shoulder-positions[64,65] and fusion at the imide oxygen with dithiafulvalene[66,67] have been successfully carried out. The substitution of acceptor groups at the axial- and core-positions of the NDI provided a promising route to synthesize novel compounds having exceptional n-type semiconductor properties. It was found that apart from lowering the LUMO level of the NDI molecules, π–π stacking and close-packing of the NDI molecules play a pivotal role in enhancing the electron mobility in the semiconductor materials.[54–57] Thus, in an innovative demonstration of molecular and supramolecular engineering, Zhu *et al.* have been able to obtain air-stable NDIs fused with 2-(1,3-dithiol-2-ylidene)malononitrile groups (NDI-DTYM2) with specific branched alkyl chains. This resulted in an electron mobility of 3.50 $cm^2\ V^{-1}\ s^{-1}$, which is an unprecedented value for ambient stable n-channel organic thin-film transistors (OTFTs).[68]

Within the core-functionalized NDIs, the tetra-cyano substituted NDI molecule [$NDI(CN)_4$] was anticipated to be one of the strongest π-acceptors, with a theoretically calculated LUMO of −4.82 eV. However, despite its wide-ranging appeal for academia as well as industry, its ambient isolation has remained elusive to date.[69] Matile's group, in a pioneering development, reported the synthesis of the core-functionalized tetra-ethylsulfonyl-NDI [$NDI(EtSO_2)_4$] (NDI-6) molecule having a LUMO of −4.74 eV and pronounced it as a super-π-acceptor while demonstrating strong anion–π interactions.[70] We recently reported core-substitution of NDI molecules with bis-phosphonium groups [$NDI(PPh_4)_2{}^{2+}) \cdot 2BF_4{}^{-}$] (NDI-8) with a LUMO of −4.90 eV, a little lower than the seminal acceptor TCNQ molecule which had a LUMO of −4.84 eV.[71] The low LUMO level of this molecule results in spontaneous generation of highly stable radical ions.

4.3.2 ET Reactions of NDI

It is important to mention here that the first report on the NDI radical anion ($NDI^{\bullet-}$) dates back to 1967 by Stephen F. Nelsen.[72] The $NDI^{\bullet-}$ was found to be persistent in solution for up to few weeks in the absence of air. In 1991, in a significant development, the group of Larry Miller demonstrated the formation of $NDI^{\bullet-}$ π-stacks in aqueous NaCl solution with the radical anion of a bis-arylsulfonate substituted NDI molecule.[73,74] Interestingly, these π-stacks were found to have similar optical and magnetic resonance properties to the conducting solid. The radical anion was produced by the reduction of the NDI unit with sodium dithionite ($Na_2S_2O_4$) (Figure 4.9a). Later, we found that mild reducing agents like primary or secondary organic amines as well as specific anions like cyanide can generate $NDI^{\bullet-}$ under ambient conditions (Figure 4.9b).[75] Interestingly, fluoride, acetate and dihydrogenphosphate were also found to produce $NDI^{\bullet-}$, albeit to a much lesser extent.[76] In the presence of fluoride, we could realize a tetra-stable system, which included the formation of NDI^{2-} and apply the one-, two- ET, and the intramolecular CT interactions to develop

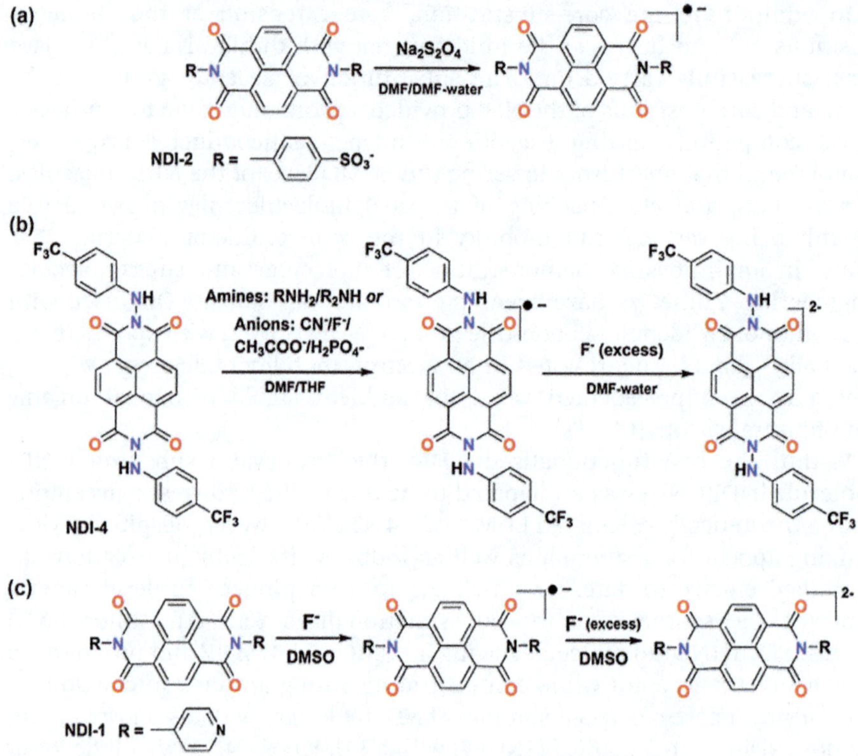

Figure 4.9 (a) One-electron reduction of NDI-2 with $Na_2S_2O_4$ to form the corresponding NDI radical anion, (b) One-electron reduction of NDI-4 with various amines (aliphatic) or cyanide/fluoride/acetate/dihydrogenphosphate anions to form the corresponding NDI radical anion and two-electron reduction of NDI-4 with excess of fluoride anion, and (c) One-electron reduction of NDI-1 with fluoride to form the corresponding NDI radical anion and dianion.

combinational logic gate systems.[77] This phenomenon was further extended by us towards the application of radical ions as a platform for sensing different analytes as described in the section of radical-based sensors of this chapter.

Subsequently, Saha *et al.* demonstrated ET from fluoride to NDI to form $NDI^{\bullet-}$ as well as the formation of the dianion NDI^{2-} with excess of fluoride in polar solvents like DMSO (Figure 4.9c).[78] In an elegant report, Saha *et al.* further showed that by positioning the HOMO of the anion above the LUMO of the NDI, the anion to NDI thermal ET pathway can be activated. While in the event of the HOMO of a weakly Lewis basic anion lying below the LUMO of an NDI but above its HOMO, the thermal ET is not possible, but light can activate an anion to ^{1*}NDI photoinduced ET pathway from the anion's HOMO to the photogenerated ^{1*}NDI's SOMO-1.[79] Therefore, electronic regulation to thermal and light-gated ET can be achievable in NDIs with diverse anions.

Although studies related to the ET reactions of NDI have gained immense attention, the ambient isolation and stabilization of the one-electron reduced $NDI^{\bullet-}$ or two-electron reduced NDI^{2-} remains a challenging task. Thus majority of the studies were confined to solution, barring few which focused on thin films. One of the earliest attempts to isolate these reduced species was by Roth *et al.* in 1989, in which the radical anion of the naphthalene-tetracarboxylicdianhydride ($NTCDA^{\bullet-}$) was isolated by electrocrystallization.[80] However, its unusual stoichiometry and insolubility precluded its detailed characterization.

4.3.3 Electron Delocalization/Stabilization by Noncovalent Means in NDI Radical Ions

Recently, we reported the synthesis and isolation of the first NDI-based radical ion $[NDI(PPh_3)_2{}^{\bullet+}(BPh_4)^-]$ as single crystals.[71] The radical ion (NDI-$8^{\bullet+}$) as well its dicationic precursor (NDI-8^{2+}) exhibited unprecedented stability to normal workup procedures, chromatography and anion metathesis in open air (Figure 4.10a). This extraordinary stability is bestowed by the noncovalent P···O interaction, hypervalency and propeller-like shape of the phosphonium groups. Natural bond orbital (NBO)[81] calculations validated the P···O contact to be a stabilizing donor–acceptor type $n_O \rightarrow \sigma_{P-C}{}^*$ orbital interaction (Figure 4.10b). Furthermore, atom in orbital (AIM)[82] analysis clearly showed a bond critical path and electron density at bond critical points between the P and O atoms, validating sharing of the unpaired electron by the phosphonium groups (Figure 4.10c). Therefore, intramolecular noncovalent interactions can provide significant opportunities as new design elements for the stabilization of reactive systems.

In a pioneering development, Stoddart *et al.* have demonstrated through-space orbital interactions between suitably oriented NDI units as an effective way for electron sharing, when NDI units are oriented in a triangular or cofacial fashion. In the case of the NDI-based triangular prism, electronic communication between the three NDI units leads to six individually accessible redox states (Figure 4.11a).[83] The electron-sharing hypothesis was corroborated by density functional theory (DFT) calculations that revealed the presence of delocalized frontier molecular orbitals in (−)-NDI-Δ. The results suggested that the unpaired electron in $[(-)\text{-NDI-}\Delta]^{\bullet-}$ is shared among all three NDI units of the triangular prism. This constitutes a rare example of an organic mixed-valence radical in which the unpaired electron is surrounding a triangular cavity. The orbital interactions between the NDIs render the cavity of the triangular prism electron deficient and an ideal platform for studying anion–π interactions (Figure 4.11c). This was nicely demonstrated by the encapsulation of I_3^- anions inside these cavities inducing π–π stacking of the chiral prisms into one-handed supramolecular helices, affording an emergent example of anion-induced self-assembly with potential as ion-channels.

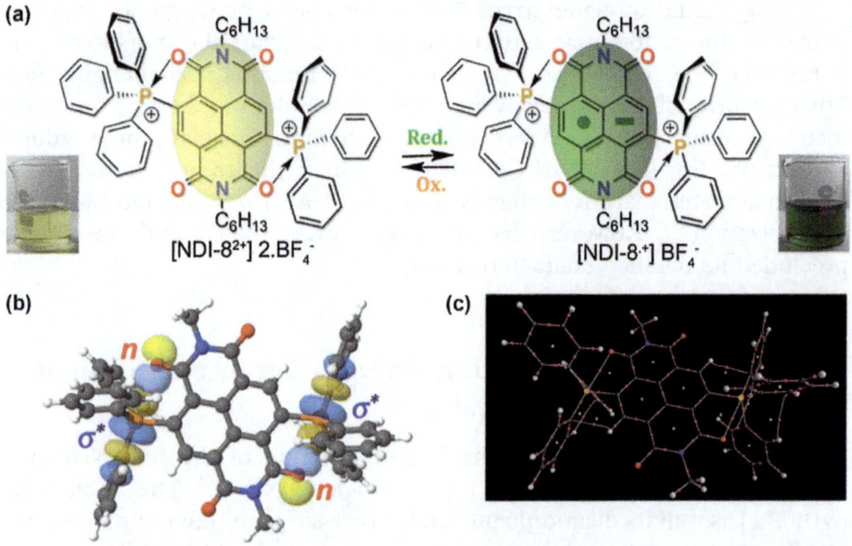

Figure 4.10 (a) Molecular structure of the NDI radical ion NDI-$8^{\bullet +}$ and its dicationic precursor NDI-8^{2+} and their reversibility to oxidation and reduction. The respective solutions remain stable under ambient condition, (b) A donor–acceptor type $n_O \rightarrow \sigma_{P-C}{}^*$ orbital interaction showing the P···O noncovalent interactions established through NBO calculations and (c) AIM analysis depicting a bond critical path and electron density at bond critical points between the P and O atoms. Colour code: C, grey; N, blue; O, red; P, orange.
Reprinted with permission from S. Kumar, M. R. Ajayakumar, G. Hundal and P. Mukhopadhyay, Extraordinary Stability of Naphthalene-diimide Radical Ion and Its Ultra-Electron-Deficient Precursor: Strategic Role of the Phosphonium Group, *J. Am. Chem. Soc.*, 2014, **136**, 12004–12010. Copyright 2014 American Chemical Society.

On the other hand, in the cofacially-oriented NDI-based cyclophane, electronic communication in the confined geometry leads to four individually accessible redox states (Figure 4.11b).[84] The neutral state was strongly exciton-coupled and has sufficient electronic coupling between the π-systems to produce a long-lived emissive excited state, which indicated extensive electron delocalization between the NDI units. The monoreduced radical anion $[2NDI]^{\bullet -}$ was found to be paramagnetic and a mixed-valence species with an absorption band centred at 1750 nm. On the other hand, the radical anion dimer (−)-$[2NDI]^{2-}$ was diamagnetic. The crystal structure of this radical anion dimer $[(-)\text{-}2NDI]^{2-}\cdot 2Bu_4N^+$ confirmed extensive π-electron delocalization between the two $NDI^{\bullet -}$ upon spin-pairing (Figure 4.11d).

Taking the noncovalent strategy for stabilizing reactive NDI radical ions forward, Zhang *et al.* reported the promotion and stabilization of $NDI^{\bullet -}$(NDI-9) in aqueous solution by shielding the radical anion with cucurbit[7]uril (CB[7]) (Figure 4.12).[85]

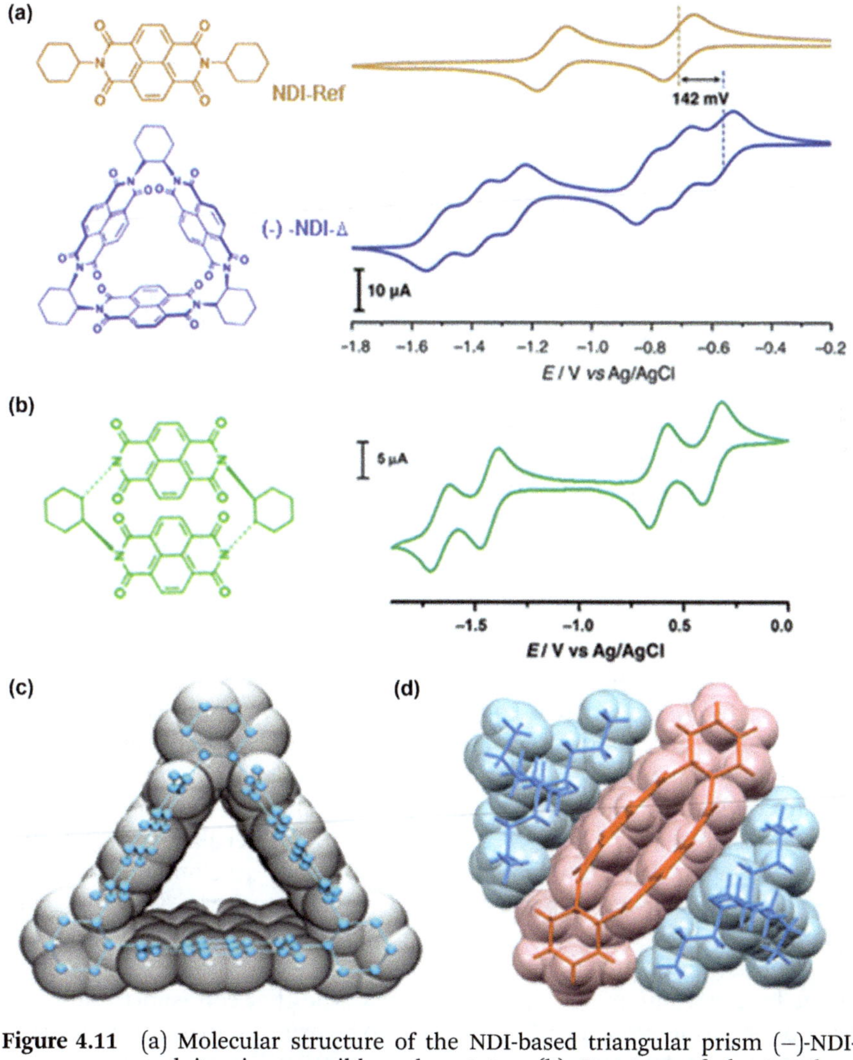

Figure 4.11 (a) Molecular structure of the NDI-based triangular prism (−)-NDI-Δ and its six accessible redox states, (b) Structure of the NDI-based cyclophane and its four individually accessible redox states, (c) Single crystal X-ray structure of the NDI triangular prism, and (d) Single crystal X-ray structure of the NDI cyclophane in radical anion dimer form $[(-)\text{-}2NDI]^{2-} \cdot 2Bu_4N^+$. Color code for 11d: NDI cyclophane in red, Bu_4N^+, in light blue.

Figures (a) and (c) Reproduced with permission from S. T. Schneebeli, M. Frasconi, Z. Liu, Y. Wu, D. M. Gardner, N. L. Strutt, C. Cheng, R. Carmieli, M. R. Wasielewski and J. F. Stoddart, Electron Sharing and Anion–π Recognition in Molecular Triangular Prisms, *Angew. Chem., Int. Ed.*, 2013, **52**, 13100–13104. Copyright 2013 Wiley-VCH Verlag GmbH & Co. KGaA, and (b) reproduced with permission from Y. Wu, M. Frasconi, D. M. Gardner, P. R. McGonigal, S. T. Schneebeli, M. R. Wasielewski and J. F. Stoddart, Electron Delocalization in a Rigid Cofacial Naphthalene-1,8:4,5-bis(dicarboximide) Dimer, *Angew. Chem., Int. Ed.*, 2014, **53**, 9476–9481. Copyright 2014 Wiley-VCH Verlag GmbH & Co. KGaA.

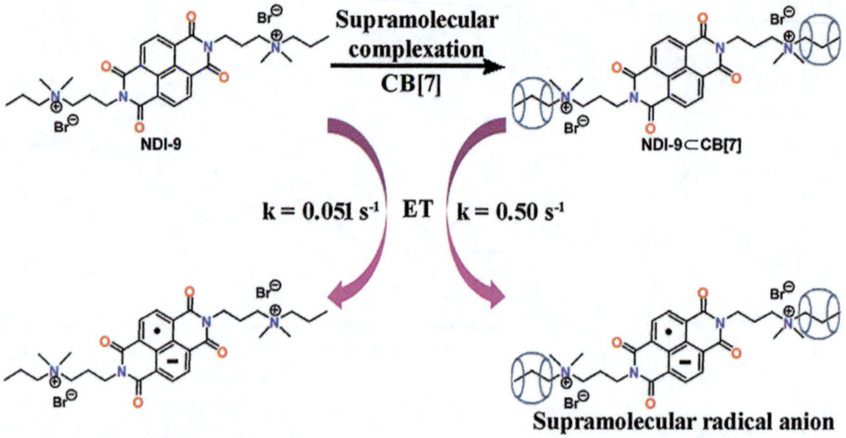

Figure 4.12 Stabilization of NDI$^{\bullet-}$ (NDI-9) by supramolecular shielding of the radical anion with CB[7] and enhancement of the photoinduced ET rate in the presence of CB[7].
Reproduced from ref. 85 with permission from the Royal Society of Chemistry.

The LUMO and HOMO energy of NDI were lowered significantly by the supramolecular complexation of the axially-substituted ammonium side chains of the NDI with two CB[7]. This efficiently promoted the photoinduced ET process between NDI and bromide anions in aqueous solution by one order of magnitude. The resulting supramolecular NDI$^{\bullet-}$ was found to foster outstanding stability in aqueous solution, even at temperatures of 60 °C. Therefore, it is anticipated that this supramolecular strategy would provide a facile way for stabilizing radicals and development of novel materials with spin-based properties and optical properties in the Vis/NIR regions.

In contrast to the general perception of NDI scaffold forming only radical anions by electronic reduction, we demonstrated that core-substituted NDI with donor amino groups NDI-10/11 can also be oxidized in a straightforward manner to give ND-centred radical cations (NDI$^{\bullet+}$) under ambient conditions.[86] We found that intramolecular H-bonding of the amino (–NH) group with the imide carbonyls along with the steric encumbrance plays a key role in enhancing the lifetime of the NDI$^{\bullet+}$. This was clearly illustrated by the fact that the NDI radical cation with the cyclohexyl amino substituents (NDI-10) had a lifetime 290-fold greater than the NDI radical cation with the piperidine substituents (NDI-11) and devoid of H-bonding groups (Figure 4.13).

The NDI$^{\bullet+}$ had appealing panchromatic optical characteristics with a strong absorption at 870 nm. Moreover, these core-substituted NDIs retained the ability to undergo chemical reduction with TBACN to form the corresponding radical anions. The strong NIR absorption characteristics could be applied to develop new generation of radical ion probes for sensing metal ions as well as anions.

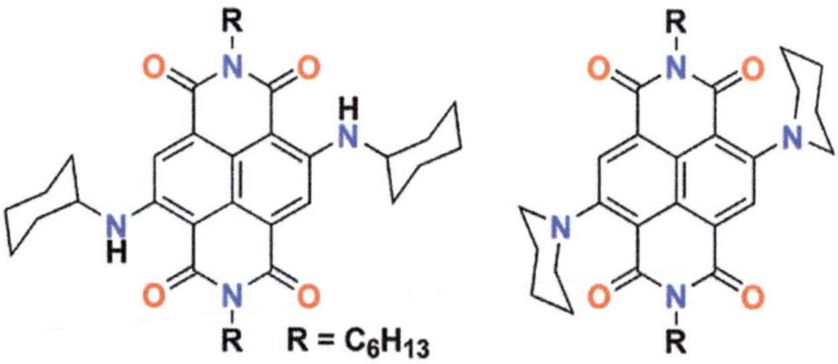

NDI-10 **NDI-11**

Figure 4.13 Molecular structures of core-substituted NDIs, NDI-10 and NDI-11 with donor amino groups.
Reprinted with permission from M. R. Ajayakumar, D. Asthana and P. Mukhopadhyay, Core-Modified Naphthalenediimides Generate Persistent Radical Anion and Cation: New Panchromatic NIR Probes, *Org. Lett.*, 2012, **14**, 4822–4825. Copyright 2012 American Chemical Society.

4.3.4 Photochemical Reactions and Photoexcitation of NDIs and NDI Radical Ions

Over the past few decades, Wasielewski's group has immensely contributed towards studies related to the photoexcitation of these radical anions.[87] Spectroscopic techniques such as transient absorption and fluorescence have been used to probe the photophysical properties of the π^* excited doublet states of these radical anions, and have shown that these properties are very different from those of the neutral parent molecule. To achieve a femtosecond optical control in a donor–acceptor molecular switch, this group reported the two-pulse femtosecond ET dynamics of a donor–acceptor(1) dyad, **ANI-NDI** (NDI-12), and a donor–acceptor(1)-acceptor(2) triad, **ANI-NDI-PMDI** (NDI-13) (Figure 4.14), in which an initial femtosecond laser flash was used to produce an ion pair, whose subsequent fate was controlled by a second femtosecond laser flash.[88]

The rate of ET from $*NDI^-$ to PMDI was found to be eight times faster than that from $*NDI^-$ to ANI^+. The larger rate constant for ET from $*NDI^-$ to PMDI was attributed to the increased electronic coupling between $*NDI^-$ and PMDI provided by the single N–N bond between them. These experiments revealed that it is possible to control the movement of electrons within a multisite donor–acceptor array on a femtosecond time scale. This work was followed by a series of intriguing studies on other donor–acceptor arrays and extended to the understanding of photosynthetic reaction-centre model.[89–91]

Multichromophoric systems having multi-step ET properties have been in focus to understand and mimic natural photosynthetic systems. However, in many cases these artificial systems do not always exhibit efficient, long-lived charge-separated (CS) states. One of the promising strategies that have emerged is that of a spin-control approach using a chromophore undergoing rapid intersystem crossing.[92] Metal complexes with M = Ru, Ir or Pt that can

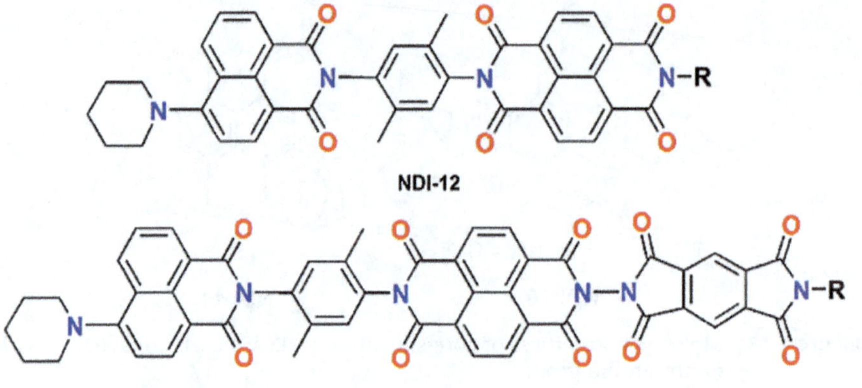

Figure 4.14 Molecular structures of donor–acceptor dyad (NDI-12) and donor-acceptor triad.
Reprinted with permission from M. P. Debreczeny, W. A. Svec, E. M. Marsh and M. R. Wasielewski, Femtosecond Optical Control of Charge Shift within Electron Donor-Acceptor Arrays: An Approach to Molecular Switches, *J. Am. Chem. Soc.*, 1996, **118**, 8174–8175. Copyright 1996 American Chemical Society.

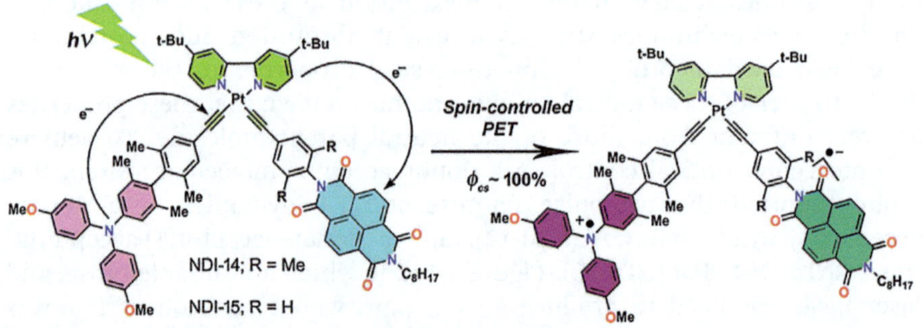

Figure 4.15 Molecular structures of donor–acceptor triads NDI-14 and NDI-15 comprising of dimethoxydimethyltriphenylamine (MTA) as the donor and a NDI/MeNDI as the acceptor.
Reprinted with permission from S. Suzuki, R. Sugimura, M. Kozaki, K. Keyaki, K. Nozaki, N. Ikeda, K. Akiyama and K. Okada, Highly Efficient Photoproduction of Charge-Separated States in Donor-Acceptor-Linked Bis(acetylide) Platinum Complexes, *J. Am. Chem. Soc.*, 2009, **131**, 10374–10375. Copyright 2009 American Chemical Society.

exhibit intense phosphorescence at room temperature *via* strong spin–orbit coupling (SOC) are necessary for this purpose.[93] Taking this idea forward, Okada *et al.* reported highly efficient photoproduction of CS states in donor–acceptor linked bis(acetylide) platinum complexes. They synthesized new triads MTA-Pt-MeNDI (NDI-14) and MTA-Pt-NDI (NDI-15) comprising of dimethoxydimethyltriphenylamine (MTA) as the donor and a NDI/MeNDI as the acceptor (Figure 4.15).[94]

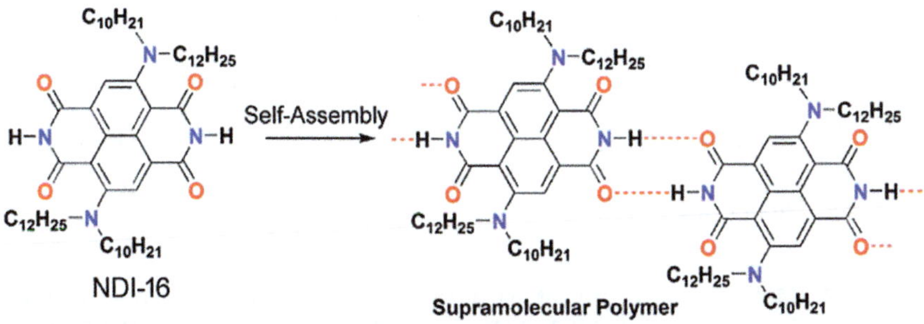

Figure 4.16 Molecular structure and self-assembly of core-substituted NDI-diamines with long alkyl chains (NDI-16).
Reproduced from ref. 95 with permission from the Royal Society of Chemistry.

These chromophores were bonded to the Pt moiety through highly twisted phenylene–ethynylene linkages. $MTA^{\bullet +}$and $NDI^{\bullet -}$ exhibited strong absorption bands in different wavelength regions enabling detection and quantification of the CS states. Thus, a highly efficient CS system with a long lifetime for the CS state using D–Pt–A complexes with twisted π linkers could be accessed through the efficient formation of D^{+}–Pt–A^{-}. Here, the SOC effect could be minimized by the elimination of the unpaired electron from the Pt moiety which resulted in a long-lived CS state.

In a recent report, Ghosh *et al.* described the synthesis and self-assembly of core-substituted NDI-diamines with long alkyl chains (NDI-16) (Figure 4.16).[95]

They demonstrated by means of different spectroscopic studies the origin and mechanistic pathway of a cooperative self-assembly, leading to nanotube formation through the intermediate sheet-like structures. The H-bonding in these systems could be modulated utilizing the two H-bond acceptors (imide oxygen) and the H-bond donor (–NH) at the axial positions. They established through time-resolved absorption and fluorescence measurements that the J-aggregated dye molecules, encapsulated in the multilayer walls of the tubes, facilitate effective delocalization of the excited states leading to prolonged excited state lifetimes. This aspect is desirable in emerging areas including photocatalysis and light harvesting.

NDIs have also been shown to exhibit a variety of photo-reactivities towards DNA, including long-range guanine oxidation[96] and preferential DNA cleavage.[97] The photochemical mechanisms of damage by hydroperoxy-substituted NDIs have been examined using laser flash photolysis.[98] Kelly *et al.* reported the synthesis, redox and photophysical properties of a series of water-soluble 1,4,5,8-naphthalene diimide derivatives (Figure 4.17).[99]

From laser flash photolysis studies, the triplet excited state of *N*,*N*′-bis[2-(*N*-pyridinium)ethyl]-1,4,5,8-naphthalene diimide (NDI-17) was found to undergo oxidative quenching with the electron donors like DABCO, tyrosine, *etc.* Interestingly, the reactivities of NDIs possessing α- and β-carboxylic acid substituents (R = $-CH_2COO^-$, $-C(CH_3)_2COO^-$, and $-CH_2CH_2COO^-$) (NDI-18-20) were strikingly different. In these compounds, the transient species

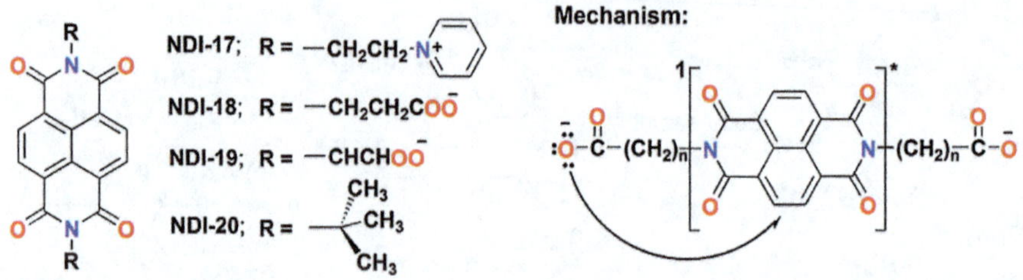

Figure 4.17 Molecular structure of NDIs (NDI-17-20) and the intramolecular ET from the carboxylate moiety to the singlet excited state of the NDI. Reprinted with permission from B. Abraham, S. McMasters, M. A. Mullan and L. A. Kelly, Reactivities of Carboxyalkyl-Substituted 1,4,5,8-Naphthalene Diimides in Aqueous Solution, *J. Am. Chem. Soc.*, 2004, **126**, 4293–4300. Copyright 2004 American Chemical Society.

produced upon excitation did not react with the electron donors, while rapidly reacting with known electron acceptors like benzyl viologen and ferricyanide. The transient spectrum of the carboxyalkyl-substituted NDIs observed after the laser pulse was nearly identical to the $NDI^{\bullet -}$. From these studies they concluded that the transient species produced upon photolysis of NDI-$(CH_2)_nCOO^-$ is the species produced upon intramolecular ET from the carboxylate moiety to the singlet excited state of NDI. The results are significant as they introduced new reactive intermediates, which can have functions as DNA and also as protein structural probes.

As an important extension to this work, Shinmyozu *et al.* recently reported photoinduced colour change of NDIs bearing pendant alkylamine moieties as axial substituents, in the solid state.[100] The colour change was attributed to the generation of $NDI^{\bullet -}$ formed through a photoinduced ET process from the alkylamine moiety to the NDI. The photosensitivity of NDIs was highly dependent on the structures of the alkylamine moieties. Upon photoirradiation, the NDI-21,22,23 and NDI-26 showed strong colour changes from yellow to black, while NDIs 27–31 did not show any colour changes (Figure 4.18).

Analysis of the photoproduct suggested that the $NDI^{\bullet -}$ was formed through an irreversible process initiated by proton abstraction between an amine radical cation and the neutral amine moiety. The broad absorption bands in NIR region suggested that the radical anions formed stacks including mixed-valence stacks and radical anion stacks. Interestingly, these photosensitive NDIs showed crystal bending upon photoirradiation, as a result of the changes in the intermolecular distance of the NDIs in the monomeric radical anions, mixed-valence or radical anion π-stacks.

4.4 Perylenediimides

4.4.1 Introduction to PDIs

Perylenediimides (PDIs), were initially applied for industrial purposes as red vat dyes. Over the years, several members of the PDI family found

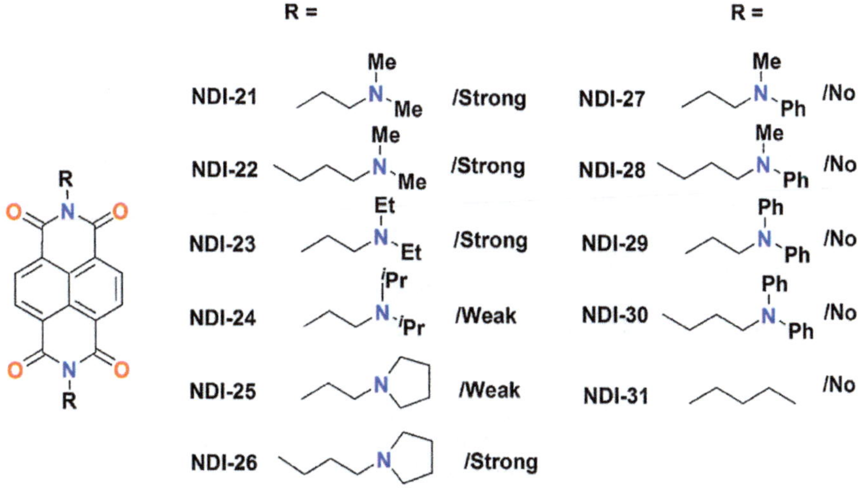

Figure 4.18 Molecular structures of NDI21-31 and their response (Strong/Weak/No) to photoirradiation.
Reproduced with permission from Y. Matsunaga, K. Goto, K. Kubono, K. Sako and T. Shinmyozu, Photoinduced Color Change and Photomechanical Effect of Naphthalene Diimides Bearing Alkylamine Moieties in the Solid State, *Chem. - Eur. J.*, 2014, **20**, 7309–7316. Copyright 2014 Wiley-VCH Verlag GmbH & Co. KGaA.

remarkable industrial applications as pigments as it could exist with hues ranging from red to violet and even black, and most significantly due to its inherent insolubility, chemical inertness, light and thermal stability.[101,102] A large volume of investigation has been carried out to understand its role in self-assembly and its effect on the opto-electronic properties.[103–107] Their potential usefulness as electron-accepting materials was reported as early as in 1986. In this pioneering work, Tang *et al.* reported an all-organic p–n junction-based photovoltaic solar cell, applying a perylenetetracarboxylic derivative as the n-type material and phthalocyanine as the p-type material.[108] In 2000, with the work of Sudhölter *et al.*, PDIs made their mark as promising candidates for n-type semiconductor materials.[109] This group reported PDI-based liquid crystalline (LC) materials with an electron mobility of 0.10 $cm^2\ V^{-1}\ s^{-1}$ in the LC phase, and 0.21 $cm^2\ V^{-1}\ s^{-1}$ for the crystalline phase. The n-type semiconductivity is related to the high electron affinity of PDIs, which makes PDIs promising for application in organic field-effect transistors. In addition, based on their unique combination of optical, redox, and stability properties, PDIs have been investigated in electrophotography. The advancement in the synthesis of the PDIs in the axial, bay and non-bay positions (Figure 4.19a) played a crucial role towards the development of the PDI-based smart organic electronic materials.

Langhal *et al.* introduced solubilizing substituents at the axial positions, which substantially increased the solubility of these classes of dyes in common organic solvents.[110,111] Later, Seybold *et al.* reported the

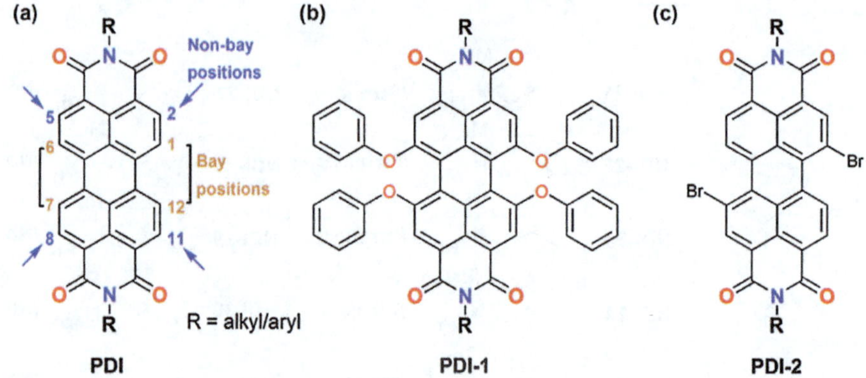

Figure 4.19 (a) Molecular structure of a PDI showing the different positions for substitution/coupling reactions, (b) PDI with phenoxy groups at the bay positions (PDI-1), and (c) structure of 1,7-dibromo PDI (PDI-2). Figure (a) reprinted with permission from C. W. Struijk, A. B. Sieval, J. E. J. Dakhorst, M. van Dijk, P. Kimkes, R. B. M. Koehorst, H. Donker, T. J. Schaafsma, S. J. Picken, A. M. van de Craats, J. M. Warman, H. Zuilhof and E. J. R. Sudhölter, Liquid Crystalline Perylene Diimides: Architecture and Charge Carrier Mobilities, *J. Am. Chem. Soc.*, 2000, **122**, 11057. Copyright 2000 American Chemical Society. Figure (c) reprinted with permission from F. Würthner, V. Stepanenko, Z. Chen, C. R. Saha-Möller, N. Kocher and D. Stalke, Preparation and Characterization of Regioisomerically Pure1,7-Disubstituted Perylene Bisimide Dyes, *J. Org. Chem.*, 2004, **69**, 7933–7939. Copyright 2004 American Chemical Society.

nucleophilic substitution with phenoxy groups at the bay positions of the PDI (PDI-1) starting from the fourfold chlorinated PDI (Figure 4.19b).[112] Subsequently, the regioisomeric purification of the 1,7-dibromo PDI (PDI-2) from the respective 1,6 regioisomer by Würthner *et al.* initiated a wide range of nucleophilic substitution and coupling reactions at the bay positions of the PDI (Figure 4.19c).[113]

4.4.2 Electron Injection and Transport With PDIs

With the accessibility to new synthetic methodologies for bay substitution at the PDI, a comprehensive study was undertaken by Würthner *et al.* to understand the effect of bay position substitution and also the different axial groups on the semiconductor property of PDIs. A diverse range of different PDIs (PDI-3-17) with up to four halogen substituents in the bay positions of the perylene core and five different fluorinated imide axial substituents were synthesized and examined for their electron transport characteristics (Figure 4.20).[114]

The influence of the substituents on the LUMO level and the solid-state packing of PDIs were examined by electrochemical and single crystal structure analyses. PDIs bearing only hydrogen or up to two fluorine substituents

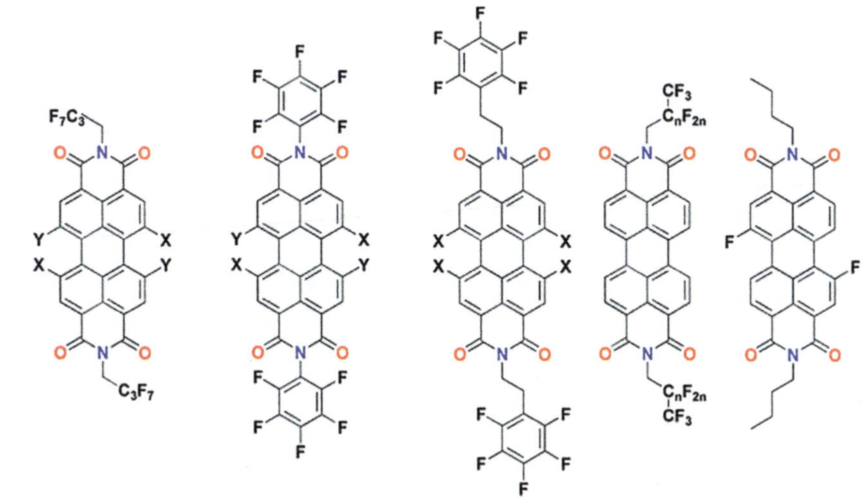

PDI-3 X = Y = H
PDI-4 X = H, Y = F
PDI-5 X = Y = F
PDI-6 X = Y = Cl
PDI-7 X = Y = Br

PDI-8 X = Y = H
PDI-9 X = H, Y = F
PDI-10 X = Y = Cl
PDI-11 X = Y = Br

PDI-12 X = H
PDI-13 X = Cl

PDI-14 n = 0
PDI-15 n = 2
PDI-16 n = 3

PDI-17

Figure 4.20 Molecular structures of n-type semiconductors PDI-3-17. Reprinted with permission from R. Schmidt, J. H. Oh, Y.-S. Sun, M. Deppisch, A.-M. Krause, K. Radacki, H. Braunschweig, M. Könemann, P. Erk, Z. Bao and F. Würthner, High-Performance Air-Stable n-Channel Organic Thin Film Transistors Based on Halogenated Perylene Bisimide Semiconductors, *J. Am. Chem. Soc.*, 2009, **131**, 6215–6228. Copyright 2009 American Chemical Society.

at the flat PDI core (PDI-3/PDI-4) afforded excellent n-channel transistors with electron mobility better than 0.5 $cm^2\ V^{-1}\ s^{-1}$ and $I_{on}/I_{off} > 10^6$. The best OTFTs were obtained for 2,2,3,3,4,4,4-heptafluorobutyl-substituted PDI (PDI-3) with electron mobilities consistently greater than 1 $cm^2\ V^{-1}\ s^{-1}$. For distorted core-tetrahalogenated (fluorine, chlorine, or bromine) PDIs (PDI-5/6/7/10/11/13), less favourable solid-state packing properties were found. High performance OTFTs were only obtained from the tetrachlorinated PDI derivative. This inferior behaviour was attributed to packing constraints arising from the distorted core, which encumber the formation of extended highly crystalline grains. The easy synthetic availability, excellent on-to-off current modulation combined with high mobility in air makes these PDIs suitable for a wide range of organic electronic devices.

To better the electron mobility of the tetra-substituted PDIs, particularly of the Cl_4-PDI, Würthner *et al.* synthesized a new octa-chloro-substituted PDI (PDI-18) that bears eight chlorine substituents at the aromatic core and free NH imide functionalities (Figure 4.21a).[115] By analysing the limitations of crystal packing in the tetra-substituted PDIs, they anticipated that by increasing the number of chlorine substituents a "brick" with sides of

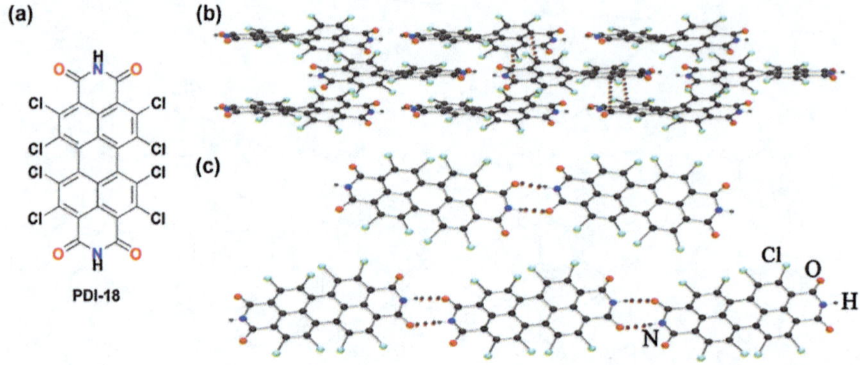

Figure 4.21 (a) Molecular structure of n-type semiconductor PDI-18, (b) Self-assembly of PDI-18 showing the π-stacking interactions, and (c) N–H · O H-bonding driven self-assembly of PDI-18. Colour code: C, dark grey; N, blue; O, red; Cl, turquoise blue; H, light grey. Reproduced with permission from 115. M. Gsänger, J. H. Oh, M. Könemann, H. W. Hffken, A.-M. Krause, Z. Bao and F. Würthner, A Crystal-Engineered Hydrogen-Bonded Octachloroperylene Diimide with a Twisted Core: An n-Channel Organic Semiconductor, *Angew. Chem., Int. Ed.*, 2010, **49**, 740–743. Copyright 2010 Wiley-VCH Verlag GmbH & Co. KGaA.

similar dimensions can be formed. The free NH imide groups were envisioned to enforce close hydrogen-bonded contacts between adjacent molecules.

The X-ray crystal structure of PDI-18 unambiguously proved the extended π–π stacking between the contorted PDI rings (Figure 4.21b) and the array of N–H · O H-bonding formed from the infinite recurrence of eight membered cyclic H-bonding pattern (Figure 4.21c). An increase in the number of chlorine substituents to eight led to lowering of the LUMO level to −4.23 eV, while the combination of hydrogen bonding and contortion of the π-core directs two-dimensional π–π-stacked percolation paths for electron transport. Thus as a consequence of judicious molecular and crystal engineering PDI-18 based OTFTs showed excellent electron mobility 1 $cm^2\ V^{-1}\ s^{-1}$ in air with an on-to-off current ratios greater than 10^6.

From our group, we reported the facile synthesis of octabromoperylene-3,4,9,10 tetracarboxylic dianhydride (Br_8-PDA), its diimides (PDI-19) (Figure 4.22a).[116] Diverse halogen bonding interactions were identified in the Br_8-PDA (Figure 4.22b) and PDI-19 (Figure 4.22c), which along with the characteristic high electron-deficient nature enhanced the electron transport characteristics compared to the bay-position-substituted di-/tetra-brominated PDIs (Br_2-PDI/Br_4-PDI).

The best electron mobilities were found with the axially butyl substituted Br_8-PDIs, which approached $\sim 2\times10^{-3}\ cm^2\ V^{-1}\ s^{-1}$. Notably, the reported electron mobilities for previously described bay-position-substituted Br_2-PDI and Br_4-PDI derivatives are $\sim 10^{-5}$–$10^{-6}\ cm^2\ V^{-1}\ s^{-1}$. The greater μ_e was

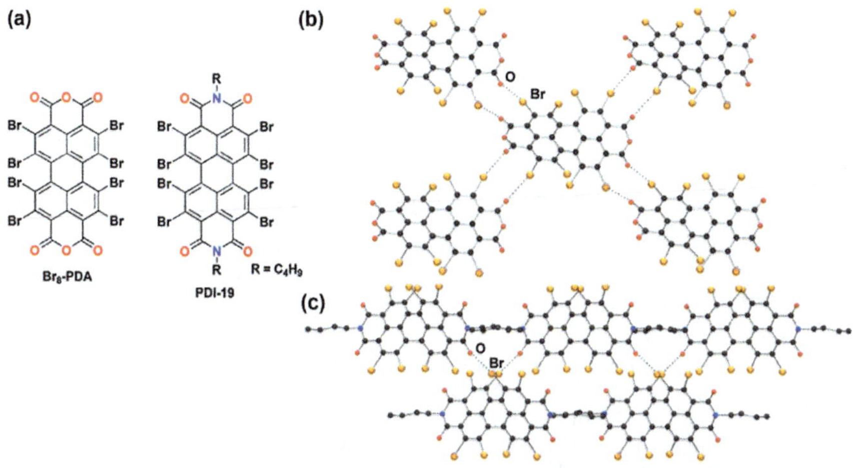

Figure 4.22 (a) Molecular structures of Br_8-PDA and PDI-19, (b) Self-assembly of Br_8-PDA using ten-membered C–Br · O halogen bonding interactions, and (c) Self-assembly of PDI-19 using C–Br · O halogen bonding interactions. Colour code: C, dark grey; N, blue; O, red; Br, orange.
Reprinted with permission from Y. Kumar, S. Kumar, S. K. Keshri, J. Shukla, S. S. Singh, T. S. Thakur, M. Denti, A. Facchetti and P. Mukhopadhyay, Synthesis of Octabromoperylene Dianhydride and Diimides: Evidence of Halogen Bonding and Semiconducting Properties, *Org. Lett.*, 2016, **18**, 472–475. Copyright 2016 American Chemical Society.

rationalized considering the close-packing due to the halogen bonding interactions and significantly reduced LUMO energy.

In an innovative molecular design, Wang *et al.* reported the synthesis of diperylenedimide having four chlorine atoms substituted at the bay positions (PDI-20) by a simple homocoupling method (Figure 4.23a).[117] The LUMO level of PDI-20 was found to be around −4.22 eV, bestowing it strong electron deficient character. The transistors fabricated based on the single crystals exhibited electron mobility of 4.65 $cm^2 V^{-1} s^{-1}$, which is one of the highest reported to date with PDI-based n-type semiconductors. Furthermore, the devices revealed excellent environmental stability. In an extension of this coupling reaction, Wang *et al.* reported a new family of hybrid rylene arrays in which a central PDI core was coupled with two NDI units (PDI-21/22/23) *via* a combination of Stille coupling and C–H transformation (Figure 4.23b).[118]

This new design concept led to a markedly broadened absorption spectrum and increased electron affinities, which facilitated electron injection and transport with ambient stability. One of these hybrid (PDI-22) derivatives exhibited excellent OTFT characteristics under ambient conditions with an electron mobility of 0.25 $cm^2 V^{-1} s^{-1}$ and a high on-to-off current ratio of 10^7.

Narayan *et al.* in an appealing demonstration have shown the function of non-planar PDIs as alternatives to fullerene electron acceptors in organic solar cells. This group identified the crystallinity of PDIs to be responsible

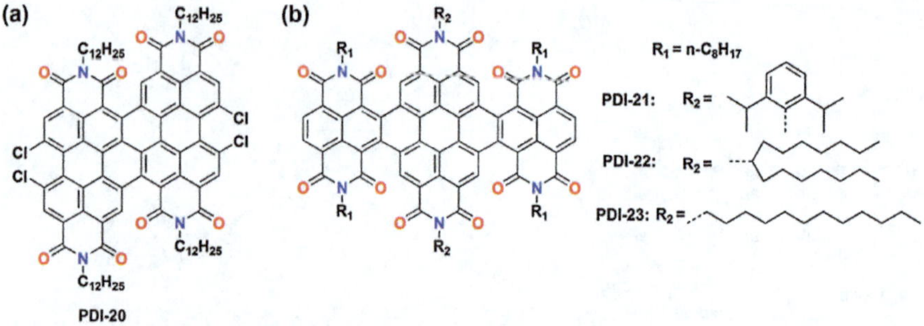

Figure 4.23 (a) Molecular structure of two-PDI coupled n-type semiconductor PDI-20. Reproduced with permission from A. Lv, S. R. Puniredd, J. Zhang, Z. Li, H. Zhu, W. Jiang, H. Dong, Y. He, L. Jiang, Y. Li, W. Pisula, Q. Meng, W. Hu and Z. Wang, High Mobility, Air Stable, Organic Single Crystal Transistors of an n-Type Diperylene Bisimide, *Adv. Mater.* 2012, **24**, 2626–2630. Copyright 2012 from the Wiley-VCH Verlag GmbH & Co. KGaA, and (b) Structure of NDI-PDI-NDI coupled n-type semiconductor PDI-21-23. Reprinted with permission from W. Yue, A. Lv, J. Gao, W. Jiang, L. Hao, C. Li, Y. Li, L. E. Polander, S. Barlow, W. Hu, S. Di Motta, F. Negri, S. R. Marder and Z. Wang, Hybrid Rylene Arrays via Combination of Stille Coupling and C-H Transformation as High-Performance Electron Transport Materials, *J. Am. Chem. Soc.*, 2012, **134**, 5770–5773. Copyright 2012 American Chemical Society.

for the low short circuit current density (J_{SC}), which is one of the primary limitations of these cells.[119] In their design concept, they applied a non-planar bis-functionalized PDI (PDI-24) which disrupts the crystallinity without adversely impacting the CT properties of PDIs (Figure 4.24).

In combination with a hole transporting polymer (Poly-1), a device efficiency of 2.77% was achieved. Furthermore, a tenfold increase in J_{SC} was observed compared to the planar PDI (PDI-25), resulting in one of the highest J_{SC} values reported for a solution-processed device featuring a PDI.

In a recent report, Marks *et al.* have further extended the application of PDIs as non-fullerene-based electron acceptors in organic photovoltaics.[120] In this work, they reported the synthesis of three non-bay-substituted PDIs having four hexyl (PDI-26) or phenyl ethyl (PDI-27) or phenyl (PDI-28) groups (Figure 4.25).

These substituted PDIs revealed slip-stacked packing, which suppressed excimer formation while preserving good π-stacking for charge transport. With the increasing steric bulk in the non-bay position and with increasing driving force for charge separation, the relative proportion of geminate recombination decreased. Bulk-heterojunction (BHJ) devices employing these three acceptors afforded high open circuit voltages ($V_{oc's}$) (>1.0 V) and relatively high fill factors. PDI-28 blended with the donor polymer (Poly-2) was found to be the best performing BHJ blend, affording power conversion efficiency (PCE) of 3.67%. Transient absorption spectroscopy revealed that the Poly-2-PDI-28 blend supported higher yields of long-lived CS species *versus* the other two PDI blends contributing to the higher PCE. In addition,

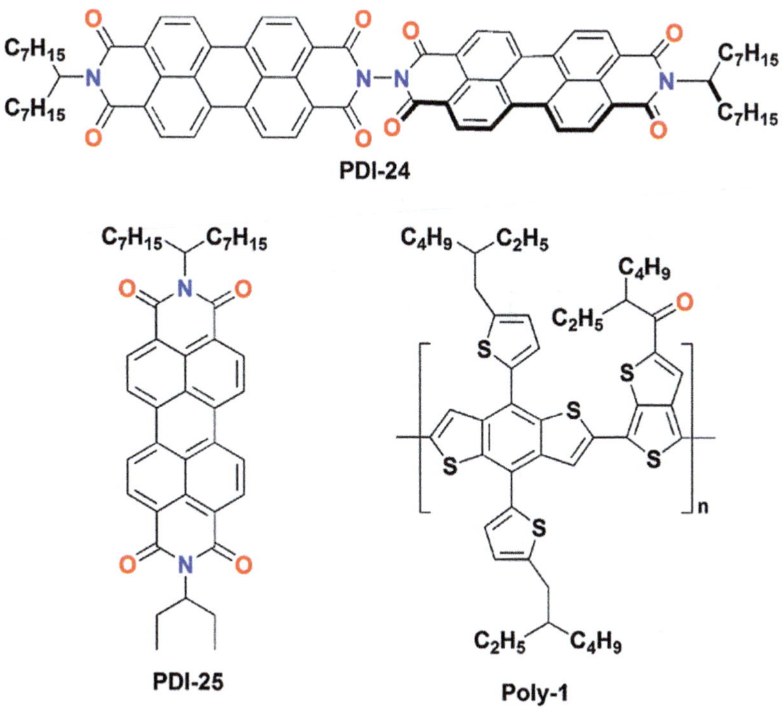

Figure 4.24 Structures of PDI-24, control compound (PDI-25) as alternatives to fullerene for solar cell application and the polymer (Poly-1). Reprinted with permission from S. Rajaram, R. Shivanna, S. K. Kandappa and K. S. Narayan, Nonplanar Perylene Diimides as Potential Alternatives to Fullerenes in Organic Solar Cells, *J. Phys. Chem. Lett.*, 2012, **3**, 2405–2408. Copyright 2012 American Chemical Society.

characterization of surface morphology of the blends suggested that moderate crystalline size with well-defined nanoscale donor–acceptor domains favoured higher PCEs.

4.4.3 ET Reactions of PDIs

PDIs akin to the other members of the rylene family undergo a sequential two-electron reversible reduction to form the corresponding radical anions and dianions. The radical anion and the dianion of PDIs are excellent chromophores having signature absorption bands in the visible and near-infrared (NIR) region. However, the PDI radical anion ($PDI^{\bullet -}$) and dianion (PDI^{2-}) are highly sensitive to ambient conditions and hence their stabilization and isolation under ambient conditions remains a highly challenging task. A diverse range of reducing agents, for example Na metal, hydrazine, sodium dithionite, *etc.*, have been used to chemically generate the anionic PDIs.[121–123] In our group, we found TBACN to be an effective

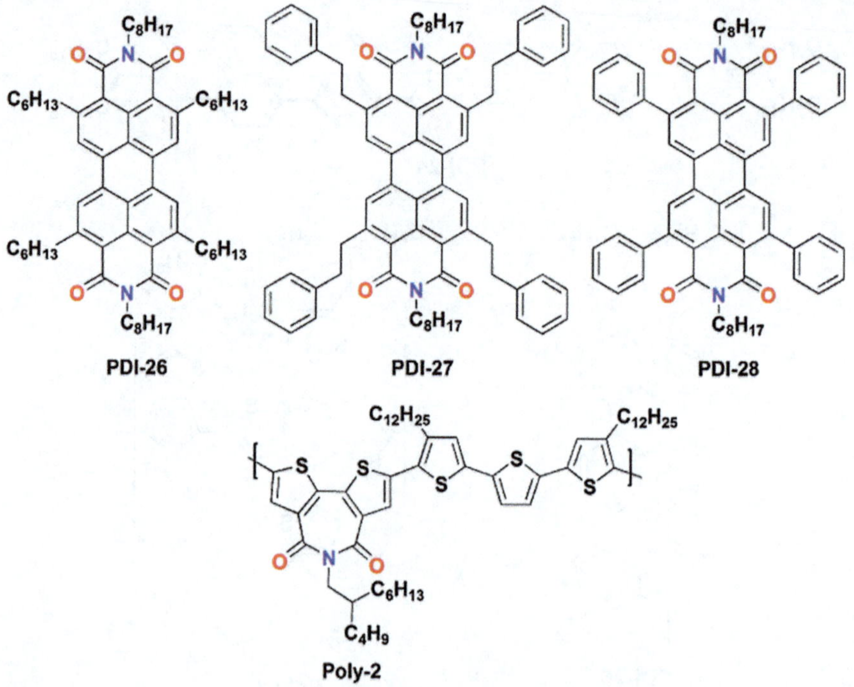

Figure 4.25 Structures of PDI-26, PDI-27 and PDI-28 as alternatives to fullerene for solar cell application and the polymer (Poly-2).
Reprinted with permission from P. E. Hartnett, A. Timalsina, H. S. S. R. Matte, N. Zhou, X. Guo, W. Zhao, A. Facchetti, R. P. H. Chang, M. C. Hersam, M. R. Wasielewski and T. J. Marks, Slip-Stacked Perylenediimides as an Alternative Strategy for High Efficiency Non-fullerene Acceptors in Organic Photovoltaics, *J. Am. Chem. Soc.*, 2014, **136**, 16345–16356. Copyright 2014 American Chemical Society.

reducing agent to reduce PDI-29 and generate the $PDI^{\bullet-}$ (Figure 4.26).[76] Although the stability of the $PDI^{\bullet-}$ increased with the tetrabutylammonium (TBA) counter-cation in solution, ambient isolation of the radical anion form still remained elusive.

Würthner *et al.* provided the first synthetic access to an ambient stable zwitterionic PDI centred radical ($PDI\text{-}30^{\bullet\pm}$).[124] The zwitterionic radical constitutes an imidazolium substituent at the *ortho*-position to the imide functionality with four chlorine atoms in the bay positions (Figure 4.27a). Interestingly, this zwitterionic radical could be reduced reversibly in the presence of tetrabutylammonium fluoride (TBAF) to its anionic state or oxidized to the corresponding cation with nitrosyl tetrafluoroborate ($NOBF_4$) as the oxidant. The radical was characterized in detail by UV/Vis absorption, EPR, and NMR spectroscopy. Furthermore, the zwitterionic radical was isolated as single crystals and characterized by X-ray crystallography (Figure 4.27b). The π-stacking interactions between the PDI rings varied

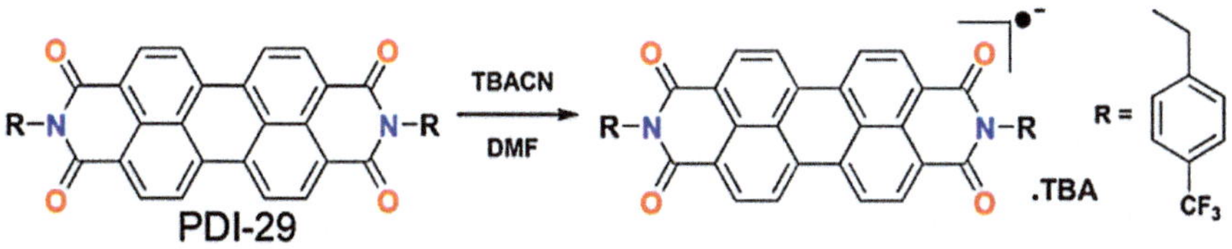

Figure 4.26 Molecular structure of PDI-29 and its ET with TBACN.
Reprinted with permission from M. R. Ajayakumar, S. Yadav, S. Ghosh and Pritam Mukhopadhyay, Single-Electron Transfer Driven Cyanide Sensing: A New Multimodal Approach, *Org. Lett.*, 2010, **12**, 2646–2649. Copyright 2010 American Chemical Society.

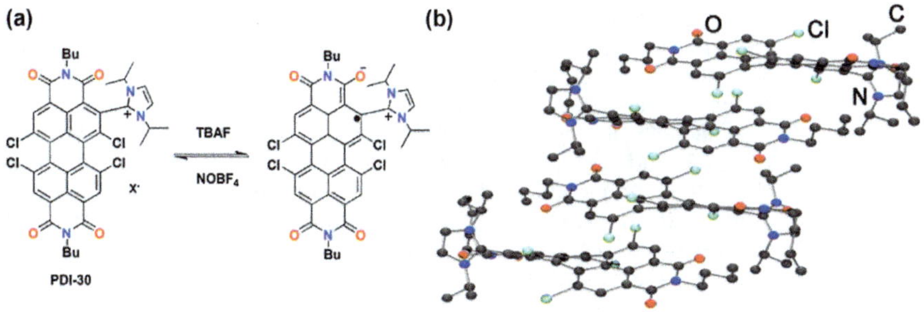

Figure 4.27 (a) Molecular structure of PDI-30 and its reduction with TBAF to form the zwitterionic radical, and its reversible oxidation with $NOBF_4$ to PDI-30, (b) π-stacking interactions of the zwitterionic radical. Colour code: C, dark grey; N, blue; O, red; Cl, turquoise blue.
Reproduced with permission from ref. 124. D. Schmidt, D. Bialas and F. Würthner, Ambient Stable Zwitterionic Perylene Bisimide-Centred Radical, *Angew. Chem., Int. Ed.*, 2015, **54**, 3611–3614. Copyright 2015 from the Wiley-VCH Verlag GmbH & Co. KGaA.

from 3.606–4.735 Å. The chemical stability of this isolable radical offers opportunities towards ambient processable n-type dopants for organic electronic devices, magnetism, *etc.*

In a major development, Rybtchinski *et al.* reported a PDI aromatic dianion (PDI^{2-}) that is almost indefinitely stable in an oxygen-free aqueous medium.[125] The PDI derivative comprised of polyethylene glycol groups (PDI-31) possessed excellent solubility in water, which allowed simple reducing agents like $Na_2S_2O_4$ and precise control over the stoichiometry of reduction (Figure 4.28a). When an aqueous solution of (PDI-31) was titrated with $Na_2S_2O_4$ under an inert atmosphere its colour changed from red to blue, and characteristic bands for $PDI^{\bullet-}$ appeared in the UV-vis spectrum. Further titration resulted in the appearance of a strong magenta colour, and signals characteristic to the PDI-31^{2-} absorption. PDI-31^{2-} was found to be EPR-silent, as typical of most aromatic dianions. Although PDI-31^{2-} was inert in water, it was extremely sensitive to oxygen and exposure to air, instantly converted PDI-31^{2-} to the neutral parent compound PDI-31. Experimental and theoretical studies revealed extensive charge delocalization and

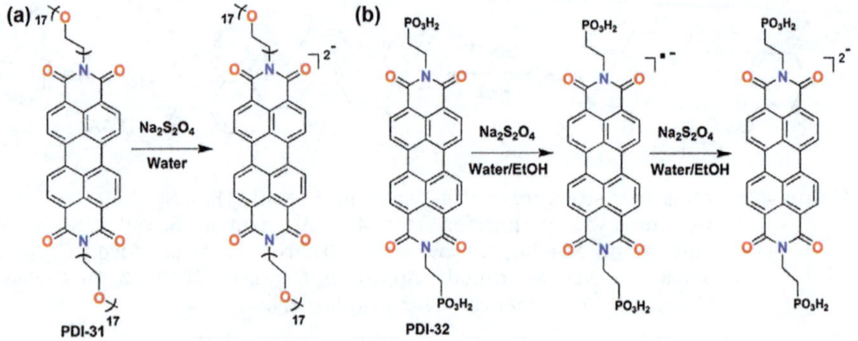

Figure 4.28 (a) Molecular structure of PDI-31 and its reduction with $Na_2S_2O_4$ in water to form the PDI-31 dianion. Reprinted with permission from E. Shirman, A. Ustinov, N. B. Shitrit, H. Weissman, M. A. Iron, R. Cohen and B. Rybtchinski, Stable Aromatic Dianion in Water, *J. Phys. Chem. B*, 2008, **112**, 8855–8858. Copyright 2008 American Chemical Society. (b) Structure of PDI-32 and its stepwise reduction with $Na_2S_2O_4$ in water-ethanol to form the PDI-32 radical anion and PDI-32 dianion. Reprinted with permission from I. Ghosh, T. Ghosh, J. I. Bardagi and B. König, Reduction of aryl halides by consecutive visible light-induced ET processes, *Science*, 2014, **346**, 725–728. Copyright 2009 American Chemical Society.

aromatic character of the PDI-31^{2-}. The reversible charging and discharging demonstrated in this PDI-31 render it useful for controlled electron storage and release in aqueous media.

This work was followed by the report of Brochsztain *et al.*, whereby they reported the synthesis of *N*,*N*′-bis(2-phosphonoethyl)-3,4,9,10-perylenediimide (PDI-32).[126] In aqueous ethanol (1 : 1 v/v), (PDI-32) existed as monomers and remained in the monomeric form in all the reduced states, *i.e.* radical anion and dianion states (Figure 4.28b).

In water, on the other hand, (PDI-32) formed dimers, which were then transformed into π-stacked aggregated form upon one-electron reduction to the PDI-$32^{\bullet-}$. The aggregates dissociated upon the second electron reduction, generating the PDI-32^{2-} in the monomeric form. Interestingly, the authors observed that the presence of CTAB micelles prevented the aggregation of PDI-$32^{\bullet-}$ in water, since the radical anions get incorporated into the micellar phase. Reduction of (PDI-32) in ethanol solution where the dye was aggregated resulted in soluble mixed stacks containing both $PDI^{\bullet-}$ and PDI^{2-}. These mixed stacks were stabilized in the radical anion form, which prevented their further reduction to the dianion.

While efforts to synthesize the PDI^{2-} in different solvent media continued, its ambient isolation remained out of reach. In a significant progress, Würthner *et al.* recently reported the synthesis of tetrachloro–tetracyano-PDI (PDI-33) (Figure 4.29a). In PDI-33, the four chloro substituents occupy the bay positions, while the four cyano groups populate the non-bay positions, neighbouring the imide carbonyl groups of the PDI.[127] PDI-33, when reduced with hydrogen in the presence of 10% (wt) palladium on activated charcoal

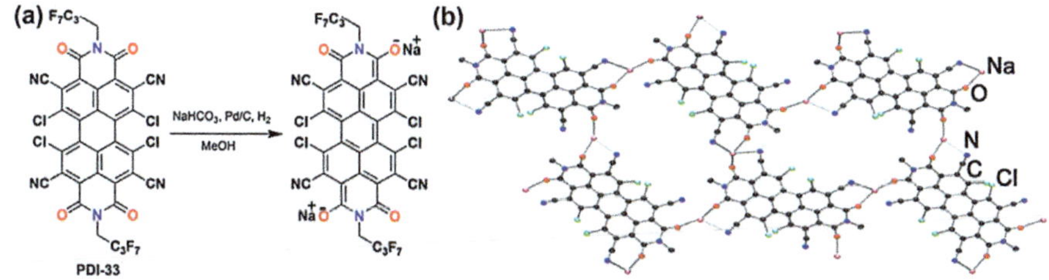

Figure 4.29 (a) Molecular structure of PDI-33 and its reduction to form the PDI-33 dianion, and (b) Single crystal structure of the PDI-33 dianion showing the assembly formed by the interaction of the imide oxygen and cyano nitrogen with the sodium ion. Colour code: C, dark grey; N, blue; O, red; Cl, turquoise blue; Na, pink.
Reproduced from ref. 127 with permission from the Royal Society of Chemistry.

(Pd/C) and an excess of sodium hydrogen carbonate, led to the generation of the PDI-33^{2-} disodium salt. A colour change of the reaction mixture from orange to dark blue was observed upon changing the nitrogen atmosphere with hydrogen. The PDI-33^{2-} disodium salt was isolated in 80% yield. The dianionic nature was fully characterized by UV-vis absorption and NMR spectroscopy.

Furthermore, X-ray crystallography studies confirmed the identity of the dianionic state (Figure 4.29b). Importantly, no spectral changes were observed when the solid PDI^{2-} was exposed for approximately 5 months to ambient conditions and upon dissolution of the solid in acetone. It is believed that the highly electron deficient nature of PDI-33 plays a pivotal role for the stabilization and electronic delocalization of the additional negative charges. From the semiquinone formation, constants examined by CV and SQV revealed that the formation of PDI^{2-} is preferred in polar solvents, whereas the generation of PDI$^{\bullet-}$ is favoured in less polar solvents.

Next, we discuss the intriguing report by König *et al.*, whereby they exemplified the reduction of aryl halides by consecutive visible light-induced ET processes.[128] This notable report described a new approach to overcome the limitations of visible light-mediated chemical photocatalysis, by using the energies of two photons in one catalytic cycle. It is known from earlier investigations that photocatalytic alkylation or arylation reactions employ a typical PET process in which the excited dye is converted into its radical anion, which activates the substituted benzyl bromides, alpha bromo carbonyl compounds, aryl iodides, *etc.* However, aryl bromides and chlorides that are less reactive as a result of more negative reduction potential and higher carbon–halide bond dissociation energy are not accessible by such photocatalysts and, visible light. To find a solution, König *et al.*, hypothesized that a second visible light excitation can be added to the process if the radical anion of the dye is stable in the ground state, and thus can be excited again by visible light. The *N*,*N*-bis(2,6-diisopropylphenyl)perylene-3,4,9,10-bis(dicarboximide) (PDI-34) (Figure 4.30a) was selected as the photocatalyst due to its good solubility in DMF and DMSO. Upon irradiation with blue

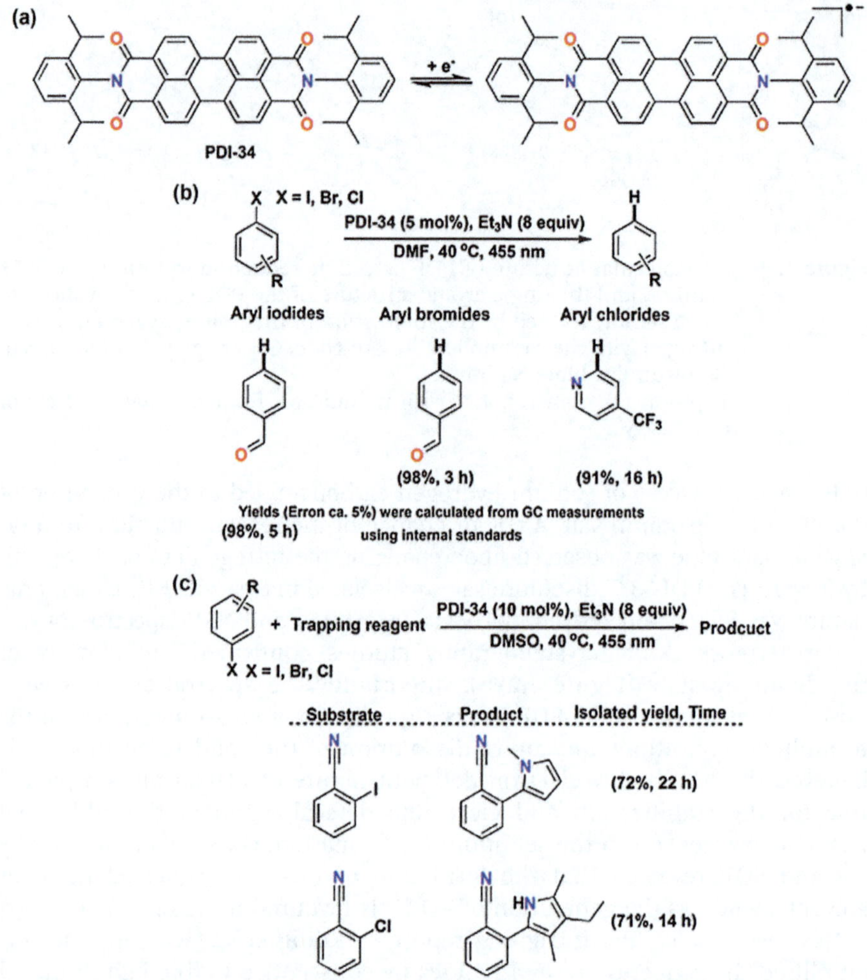

Figure 4.30 (a) Molecular structure of PDI-34 as the photocatalyst, upon irradiation with blue light (455 nm) in the presence of Et_3N, PDI-34 formed PDI-34$^{\bullet-}$, which could be excited by visible light, (b) Application of the catalytic system towards the photoreduction of aryl halides, including aryl chlorides, using visible light irradiation, and (c) Application of the catalytic system towards C–C bond-forming arylation reactions, the arylated products were obtained by irradiating the aryl halides in the presence of *N*-methylpyrrole and catalytic amounts of PDI. Reproduced with permission from I. Ghosh, T. Ghosh, J. I. Bardagi and B. König, Reduction of aryl halides by consecutive visible light-induced ET processes, *Science*, 2014, **346**, 725–728. Copyright 2014, American Association for the Advancement of Science.

light (455 nm) in the presence of an electron donor (Et_3N), PDI-34 formed a coloured radical anion PDI-34$^{\bullet-}$, which could again be excited by visible light. This catalytic system could be synthetically applied to the

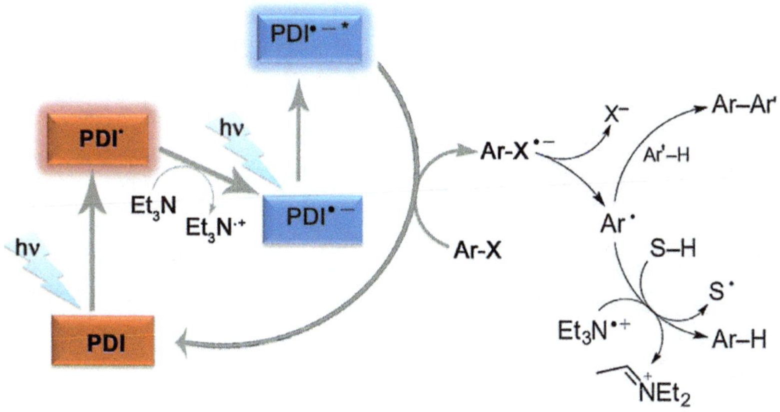

Figure 4.31 Mechanism of the photocatalytic redox process. Reproduced with permission from I. Ghosh, T. Ghosh, J. I. Bardagi and B. König, Reduction of aryl halides by consecutive visible light-induced ET processes, *Science*, 2014, **346**, 725–728. Copyright 2014, American Association for the Advancement of Science.

photoreduction of aryl halides, including aryl chlorides, using visible light irradiation (Figure 4.30b).

A total of seventeen products were obtained applying this photocatalytic protocol. Furthermore, the authors extended these PDI radical intermediates for C–C bond-forming arylation reactions (Figure 4.30c). The arylated products were obtained by irradiating the aryl halides in the presence of *N*-methylpyrrole and catalytic amounts of PDI. The tables show only few representative examples. A total of twelve arylated products were reported by following this photocatalytic system. Figure 4.31 shows the detailed mechanism of the photocatalytic redox process.

4.4.4 ET in PDI-based Supramolecular Systems

Applying a metallo-supramolecular coordination strategy, Würthner *et al.* reported a PDI derivative having pyridyl groups as axial substituents and four ferrocenyl moieties linked through the bay positions (PDI-35), which assembles in the presence of Pt(ɪɪ) to form a molecular square box (PDI-Sq-1) containing 20 redox-active units (Figure 4.32).[129]

CV studies of (PDI-Sq-1) revealed two reversible four-electron reductive processes due to the inner scaffold of the four PDIs, which are negligibly influenced by this supramolecular arrangement. On the other hand, it was found that the redox properties of the 16 peripheral ferrocenyl subunits are affected by the steric constraints imposed by the core-superstructure. In addition, chemical oxidation of ferrocene-functionalized PDI-box (PDI-Sq-1) could be realized and the redox process could be conveniently monitored by spectrophotometry. The study showed that metal-directed self-assembly of square structures could organize multiple redox-active functional units in

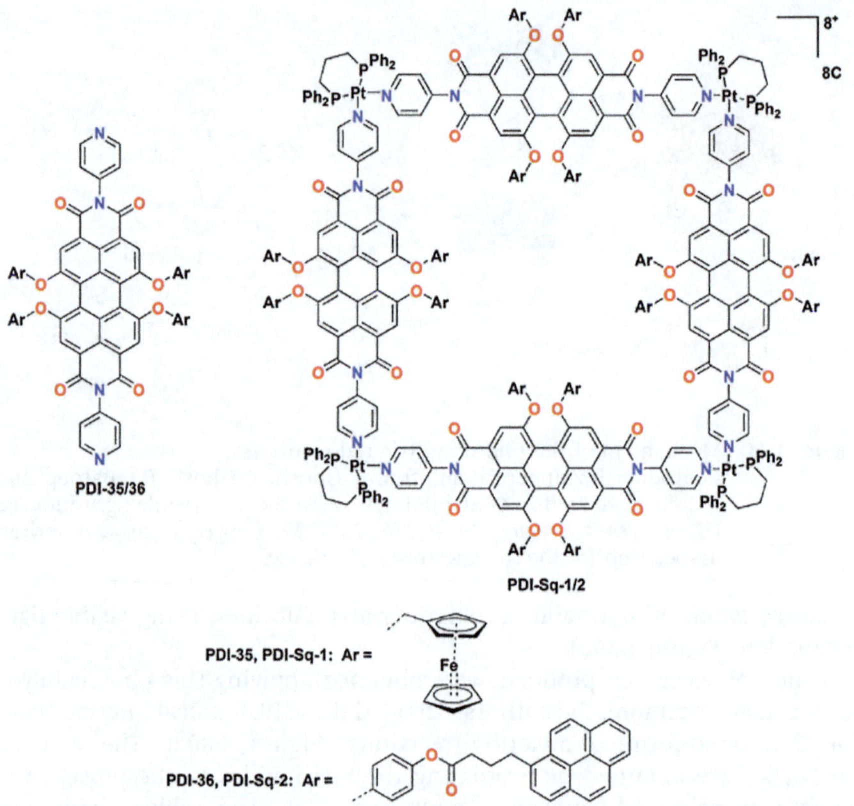

Figure 4.32 Molecular structure of PDI-35 and PDI-36 and the PDI-square box, PDI-Sq-1 and PDI-Sq-2.
Reprinted with permission from C.-C. You and F. Würthner, Self-Assembly of Ferrocene-Functionalized Perylene Bisimide Bridging Ligands with Pt(II) Corner to Electrochemically Active Molecular Squares, *J. Am. Chem. Soc.*, 2003, **125**, 9716–9725. Copyright 2003 American Chemical Society and A. Sautter, B. K. Kaletasü, D. G. Schmid, R. Dobrawa, M. Zimine, G. Jung, I. H. M. van Stokkum, L. De Cola, R. M. Williams and F. Würthner, Ultrafast Energy-Electron Transfer Cascade in a Multichromophoric Light-Harvesting Molecular Square, *J. Am. Chem. Soc.*, 2005, **127**, 6719–6729, Copyright 2005 American Chemical Society.

three-dimensional space and ET processes can be controlled in such a supramolecular architecture.

Subsequent to this report, Würthner *et al.* reported the synthesis of a molecular square box (PDI-Sq-2) incorporating sixteen pyrene chromophores attached to four ditopic bay-functionalized PDI units (PDI-36) and coordinated to four Pt(II)-phosphine units (Figure 4.32).[130] A highly efficient (>90%) and fast photoinduced energy transfer ($k_{En} \sim 5.0 \times 10^9$ s^{-1}) from the pyrene to the PDI chromophores and a very fast and efficient ET (>94%, $k_{ET} \sim 5 \times 10^{11}$ up to 43×10^{11} s^{-1}) was revealed from the steady-state and time-resolved emission as well as femtosecond transient absorption studies. The extremely

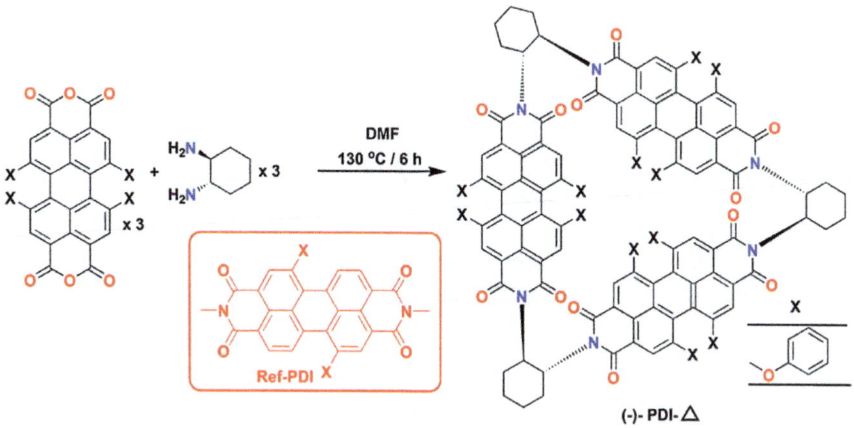

Figure 4.33 Molecular structure of the PDI triangle, (−)-PDI-Δ. Reprinted with permission from Y. Wu, R. M. Young, M. Frasconi, S. T. Schneebeli, P. Spenst, D. M. Gardner, K. E. Brown, F. Würthner, J. F. Stoddart and M. R. Wasielewski, Ultrafast Photoinduced Symmetry-Breaking Charge Separation and Electron Sharing in Perylenediimide Molecular Triangles, *J. Am. Chem. Soc.*, 2015, **137**, 13236–13239. Copyright 2015 American Chemical Society.

fast ET process, which was significantly faster and efficient than the free ligand, is generally observed only in solid materials. Therefore, this molecular square can be considered as a monodisperse nanoaggregate of oriented donor–acceptor chromophores that partly behaves like a solid material.

Photoexcitation of the PDI units in a constrained macrocycle offers a great prospect to understand the CS states and their rate of recombination by transient spectroscopic techniques. In this context, Wasielewski *et al.* reported a visible-light-absorbing chiral molecular triangle composed of three covalently linked 1,6,7,12-tetra(phenoxy)perylene-3,4:9,10-bis-(dicarboximide) (PDI) units (Figure 4.33).[131]

The constrained triangular arrangement of the PDI units reduced the π interaction and electronic coupling between the PDIs. As a result, ultrafast symmetry-breaking charge separation (SB-CS) was kinetically favoured over intramolecular excimer formation, which was revealed by femtosecond transient absorption spectroscopy. Photoexcitation of the PDI triangle dissolved in DCM, gave $PDI^{\bullet+}$-$PDI^{\bullet-}$ with a τ_{CS} of ~12 ps and the SB-CS state decays by charge recombination (CR) to the ground state in τ_{CR}~1.1 ns with no evidence of triplet excited state. The authors probed the effect of solvent polarity on SB-CS, and in chloroform the charge separation significantly slows down to τ_{CS} of ~59 ps and the SB-CS lifetime was extended to τ_{CR}~4.3 ns, in line with the Marcus theory. The much slower CR relative to charge separation, even though the former process is quite exothermic, can be explained by the fact that the recombination lies far into the Marcus inverted region. In non-polar toluene, the SB-CS process is switched off because the ion pair state has a higher energy than the excited singlet state.

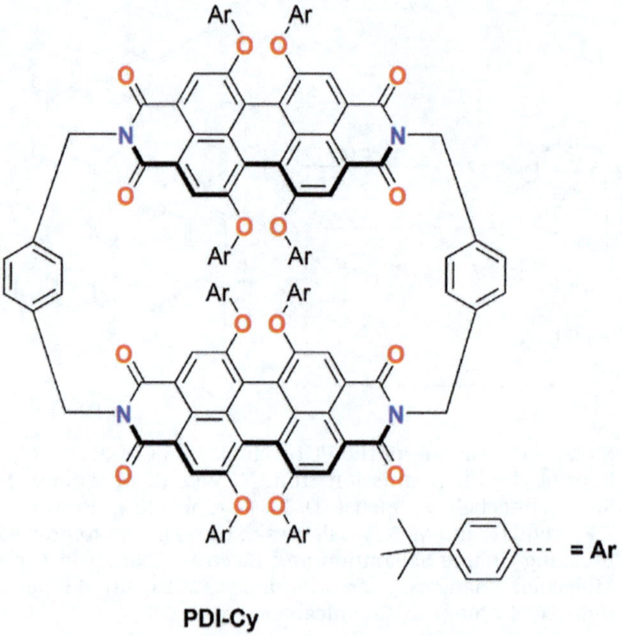

Figure 4.34 Molecular structure of the PDI cyclophane, PDI-Cy. Reproduced from ref. 132 with permission from the Royal Society of Chemistry.

One-electron reduction of the PDI triangle results in complete electron sharing among the three PDIs, as observed earlier with the geometrically constrained NDI units.

Würthner *et al.* reported a PDI-based cyclophane (PDI-Cy), in which the PDI units are cofacially positioned with a distance of 6.5 Å (Figure 4.34).[132]

The rigid cyclophane structure prohibits the common excimer formation and directs photoexcited singlet state relaxation towards intramolecular SB-CS with a τ_{CS} of ~161 ps in DCM, which then recombine slowly in τ_{CR}~8.90 ns to form the PBI triplet excited state. This can be used to generate singlet oxygen in 27% quantum yield. This sequence of events can be switched off by dissolving the PDI-Cy in non-polar toluene, where only excited singlet state decay occurs. In contrast, complexation of electron-rich aromatic hydrocarbons by the PDI-Cy followed by photoexcitation results in ultrafast ET (<10 ps) from the guest to the PDI in DCM. The authors observed that the rate constants for charge separation and recombination increase as the guest molecules become easier to oxidize. This demonstrated that charge separation occurs close to the peak of the Marcus curve, while the recombination lies far into the Marcus inverted region.

Rybtchinski *et al.* demonstrated that in aqueous media (PDI-37) self-assembles into extended supramolecular fibres forming a robust three-dimensional network structure, which results in gelation (Figure 4.35a).[133]

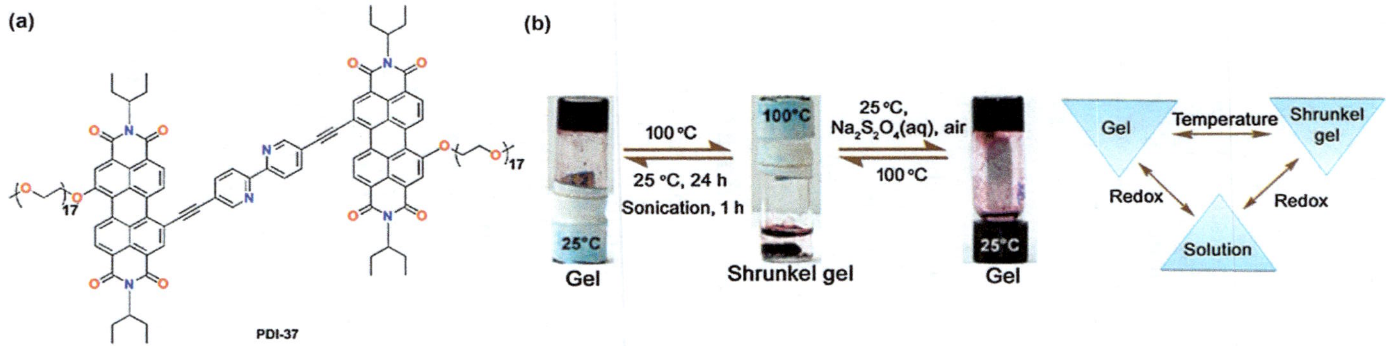

Figure 4.35 (a) Molecular structure of PDI-37, and (b) Shrinkage of the PDI-37 gel in response to high temperature and the reversal to the swollen gel after 24 h, or by sonication, or by addition of one equiv. of $Na_2S_2O_4$ in air. Right scheme: Switching between swollen gel, shrunken gel, and solution in response to temperature and reversible charging.
Reprinted with permission from E. Krieg, E. Shirman, H. Weissman, E. Shimoni, S. G. Wolf, I. Pinkas and B. Rybtchinski, Supramolecular Gel Based on a Perylene Diimide Dye: Multiple Stimuli Responsiveness, Robustness, and Photofunction, *J. Am. Chem. Soc.*, 2009, **131**, 14365–14373. Copyright 2009 American Chemical Society.

This supramolecular gel exhibited multiple stimuli responsiveness. Interestingly, reversible charging/discharging of (PDI-37) allowed for switching between the gel state and fluid solution. After the reduction with $Na_2S_2O_4$ (charging process), the radical anions lead to the destruction of the gel network, probably due to electrostatic repulsion and better solvation of the more polar anion species resulting in fission of the nanofibers. No evidence of decomposition was found after the reduction/oxidation cycle, which corroborated that the charging/discharging sequence to be fully reversible. In addition, high temperature (100 °C) triggered shrinkage of the gel leading to the expulsion of a large fraction of the aqueous solvent, which reversibly returns back to the gel state while standing at room temperature for 24 hours or sonication for one hour (Figure 4.35b). Further, addition of one equivalent of aqueous $Na_2S_2O_4$ at room temperature in air to the solution of the shrunken gel creates a fluid solution of reduced PDI-37, followed by its gelation within several minutes due to oxidation by atmospheric oxygen. This multiple sensibility towards chemical reduction, temperature and sound with interrelated response to each of these stimuli allowed adaptive behaviour and property switching. In addition, the authors demonstrated excellent light absorption and exciton mobility characteristics of the gel, rendering it advantageous for light harvesting applications.

4.5 Introduction to Corannulenes

In the last two decades bowl-shaped polyaromatic hydrocarbons have been investigated, which unearthed their unique coordination and redox properties. These systems have been found to readily accept multiple electrons in stepwise reduction reactions to form reduced non-planar polyaromatic carbanions. In addition, the multi-electron reduction processes of curved carbon-rich compounds attract attention due to their thermoelectric, superconductive properties and their performance in Li-ion batteries. The smallest non-planar fragment of the C_{60}-fullerene, corannulene ($C_{20}H_{10}$), has a doubly degenerate LUMO and thus accepts up to four electrons. Figure 4.36 depicts the hub, rim, spoke and flank regions of corannulene. Interestingly, corannulene (Cor) exhibits a high reversible lithium capacity, 602 mAh g^{-1}, which is significantly higher compared to the theoretical capacity of the fully-lithiated planar graphite material (LiC_6, 372 mAh g^{-1}).[134]

With a variety of alkali metals, corannulene readily acquires up to two electrons,[135] while a trianionic state was detected electrochemically.[136] The highest reduction state of corannulene, $C_{20}H_{10}{}^{4-}$ (Cor^{4-}) has been clearly identified in reduction with Li metal, where the stabilization of the tetraanion was achieved through formation of supramolecular aggregates with lithium ions. The tetrareduced corannulene Li_5 sandwich complex is shown in Figure 4.36. The supramolecular aggregation of corannulene anions has been investigated by the groups of Scott and Rabinovitz using solution NMR spectroscopy.[137,138] Stable ten-component sandwich compounds were characterized in which four lithium ions reside between two Cor^{4-} derived from corannulene or its

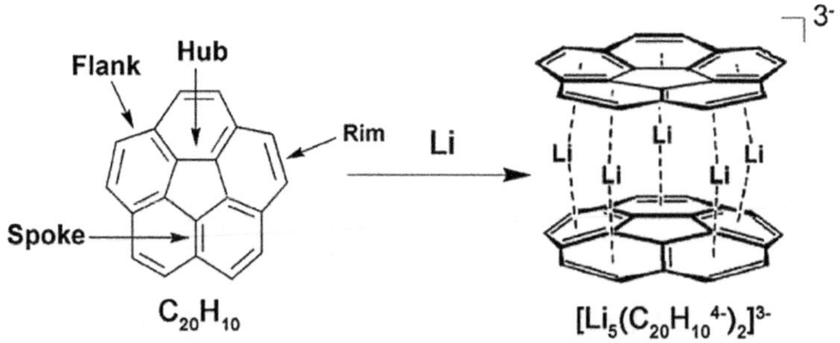

Figure 4.36 Molecular structure of corannulene, its different regions and its tetra-reduction in the presence of Li.
Reproduced with permission from A. S. Filatov, A. V. Zabula, S. N. Spisak, A. Y. Rogachev and M. A. Petrukhina, Clamshell Opening in the Mixed-Metal Supramolecular Aggregates Formed by Fourfold Reduced Corannulene for Maximizing Intercalated Metal Content, *Angew. Chem., Int. Ed.*, 2014, **53**, 140–145. Copyright 2014 from the Wiley-VCH Verlag GmbH & Co. KGaA.

alkyl-substituted derivatives and four additional lithium ions decorate the exterior. It was demonstrated that in tetrahydrofuran solution, the four lithium ions inside the sandwich can exchange environments with the four external lithium atoms, but the two tetraanion decks of the sandwich do not separate on the time scale of nuclear magnetic resonance. Therefore, although detailed characterization of the corannulene anions and their supramolecular aggregates was possible in solution, however, structural identification of these tetraanions in the solid state remained unavailable until 2011.

4.5.1 ET and Solid-state Characterization of Anionic Corannulenes

One of the significant aspects of the reduction of corannulene is that the Cor^{4-} bears one electron per five carbon atoms and therefore is substantially more electron-rich than the fullerene-hexaanion ($C_{60}{}^{6-}$), which bears one electron per ten carbon atoms in $C_{60}{}^{6-}$. However, the extreme sensitivity of Cor^{4-} towards air and moisture made it a real challenge to isolate these tetranions and obtain the solid state structural information.

In a major development, recently, Petrukhina *et al.* successfully crystallized the Li salt of Cor^{4-} under strictly anaerobic conditions.[139] X-ray structural analysis confirmed the sandwich-type assembly for Cor^{4-}, as expected from the earlier NMR studies, and the product was formulated as $Li(THF)_4]^+\{[Li(THF)_2]//[Li_5(Cor)_2]//[Li(THF)_3]\}^-$ (Figure 4.37).

The structure comprised of five Li ions sandwiched between two anionic corannulene decks, two more Li^+ ions were bound to the external surface of the sandwich, while the remaining Li^+ ion was found to be solvent (THF)-separated (Figure 4.37a–b). It is noteworthy that the sandwich was found to

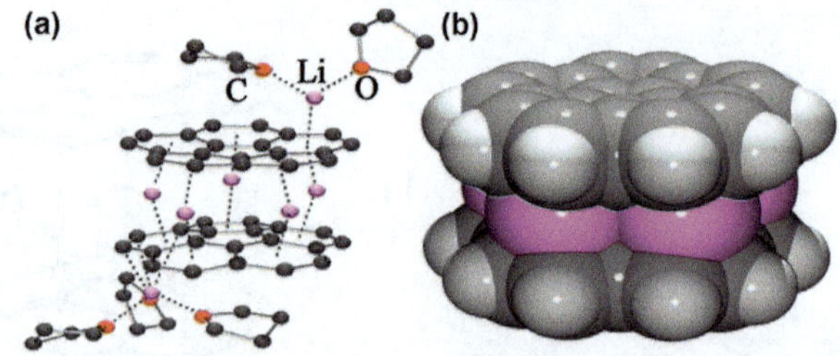

Figure 4.37 (a) Single crystal structure of the sandwich in monoanionic form and $[Li(THF)_4]^+$ as the counterion, and (b) Space-filling model show top view of the sandwich complex. Colour Code: C, dark grey; O, red; Li, purple-pink.
Reproduced from A. V. Zabula, A. S. Filatov, S. N. Spisak, A. Y. Rogachev and M. A. Petrukhina, A Main Group Metal Sandwich: Five Lithium Cations Jammed Between Two Corannulene Tetraanion Decks, *Science*, 2011, **333**, 1008–1011. Copyright 2011. Reprinted with permission from AAAS.

be robust in solution and solid states even in the presence of crown ethers, which competed for the Li^+ ions.

Subsequently, Petrukhina *et al.* reported a new class of supramolecular mixed-alkali-metal cluster cores, LiK_5 and Li_3K_3, which are sandwiched between two Cor^{4-} decks.[140] The triple-decker supramolecular anions, $[(Cor^{4-})\text{-}(LiK_5)^{6+}(Cor^{4-})]^{2-}$ and $[(Cor^{4-})(Li_3K_3)^{6+}(Cor^{4-})]^{2-}$, illustrated the charged corannulene to provide all its sites for binding of six alkali metal ions. Figure 4.38 shows the single crystal structure of mixed-metal sandwich complex $[(Cor^{4-})\text{-}(LiK_5)^{6+}(Cor^{4-})]^{2-}$.

This particular binding mode which utilizes five benzene rings and the central five-membered ring (hub-site) for the binding process is a record in itself. This newly-recognized engagement of the hub-site of Cor^{4-} in Li binding was accompanied by an extraordinary shift of up to −24 ppm in ^{7}Li NMR spectrum. Theoretical calculations of the mixed Li/K sandwiches with variable core compositions were carried out, starting from the Li_5 cluster. Calculations revealed that substitution of small Li^+ ions by K^+ ions along the periphery gradually leads to the sandwich opening, simulating the opening of a clamshell. This clamshell-like opening was quantified by measuring the angles between the mean planes passing through the rim or hub carbon atoms of Cor^{4-} within the sandwich.

Extending this work further, this group recently reported the isolation of triple-decker organometallic aggregates having a mixed-alkali-metal Li-Rb-based core, for example, Li_3Rb_2, Li_3Rb_3, and $LiRb_5$ sandwiched between two Cor^{4-} decks.[141] In these sandwiches the central cavity located in between the hub-sites of two Cor^{4-} decks is occupied by an internal Li^+ ion that exhibits one of the highest negative shifts, of up to −25 ppm in ^{7}Li NMR spectrum. Theoretical calculations predicted that the coupling of the two Cor^{4-} bowls results in a shared region of high negative electron density around the central Li^+ ion, which is responsible for this record shielding effect.

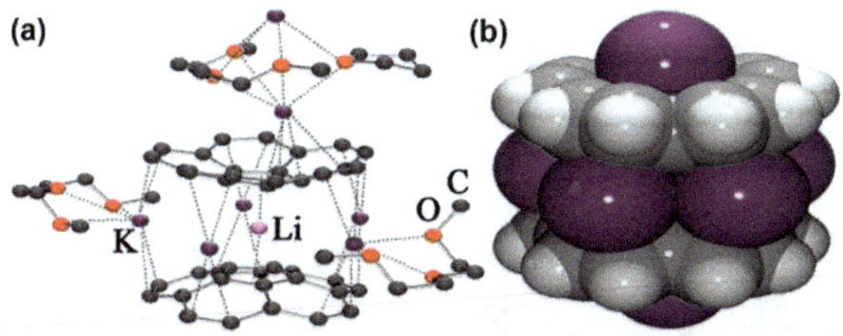

Figure 4.38 (a) Single crystal structure of mixed-metal sandwich complex $[(Cor^{4-})(LiK_5)^{6+}(Cor^{4-})]^{2-}$, and (b) Space-filling model show top view of the sandwich complex. Colour Code: C, dark grey; O, red; Li, purple-pink; K, Purple.
Reproduced with permission from A. S. Filatov, A. V. Zabula, S. N. Spisak, A. Y. Rogachev and M. A. Petrukhina, Clamshell Opening in the Mixed-Metal Supramolecular Aggregates Formed by Fourfold Reduced Corannulene for Maximizing Intercalated Metal Content, *Angew. Chem., Int. Ed.*, 2014, **53**, 140–145. Copyright 2014 from the Wiley-VCH Verlag GmbH & Co. KGaA.

4.6 Introduction to Bipyridinium Systems

The bis-quaternary salt of the *N*,*N*′2-dimethyl-4,4′-bipyridinium dication, commonly referred to as methylviologen (MV^{2+}), has been known as an herbicide and acts as an important electron mediator in water-splitting systems as well as in photochemical, radiolytic and electrochemical applications. Its strong binding to supramolecular hosts like cucurbiturils and to some extent cyclodextrins is also well documented. The reduction of MV^{2+} was discovered by Michaelis way-back in 1932 and led to its realization as the prototypical ET reagent.[142,143] The key to its usefulness as a ET mediator is its facile one-electron reduction to its cation radical ($MV^{\bullet+}$) in aqueous solutions (Figure 4.39a). However the $MV^{\bullet+}$ was found to be extremely sensitive to air and its isolation as a crystalline salt remained elusive for a long period of time. In addition, the reduction of MV^{2+} with sodium dithionite in aqueous solution yields the neutral dihydrobipyridyl derivative (MV^0), corresponding to an overall two-electron reduction. Similar to $MV^{\bullet+}$, MV^0 undergoes ready oxidation upon exposure to air. Although the singly and doubly reduced species were examined through various spectroscopic techniques, in solution, thin films, and zeolites, their structural details remained unavailable for a long time.

4.6.1 ET Reactions of Bipyridinium-based Systems

In a vital finding, Kochi *et al.* in 1990 reported the first isolation and structural characterization of the crystalline forms of $MV^{\bullet+}$ and MV^0.[144] The use of aprotic solvents greatly aided the stabilization and characterization of the reduced species. The structural parameters of MV^0 indicated it to be a nonaromatic, polyene-like dihydrobipyridine with excellent electron donor

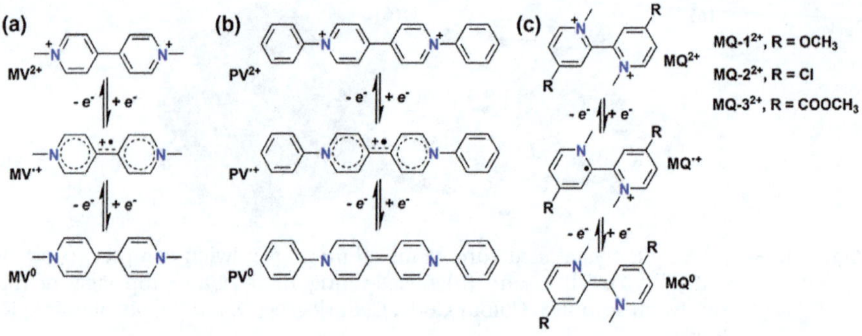

Figure 4.39 Stepwise two-electron reduction reactions of (a) MV^{2+}, (b) PV^{2+} and (c) MQ^{2+}.
(a) Reprinted with permission from T. M. Bockman and J. K. Kochi, Isolation and Oxidation-Reduction of Methylviologen Cation Radicals. Novel Disproportionation in Charge-Transfer Salts by X-ray Crystallography, *J. Org. Chem.*, 1990, **55**, 4127–4135. Copyright 1990 American Chemical Society. (b) Isolation and Characterization of Phenyl Viologen as a Radical Cation and Neutral Molecule, William W. Porter, III and Thomas P. Vaid, *J. Org. Chem.*, 2005, **70**, 5028–5035. Copyright 2005 American Chemical Society. (c) D. Zhang, E. J. Dufek and E. L. Clennan, Syntheses, Characterizations, and Properties of Electronically Perturbed 1,1′-Dimethyl-2,2′-bipyridinium Tetrafluoroborates, *J. Org. Chem.*, 2006, **71**, 315–319. Copyright 2006 American Chemical Society.

properties. The crystalline $MV^{\bullet+}$ as the 1 : 1 PF_6^- salt was found in the form of an infinite stack of two structurally distinct cationic moieties, which were designated as MV-I and MV-II. Structural parameters from X-ray crystallography demonstrated the MV-I component to be related to the oxidized MV^{2+}, and MV-II to the reduced MV^0. The alternate stacking of MV-I and MV-II in $MV^{\bullet+}$. PF_6^- was attributed to an intracrystalline disproportionation originating from strong CT interactions inherent to the $MV^{\bullet+}$.

Subsequently, Vaid *et al.* reported the synthesis of phenyl viologen (PV^{2+}) and importantly isolation and characterization of its three redox-active forms, namely the dication (PV^{2+}), the radical cation ($PV^{\bullet+}$), and the neutral species (PV^0) (Figure 4.39b).[145] Interestingly, X-ray crystallography revealed a structural dichotomy between the PV^{2+} and $PV^{\bullet+}$ twisted *versus* coplanar central rings, which was not clearly illustrated previously in other viologen systems. The $PV^{\bullet+}$ exhibited characteristics similar to other related radical cations, including π-dimerization in the solid state and reversible π-dimerization in the solution. One of the most significant as well as unusual finding of this work was the demonstration of the electrical conductivity of PV^0 in the solid state, which was three orders of magnitude larger than that of the $PV^{\bullet+}$. The plausible explanation provided by the authors was based on the fact that during hole-transfer, there is a significant geometric change between the radical cation and dication, while the neutral molecule and radical cation have similar structures.

In 2005, Clennan *et al.* extended this area of study and reported the synthesis of new 2,2′-bipyridinium salts with electronically communicating substituents directly attached to the pyridinium core.[146] Importantly, 2,2′-bipyridinium ions with the carbomethoxy substituents ($MQ\text{-}3^{2+}$) at the 4,4′-positions showed two reversible one-electron reductions to $MQ\text{-}3^{\bullet +}$ and $MQ\text{-}3^{0}$ reminiscent of the viologens, while with methoxy ($MQ\text{-}1^{2+}$) or halogen ($MQ\text{-}2^{2+}$) substituents such well-defined reversible reduction was not achieved (Figure 4.39c). The carbomethoxy groups possibly provide sufficient resonance interaction to produce a radical cation ($MQ\text{-}3^{\bullet +}$) stable on the cyclic voltammetry timescale. Optical studies of these new compounds resulted in the first example of triplet–triplet absorption spectra. However, their absorption maxima are substantially blue-shifted compared to the more widely used pyrylium salts, decreasing their utility as ET sensitizers.

4.6.2 Supramolecular Stabilization of the MV Radical Cation

The designing and function of the new generation of the advanced molecular devices has been largely possible due to the understanding of the principles governing the interplay between redox events and molecular recognition in supramolecular host–guest systems. In this regard, a number of host–guest systems involving redox-active guests and different macrocycles have been studied. The binding and voltammetric behaviour of host–guest systems with viologens such as cyclodextrin (CD)-viologen[147] and calixarene-viologen[148] have been studied.

Kim *et al.* studied the inclusion behaviour coupled with the redox chemistry of MV^{2+} with cucurbit[7]uril (CB[7]) and found that CB[7] prefers the MV^{2+} charged species and the cation radical ($MV^{\bullet +}$) driven by the ion–dipole interactions, compared to the fully reduced neutral (MV^{0}) species as guests.[149] Furthermore, dimerization of $MV^{\bullet +}$ was suppressed by forming a stable complex with CB[7] in aqueous solution. Interestingly, the redox process ($MV^{2+}/MV^{\bullet +}$) of the MV^{2+}-CB[7] complex occurred predominantly *via* a direct ET pathway, while the second redox process ($MV^{+\bullet}/MV^{0}$) ensued *via* both the direct and indirect pathway because of the low affinity of the fully reduced species MV^{0} to CB[7] (Figure 4.40).

Subsequently, Kim *et al.* reported that CB[8] forms a 1:1 host–guest complex with MV^{2+} and was confirmed by ^{1}H NMR and mass spectrometry, even though CB[8] offered a large cavity to accommodate two MV^{2+} molecules.[150] The exclusive formation of the 1:1 complex was also confirmed by UV-visible spectroscopy. The formation constant for the 1:1 complex of the MV^{2+}-CB[8] was found to be $\sim 1\times 10^{5}$ M^{-1} at 25 °C in water. In a very interesting finding, Kim's group later on reported that the binding stoichiometry of this host–guest complex could be effectively controlled by the redox chemistry of the MV^{2+} guest.[151] The 1:1 MV^{2+}-CB[8] inclusion complex in CB[8] converted completely and reversibly to a 2:1 inclusion complex of the cation radical ($MV^{+\bullet}$) in CB[8] upon the reduction of the guest.

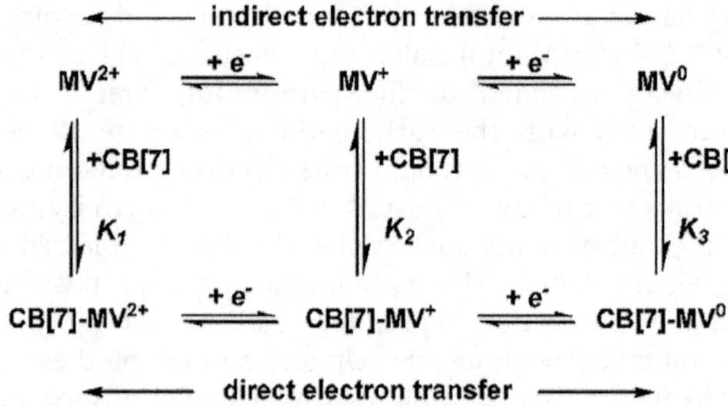

Figure 4.40 A schematic diagram depicting the indirect and direct ET of MV^{2+}/MV^{+} without CB[7] and with CB[7], respectively.
Reproduced with permission from H.-J. Kim, W. S. Jeon, Y. H. Ko and K. Kim, Inclusion of methylviologen in cucurbit[7]uril, *Proc. Natl. Acad. Sci.*, 2002, **99**, 5007–5011. Copyright 2002 from the PNAS, USA.

4.6.3 Multi-ET Reactions/Electron Reservoirs

Inorganic materials, especially the metal oxides, for example MnO_2, NiO, IrO_2, RuO_2, *etc.* are capable of multiple electronic charge storage. Their electronic band shape, lattice defect structure, and mixed-valence state formation provides means to support mass charge accumulation.[152] However, the major problem often encountered in these systems is their slow rate of charging. In comparison to the inorganic materials, multiple electronic charge storage within a single organic molecule and at a reasonable thermodynamic potential becomes an arduous task to achieve. The advantage in organic molecules, however, is the generally observed fast rate of charging/discharging.[153] The most well-known organic system capable of storing multiple electrons is that of the seminal C_{60} molecule, which stores up to six electrons in solution within a potential window of -3.5 V (*vs.* Fc^{+}/Fc).[154] A few other organic compounds for multiple electron storage have been synthesized, which include the dithiines,[155] and polymeric structures and surfaces grafted with electroactive subunits.[156] In this context, it has recently been realized that viologens possess an attractive π-conjugated scaffold and appealing electronic properties, ideal for studies related to multi-electron reservoirs.

In a recent report, Harrington *et al.* reported the synthesis of a disulfide-strapped viologen derivative (DS-1) capable of storing multiple electrons (Figure 4.41a).[157] Incorporation of the disulfur-bridge results into the formation viologen dication, which accepts up to four electrons at a potential of less than -1.6V *vs.* Ag/AgCl. Furthermore, it was shown that the addition of the first two electrons is extremely facile, making DS-1 one of the easiest-to-reduce viologen derivative, to date. The authors found that to alleviate electrostatic repulsion between the two thiolate ions the molecular system twists or "springs open" to accommodate the final two electrons. This unique redox property of the disulfide-strapped viologen and its multiple charge accumulation should open up opportunities for applications in

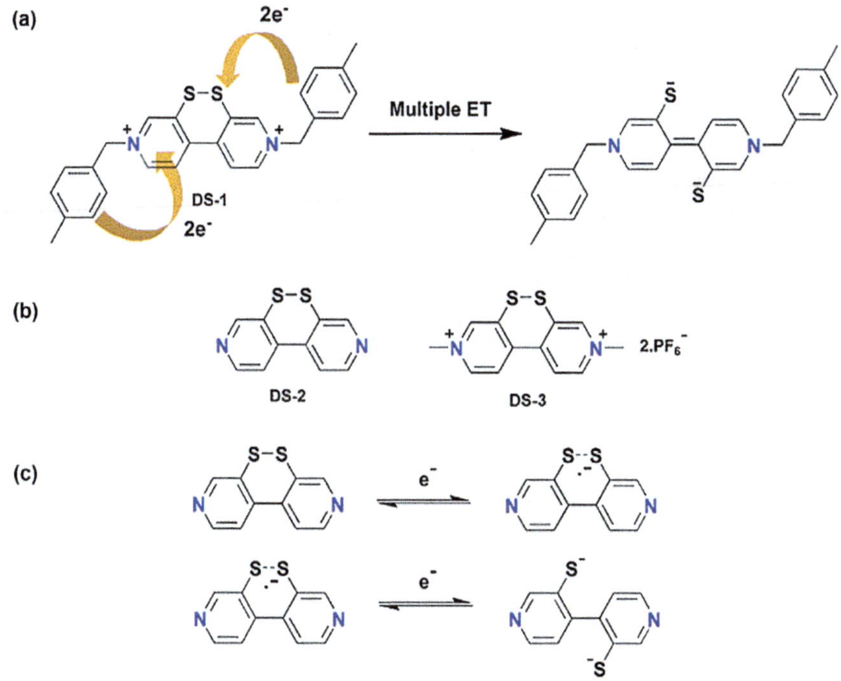

Figure 4.41 (a) Molecular structure of DS-1 and its multiple-ET reactions, (b) Structure of DS-2 and DS-3, and, (c) Mechanism of S–S bond cleavage through a two-electron reduction process.
(a) Reprinted with permission from A. C. Benniston, J. Hagon, X. He, S. Yang and R. W. Harrington, Spring Open Two-plus-Two Electron Storage in a Disulfide-Strapped Methyl Viologen Derivative, *Org. Lett.*, 2012, **14**, 506–509. Copyright 2012 American Chemical Society. (b) and (c) Reprinted with permission from G. B. Hall, R. Kottani, G. A. N. Felton, T. Yamamoto, D. H. Evans, R. S. Glass and D. L. Lichtenberger, Intramolecular Electron Transfer in Bipyridinium Disulfides, *J. Am. Chem. Soc.*, 2014, **136**, 4012–4018. Copyright 2014 American Chemical Society.

artificial photosynthesis, molecular memory devices and electron reservoir-based materials.

To gain further insight on whether the S–S bond cleavage occurs stepwise or in a concerted fashion upon ET, Lichtenberger *et al.* reported the synthesis and detailed electrochemical and computational studies of 4,4′-bipyridyl-3,3′-disulfide (DS-2) and the di-*N*-methylated derivative (DS-3) (Figure 4.41b).[158]

DS-2 was found to undergo reduction directly at the disulfide moiety with a substantially more negative potential for the first electron than for the second electron, resulting in an overall two-electron reduction and rapid cleavage of the S–S bond (Figure 4.41c). In sharp contrast, DS-3 was reduced at less negative potential than DS-2 and at the dimethyl bipyridinium moiety rather than at the disulfide moiety. Importantly, the second reduction of the bipyridinium moiety resulted in a fast and reversible intramolecular two-ET to reduce the S–S bond and form the dithiolate. Therefore, it was concluded that the redox-active bipyridinium moiety provides a low energy pathway for reductive cleavage of the S–S bond, avoiding the highly negative potential for the first direct electron reduction. Subsequent to the intramolecular two-ET and cleavage of the S–S bond the bipyridinium moiety undergoes two additional reversible reductions at more negative potentials.

The cyclo-paraquat-*p*-phenylene ring systems provide ideal architecture to study multiple-ET reactions. Stoddart *et al.*, in a pioneering work, reported the intricate multi-step synthesis of a homo[2]catenane comprising of two rigid and fixed cyclobis (paraquat-*p*-phenylene) rings capable of storing of up to eight electrons (Figure 4.42a).[159]

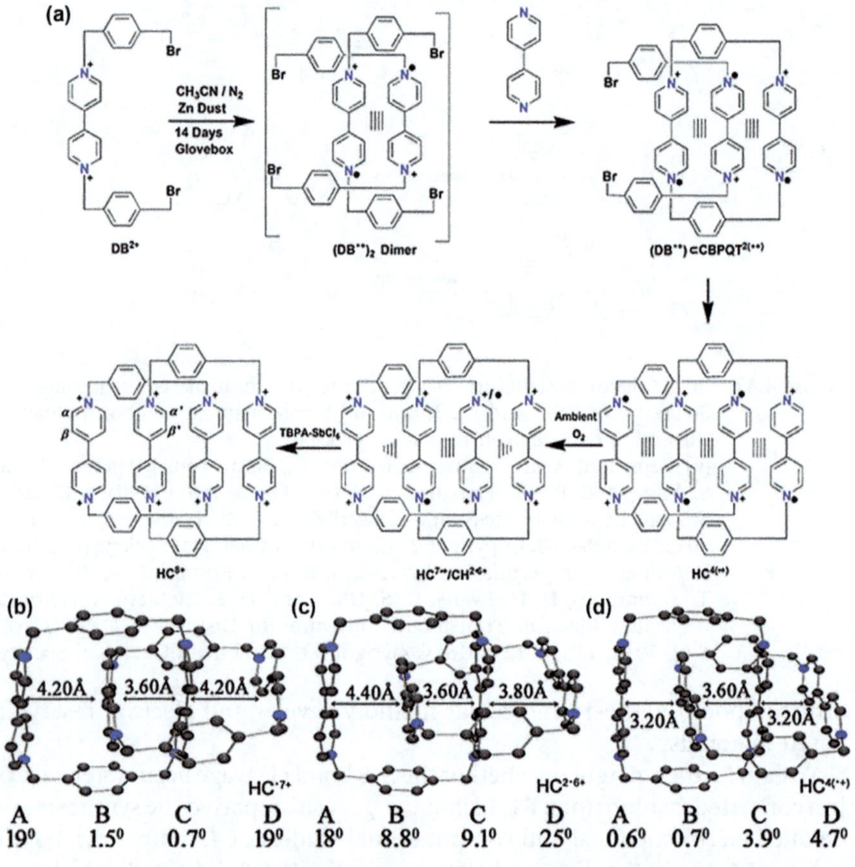

Figure 4.42 (a) Stepwise synthesis of HC^{8+}, (b) Single crystal structure of $HC^{\bullet 7+}$, (c) Single crystal structure of $HC^{2.6+}$and, (d) Single crystal structure of $HC^{4(\bullet+)}$. The dihedral angles between the pyridinium rings (A, B, C and D) are listed below the crystal structures. Colour Code: C, dark grey; N, blue. Reproduced from J. C. Barnes, A. C. Fahrenbach, D. Cao, S. M. Dyar, M. Frasconi, M. A. Giesener, D. Benítez, E. Tkatchouk, O. Chernyashevskyy, W. H. Shin, H. Li, S. Sampath, C. L. Stern, A. A. Sarjeant, K. J. Hartlieb, Z. Liu, R. Carmieli, Y. Y. Botros, J. W. Choi, A. M. Z. Slawin, J. B. Ketterson, M. R. Wasielewski, W. A. Goddard III and J. F. Stoddart, A Radically Configurable Six-State Compound, *Science*, 2013, **339**, 429–433. Copyright 2013. Reprinted with permission from AAAS.

The octacationic homo[2]catenane could be configured reversibly, both chemically and electrochemically, between each one of the six experimentally accessible redox states *i.e.* 0, 2^+, 4^+, 6^+, 7^+, and 8^+ from within the total of nine states evaluated by quantum mechanical methods. All six of the observable redox states were identified by electrochemical techniques and three states, namely 4^+, 6^+, and 7^+, were characterized by X-ray crystallography (Figure 4.42b). Most significantly, this novel synthesis by radical templation generated a class of air- and water-stable organic radicals, trapped within the homo[2]catenane.

Stoddart *et al.* in a recent report described the synthesis of a series of homologous oligoviologens ($2MV^{2+}$, $3MV^{2+}$, $4MV^{2+}$, $5MV^{2+}$ and $12MV^{2+}$) (Figure 4.43a), in which different numbers of MV^{2+} subunits are linked by *p*-xylylene bridges, to investigate how their radical cationic forms self-assemble both in solution and in the solid state.[160]

Interestingly, the strong radical–radical interactions between the radical cationic forms in these oligoviologens induce intra- or intermolecular folding of these homologues. UV/Vis/NIR spectroscopic studies and DFT quantum mechanics indicate that the folding of the shorter oligoviologens is dominated by intermolecular radical–radical interactions. In addition, strong intramolecular radical–radical interactions give rise to an NIR absorption band at 900 nm and play a crucial role in governing the folding of the longer oligoviologens. The single crystal structure of the oligoviologen with three MV^{2+} units revealed that two inter-twining chains fold together to form a dimer, stabilized by intermolecular radical–radical interactions (Figure 4.43b–c). These dimers stack in an infinite column through intermolecular radical–radical interactions between them. This artificial biomimetic system bears intricate secondary and tertiary structures, reminiscent to that of observed in nucleic acids and proteins.

4.7 Applications

4.7.1 Sensors

Chromogenic molecular sensors provide the most inexpensive and rapid route to sense a variety of analytes.[161–165] There has been tremendous progress in the development of smart chromogenic sensors for a diverse range of metal ions and anionic analytes. On the other hand, there have been fewer investigations on sensing of neutral organic analytes. The chromogenic sensors discovered to date, and excluding that of the Au-NP based sensors, generally function through intramolecular charge transfer (ICT) as the primary signal transduction unit (STU). In these sensors, the donor and acceptor unit is separated through a π-conjugated pathway, which acts as the CT conduit, having a recognition unit (for the analyte) either integrated with the donor/acceptor unit or positioned separately. In these sensors, the emanating absorption bands due to such STUs are generally broad and suffer from weak molar extinction coefficients. This significantly limits its

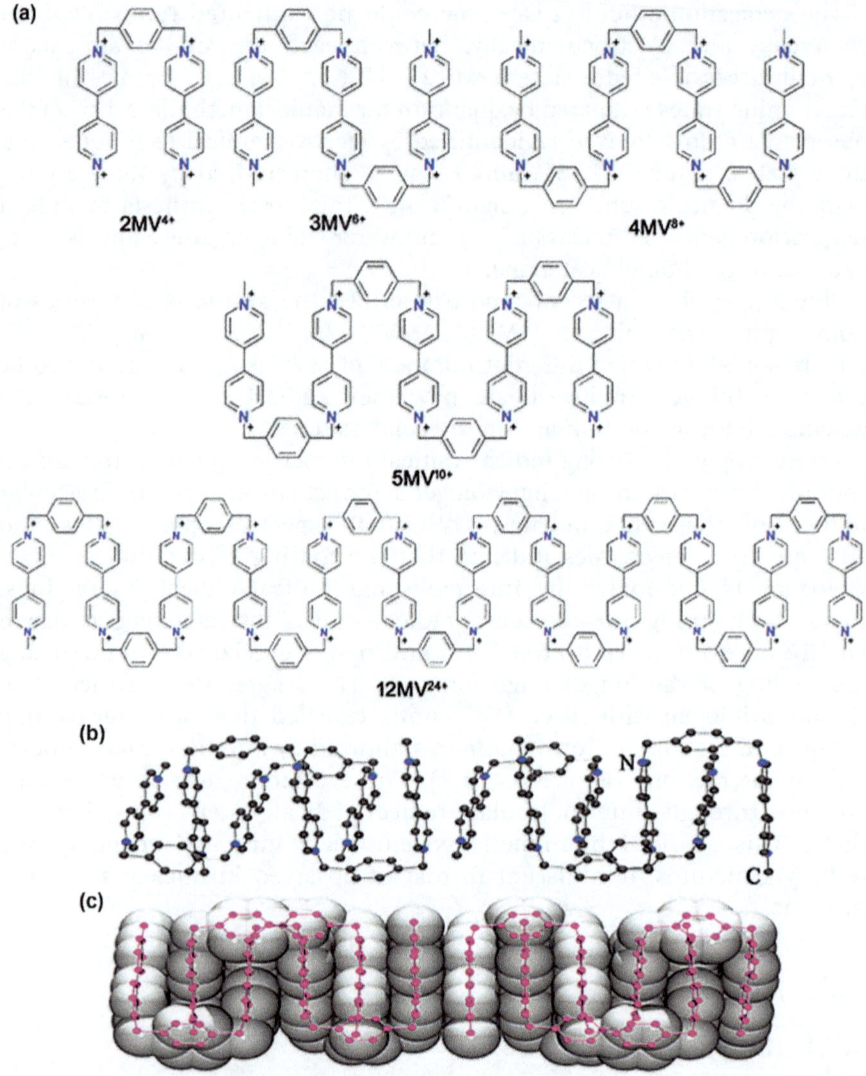

Figure 4.43 (a) Molecular structures of the MV^{2+} oligomers, (b) Single crystal structure of $3MV^{6+}$ and (c) space-filling model of the folded self-assembled dimers of $3MV^{6+}$ with superimposed ball and stick structure of $3MV^{6+}$. Colour Code: C, dark grey; N, blue.
Reprinted with permission from Y. Wang, M. Frasconi, W.-G. Liu, Z. Liu, A. A. Sarjeant, M. S. Nassar, Y. Y. Botros, W. A. Goddard, III and J. F. Stoddart, Folding of Oligoviologens Induced by Radical – Radical Interactions, *J. Am. Chem. Soc.*, 2015, **137**, 876–885. Copyright 2015 American Chemical Society.

applicability to demarcate closely related analytes having small differences in shape, size, isomerism and electronic factors. The other major drawback with such conventional sensors is the interference of their signal read-out

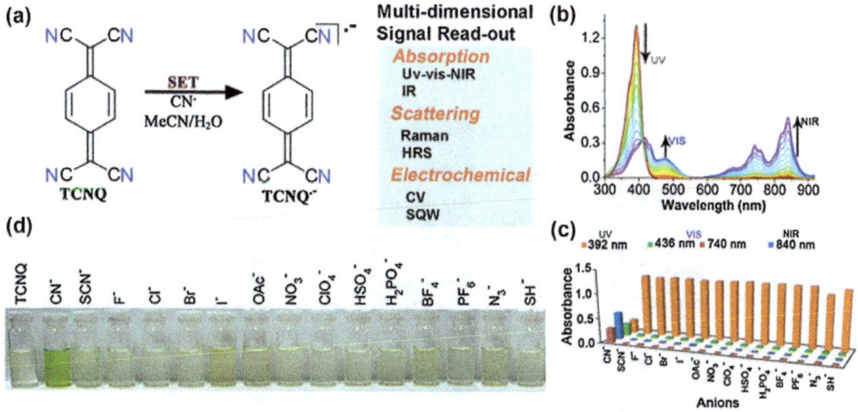

Figure 4.44 (a) Single ET reaction of TCNQ with cyanide anion and the multidimensional signal read-out of the TCNQ radical anion, (b) UV-Vis-NIR absorption spectrum showing the formation of the TCNQ radical anion upon addition of cyanide anion, (c) Selectivity profile of TCNQ with other anions, and (d) Colourimetric response of TCNQ in the presence of different anions.
Reprinted with permission from M. R. Ajayakumar, K. Mandal, K. Rawat, D. Asthana, R. Pandey, A. Sharma, S. Yadav, S. Ghosh and P. Mukhopadhyay, Single Electron Transfer-Driven Multi-Dimensional Signal Read-out Function of TCNQ as an "Off-the-Shelf" Detector for Cyanide, *ACS Appl. Mater. Interfaces*, 2013, **5**, 6996–7000. Copyright 2013 American Chemical Society.

with common biological chromophores. All these necessitate an unconventional approach to sense demanding analytes.

As a proof-of-concept, we recently applied a kinetically selective ET from a cyanide anion to deep-lying LUMO orbitals of TCNQ to generate a persistently stable radical anion ($TCNQ^{\bullet-}$) under ambient conditions.[166] In contrast to the known cyanide sensors that operate with limited signal outputs, $TCNQ^{\bullet-}$ offers a unique multiple signalling platform (Figure 4.44a). The signal readability is facilitated through multichannel absorption in the UV-vis-NIR region and scattering-based Raman spectroscopy and hyper Rayleigh scattering techniques. Particularly notable is the application of the intense 840 nm NIR absorption band (along with the UV and visible bands) to detect cyanide, which can be useful for avoiding background interference in the UV-vis region of biological samples (Figure 4.44b–c).

Figure 4.44d shows the colorimetric response of TCNQ on the addition of cyanide anions and other anions. Furthermore, we demonstrated the fabrication of an electronic device with TCNQ as a detector. The device generates multi-order enhancement in current with cyanide because of the formation of the conductive $TCNQ^{\bullet-}$.

We also found that NDI-4 and PDI-38, having low-lying LUMO levels, can accept electrons selectively from cyanide anions (Figure 4.45a–b). The characteristic signature absorption bands for $NDI^{\bullet-}$ and $PDI^{\bullet-}$ upon ET

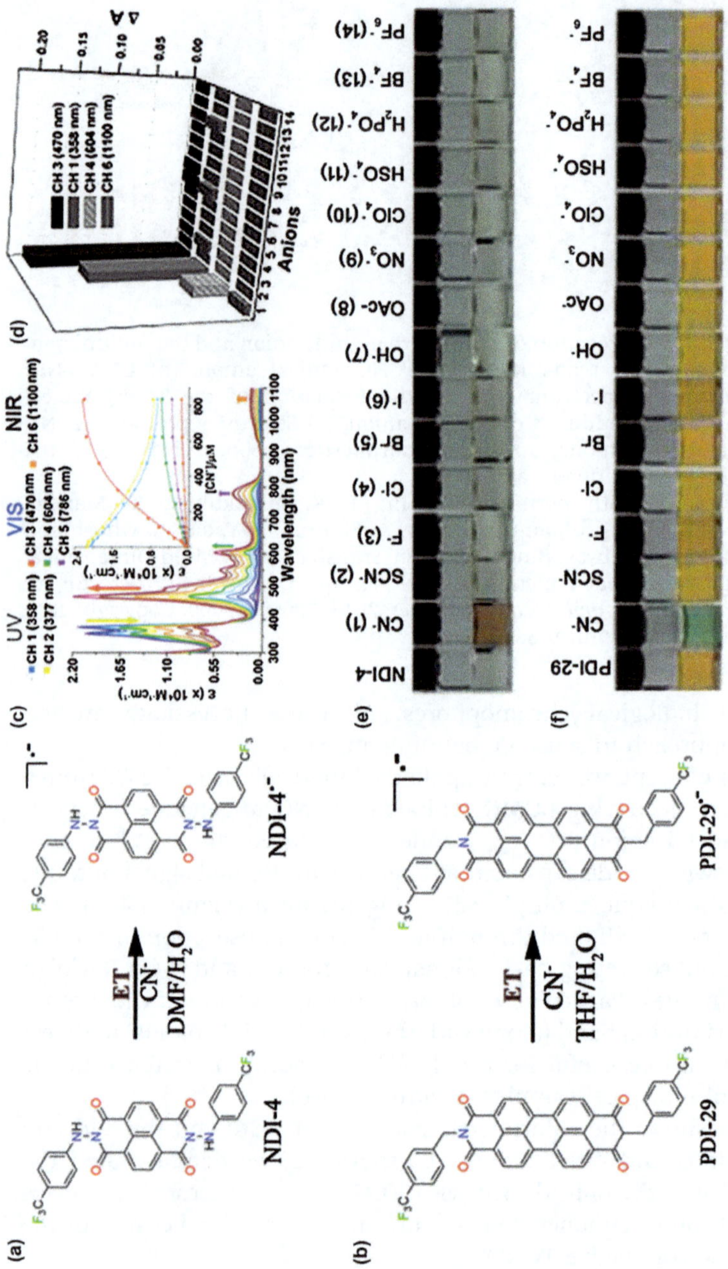

Figure 4.45 (a) and (b) Single ET reaction of NDI-4 and PDI-29 with cyanide anion, (c) UV-Vis-NIR absorption spectrum showing the formation of the NDI-4 radical anion in the presence of cyanide anion, (d) Selectivity profile of NDI-4 with different anions, (e) and (f) Colourimetric response of NDI-4 and PDI-29 with different anions.
Reprinted with permission from M. R. Ajayakumar, S. Yadav, S. Ghosh and Pritam Mukhopadhyay, Single-Electron Transfer Driven Cyanide Sensing: A New Multimodal Approach, *Org. Lett.*, 2010, **12**, 2646–2649. Copyright 2010 American Chemical Society.

from cyanide anions could be observed in a polar solvent medium like DMF, THF, or mixed aqueous solution.[76] The absorption bands in the UV-vis and NIR region (Figure 4.45c) provides multiple channels for independent investigation of the particular anion. This is evident from Figure 4.45d, which shows the selectivity pattern of the different anions. From this study, it became clear that other basic anions like fluoride, acetate and dihydrogenphospahate could also exhibit ET, albeit to a significantly lower extent.

This ET reaction-based system was found to be quite sensitive (0.2–16 μM) to cyanide. The neutral NDI-4/PDI-38 molecules could be regenerated using $NOBF_4$. Importantly, utilizing the conductive nature of these radical anion probes, an electronic device could be fabricated for sensing the cyanide anion.

Subsequent to our report, Saha *et al.* reported the sensing of the fluoride anion by axially pyridine-substituted NDI-1 and pre-organized receptor molecules comprising of two NDIs linked with bisamide linker (NDI-32/33) (Figure 4.46).[78]

With smaller equivalents of the fluoride anion in DMSO, the one-electron reduction process was observed with a concomitant orange coloration and formation of the $NDI^{\bullet-}$. In a significant finding, with a large excess of the fluoride anion, a two-electron reduction process leading to the NDI^{2-} was confirmed by various spectroscopic techniques with the solution turning into pink. From the ESI-MS results, the authors concluded that supramolecular fluoride-NDI (anion–π) interaction drove the ET and the one-electron reduction process. The ET was reported to be selective to fluoride, although the effect of the cyanide anion on the ET was not discussed. Importantly, the pre-organized receptor NDI-32 exhibited nM range sensitivity compared to μM sensitivity in case of NDI-1. In addition, regeneration of the receptors could be achieved utilizing oxidants, showing the promise and usability of an ET-based sensing approach.

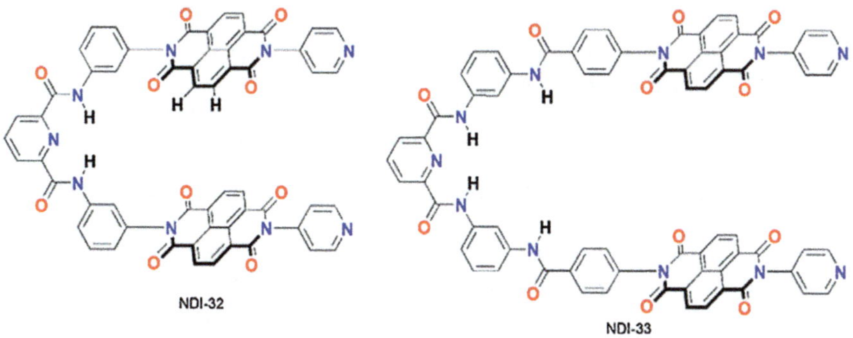

Figure 4.46 Molecular structure of NDI-32 and NDI-33 as acceptors of electrons from TBAF.
Reprinted with permission from Samit Guha and Sourav Saha, Fluoride Ion Sensing by an Anion-π Interaction, *J. Am. Chem. Soc.*, 2010, **132**, 17674–17677. Copyright 2010 American Chemical Society.

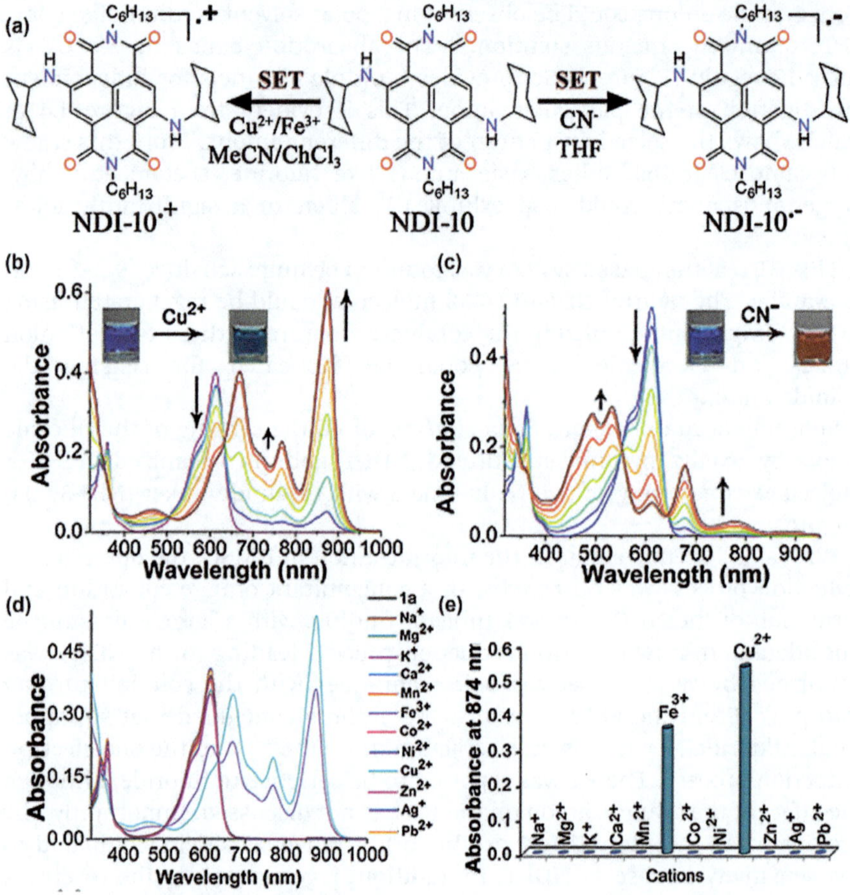

Figure 4.47 (a) Single ET oxidation and reduction of NDI-10 with Cu^{2+}/Fe^{3+} and cyanide anion, respectively, (b) UV-Vis-NIR absorption spectrum showing the formation of NDI-10$^{\bullet+}$, (c) UV-Vis-NIR absorption spectrum showing the formation of NDI-10$^{\bullet-}$, (d) UV-Vis-NIR absorption spectrum of NDI-10 with different cations, and (e) Selectivity profile of NDI-10 with different cations.
Reprinted with permission from M. R. Ajayakumar, D. Asthana and P. Mukhopadhyay, Core-Modified Naphthalenediimides Generate Persistent Radical Anion and Cation: New Panchromatic NIR Probes, *Org. Lett.*, 2012, **14**, 4822–4825. Copyright 2012 American Chemical Society.

Continuing with our interest in sensing of diverse ions through radical ion probes, we reported the formation of a NDI-based radical cation (NDI-10$^{\bullet+}$) as well as NDI-based radical anion (NDI-10$^{\bullet-}$) under ambient conditions (Figure 4.47b). The NDI-10 having its core functionalized with the amino donor groups formed the radical cation in the presence of Cu^{2+} or Fe^{3+} in MeCN, under ambient conditions.[86] The intense deep-blue colour of NDI-10 immediately changed to turquoise blue as a result of formation of the radical cation (Figure 4.47b). Upon gradual addition of Cu^{2+} to NDI-10, new bands appeared at 290, 464, 669, 768, and 874 nm with a concomitant decrease of the 346 and 361 nm (π–π*) bands and the 613 nm ICT band. This radical cation formation was selective to Cu^{2+} and Fe^{3+} (Figure 4.47d–e).

Notably, the intense narrow band at 874 nm offers potential for developing these radical cations as NIR probes. In addition, in the presence of cyanide anions, a red-coloured radical anion formation was observed in THF medium (Figure 4.47c). Therefore, a single donor-substituted NDI molecule can act as panchromatic and multichannel (UV-VIS-NIR) optical probes for sensing of a wide range of cations as well as anions.

From our group, we have also explored the application of NDI-4 radical anions as probes (Figure 4.48a) along with a conventional ICT-based signal transduction process, as a dual-mode of optical sensing of a library of amines (Figure 4.48b).[75] An information-rich colour palette could be generated, which aids in sensing a particular amine within a library of amines having small differences in pK_a, lipophilicity, *etc.* Thus, NDI-4in THF–H_2O (9 : 1) and (1 : 1) with various amines (**1–19**) shows a new set of bands at 470, 609, 690 and 765 nm (Figure 4.48c–e), which matched with the signature peaks of radical anions of the NDI moiety.

Interestingly, the amines **7**, **13** and **20**, which remained difficult to recognize in dry THF as a result of non-colouration (non-operation of the ICT process) could easily be identified in THF–H_2O (9 : 1) and (1 : 1) as a result of the distinct colour changes through the radical anion probe. Similarly, the amines belonging to C-4 isomers **3**, **13** and **7**, C-8 isomers like, **5** and **15** could be differentiated with the aid of this colour palette. Therefore, a straightforward ET process from the amine analytes to an appropriately designed molecule can generate absorption bands with multiple read-out channels in the UV, visible and NIR region. Depending on the efficacy of the ET from the analyte, the intensity of the doublet $D_0 \rightarrow D_n$ transitions varied, providing a simple and easy way to follow a particular amine.

Recently Banerjee *et al.*, in an interesting demonstration, reported the synthesis of a Mg(II) metal organic framework (MOF) involving the NDI unit (Figure 4.49a).[167]

This Mg-MOF exhibited instant and reversible solvatochromic behaviour in presence of various solvents with different polarity (Figure 4.49b). Upon sunlight irradiation, the Mg-MOF showed photochromism, changing its colour from yellow to black in a rapid and reversible manner. The authors concluded the process to be *via* the radical ion formation based on the UV-Vis and EPR spectroscopic results. With different aromatic systems, the Mg-MOF aniline showed a differential response resulting in dark brown coloration, while phenol, nitro-phenol, toluene, *etc.* remained silent. Furthermore, the presence of the electron deficient NDI unit in the Mg-MOF drives selective sensing of organic amines (aniline, hydrazine, ethylene diamine, triethylamine, dimethylamine, 1,3-propanediamine, ethylamine and methylamine) in the solid state (Figure 4.49c). The detection of organic amines was confirmed by fluorescence quenching experiments and visual colour change. The fluorescence quenching was attributed to the ET from

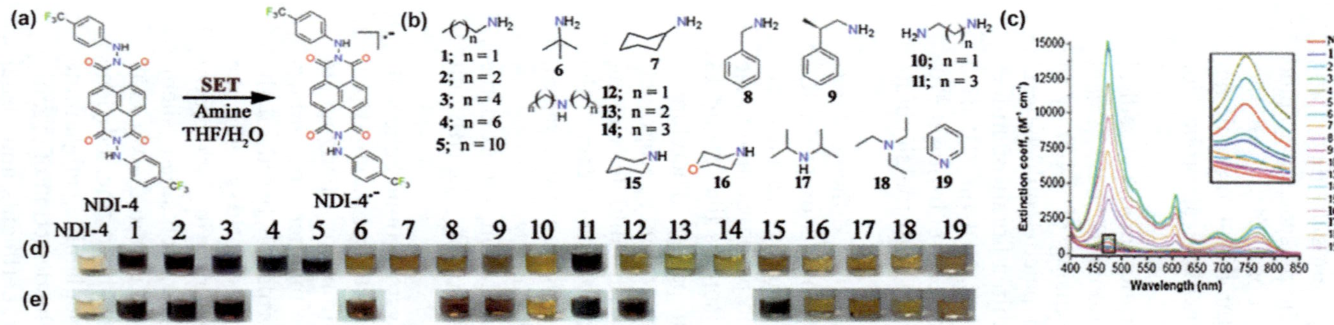

Figure 4.48 (a) Single ET reaction of NDI-4 with organic amines, (b) molecular structures of amines, (c) UV-Vis-NIR absorption spectrum showing the differential response and formation of the NDI-4 radical anion in the presence of amines, and Colourimetric response of NDI-4 with different amines in (d) THF : water (9 : 1) and (e) THF : water (1 : 1).
Reproduced with permission from ref. 75. Copyright 2009 Royal Society of Chemistry.

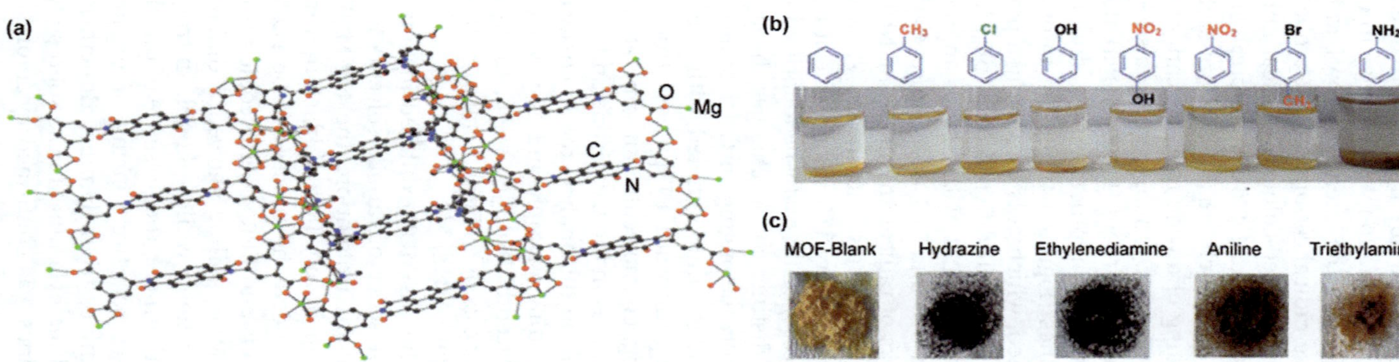

Figure 4.49 (a) Single crystal structure of the Mg-MOF, (b) Colourimetric response of the Mg-MOF in the presence of different aromatic systems, (c) Colourimetric response of the Mg-MOF in the presence of electron-rich systems. Colour Code: C, dark grey; N, blue; O, Red; Mg, Green.
Reproduced with permission from ref. 167. Copyright 2015 Royal Society of Chemistry.

the amines to the electron deficient NDI unit. This was also validated by frontier orbital calculations of the corresponding donor–acceptor units. Therefore, it was shown that suitably oriented NDI units in a MOF can be an effective scaffold to propel ET-driven photochromism and sensing of electron-rich analytes.

4.7.2 Organic Battery

Presently, the portable electronics market is dominated by rechargeable Li-ion batteries. These batteries offer the largest energy density and output voltage compared to other rechargeable battery technologies available to date. Additionally, Li-ion batteries are well-placed to take over the markets of electric vehicles. In the current commercial Li-ion batteries, the electrode materials are largely procured through mining, which makes their procurement energetically demanding and involves significant amount of carbon emissions. Therefore, it is desirable to consider all aspects of the manufacturing life-cycle, for example, scarcity of materials, carbon foot print, pollution, toxicity, *etc.* for designing advanced Li-ion batteries.

From this perspective, organic molecules are attractive candidates for electrode materials as the large expanse of easily available organic synthetic methodologies allows for control over structure and functional groups. Also the organic compounds are potentially renewable as these can be synthesized from natural products and biomass. The initial attempts to incorporate organic molecules into batteries were focused around conjugated polymers. However, electrodes manufactured from conducting polymers suffered from low specific energies, high costs of manufacturing, and experienced batch-to-batch variations in their efficiency.

Wudl *et al.* reported the synthesis and characterization of an organic salt, lithium 2,6-bis(ethoxycarbonyl)-3,7-dioxo-3,7-dihydro-*s*-indacene-1,5-bis(olate) (PMDE 2. Li^+), through a multi-step synthesis starting from the commercially available PMDA (Figure 4.50).[168]

They evaluated the device performance, and found it capable of reversibly intercalating Li with minimal polarization of 1.8 Li per unit formula with an overall capacity of 125 mAh g^{-1}. ^{1}H NMR and *in situ* XRD analyses of the battery cycling *versus* Li at room temperature was carried out to have an insight on the Li insertion–deinsertion process. The studies revealed that the insertion–deinsertion process is fully reversible with dips in the voltage-composition traces, probably associated with changes in the 3-D structural packing of the electrochemically active molecules. These findings demonstrated the opulence of organic chemistry in producing organic electrodes capable of reversibly reacting with Li, and soliciting the ester function as the redox centre.

In an ensuing report, Poizot *et al.* reported that the PMDI-dilithium salt (PMDI 2.Li^+) can also reversibly insert nearly two Li^+ per formula unit leading to a stable capacity value of 200 mAh g^{-1} (Figure 4.50).[169] It was demonstrated that the insertion process occurs *via* a series of steps involving

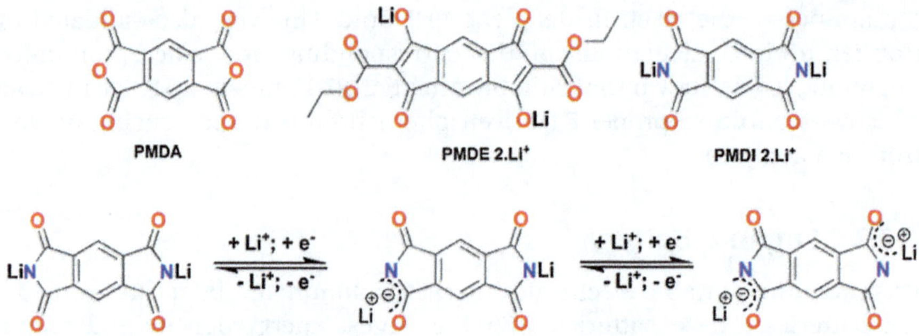

Figure 4.50 Top: Molecular structures of PMDA, PMDE $2.Li^+$, and PMDI $2.Li^+$. Below: Stepwise ET reaction of PMDI $2.Li^+$.
Reprinted with permission from W. Walker, S. Grugeon, O. Mentre, S. Laruelle, J.-M. Tarascon and F. Wudl, Ethoxycarbonyl-Based Organic Electrode for Li-Batteries, *J. Am. Chem. Soc.*, 2010, **132**, 6517–6523. Copyright 2010 American Chemical Society.

two major plateaus. This study again highlighted the potential of using negatively charged organic structures to decrease both the redox potential and the solubility issues in electrolyte.

Dunn *et al.* reported the synthesis of a family of NDI derivatives (NDI-34-41) as electroactive materials for organic Li-ion batteries.[170] The NDI core was chosen with the anticipation that the extended aromatic core of NDI as compared to that of PMDI should serve the dual purpose of decreasing the solubility of the crystalline material while providing opportunities to modify the LUMO levels of the parent hydrocarbon structure. The authors reported synthesis of two sets of structures, with $R = C_6H_{13}$ (NDI-34-37) and without $R = H$ (NDI-38-41) (Figure 4.51a).

The NDI core was also varied, which comprised of a hydrocarbon aromatic core ($X = H$, NDI-34/38), as well as molecules with electron donating $X = NMe_2$ (NDI-35/39) and electron withdrawing $X = F$ (NDI-36/40) and $X = CN$ (NDI-37/41) substituents. Changes in the electronic characteristics of the aryl substituents resulted in materials with discharge potentials that vary from 2.3 (NDI-35) to 2.9 V (NDI-41) *vs.* Li/Li^+. The discharge capacities varied from 34 (NDI-41) to as high as 121 mAh g^{-1} for molecule NDI-38.

In a recent report, Krishnamoorthy *et al.* reported the synthesis of a benzoicacid-PDI (PDI-39), to deliberately induce insolubility in common organic solvents and particularly in the electrolytes used in lithium batteries (Figure 4.51b).[171] Although PDI-39 was found to be insoluble in the electrolyte solution, the lithium battery exhibited poor performance due to slow redox rates of the PDI unit. The authors hypothesized that a chemical reduction to PDI-39$^{\bullet -}$ could precondition the molecules for faster Li-ion association and disassociation and n-type doping may increase the conductivity of these molecules. Indeed, chemical reduction increased the rate of redox process as well as the conductivity of the molecule, which enhanced the overall performance of the Li-ion battery. This PDI-39-based battery

(a)

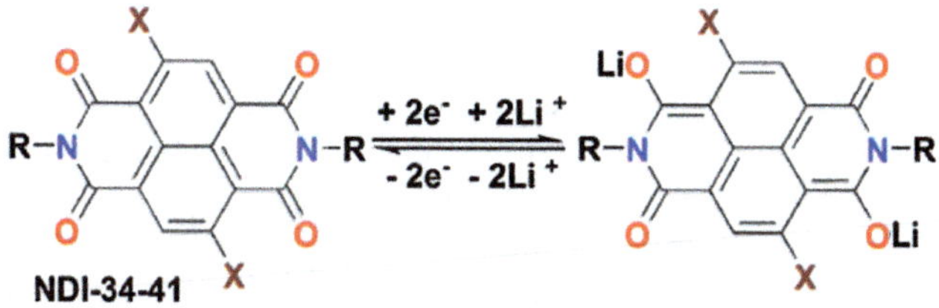

NDI-34-41

Compound	X	R
NDI-34	H	Hexyl
NDI-35	NMe_2	Hexyl
NDI-36	F	Hexyl
NDI-37	CN	Hexyl
NDI-38	H	H
NDI-39	NMe_2	H
NDI-40	F	H
NDI-41	CN	H

(b)

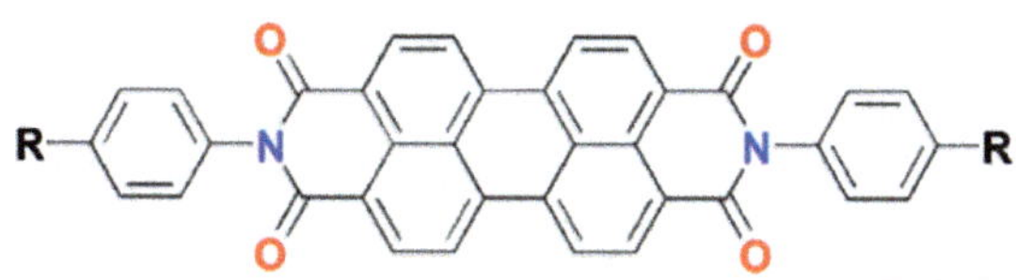

PDI-38; R= COOH
PDI-39; R= H

Figure 4.51 (a) Molecular structures of NDI-34-41 and its ET reaction in presence of Li^+ ions and (b) Structures of PDI-38 and control compound, PDI-39. (a) Reprinted with permission from G. S. Vadehra, R. P. Maloney, M. A. Garcia-Garibay and B. Dunn, Naphthalene Diimide Based Materials with Adjustable Redox Potentials: Evaluation for Organic Lithium-Ion Batteries, *Chem. Mater.*, 2014, **26**, 7151–7157. Copyright 2014 American Chemical Society. (b) Reprinted with permission from M. E. Bhosale and K. Krishnamoorthy, Chemically Reduced Organic Small-Molecule-Based Lithium Battery with Improved Efficiency, *Chem. Mater.*, 2015, **27**, 2121–2126. Copyright 2015 American Chemical Society.

exhibited 100% Coulombic efficiency, while the specific power and energy of the battery was found to be 8548 W kg^{-1} and 213 Wh kg^{-1}, respectively. When the lithium batteries were fabricated with the control compound, *i.e.* the reduced phenyl-PDI-40, the specific capacity was found to rapidly decrease in subsequent cycles and reached 40 mAh g^{-1} in the 50th cycle, which pointed to the dissolution of phenyl-PDI in the electrolyte solution. Therefore, chemical reduction of the electron deficient PDI unit in PDI-39 and the judicious choice of functional groups to drive insolubility, dramatically improved multiple aspects of this Li-ion battery.

4.8 Conclusion

In summary, an effort is made through this chapter to describe some of the most comprehensively investigated π-conjugated scaffolds, which can accept one or more electrons, in recent times. Our intention has been to highlight the new developments in the direction of design and synthesis of ambient stable open-shell, multi-anionic and multi-radical-based systems and their applications. The immense significance and potential of weak intra-molecular and intermolecular noncovalent interactions towards the de-localization of the unpaired electron and stability of open-shell system becomes evident from this chapter. The photo-controlled intramolecular ET reactions from the electron-rich pendant to the electron deficient scaffolds is highlighted, which shows the potential to harness light-driven switchable chromatic systems. The extensive use of molecular and crystal engineering principles is discussed, in the light of having efficient electron transport and high charge mobility in n-type semiconductors. The recent advances in molecular design approaches towards non-fullerene based electron acceptors for organic solar cells are also illustrated. The application of the photoexcited radical ions towards C–C bond formation reactions and aryl halide reduction is elucidated. Furthermore, the applications of the radical ions as probes to sense toxic analytes like cyanide anions and the application of their attractive multichannel absorption properties are explained. The tremendous potential of having porous electroactive materials to sense volatile electron-rich species is elucidated. Finally, we examined the recent progress towards the development of π-conjugated electroactive materials for organic Li-ion batteries.

Abbreviations

Electron Transfer	ET
Charge Transfer	CT
Charge Separation	CS
Cyclic Voltammetry	CV
Square Wave Voltammetry	SQV
Bulk Hetero Junction	BHJ
Open Circuit Voltage	V_{OC}
Power Conversion Efficiency	PCE
Highest Occupied Molecular Orbital	HOMO
Lowest Unoccupied Molecular Orbital	LUMO
Singly Occupied Molecular Orbital	SOMO
Cucurbit[*n*]uril	CB[*n*] ($n = 7$ or 8)
Methylviologen	MV
Phenylviologen	PV
Naphthalenediimide	NDI
Perylenediimide	PDI

Pyromelliticdiimide	PMDI
Tetracyanoquinodimethane	TCNQ
Cetyl trimethylammonium bromide	CTAB
Acetonitrile	MeCN
Dimethylsulfoxide	DMSO
N,*N*-Dimethylformamide	DMF
Tetrahydrofuran	THF
Tetrabutylammonium Fluoride	TBAF
Tetrabutylammonium Cyanide	TBACN
Nitrosyl tetrafluoroborate	$NOBF_4$
Electron Paramagnetic Resonance	EPR
Nuclear Magnetic Resonance	NMR
X-Ray Diffraction	XRD

References

1. J. T. Govindjee, H. Beatty and J. F. Gest, Allen, *Discoveries in Photosynthesis*, *Advances in Photosynthesis and Respiration*, Berlin, Springer, 2006, p. 20.
2. R. E. Blankenship, M. T. Madigan and C. E. Bauer, *Anoxygenic Photosynthetic Bacteria*, Kluwer Academic Publishers, Dordrecht, 1995.
3. T. M. F. Duarte and K. Müllen, Pyrene-Based Materials for Organic Electronics, *Chem. Rev.*, 2011, **111**, 7260–7314.
4. C. Li, M. Liu, N. G. Pschirer, M. Baumgarten and K. Müllen, Polyphenylene-Based Materials for Organic Photovoltaics, *Chem. Rev.*, 2010, **110**, 6817–6855.
5. N. Zhang, S. R. Samanta, B. M. Rosen and V. Percec, Single Electron Transfer in Radical Ion and Radical-Mediated Organic Materials and Polymer Synthesis, *Chem. Rev.*, 2014, **114**, 5848–5958.
6. J.-M. Lehn, Perspectives in Supramolecular Chemistry-From Molecular Recognition towards Molecular Information Processing and Self-Organization, *Angew. Chem., Int. Ed. Engl.*, 1990, **29**, 1304–1319.
7. J.-M. Lehn, *Perspectives in Chemistry-Steps towards Complex Matter*, 2013, vol. 52, pp. 2836–2850.
8. G. R. Desiraju, Supramolecular Synthons in Crystal Engineering-A New Organic Synthesis, *Angew. Chem., Int. Ed. Engl.*, 1995, **34**, 2311–2327.
9. G. R. Desiraju, The C–H···O Hydrogen Bond: Structural Implications and Supramolecular Design, *Acc. Chem. Res.*, 1996, **29**, 441–449.
10. G. R. Desiraju, Crystal Engineering: From Molecule to Crystal, *J. Am. Chem. Soc.*, 2013, **135**, 9952–9967.
11. J. Jens Wolff, Crystal Packing and Molecular Geometry, *Angew. Chem., Int. Ed. Engl.*, 1996, **35**, 2195–2197.
12. J. Bernstein, R. E. Davis, L. Shimoni and N.-L. Chang, Patterns in Hydrogen Bonding: Functionality and Graph Set Analysis in Crystals, *Angew. Chem., Int. Ed. Engl.*, 1995, **34**, 1555–1573.

13. T. Steiner, The Hydrogen Bond in the Solid State, *Angew. Chem., Int. Ed.*, 2002, **41**, 48–76.
14. L. M. Salonen, M. Ellermann and F. Diederich, *Angew. Chem., Int. Ed.*, 2011, **50**, 4808–4842.
15. J. S. Park, E. Karnas, K. Ohkubo, P. Chen, K. M. Kadish, S. Fukuzumi, C. W. Bielawski, T. W. Hudnall, V. M. Lynch and J. L. Sessler, Ion-mediated electron transfer in a supramolecular donor-acceptor ensemble, *Science*, 2010, **329**, 1324–1327.
16. S. Fukuzumi, K. Ohkubo, Y. Kawashima, D. S. Kim, J. S. Park, A. Jana, V. M. Lynch, D. Kim and J. L. Sessler, Ion-controlled on-off switch of electron transfer from tetrathiafulvalene calix[4]pyrroles to Li + @C60, *J. Am. Chem. Soc.*, 2011, **133**, 15938–15941.
17. R. Rathore, S. H. Abdelwahed and I. A. Guzei, Synthesis of a Calix[4]-arene Derivative for Isolation of a Stable Cation Radical Salt for Use as a Colorimetric Sensor of Nitric Oxide, *J. Am. Chem. Soc.*, 2004, **126**, 13582–13583.
18. M. M. Torrent, N. Crivillers, C. Rovira and J. Veciana, Attaching Persistent Organic Free Radicals to Surfaces: How and Why, *Chem. Rev.*, 2012, **112**, 2506–2527.
19. P. P. Power, Persistent and Stable Radicals of the Heavier Main Group Elements and Related Species, *Chem. Rev.*, 2003, **103**, 789–810.
20. R. C. Haddon, Conducting charge-transfer salts based on neutral π-radicals, *Nature*, 1975, **256**, 821–823.
21. Y. Morita and S. Nishida, Stable Radicals: Fundamental and Applied Aspects of Odd-Electron Compounds; ed. R. Hicks, Wiley-Blackwell, New York, 2010.
22. J. S. Miller and M. Drillon, *Magnetism: Molecules to Materials II*, Wiley-VCH: Weinheim, Germany, 2001.
23. R. A. Marcus, On the Theory of Oxidation-Reduction Reactions Involving Electron Transfer-I, *J. Chem. Phys.*, 1956, **24**, 966–978.
24. R. A. Marcus, On the Theory of Oxidation-Reduction Reactions Involving Electron Transfer-II, *J. Chem. Phys.*, 1957, **26**, 867.
25. R. A. Marcus, Chemical and Electrochemical Electron-Transfer Theory, *Annu. Rev. Phys. Chem.*, 1964, **15**, 155–196.
26. R. A. Marcus and N. Sutin, Electron Transfers in Chemistry and Biology, *Biochim. Biophys. Acta.*, 1985, **811**, 265–322.
27. R. A. Marcus, Electron Transfer Reactions in Chemistry: Theory and Experiment (Nobel Lecture), *Angew. Chem., Int. Ed. Engl.*, 1993, **32**, 1111–1121.
28. D. S. Acker, R. J. Harder, W. R. Hertler, W. Mahler, L. R. Melby, R. E. Benson and W. E. Mochel, 7,7,8,8-Tetracyanoquinodimethane and its Electrically Conducting Anion-Radical Derivatives, *J. Am. Chem. Soc.*, 1960, **82**, 6408–6409.
29. D. S. Acker and W. R. Hertler, Substituted Quinodimethans. I. Preparation and Chemistry of 7,7,8,8-Tetracyanoquinodimethane, *J. Am. Chem. Soc.*, 1962, **84**, 3370–3374.

30. N. Martín, J. L. Segura and C. Seoane, Design and synthesis of TCNQ and DCNQI type electron acceptor molecules as precursors for 'organic metals', *J. Mater. Chem.*, 1997, **7**, 1661–1676.
31. J. B. Torrance, The difference between metallic and insulating salts of tetracyanoquinodimethone (TCNQ): how to design an organic metal, *Acc. Chem. Res.*, 1979, **12**, 79–86.
32. R. A. Heintz, H. Zhao, X. Ouyang, G. Grandinetti, J. Cowen and K. R. Dunbar, New Insight into the Nature of Cu(TCNQ): Solution Routes to Two Distinct Polymorphs and Their Relationship to Crystalline Films That Display Bistable Switching Behavior, *Inorg. Chem.*, 1999, **38**, 144–156.
33. N. Lopez, H. Zhao, A. Ota, A. V. Prosvirin, E. W. Reinheimer and K. R. Dunbar, Unprecedented Binary Semiconductors Based on TCNQ: Single-Crystal X-ray Studies and Physical Properties of $Cu(TCNQX_2)$ X = Cl, Br, *Adv. Mater.*, 2010, **22**, 986–989.
34. J. Ferraris, D. O. Cowan, V. Walatka and J. H. Perlstein, Electron transfer in a new highly conducting donor-acceptor complex, *J. Am. Chem. Soc.*, 1973, **95**, 948–949.
35. M. B. Rivas, A. Ota, E. Reinheimer, A. Prosvirin, J. V. Martinez and K. R. Dunbar, Highly Conducting Coordination Polymers Based on Infinite M(4,4'-bpy) Chains Flanked by Regular Stacks of Non-Integer TCNQ Radicals, *Angew. Chem., Int. Ed.*, 2011, **50**, 9703–9707.
36. X. Qu, J. Lu, C. Zhao, J. F. Boas, B. Moubaraki, K. S. Murray, A. Siriwardana, A. M. Bond and L. L. Martin, $(Pro_2H^+)_2(TCNQC^{\bullet -})_2 \cdot TCNQ$: An Amino Acid Derived Semiconductor, *Angew. Chem., Int. Ed.*, 2011, **50**, 1589–1592.
37. H. Fukunaga and H. Miyasaka, Magnet Design by Integration of Layer and Chain Magnetic Systems in a π-Stacked Pillared Layer Framework, *Angew. Chem., Int. Ed.*, 2015, **54**, 569–573.
38. S. Shimomura, R. Matsuda, T. Tsujino, T. Kawamura and S. Kitagawa, TCNQ Dianion-Based Coordination Polymer Whose Open Framework Shows Charge-Transfer Type Guest Inclusion, *J. Am. Chem. Soc.*, 2006, **128**, 16416–16417.
39. T. A. Hudson and R. Robson, A New Class of TCNQ Derivatives Easily Generated from $TCNQH_2$ Containing Discrete $TCNQ^{2-}$ Anions and Noncoordinating Cations, *Cryst. Growth Des.*, 2009, **9**, 1658–1662.
40. A. L. Sutton, B. F. Abrahams, D. M. D'Alessandro, R. W. Elliott, T. A. Hudson, R. Robson and P. M. Usov, Structural and optical investigations of charge transfer complexes involving the F_4TCNQ dianion, *CrystEngComm.*, 2014, **16**, 5234–5243.
41. S. V. Bhosale, C. H. Jani and S. J. Langford, Chemistry of Naphthalene diimides, *Chem. Soc. Rev.*, 2008, **37**, 331–342.
42. S.-L. Suraru and F. Würthner, Strategies for the Synthesis of Functional NaphthaleneDiimides, *Angew. Chem., Int. Ed.*, 2014, **53**, 7428–7448.

43. R. S. Lokey and B. L. Iverson, Synthetic molecules that fold into a pleated secondary structure in solution, *Nature*, 1995, **375**, 303–305.
44. G. J. Gabriel, S. Sorey and B. L. Iverson, Altering the Folding Patterns of Naphthyl Trimers, *J. Am. Chem. Soc.*, 2005, **127**, 2637–2640.
45. H. Y. Au-Yeung, G. D. Pantoş and J. K. M. Sanders, Amplifying Different [2]Catenanes in an Aqueous Donor-Acceptor Dynamic Combinatorial Library, *J. Am. Chem. Soc.*, 2009, **131**, 16030–16032.
46. P. Mukhopadhyay, Y. Iwashita, M. Shirakawa, S.-I. Kawano, N. Fujita and S. Shinkai, Spontaneous Colorimetric Sensing of the Positional Isomers of Dihydroxynaphthalene in a 1D Organogel Matrix, *Angew. Chem., Int. Ed.*, 2006, **45**, 1592–1595.
47. S. Ghosh and S. Ramakrishnan, Aromatic Donor-Acceptor Charge-Transfer and Metal-Ion-Complexation-Assisted Folding of a Synthetic Polymer, *Angew. Chem., Int. Ed.*, 2004, **43**, 3264–3268.
48. V. Gorteau, G. Bollot, J. Mareda, A. P. Velasco and S. Matile, Rigid Oligonaphthalenediimide Rods as Transmembrane Anion-π Slides, *J. Am. Chem. Soc.*, 2006, **128**, 14788–14789.
49. R. E. Dawson, A. Hennig, D. P. Weimann, D. Emery, V. Ravikumar, J. Montenegro, T. Takeuchi, M. J. Mareda, C. A. Schalley and S. Matile, Experimental evidence for the functional relevance of anion-pi interactions, *Nat. Chem.*, 2010, **2**, 533–538.
50. A. L. Sisson, N. Sakai, N. Banerji, A. Fürstenberg, E. Vauthey and S. Matile, Zipper Assembly of Vectorial Rigid-Rod p-Stack Architectures with Red and Blue Naphthalenediimides: Toward Supramolecular Cascade n/p-Heterojunctions, *Angew. Chem., Int. Ed.*, 2008, **47**, 3727–3729.
51. S. Guha, F. S. Goodson, R. J. Clark and S. Saha, Deciphering anion–π-acceptor interactions and detecting fluoride using a naphthalenediimide-based Pd(II) coordination polymer, *CrystEngComm.*, 2012, **14**, 1213–1215.
52. Y. Domoto, E. Orentas, C. Beuchat, D. Emery, J. Mareda, N. Sakai and S. Matile, Catalysis with anion-π interactions, *Angew. Chem., Int. Ed.*, 2013, **52**, 9940–9943.
53. Y. Zhao, C. Beuchat, Y. Domoto, J. Gajewy, A. Wilson, J. Mareda, N. Sakai and S. Matile, Anion-π Catalysis, *J. Am. Chem. Soc.*, 2014, **136**, 2101–2111.
54. H. E. Katz, A. J. Lovinger, C. Kloc, T. Siegrist, W. Li, Y.-Y. Linand and A. Dodabalapur, A soluble and air-stable organic semiconductor with high electron mobility, *Nature*, 2000, **404**, 478–481.
55. B. A. Jones, A. Facchetti, M. R. Wasielewski and T. J. Marks, Tuning Orbital Energetics in Arylene Diimide Semiconductors. Materials Design for Ambient Stability of n-Type Charge Transport, *J. Am. Chem. Soc.*, 2007, **129**, 15259–15278.
56. Z. Chen, Y. Zheng, H. Yan and A. Facchetti, Naphthalenedicarboximide- vs Perylenedicarboximide-Based Copolymers. Synthesis and

Semiconducting Properties in Bottom-Gate N-Channel Organic Transistors, *J. Am. Chem. Soc.*, 2009, **131**, 8–9.
57. T. Earmme, Y.-J. Hwang, N. M. Murari, S. Subramaniyan and S. A. Jenekhe, All-Polymer Solar Cells with 3.3% Efficiency Based on Naphthalene Diimide-Selenophene Copolymer Acceptor, *J. Am. Chem. Soc.*, 2013, **135**, 14960–14963.
58. C. Thalacker, C. Röger and F. Würthner, *J. Org. Chem.*, 2006, **71**, 8098–8105.
59. X. Gao, W. Qiu, X. Yang, Y. Liu, Y. Wang, H. Zhang, T. Qi, Y. Liu, K. Lu, C. Du, Z. Shuai, G. Yu and D. Zhu, First Synthesis of 2,3,6,7-Tetrabromonaphthalene Diimide, *Org. Lett.*, 2007, **9**, 3917–3920.
60. C. Röger and F. Würthner, Core-Tetrasubstituted Naphthalene Diimides: Synthesis, Optical Properties, and Redox Characteristics, *J. Org. Chem.*, 2007, **72**, 8070–8075.
61. R. S. K. Kishore, V. Ravikumar, G. Bernardinelli, N. Sakai and S. Matile, Rapid and Mild Synthesis of Functionalized Naphthalenediimides, *J. Org. Chem.*, 2008, **73**, 738–740.
62. L. E. Polander, A. S. Romanov, S. Barlow, D. K. Hwang, B. Kippelen, T. V. Timofeeva and S. R. Marder, Stannyl Derivatives of Naphthalene Diimides and Their Use in Oligomer Synthesis, *Org. Lett.*, 2012, **14**, 918–921.
63. V. V. Roznyatovskiy, D. M. Gardner, S. W. Eaton and M. R. Wasielewski, Radical Anions of Trifluoromethylated Perylene and Naphthalene Imide and Diimide Electron Acceptors, *Org. Lett.*, 2014, **16**, 696–699.
64. L. E. Polander, L. Pandey, A. Romanov, A. Fonari, S. Barlow, B. M. Seifried, T. V. Timofeeva, J. Luc Brédas and S. R. Marder, 2,6-Diacylnaphthalene-1,8:4,5-Bis(dicarboximides): Synthesis, Reduction Potentials, and Core Extension, *J. Org. Chem.*, 2012, **77**, 5544–5551.
65. Y. Li, G. Zhang, G. Yang, Y. Guo, C. Di, X. Chen, Z. Liu, H. Liu, Z. Xu, W. Xu, H. Fu and D. Zhang, Extended π-Conjugated Molecules Derived from Naphthalene Diimides toward Organic Emissive and Semiconducting Materials, *J. Org. Chem.*, 2013, **78**, 2926–2934.
66. D. Asthana, M. R. Ajayakumar, R. P. Pant and P. Mukhopadhyay, NTCDA-TTF first axial fusion: emergent panchromatic, NIR optical, multi-state redox and high optical contrast photooxidation, *Chem. Commun.*, 2012, **48**, 6475–6477.
67. D. Asthana, PhD thesis, Jawaharlal Nehru University, 2013.
68. Zhang, Y. Hu, T. Schuettfort, C. Di, X. Gao, C. R. McNeill, L. Thomsen, S. C. B. Mannsfeld, W. Yuan, H. Sirringhaus and D. Zhu, Critical Role of Alkyl Chain Branching of Organic Semiconductors in Enabling Solution-Processed N-Channel Organic Thin-FilmTransistors with Mobility of up to 3.50 $cm^2 V^{-1} s^{-1}$, *J. Am. Chem. Soc.*, 2013, **135**, 2338–2349.
69. R. E. Dawson, A. Hennig, D. P. Weimann, D. Emery, V. Ravikumar, J. Montenegro, T. Takeuchi, S. Gabutti, M. Mayor, J. Mareda, C. A. Schalley and S. Matile, Experimental evidence for the functional relevance of anion-π interactions, *Nat. Chem.*, 2010, **2**, 533–538.

70. J. Míšek, A. V. Jentzsch, S.-I. Sakurai, D. Emery, J. Mareda and S. Matile, A Chiral and Colorful Redox Switch: Enhanced π Acidity in Action, *Angew. Chem., Int. Ed.* 2010, **49**, 7680–7683.
71. S. Kumar, M. R. Ajayakumar, G. Hundal and P. Mukhopadhyay, Extraordinary Stability of Naphthalenediimide Radical Ion and Its Ultra-Electron-Deficient Precursor: Strategic Role of the Phosphonium Group, *J. Am. Chem. Soc.*, 2014, **136**, 12004–12010.
72. S. F. Nelsen, Heterocyclic radical anions. II. Naphthalic and 1,4,5,8-naphthalenetetracarboxylic acid derivatives, *J. Am. Chem. Soc.*, 1967, **89**, 5925–5931.
73. J.-F. Penneau and L. L. Miller, An Imide Radical Anion which Assembles into n-Stacks in Solution, *Angew. Chem., Int. Ed.*, 1991, **30**, 986–987.
74. C. J. Zhong, W. Sum, V. Kwan and L. L. Miller, Self-Assembly of Delocalized π-Stacks in Solution. Assessment of Structural Effects, *Chem. Mater.*, 1992, **4**, 1423–1428.
75. M. R. Ajayakumar and P. Mukhopadhyay, Naphthalene-bis-hydrazimide: radical anions and ICT as new bimodal probes for differential sensing of a library of amines, *Chem. Commun.*, 2009, 3702–3704.
76. M. R. Ajayakumar, S. Yadav, S. Ghosh and P. Mukhopadhyay, Single-Electron Transfer Driven Cyanide Sensing: A New Multimodal Approach, *Org. Lett.*, 2010, **12**, 2646–2649.
77. M. R. Ajayakumar, G. Hundal and P. Mukhopadhyay, *Chem. Commun.*, 2013, **49**, 7684–7686.
78. S. Guha and S. Saha, Fluoride Ion Sensing by an Anion-π Interaction, *J. Am. Chem. Soc.*, 2010, **132**, 17674–17677.
79. S. Guha, F. S. Goodson, S. Roy, L. J. Corson, C. A. Gravenmier and S. Saha, Electronically Regulated Thermally and Light-Gated Electron Transfer from Anions to Naphthalenediimides, *J. Am. Chem. Soc.*, 2011, **133**, 15256–15259.
80. G. Heywang, L. Born, H.-G. Fitzky, T. Hassel, J. Hocker, H.-K. Muller, B. Pittel and S. Roth, *Angew. Chem., Int. Ed. Engl.*, 1989, **28**, 483–485.
81. E. D. Glendening, J. K. Badenhoop, A. E. Reed, J. E. Carpenter, J. A. Bohmann, C. M. Morales, C. R. Landis and F. Weinhold, NBO 6.0; Theoretical Chemistry Institute, University of Wisconsin, Madison, WI, 2013.
82. R. F. W. Bader, *Atoms in Molecules: A Quantum Theory*, Clarendon Press, Oxford, U.K., 1990.
83. S. T. Schneebeli, M. Frasconi, Z. Liu, Y. Wu, D. M. Gardner, N. L. Strutt, C. Cheng, R. Carmieli, M. R. Wasielewski and J. F. Stoddart, Electron Sharing and Anion–p Recognition in Molecular Triangular Prisms, *Angew. Chem., Int. Ed.*, 2013, **52**, 13100–13104.
84. Y. Wu, M. Frasconi, D. M. Gardner, P. R. McGonigal, S. T. Schneebeli, M. R. Wasielewski and J. F. Stoddart, Electron Delocalization in a Rigid Cofacial Naphthalene-1,8:4,5-bis(dicarboximide) Dimer, *Angew. Chem., Int. Ed.*, 2014, **53**, 9476–9481.

85. Q. Song, L. Fei, Z. Wang and X. Zhang, , A Supramolecular Strategy for Tuning the Energy Level of Naphthalenediimide: Promoted Formation of Radical Anions with Extraordinary Stability, *Chem. Sci.*, 2015, **6**, 3342–3346.
86. M. R. Ajayakumar, D. Asthana and P. Mukhopadhyay, Core-Modified Naphthalenediimides Generate Persistent Radical Anion and Cation: New Panchromatic NIR Probes, *Org. Lett.*, 2012, **14**, 4822–4825.
87. D. Gosztola, M. P. Niemczyk, W. Svec, A. S. Lukas and M. R. Wasielewski, Excited Doublet States of Electrochemically Generated Aromatic Imide and Diimide Radical Anions, *J. Phys. Chem. A*, 2000, **104**, 6545–6551.
88. M. P. Debreczeny, W. A. Svec, E. M. Marsh and M. R. Wasielewski, Femtosecond Optical Control of Charge Shift within Electron Donor-Acceptor Arrays: An Approach to Molecular Switches, *J. Am. Chem. Soc.*, 1996, **118**, 8174–8175.
89. K. Hasharoni, H. Levanon, S. R. Greenfield David, J. Gosztola, W. A. Svec and M. R. Wasielewski, Radical Pair and Triplet State Dynamics of a Photosynthetic Reaction-Center Model Embedded in Isotropic Media and Liquid Crystals, *J. Am. Chem. Soc.*, 1996, **118**, 10228–10235.
90. H. Levanon, T. Galili, A. Regev, G. P. Wiederrecht, W. A. Svec and M. R. Wasielewski, Determination of the Energy Levels of Radical Pair States in Photosynthetic Models Oriented in Liquid Crystals with Time-Resolved Electron Paramagnetic Resonance, *J. Am. Chem. Soc.*, 1998, **120**, 6366–6373.
91. S. E. Miller, A. S. Lukas, E. Marsh, P. Bushard and M. R. Wasielewski, Photoinduced Charge Separation Involving an Unusual Double Electron Transfer Mechanism in a Donor-Bridge-Acceptor Molecule, *J. Am. Chem. Soc.*, 2000, **122**, 7802–7810.
92. J. W. Verhoeven, *J. Photochem. Photobiol., C*, 2006, **7**, 40–60.
93. R. C. Evans, P. Douglas and C. J. Winscom, *Coord. Chem. Rev.* 2006, **250**, 2093–2126.
94. S. Suzuki, R. Sugimura, M. Kozaki, K. Keyaki, K. Nozaki, N. Ikeda, K. Akiyama and K. Okada, Highly Efficient Photoproduction of Charge-Separated States in Donor-Acceptor-Linked Bis(acetylide) Platinum Complexes, *J. Am. Chem. Soc.*, 2009, **131**, 10374–10375.
95. H. Kar, D. W. Gehrig, N. K. Allampally, G. Fernández, F. Laquai and S. Ghosh, Cooperative supramolecular polymerization of an amine-substituted naphthalene-diimide and its impact on excited state photophysical properties, *Chem. Sci.*, 2016, **7**, 1115–1120.
96. M. E. Nunez, K. T. Noyes, D. A. Gianolio, L. W. McLaughlin and J. K. Barton, Long-range guanine oxidation in DNA restriction fragments by a triplex-directed naphthalene diimide intercalator, *Biochemistry*, 2000, **39**, 6190–6199.
97. S. Matsugo, S. Kawanishi, K. Yamamoto, H. Sugiyama, T. Matsuura and I. Saito, Bis(hydroperoxy)naphthaldiimide as a "Photo-Fenton

Reagent": Sequence-Specific Photocleavage of DNA, *Angew. Chem., Int. Ed. Engl.*, 1991, **30**, 1351–1353.

98. B. A. Aveline, S. Matsugo and R. W. Redmond, Photochemical Mechanisms Responsible for the Versatile Application of Naphthalimides and Naphthaldiimides in Biological Systems, *J. Am. Chem. Soc.*, 1997, **119**, 11785–11795.
99. B. Abraham, S. McMasters, M. A. Mullan and L. A. Kelly, Reactivities of Carboxyalkyl-Substituted 1,4,5,8-Naphthalene Diimides in Aqueous Solution, *J. Am. Chem. Soc.*, 2004, **126**, 4293–4300.
100. Y. Matsunaga, K. Goto, K. Kubono, K. Sako and T. Shinmyozu, Photoinduced Color Change and Photomechanical Effect of Naphthalene Diimides Bearing Alkylamine Moieties in the Solid State, *Chem. – Eur. J.*, 2014, **20**, 7309–7316.
101. H. Zollinger, *Color Chemistry*, VCH, Weinheim, 3rd edn, 2003.
102. W. Herbst and K. Hunger, *Industrial Organic Pigments: Production, Properties, Applications*, WILEY-VCH, Weinheim, 2nd edn., 1997.
103. P. Jonkheijm, N. Stutzmann, Z. Chen, D. M. de Leeuw, E. W. Meijer, A. P. H. J. Schennin and F. Würthner, Control of Ambipolar Thin Film Architectures by Co-Self-Assembling Oligo(p-phenylenevinylene)s and Perylene Bisimides, *J. Am. Chem. Soc.*, 2006, **128**, 9535–9540.
104. T. E. Kaiser, V. Stepanenko and F. Würthner, Fluorescent J-Aggregates of Core-Substituted Perylene Bisimides: Studies on Structure-Property Relationship, Nucleation-Elongation Mechanism, and Sergeants-and-Soldiers Principle, *J. Am. Chem. Soc.*, 2009, **131**, 6719–6732.
105. F. Fennel, S. Wolter, Z. Xie, P.-A. Plötz, O. Kühn, F. Würthner and S. Lochbrunner, Biphasic Self-Assembly Pathways and Size-Dependent Photophysical Properties of Perylene Bisimide Dye Aggregates, *J. Am. Chem. Soc.*, 2013, **135**, 18722–18725.
106. G. Sforazzini, E. Orentas, A. Bolag, N. Sakai and S. Matile, Toward Oriented Surface Architectures with Three Coaxial Charge-Transporting Pathways, *J. Am. Chem. Soc.*, 2013, **135**, 12082–12090.
107. K. Mahata, P. D. Frischmann and F. Würthner, Giant Electroactive M_4L_6 Tetrahedral Host Self-Assembled with Fe(II) Vertices and Perylene Bisimide Dye Edges, *J. Am. Chem. Soc.*, 2013, **135**, 15656–15661.
108. C. W. Tang, Two-layer organic photovoltaic cell, *Appl. Phys. Lett.*, 1986, **48**, 183–185.
109. C. W. Struijk, A. B. Sieval, J. E. J. Dakhorst, M. van Dijk, P. Kimkes, R. B. M. Koehorst, H. Donker, T. J. Schaafsma, S. J. Picken, A. M. van de Craats, J. M. Warman, H. Zuilhof and E. J. R. Sudhölter, Liquid Crystalline Perylene Diimides: Architecture and Charge Carrier Mobilities, *J. Am. Chem. Soc.*, 2000, **122**, 11057.
110. A. Rademacher, S. Markle and H. Langhals, Lösliche Perylen-Fluoreszenzfarbstoffe mit hoher Photostabilität, *Chem. Ber.*, 1982, **115**, 2927–2934.
111. S. Demmig and H. Langhals, Leichtlösliche, lichtechte Perylen-Fluoreszenzfarbstoffe, *Chem. Ber.*, 1988, **121**, 225–230.

112. G. Seybold and A. Stange (BASF AG), *Ger. Pat.*, DE 35 45 004, 1987.
113. F. Würthner, V. Stepanenko, Z. Chen, C. R. Saha-Möller, N. Kocher and D. Stalke, Preparation and Characterization of Regioisomerically Pure1,7-Disubstituted Perylene Bisimide Dyes, *J. Org. Chem.*, 2004, **69**, 7933–7939.
114. R. Schmidt, J. H. Oh, Y.-S. Sun, M. Deppisch, A.-M. Krause, K. Radacki, H. Braunschweig, M. Könemann, P. Erk, Z. Bao and F. Würthner, High-Performance Air-Stable n-Channel Organic Thin Film Transistors Based on Halogenated Perylene Bisimide Semiconductors, *J. Am. Chem. Soc.*, 2009, **131**, 6215–6228.
115. M. Gsänger, J. H. Oh, M. Könemann, H. W. Hffken, A.-M. Krause, Z. Bao and F. Würthner, A Crystal-Engineered Hydrogen-Bonded Octachloroperylene Diimide with a Twisted Core: An n-Channel Organic Semiconductor, *Angew. Chem., Int. Ed.*, 2010, **49**, 740–743.
116. Y. Kumar, S. Kumar, S. K. Keshri, J. Shukla, S. S. Singh, T. S. Thakur, M. Denti, A. Facchetti and P. Mukhopadhyay, Synthesis of Octabromoperylene Dianhydride and Diimides: Evidence of Halogen Bonding and Semiconducting Properties, *Org. Lett.*, 2016, **18**, 472–475.
117. A. Lv, S. R. Puniredd, J. Zhang, Z. Li, H. Zhu, W. Jiang, H. Dong, Y. He, L. Jiang, Y. Li, W. Pisula, Q. Meng, W. Hu and Z. Wang, High Mobility, Air Stable, Organic Single Crystal Transistors of an n-Type Diperylene Bisimide, *Adv. Mater.*, 2012, **24**, 2626–2630.
118. W. Yue, A. Lv, J. Gao, W. Jiang, L. Hao, C. Li, Y. Li, L. E. Polander, S. Barlow, W. Hu, S. Di Motta, F. Negri, S. R. Marder and Z. Wang, Hybrid Rylene Arrays via Combination of Stille Coupling and C-H Transformation as High-Performance Electron Transport Materials, *J. Am. Chem. Soc.*, 2012, **134**, 5770–5773.
119. S. Rajaram, R. Shivanna, S. K. Kandappa and K. S. Narayan, Nonplanar Perylene Diimides as Potential Alternatives to Fullerenes in Organic Solar Cells, *J. Phys. Chem. Lett.*, 2012, **3**, 2405–2408.
120. P. E. Hartnett, A. Timalsina, H. S. S. R. Matte, N. Zhou, X. Guo, W. Zhao, A. Facchetti, R. P. H. Chang, M. C. Hersam, M. R. Wasielewski and T. J. Marks, Slip-Stacked Perylenediimides as an Alternative Strategy for High Efficiency Nonfullerene Acceptors in Organic Photovoltaics, *J. Am. Chem. Soc.*, 2014, **136**, 16345–16356.
121. T. Kircher and H.-G. Löhmannsröben, Photoinduced charge recombination reactions of a perylene dye in acetonitrile, *Phys. Chem. Chem. Phys.*, 1999, **1**, 3987–3992.
122. Y. Che, A. Datar, X. Yang, T. Naddo, J. Zhao and L. Zang, Enhancing One-Dimensional Charge Transport through Intermolecular -Electron Delocalization: Conductivity Improvement for Organic Nanobelts, *J. Am. Chem. Soc.*, 2007, **129**, 6354–6355.
123. Y. Che, A. Datar, K. Balakrishnan and L. Zang, Ultralong Nanobelts Self-Assembled from an Asymmetric Perylene Tetracarboxylic Diimide, *J. Am. Chem. Soc.*, 2007, **129**, 7234–7235.

124. D. Schmidt, D. Bialas and F. Würthner, Ambient Stable Zwitterionic Perylene Bisimide-Centered Radical, *Angew. Chem., Int. Ed.*, 2015, **54**, 3611–3614.
125. E. Shirman, A. Ustinov, N. B. Shitrit, H. Weissman, M. A. Iron, R. Cohen and B. Rybtchinski, Stable Aromatic Dianion in Water, *J. Phys. Chem. B*, 2008, **112**, 8855–8858.
126. R. O. Marcon and S. Brochsztain, Aggregation of 3,4,9,10-Perylenediimide Radical Anions and Dianions Generated by Reduction with Dithionite in Aqueous Solutions, *J. Phys. Chem. A*, 2009, **113**, 1747–1752.
127. S. Seifert, D. Schmidt and F. Würthner, An ambient stable core-substituted perylene bisimide dianion: isolation and single crystal structure analysis, *Chem. Sci.*, 2015, **6**, 1663–1667.
128. I. Ghosh, T. Ghosh, J. I. Bardagi and B. König, Reduction of aryl halides by consecutive visible light-induced electron transfer processes, *Science*, 2014, **346**, 725–728.
129. C.-C. You and F. Würthner, Self-Assembly of Ferrocene-Functionalized Perylene Bisimide Bridging Ligands with Pt(II) Corner to Electrochemically Active Molecular Squares, *J. Am. Chem. Soc.*, 2003, **125**, 9716–9725.
130. A. Sautter, B. K. Kaletasü, D. G. Schmid, R. Dobrawa, M. Zimine, G. Jung, I. H. M. van Stokkum, L. De Cola, R. M. Williams and F. Würthner, Ultrafast Energy-Electron Transfer Cascade in a Multichromophoric Light-Harvesting Molecular Square, *J. Am. Chem. Soc.*, 2005, **127**, 6719–6729.
131. Y. Wu, R. M. Young, M. Frasconi, S. T. Schneebeli, P. Spenst, D. M. Gardner, K. E. Brown, F. Würthner, J. F. Stoddart and M. R. Wasielewski, Ultrafast Photoinduced Symmetry-Breaking Charge Separation and Electron Sharing in Perylenediimide Molecular Triangles, *J. Am. Chem. Soc.*, 2015, **137**, 13236–13239.
132. P. Spenst, R. M. Young, M. R. Wasielewski and F. Würthner, Guest and solvent modulated photo-driven charge separation and triplet generation in a perylene bisimide cyclophane, *Chem. Sci.*, 2016, 7, 5428–5434.
133. E. Krieg, E. Shirman, H. Weissman, E. Shimoni, S. G. Wolf, I. Pinkas and B. Rybtchinski, Supramolecular Gel Based on a Perylene Diimide Dye: Multiple Stimuli Responsiveness, Robustness, and Photofunction, *J. Am. Chem. Soc.*, 2009, **131**, 14365–14373.
134. R. E. Gerald II, C. S. Johnson, J. W. Rathke, R. J. Klingler, G. Sandí and L. G. Scanlon, ^{7}Li NMR study of intercalated lithium in curved carbon lattices, *J. Power Sources*, 2000, **89**, 237–243.
135. M. Baumgarten, J. L. Gherghel, M. Wagner, A. Weitz, M. Rabinovitz, P.-C. Cheng and L. T. Scott, Corannulene Reduction: Spectroscopic Detection of All Anionic Oxidation States, *J. Am. Chem. Soc.*, 1995, **117**, 6254–6257.
136. C. Bruno, R. Benassi, A. Passalacqua, F. Paolucci, C. Fontanesi, M. Marcaccio, E. A. Jackson and L. T. Scott, Electrochemical and

Theoretical Investigation of Corannulene Reduction Processes, *J. Phys. Chem. B*, 2009, **113**, 1954–1962.
137. A. Ayalon, M. Rabinovitz, P.-C. Cheng and L. T. Scott, Corannulene Tetraanion: A Novel Species with Concentric Anionic Rings, *Angew. Chem., Int. Ed. Engl.*, 1992, **31**, 1636–1637.
138. A. Ayalon, A. Sygula, P.-C. Cheng, M. Rabinovitz, P. W. Rabideau and L. T. Scott, *Stable High-Order Molecular Sandwiches: Hydrocarbon Polyanion Pairs with Multiple Lithium Ions Inside and Out, Science*, 1994, vol. 265, pp. 1065–1067.
139. A. V. Zabula, A. S. Filatov, S. N. Spisak, A. Y. Rogachev and M. A. Petrukhina, A Main Group Metal Sandwich: Five Lithium Cations Jammed Between Two Corannulene Tetraanion Decks, *Science*, 2011, **333**, 1008–1011.
140. A. S. Filatov, A. V. Zabula, S. N. Spisak, A. Y. Rogachev and M. A. Petrukhina, Clamshell Opening in the Mixed-Metal Supramolecular Aggregates Formed by Fourfold Reduced Corannulene for Maximizing Intercalated Metal Content, *Angew. Chem., Int. Ed.*, 2014, **53**, 140–145.
141. A. S. Filatov, S. N. Spisak, A. V. Zabula, J. McNeely, R. A. Yu and M. A. Petrukhina, Self-assembly of tetrareduced corannulene with mixed Li–Rb clusters: dynamic transformations, unique structures and record ^{7}Li NMR shifts, *Chem. Sci.*, 2015, **6**, 1959–1966.
142. L. Michaelis, Ein Reduktionsindikator im Bereich der Wasserstoffüberspannung, *Biochem. Z.*, 1932, **250**, 564.
143. L. Michaelis and E. S. Hill, The Viologen Indicators, *J. Gen. Physiol.*, 1933, **16**, 859–873.
144. T. M. Bockman and J. K. Kochi, Isolation and Oxidation-Reduction of Methylviologen Cation Radicals. Novel Disproportionation in Charge-Transfer Salts by X-ray Crystallography, *J. Org. Chem.*, 1990, **55**, 4127–4135.
145. W. W. Porter III and T. P. Vaid, Isolation and Characterization of Phenyl Viologen as a Radical Cation and Neutral Molecule, *J. Org. Chem.*, 2005, **70**, 5028–5035.
146. D. Zhang, E. J. Dufek and E. L. Clennan, Syntheses, Characterizations, and Properties of Electronically Perturbed 1,1′-Dimethyl-2,2′-bipyridinium Tetrafluoroborates, *J. Org. Chem.*, 2006, **71**, 315–319.
147. A. Mirzoian and A. E. Kaifer, Reactive Pseudorotaxanes: Inclusion Complexation of Reduced Viologens by the Hosts β-Cyclodextrin and Heptakis(2,6-di-o-Methyl)-β-Cyclodextrin, *Chem. – Eur. J.*, 1997, **3**, 1052–1058.
148. A. R. Bernardo, T. Lu, E. Córdova, L. Zhang, G. W. Gokel and A. E. Kaifer, Host–guest complexation at the electrode/solution interface: the inclusion of an amphiphilic viologen guest by an amphiphilic calix[6]arene host, *J. Chem. Soc. Chem. Commun.*, 1994, 529–530.
149. H.-J. Kim, W. S. Jeon, Y. H. Ko and K. Kim, Inclusion of methylviologen in cucurbit[7]uril, *Proc. Natl. Acad. Sci.*, 2002, **99**, 5007–5011.
150. H.-J. Kim, J. Heo, W. S. Jeon, E. Lee, J. Kim, S. Sakamoto, K. Yamaguchi and K. Kim, Selective Inclusion of a Hetero-Guest Pair in a Molecular

Host: Formation of Stable Charge-Transfer Complexes in Cucurbit[8]-uril, *Angew. Chem., Int. Ed.*, 2001, **40**, 1526–1529.
151. W. S. Jeon, H.-J. Kim, C. Lee and K. Kim, Control of the stoichiometry in host-guest complexation by redox chemistry of guests: Inclusion of methylviologen in cucurbit[8]uril, *Chem. Commun.*, 2002, 1828–1829.
152. W. Wei, X. Cui, W. Chen and D. G. Ivey, Manganese oxide-based materials as electrochemical supercapacitor electrodes, *Chem. Soc. Rev.*, 2011, **40**, 1697–1721.
153. M. Fukai, A. Kitani, C. Degrand and L. L. Miller, Propagation of a redox reaction through a quinoid polymer film on an electrode, *J. Am. Chem.Soc.*, 1982, **104**, 28–33.
154. L. Echegoyen and L. E. Echegoyen, Electrochemistry of Fullerenes and Their Derivatives, *Acc. Chem. Res.*, 1998, **31**, 593–601.
155. T. Inamasu, D. Yoshitoku, Y. Sumi-Otorii, H. Tani and N. Ono, Electrochemical Behaviors of Naphtho[1,8-cd][1,2]dithiol,Dibenzo[c,e][1,2]-dithiin, and Naphtho[1,8-cd:4,5 c′d′]bis[1,2]dithiol, *J. Electrochem. Soc.*, 2003, **150**, A128–A132.
156. B. Fabre, Ferrocene-Terminated Monolayers Covalently Bound to Hydrogen-Terminated Silicon Surfaces. Toward the Development of Charge Storage and Communication Devices, *Acc. Chem. Res.*, 2010, **43**, 1509–1518.
157. A. C. Benniston, J. Hagon, X. He, S. Yang and R. W. Harrington, Spring Open Two-plus-Two Electron Storage in a Disulfide-Strapped Methyl Viologen Derivative, *Org. Lett.*, 2012, **14**, 506–509.
158. G. B. Hall, R. Kottani, G. A. N. Felton, T. Yamamoto, D. H. Evans, R. S. Glass and D. L. Lichtenberger, Intramolecular Electron Transfer in Bipyridinium Disulfides, *J. Am. Chem. Soc.*, 2014, **136**, 4012–4018.
159. J. C. Barnes, A. C. Fahrenbach, D. Cao, S. M. Dyar, M. Frasconi, M. A. Giesener, D. Benítez, E. Tkatchouk, O. Chernyashevskyy, W. H. Shin, H. Li, S. Sampath, C. L. Stern, A. A. Sarjeant, K. J. Hartlieb, Z. Liu, R. Carmieli, Y. Y. Botros, J. W. Choi, A. M. Z. Slawin, J. B. Ketterson, M. R. Wasielewski, W. A. Goddard III and J. F. Stoddart, A Radically Configurable Six-State Compound, *Science*, 2013, **339**, 429–433.
160. Y. Wang, M. Frasconi, W.-G. Liu, Z. Liu, A. A. Sarjeant, M. S. Nassar, Y. Y. Botros, W. A. Goddard III and J. F. Stoddart, Folding of Oligoviologens Induced by Radical – Radical Interactions, *J. Am. Chem. Soc.*, 2015, **137**, 876–885.
161. D. Leung, S. O. Kang and E. V. Anslyn, Rapid determination of enantiomeric excess: a focus on optical approaches, *Chem. Soc. Rev.*, 2012, **41**, 448–479.
162. R. M. Duke, E. B. Veale, F. M. Pfeffer, P. E. Kruger and T. Gunnlaugsson, Colorimetric and fluorescent anion sensors: an overview of recent developments in the use of 1,8-naphthalimide-based chemosensors, *Chem. Soc. Rev.*, 2010, **39**, 3936–3953.

163. J. R. Askim, M. Mahmoudi and K. S. Suslick, Optical sensor arrays for chemical sensing: the optoelectronic nose, *Chem. Soc. Rev.*, 2013, **42**, 8649–8682.
164. Y. Salinas, R. M. Máñez, M. D. Marcos, F. Sancenón, A. M. Costero, M. Parra and S. Gil, Optical chemosensors and reagents to detect explosives, *Chem. Soc. Rev.*, 2012, **41**, 1261–1296.
165. L. E. S. Figueroa, M. E. Moragues, E. Climent, A. Agostini, R. M. Máñez and F. Sancenón, Chromogenic and fluorogenic chemosensors and reagents for anions. A comprehensive review of the years 2010–2011, *Chem. Soc. Rev.*, 2013, **42**, 3489–3613.
166. M. R. Ajayakumar, K. Mandal, K. Rawat, D. Asthana, R. Pandey, A. Sharma, S. Yadav, S. Ghosh and P. Mukhopadhyay, Single Electron Transfer-Driven Multi-Dimensional Signal Read-out Function of TCNQ as an "Off-the-Shelf" Detector for Cyanide, *ACS Appl. Mater. Interfaces*, 2013, **5**, 6996–7000.
167. A. Mallick, B. Garai, M. A. Addicoat, P. St. Petkov, T. Heinec and R. Banerjee, Solid state organic amine detection in a photochromic porous metal organic framework, *Chem. Sci.*, 2015, **6**, 1420–1425.
168. W. Walker, S. Grugeon, O. Mentre, S. Laruelle, J.-M. Tarascon and F. Wudl, Ethoxycarbonyl-Based Organic Electrode for Li-Batteries, *J. Am. Chem. Soc.*, 2010, **132**, 6517–6523.
169. S. Renault, J. Geng, F. Dolhem and P. Poizot, Evaluation of polyketones with N-cyclic structure as electrode material for electrochemical energy storage: case of pyromellitic diimide dilithium salt, *Chem. Commun.*, 2011, **47**, 2414–2416.
170. G. S. Vadehra, R. P. Maloney, M. A. Garcia-Garibay and B. Dunn, Naphthalene Diimide Based Materials with Adjustable Redox Potentials: Evaluation for Organic Lithium-Ion Batteries, *Chem. Mater.*, 2014, **26**, 7151–7157.
171. M. E. Bhosale and K. Krishnamoorthy, Chemically Reduced Organic Small-Molecule-Based Lithium Battery with Improved Efficiency, *Chem. Mater.*, 2015, **27**, 2121–2126.

CHAPTER 5

Influence of Supramolecular Interactions on the Solid-state Photoreactivity of Metal Complexes

HONG SHENG QUAH AND JAGADESE J. VITTAL*

Department of Chemistry, National University of Singapore, 117543 Singapore
*Email: chmjjv@nus.edu.sg

5.1 Introduction

Reactivity of a molecule is a fundamental chemical property. The reactions are usually conducted in solution due to free movements of the reactants. The solid-state reactions are the least preferred by chemists as their restricted movement is often the consideration when designing reactive compounds, even though the solvent-less approach is environmentally friendly.[1–4] If, however, the reactive functional groups are already aligned closer to the transition state of the reaction, then the solid-state reaction can easily be carried out. Due to restricted movements of the predisposed reactive functions groups, reactions in the solid state can yield regioselective and stereoselective products easily.[5,6] Unlike solution-state reactions, solid-state reactions are more sluggish due to the absence of homogeneity. Moreover, solid-state reactivity is very much dependent on the periodic

Monographs in Supramolecular Chemistry No. 22
Functional Supramolecular Materials: From Surfaces to MOFs
Edited by Rahul Banerjee

Published by the Royal Society of Chemistry, www.rsc.org

assemblies of molecules in crystals while reactivity in solution is largely a molecular property.[7] Thus, bringing the functional groups in the molecules closely in the correct alignment and orientation for the reaction to occur upon an external stimulus requires crystal engineering. Crystal engineering as defined by Desiraju laid the principles for modern crystal design and synthesis. In the context of *supramolecular chemistry*, crystal engineering is the synthesis of solid-state supramolecular structures.[8] Researchers can elegantly and intelligently exploit various supramolecular interactions, tectons and synthons to create reactive solid and obtain their desired products with ease.

Among the solid-state reactions, [2 + 2] photocycloaddition reactions have been studied quite extensively since the classical work of Schmidt and co-workers in 1960s.[9,10] Since then a number of excellent reviews on organic solid-state photoreactions have appeared.[11–23] Further structural dynamics, transformations and reactivity of coordination polymers and metal-organic frameworks have also been well reviewed.[24–31] Due to the advancement of technology, single crystal X-ray crystallography (SXRD) is becoming an increasingly common tool for solid-state chemists to follow the course of reactions. While it is complimented by other spectroscopic methods, single-crystal-to-single-crystal (SCSC) transformation is often much desired as it provides the direct proof of the reaction that occurred.[32] Since metal-mediated reactivity in the organic solid state was reviewed by Georgiev and MacGillivray in 2008,[33] we will discuss mainly the photoreactive metal complexes which are able to undergo [2 + 2] photocyclization reactions in this book chapter. The metal complexes engineered and assembled for photochemical reactions in the solid state have yielded interesting new dimers, oligomers and polymers which would otherwise be inaccessible in the solution state. We also attempt to understand how supramolecular interactions present in the solid state influence the photoreactivity and nature of the photoproducts. Although the literature covered is comprehensive, it is by no means complete. We apologize for any inadvertent omission of any interesting literature.

5.2 Metallophilic Interactions to Align Terminal Ligands Bearing Olefin Bonds

The most common oxidation state of silver or gold is +1. Depending on the ligand, silver(I) atoms generally adopt tetrahedral, T-shaped or linear geometrical structures. On the hand, Au(I) prefers to have a linear geometry. The well-established metallophilic (argentophilic or aurophilic) interactions[34–37] can be commonly employed as a strategy to bring two ligands close together for photochemical reactions to occur between two olefin bonds. Mononuclear silver(I) complexes with two linear coordination geometries tend to assemble *via* argentophilic interactions forming a dimer. For illustration, the Ag(I) complex can coordinate to pyridyl ligands containing olefin bonds,

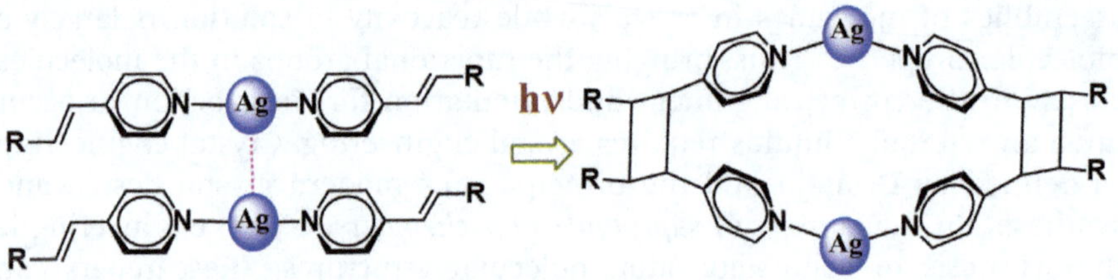

Figure 5.1 Face-to-face alignment of the asymmetrical ligands due to argentophilic interactions and linear geometry of Ag(I) and the resultant photoproduct.

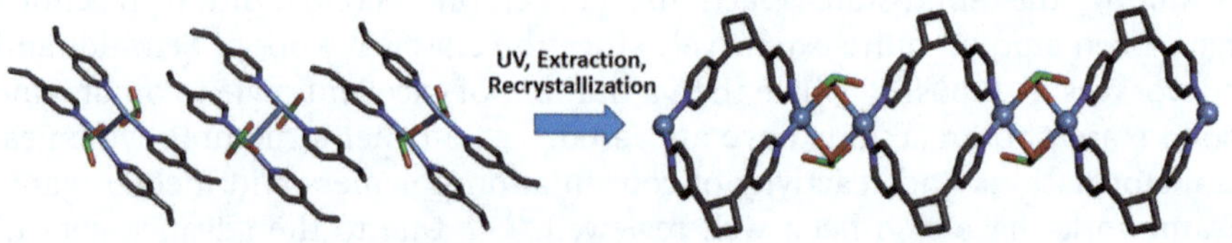

Figure 5.2 Alignment of 4-vinyl pyridine ligands in a Ag(I) complex and its photoproduct after recrystallization.
(Adapted from ref. 38 with permission from the Royal Society of Chemistry.)

forcing a face-to-face stacking as depicted in Figure 5.1. Since the Ag–Ag separation is less than the sum of the van der Waals radii of silver, it is expected that Ag···Ag interaction is present to aid the olefin ligands in packing closer. In many cases reported, the anion may or may not have any influence on this molecular interaction. The loss of argentophilic interaction after the photodimerization reaction is inevitable due to the formation of sterically-strained cyclobutane derivatives on both sides of the Ag(I) atoms as the dimer puckers about the cyclobutane.

5.2.1 Role of Argentophilic Interactions Influencing the Photoreactivity in Ag(I) Complexes

As expected 4-vinyl pyridine (4-vpy) ligand pairs can be aligned in $[Ag_2(4\text{-vpy})_4][ClO_3]_2$ as shown in Figure 5.2 with the chlorate anions weakly bridging the dimer. Here, the argentophilic interaction favours π–π contact between the pyridyl groups in the dimer. Although the pair of olefins are proximate at 3.96 Å, the olefin pairs are found to be aligned crisscrossed. Nonetheless, the solid undergoes photodimerization reaction when a broadband UV light of 450 W generated from a medium-pressure Hg lamp was used. The transformation was found to occur in a quantitative manner by monitoring with ^{1}H-NMR spectroscopy as the single crystallinity was not retained. The ligand was separated and extracted after the reaction and further reacted with $AgClO_3$ to confirm the stereospecific generation of *cis*-1,2-bis(4-pyridyl)cyclobutane (1,2-bpcb) by SXRD. Interestingly, the recrystallized product afforded a one-dimensional coordination polymer (1D CP) sustained by the bridging of $[Ag_2(1,2\text{-bpcb})_2]^{2+}$ with the chlorate anions.[38]

By substituting the anion with trifluoroacetate (TFA), the silver complex, which consists of $[Ag_2(4\text{-vpy})_4]^{2+}$ cation and TFA anions, undergoes photocycloaddition reaction under visible laser irradiation ($\lambda = 633$ nm, 3 mW) after melting, as monitored by *in situ* Raman spectroscopy complemented by NMR techniques. The power and wavelength have been found to affect the reaction kinetics. Although such a photoreaction is usually triggered by a UV source due to the relatively large HOMO–LUMO gap of the complex, this reaction could have possibly occurred *via* an upconversion process.[39]

Replacing the 4-vpy with the 4-spy ligand, the silver(I) complex $[Ag_2(4\text{-spy})_4](CO_2CF_3)_2$ (4-spy = *trans*-4-styrylpyridine) contained a pair of crisscross aligned 4-spy as shown in Figure 5.3. Nonetheless, the C=C bond pairs undergo pedal motion under UV radiation resulting in 100% isolation of *rtct*-1,2-bis(4-pyridyl)-3,4-bis(phenyl)cyclobutane isomer in an SCSC fasion.[40] Such pedal motions have been well-documented in the literature.[41] Additionally, the nanocrystals of this dimerized complex displayed higher electrical conductivity compared to the monomer. This has been attributed to higher contribution from the Ag(I) ions to the valence band in the photodimerized product.[42]

As acid and amides tend to form hydrogen-bonded aggregates,[43,44] the well-known acid-dimer synthon was used to connect the dimers formed through argentophilic interaction, forming a hydrogen-bonded aggregate. As such, introduction of acid and amide groups to 4-vpy ligand-generated photoreactive Ag(I) complexes with *trans*-3-(4′-pyridyl)acrylic acid (4-paH),

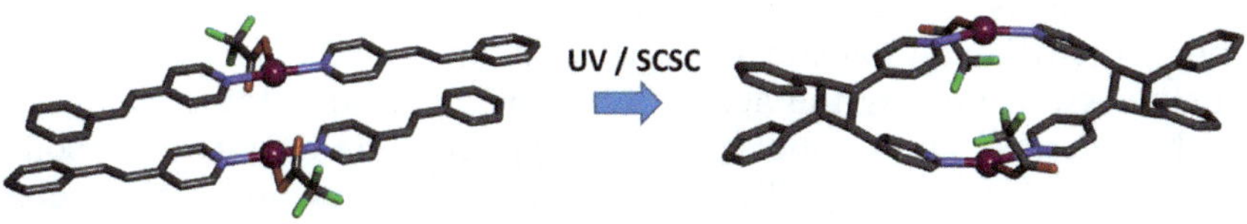

Figure 5.3 An illustration showing the crisscrossed 4-spy pairs in the Ag(I) undergoing SCSC photoreaction.

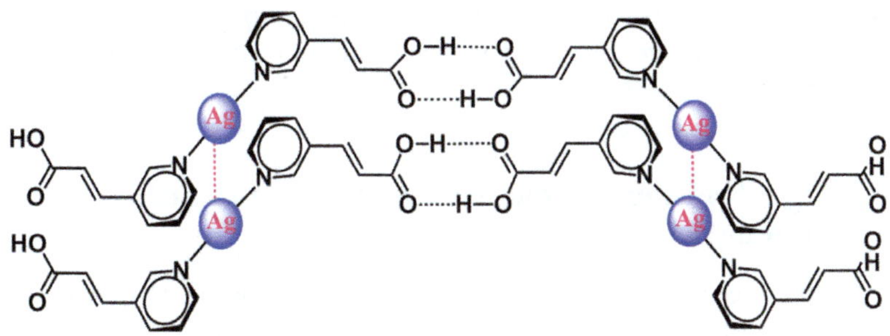

Figure 5.4 A schematic diagram showing the alignment of the 3-paH ligands in the hydrogen-bonded Ag(I) complex.

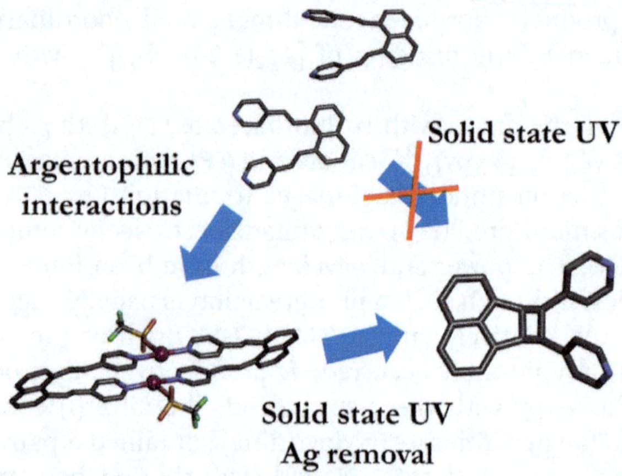

Figure 5.5 A schematic diagram showing the photoreactivity of the napthalene-based ligand and its Ag(I) complex.
(Adapted with permission from R. C. Laird, M. A. Sinnwell, N. P. Nguyen, D. C. Swenson, S. V. S. Mariappan and L. R. MacGillivray, Intramolecular [2 + 2] Photodimerization Achieved in the Solid State via Coordination-Driven Self-Assembly, *Org. Lett.*, 2015, **17**, 3233–3235. Copyright 2015 American Chemical Society.)

trans-3-(4′-pyridyl)acrylamide (4-pam) and *trans*-3-(3′-pyridyl)acrylic acid (3-paH). The photoreactive hydrogen-bonded dimer aggregate is illustrated in Figure 5.4. It is worth noting that the problem of crisscrossed alignment of olefin pairs has been eliminated in these supramolecular structures.

Although it is possible to structurally lock and align two 4-vpy groups at the 1,8-positions of a naphthalene ring in a single ligand,[45] (1,8-bis[(*E*)-2-(4-pyridyl)ethenyl] naphthalene), it was surprisingly found to be photostable. Nonetheless, by making the Ag(I) complex of this ligand, the cyclobutane readily undergoes an intramolecular [2 + 2] photodimerization in the solid state mediated by argentophilic forces within the self-assembled rectangles as shown in Figure 5.5. Of the three silver salts used to react with this ligand, only the chlorate salt provided a quantitative photoproduct as monitored by the ^{1}H-NMR spectroscopy. The geometries of the resulting cyclobutanes may allow the intramolecular reaction to act as a reliable covalent synthon to generate unsymmetrical molecular architectures in organic solids.

5.2.2 Argentophilic Interactions *versus* Donor–Acceptor Interactions

The donor–acceptor interaction between perflurophenyl-phenyl ($C_6F_5\cdots C_6H_5$) groups has been successfully utilized by Grubbs *et al.* to align olefin pairs in a number of organic molecules.[46] However in a Ag(I) complex, $[Ag_2(5F\text{-}4spy)_4][ClO_3]_2$ (where 5F-4spy = *trans*-pentafluorostyrl pyridine),

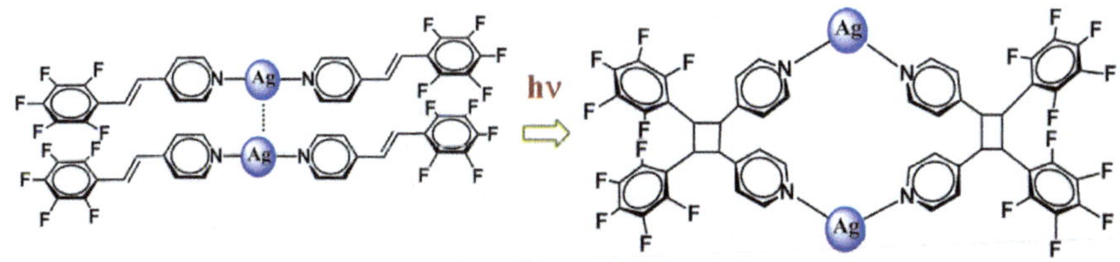

Figure 5.6 A diagram showing the unusual face-to-face $C_6F_5\cdots C_6F_5$ interactions forced by dominant Ag···Ag interactions leading to a surprising photoproduct.

MacGillivray *et al.* found that argentophilic interaction prevails over $C_6F_5\cdots NC_5H_5$ interactions.[47] The 5F-4spy ligands are aligned in a head-to-head (HH) fashion with $C_6F_5\cdots C_6F_5$ interactions as shown in Figure 5.6. The Ag···Ag distance of 3.10 Å and centroid distances between C_6F_5 pairs and pyridyl pairs are 3.37 and 3.97 Å respectively. The closest distance between the reactive C=C bonds has been found to be 3.89 Å and it undergoes photodimerization in a HH fashion quantitatively. In this complex, the argentophilic interaction is dominant over donor–acceptor ($NC_5H_5\cdots C_6F_5$) interaction in preventing the ligands in the metal complex to pack in a head-to-tail (HT) manner. The photoproduct was recrystallized to confirm its structure. Such similar $C_6F_5\cdots C_6F_5$ interactions have been observed and documented in organic[48,49] and organometallic compounds.[50]

5.2.3 Intra *versus* Intermolecular Photoreaction: Influence of C–H···π Interactions

Physical properties and chemical reactivity of solids are greatly influenced by their molecular packing. Apart from the molecular shape and steric crowdedness, the packing is governed by weak intermolecular interactions. Even weak non-directional supramolecular interactions, such as van der Waals interactions, can play a crucial role mainly due to their sheer number, as illustrated in the climbing of geckos on flat walls.[51,52] These interactions are also utilized to retain highly volatile Freon and methane molecules well above their boiling points in non-porous organic crystals.[53–58] On the other hand, directionality of hydrogen bonds have been well exploited to make photoreactive solids.[6,20,33] The C–H···π interaction was shown to play a decisive role in promoting intramolecular photodimerization in a HH manner over intermolecular cyclization of a pair of olefin bonds in an inorganic co-crystal, $[Ag_2(dppm)(4\text{-spy})_2][Ag_2(dppm)_2]_{0.5}(BF_4)_3\cdot 2THF$ (where dppm = diphenylphosphinomethane).[59] The solid contains two $[Ag_2(dppm)(4\text{-spy})_2](BF_4)_2$ and one $[Ag_2(dppm)_2]$ $(BF_4)_2$ forming an unusual inorganic co-crystal. In the $[Ag_2(dppm)(4\text{-spy})_2]^{2+}$ cation, $(dppm)Ag_2$ acts as a clip to bring the two 4-spy closer in a HH alignment. When the co-crystal is closely examined, it revealed that one $[Ag_2(dppm)(4\text{-spy})_2]^{2+}$ complex is stacked above the other alternately in a HT manner. These led to the formation of two different pairs of 4-spy that are aligned in HH (intramolecular) and HT (intermolecular) fashion. Comparatively, the intramolecular HH olefin pairs are separated by 4.30 Å while the intermolecular pairs HT are separated by 4.44 Å. In the intramolecular pair, the olefin bond pairs are crisscrossed while the intermolecular olefin pairs are slip-stacked. In both cases, their distances are greater than the distance limit of 4.2 Å by Schmidt.[10] It has been found that one of the phenyl rings of dppm exerts C–H···π interaction on one 4-spy ligand in a complementary manner. The hydrogen bonding is postulated to hinder the lateral movements of the 4-spy ligands and prevent the HT intermolecular photocycloaddition reaction. Indeed, the single crystals of this inorganic co-crystal undergo 100% photoreaction in an SCSC fashion under UV light to give the intramolecular product $[Ag_2(dppm)(4,4'\text{-bpbpcb})][Ag_2(dppm)_2]_{0.5}(BF4)_3 \cdot 2THF$ which contains the HH photo product 4,4′-bpbpcb, 1,2-bisphenyl-3,4-bis-(4′-pyridyl) cyclobutane) as shown in Figure 5.7. The C–H···π interaction became shorter and hence stronger after the photoreaction. It is likely that these 4-spy ligand pairs were under constant grip by this interaction during the photoreaction favouring the HH photoproduct. Such influence of weak interactions on the physical and chemical properties has been noted before.[60]

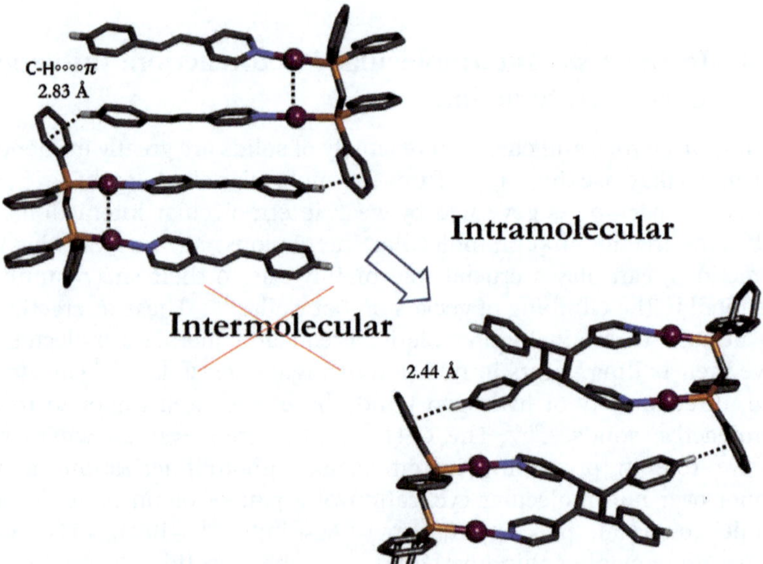

Figure 5.7 The SCSC reaction in a Ag(I) complex showing the role of C–H···π interactions in preventing intermolecular photodimerization reaction. (Adapted from ref. 59 with permission from the Royal Society of Chemistry.)

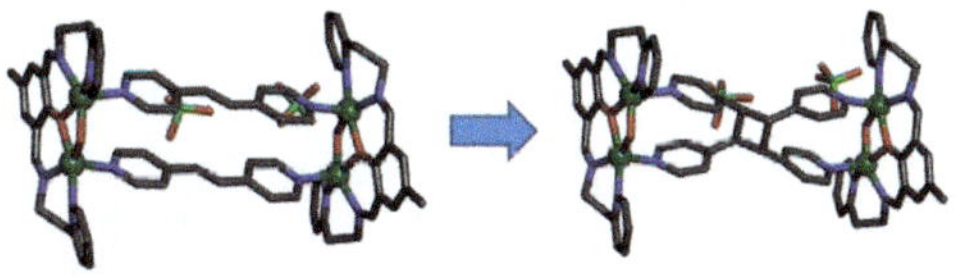

Figure 5.8 Alignment of a pair of bpe ligands in a Zn(II) complex. (Adapted with permission from G. S. Papaefstathiou, Z. Zhong, L. Geng and L. R. MacGillivray, Coordination-Driven Self-Assembly Directs a Single-Crystal-to-Single-Crystal Transformation that Exhibits Photocontrolled Fluorescence, *J. Am. Chem. Soc.*, 2004, **126**, 9158–9159. Copyright 2004 American Chemical Society.)

5.3 Ligand Design in Controlling Photoreactivity of 1,2-Bi(4-pyridyl)ethylene (bpe) Pairs

There are certainly many innovative ways of designing the ligand system by controlling the geometry and arrangement of donor atoms that bind to metal ions. The metal complexes can then have increased stability and reactivity, fascinating structural features, ability to activate small molecules, interesting catalytic and magnetic properties.[61] Surprisingly, such ligand design has not been widely exploited to control the photoreactivity of metal complexes in the solid state. Here a well-known multidentate ligand[62] L_1 has been successfully used to form a dinuclear zinc(II) complex $[Zn_2(L_1)(OH)]$ (where L_1 = 2,6-bis[*N*-(2-pyridylethyl)formimidoyl]-4-methylphenolate) which can have a custom-made Zn-Zn distance of 3.5 Å to align a pair bpe ligands through axial coordination.[63] The tetranuclear molecular rectangle $[Zn_4(L^1)_2(OH)_2(4,4'\text{-bpe})_2]^{4+}$ shown in Figure 5.8 has the olefin pairs of the bpe ligands aligned at 3.64 Å. The [2 + 2] cycloaddition reaction occurs in a SCSC manner. The same metal complex has also been successfully employed to align infinite array of bpe pairs in a photoreactive 1D CP.[64]

5.4 Diphosphine Ligands as Clipping Agents

In section 5.2.3, we have seen the utility of $\{(dppm)Ag_2\}$ to align two linear ligands in parallel. Indeed, the diphosphine ligands dppm and dppe (where dppe = diphenylphosphinoethane) are well-known clipping agents to align the metal ions in *cis* geometry.[65] The Ag(I) and Au(I) complexes of dppm and dppe have been effectively employed to arrange a pair of bpe ligands in parallel forming $[Ag_4(\mu\text{-dppm})_2(\mu\text{-bpe})_2](O_2CCF_3)_4$ and $[Au_4(\mu\text{-dppe})_2(\mu\text{-bpe})_2](O_2CCF_3)_4)$ by Puddephatt and others.[66–69] Since the bpe pairs are separated well within the Schimidt's criteria,[10] these 'molecular rectangles' were tested for their photoreactivity (Figure 5.9).[23] These complexes undergo quantitative photoconversion in the solid state as predicted. Interestingly,

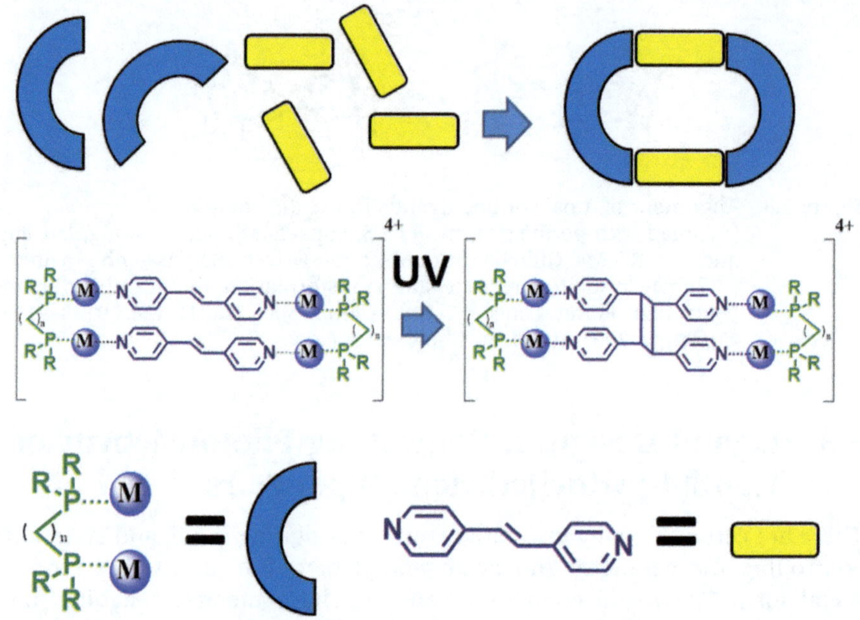

Figure 5.9 Photoreactive molecular rectangles formed by clipping agents with well-aligned pairs of bpe ligands.

the Au(I) complex will also undergo 100% photodimerization in d_6-DMSO solution. The formation of stereospecific single products, namely, *regio cis, trans, trans*-tetrapyridyl cyclobutane (*rctt*-tpcb) in quantitative yield as followed by ^{1}H NMR spectroscopy, inferring the robustness of the Au(I) complex cation in maintaining its structure in solution.[23]

5.4.1 Photoreactive Gold(I) Macrocycles

[(dppm)/(dppe)M_2] (M = Au(I) or Ag(I)) clipping agents are very useful molecular directors that can also be used to manipulate conjugated dicarboxylate ligands into proximity. In two photoreactive neutral gold(I) macrocycles, namely [Au_4(dppm)$_2$(muco)$_2$]·2MeOH and [Au_4(dppe)$_2$(muco)$_2$]·2CH_2Cl_2·MeOH, a pair of *trans,trans*-muconate ligands were aligned for photodimerization.[70] The two (P–P)Au_2 groups clip two muconate ligands in parallel, aligning face-to-face with perfect alignments of the conjugated C=C bonds for photochemical cycloaddition reactions. As expected, the photoconversion was possible under UV light in the solid state as well as in solution. However, as there are two conjugated olefins on a single muconate ligand, it is possible that various products can be expected to be formed as shown in Figure 5.10. However, the reaction proceeds through a concerted [4 + 4] cycloaddition forming 1,5-cyclooctadiene[71–73] and not through a possible Cope rearrangement after a single [2 + 2] cycloaddition as evidenced from ^{1}H-NMR spectroscopy, to form the [4 + 4] product exclusively.

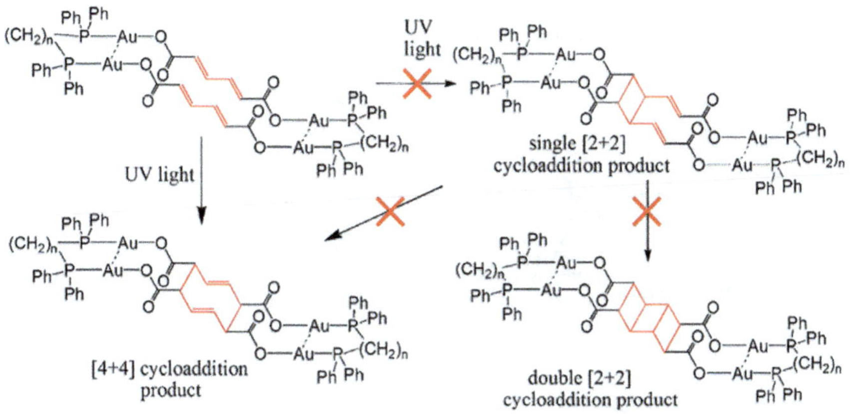

Figure 5.10 A reaction scheme showing all the possible photoproducts from the two Au(I) macrocycles under UV light.
(Reproduced from ref. 70 with permission from the Royal Society of Chemistry.)

5.5 Photoreactive Organometallic Cages and Molecular Rectangles

In many metalla macrocycles, two dipyridyl spacer ligands bridging two organometallic compounds tend to align closely in parallel. In some macrocycles, bpe ligands were used for photoreaction. The details of such photoreactive macrocycles are presented in this section.

5.5.1 Organometallic Rectangles Containing Bisimidazolate Derivatives

Recently, a number of organometallic macrocycles involving bpe ligands have been reported as photoreactive by Lu, Jin, Hahn and others. This has been reviewed by Han and Jin recently.[74] Taking advantage of the rhenium(I) tendency to form *cis* geometry with another bridging ligand, two neutral rectangles have been formed by linking a pair of bpe ligands in parallel as shown in Figure 5.11. Both $[Re_2(CO)_6(L)_2(bpe)]_2$ (L = 2,2′-bisimidazolate (biim) and 2,2′-bisbenzimidazolate (bbim)) compounds are photoreactive in an SCSC manner. In contrast, a similar rectangle formed by 2,2′-bis(4,5-dimethylimidazolate) (bdmim) failed to be photoreactive. Although the photo inertness is not fully understood, it is apparent that the methyl substitution at the backbone of the ligand is able to regulate the photoreactivity.[75] In another Re(I) rectangle, $[Re_2(CO)_6(OR)_2(L)]_2$, (where R = *n*-butyl group and L = bpe and bpeb) by Lu *et al.*, bpe and bpeb ligand pairs were nicely aligned. Unfortunately, their photoreactivity was not investigated.[76]

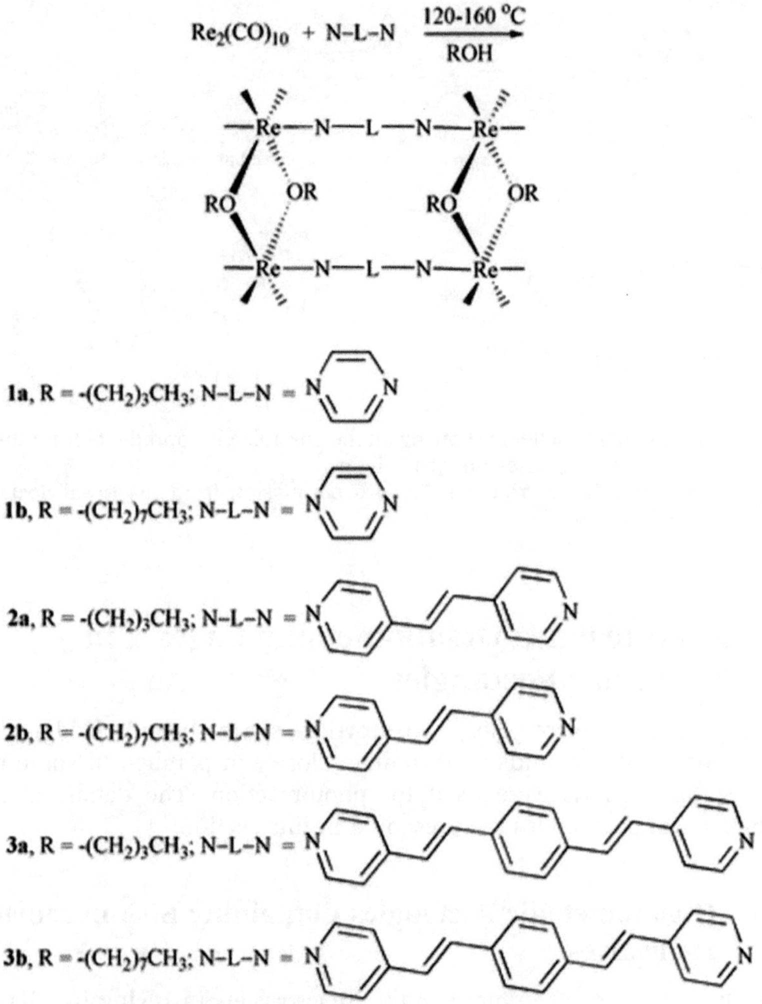

Figure 5.11 Photoreactivity of Re(I) organometallic macrocycle. (Reproduced from ref. 76 with permission from the Royal Society of Chemistry.)

5.5.2 Photoreactive Organometallic Rectangles Containing Oxalate Ligands

The oxalato-bridged organometallic dimers, $[Cp^*{}_2M_2(\mu\text{-}\eta^2\text{-}\eta^2\text{-}C_2O_4)]^{2+}$ (M = Rh, Ir) can act as a clip to pair the double bonds of bpe resulting in the formation of $[Cp^*{}_4M_4(\mu\text{-bpe})_2(\mu\text{-}\eta^2\text{-}\eta^2\text{-}C_2O_4)](OTf)_4$ [M = Rh, Ir].[77] The olefin pairs are separated by 3.23 Å and the complexes were found to be photoreactive. The powdered sample of the Ir(I) rectangle took 25 h under a Hg lamp to complete the formation of *rctt*-tpcb as monitored by ^1H and ^{13}C NMR spectroscopy. The complex can also proceed in a SCSC manner as

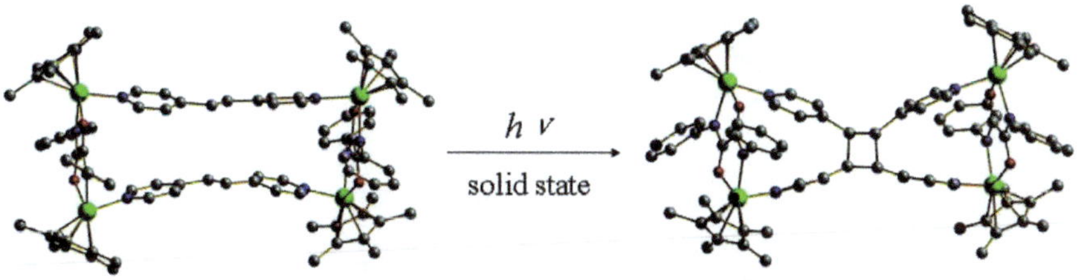

Figure 5.12 SCSC photodimerization of bpe pairs in Ir(I) organometallic rectangles. (Adapted from ref. 77 with permission from the Royal Society of Chemistry.)

shown in Figure 5.12. On the other hand, the Rh(I) compound yielded only 95% of the photoproduct in 50 h.

5.5.3 Unusual Formation of tpcb Isomer from the Crisscrossed Alignment of bpe in Photoreactive Organometallic Cages

Bis(aryl)oxamidato ligand containing organometallic dimers can also generate similar photoreactive molecular rectangles namely $[Cp^*{}_4M_4(\mu\text{-}N,N'\text{-bis(aryl)oxamidato})_2(\mu\text{-}4,4'\text{-bpe})_2](OTf)_4$ (M = Ir, Rh and aryl = Ph and C_6H_4-*p*-Me) in which the olefin bond pairs are juxtaposed in a crisscross manner.[78] Although 100% photoconversion occurred in 30 h under UV light, the reaction gave a mixture of 64.5% *rctt*-tpcb isomer and 35.5% *rctt*-isomer for the Ir(I) rectangle with aryl = Ph as evident by ^1H NMR spectroscopy in $CDCl_3$ solution and depicted in Figure 5.13. Other rectangles were incompletely photoconverted and gave *rctt* and *rtct* mixtures in the solid state. Incomplete pedal motion during the photoreaction is thought to account for the formation of the *rtct*-isomer but still dimerization of crisscrossed C=C bonds is very rare and unusual.[79]

5.5.4 Solid-state and Solution Photoreactivity of an Organometallic Rectangle with Muconate Bridging

The methodology described before has been extended to muconate bridging ligand. The photoreactivity of the tetranuclear rectangular cage, $[\{Cp^*{}_2Ir_2(\text{muconate})\}_2(\text{bpe})_2]$ has been investigated in both methonolic solution and solid-state.[80] The reaction rate is faster when the complex is dissolved in methanol than when in the solid state. Under UV light, 100% conversion has been observed in 24 h, as monitored by ^1H NMR spectroscopy. In the solid state, the reaction is relatively faster under UV light (48 h) than under sunlight (two weeks). The structure of the photoproduct has been obtained after recrystallization of the photoproduct from methanolic solution. Figure 5.14 illustrates the formation cyclobutane rings from the molecular rectangle *via* [2 + 2] cycloaddition reaction.

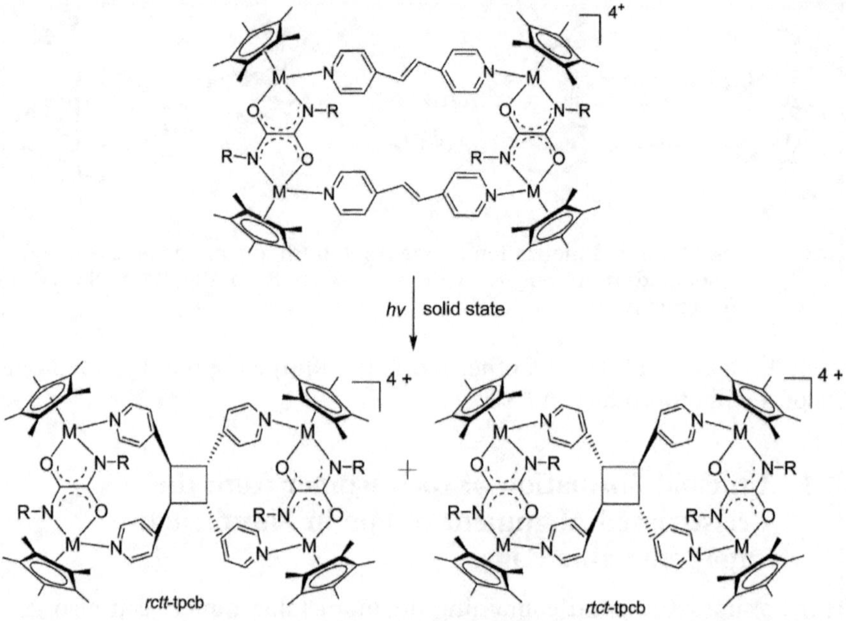

Figure 5.13 Crisscrossed olefin bonds photodimerized to give a mixture of *rctt*- and *rtct*-isomers in organometallic rectangles.
(Reproduced with permission from W. Z. Zhang, Y. F. Han, Y. J. Lin, W. G. Jia, G. L. Wang and G. X. Jin, [2 + 2] Photodimerization in the Solid State Aided by Molecular Templates of Rectangular Macrocycles Bearing Oxamidato Ligands, *Organometallics*, 2010, **29**, 2842–2849. Copyright 2010 American Chemical Society.)

5.5.5 Photoreactive NHC-based Metalla Macrocycles

N-heterocyclic carbene (NHC)-based photoreactive molecular rectangles have been designed by Jin and Hahn.[81] The ligand bis(imidazolium)-substituted stilbene derivatives were synthesized as their PF_6 salt before reacting with Ag_2O in acetonitrile to yield the disilver tetracarbene as shown in Figure 5.15. Photodimerization experiments were conducted in d_6-DMSO and acetonitrile solutions to obtain the photodimerized products. Due to the lack of precursor single crystals, SCSC conversion could not be accomplished in the solid state. The silver rectangle can be further transmetallated into its gold analogue which furnished quantitative yield in the solid state in 12 h under mercury lamp with $\lambda = 365$ nm).

5.5.6 An Unusual Photodimerization of Coumarin Derivatives in Metalla Macrocycles

A series of interesting dicarbene-based molecular rectangles with photoreactive coumarin groups as pendant have been reported in 2015.[82] The coumarin substituted diazolium salts were synthesized first and reacted with

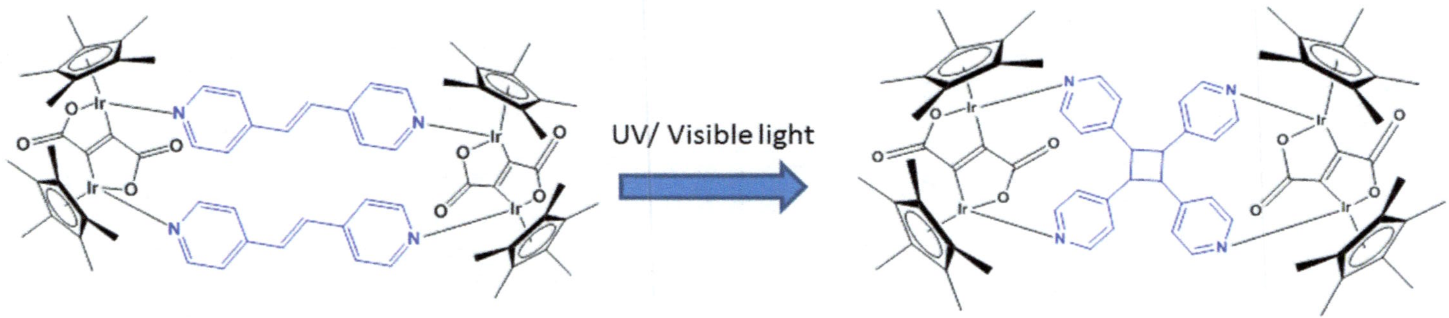

Figure 5.14 Photoreactivty of an iridium rectangle in the solid state and solution. (Adapted with permission from W. B. Yu, Y. F. Han, Y. J. Lin and G. X. Jin, Construction of Tetranuclear Macrocycles through C-H Activation and Structural Transformation Induced by [2 + 2] Photocycloaddition Reaction, *Chem. – Eur. J.*, 2011, **17**, 1863–1871. Copyright © 2011 WILEY-VCH Verlag GmbH & Co. KGaA, Weinheim.)

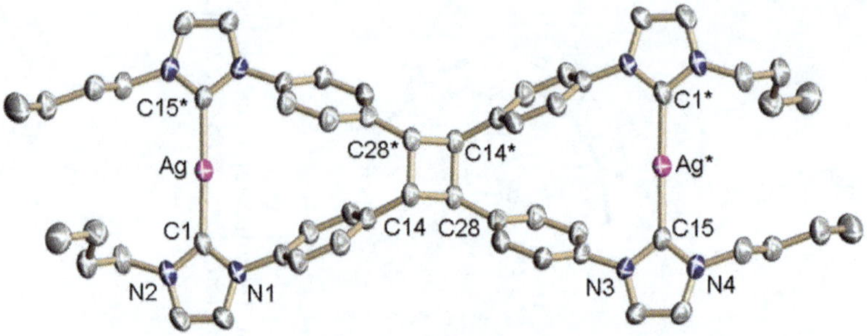

Figure 5.15 Photoproduct of NHC Ag(I)-based macrocycles.
(Reproduced with permission from Y. F. Han, G. X. Jin and F. E. Hahn, Postsynthetic Modification of Dicarbene-Derived Metallacycles via Photochemical [2+2] Cycloaddition, *J. Am. Chem. Soc.*, 2013, **135**, 9263–9266. Copyright 2013 American Chemical Society.)

Ag_2O or Cu_2O in acetonitrile yielding the corresponding dimetallic(I) tetracarbene salt. Transmetallation of the Ag(I) compound with [AuCl(tht)], (tht = tetrahydrothiophene) furnished the corresponding Au(I) rectangles. SXRD revealed that the intramolecular olefin bonds of the terminal coumarin ligands in the Au(I) compound are aligned to satisfy Schmidt's criteria for photodimerization.[10] Predictably, all these molecular rectangles of Cu(I), Ag(I) and Au(I) were shown to be photoreactive. Irradiation of the compounds under UV light with $\lambda = 365$ nm in d_6-DMSO and acetonitrile solutions led to the formation of HH photoproducts in 16–36 h, as established by ^{1}H-NMR spectroscopy as illustrated in Figure 5.16. Futhermore, the HH coumarin pendant dimer can be cleaved in solution by using a UV light of shorter wavelength ($\lambda = 254$ nm).

These results are quite significant for the following reasons. This is the first time that coumarin groups have been made photoreactive in metal complexes[15] and the dimerization is reminiscent of HH alignment by the argentophilic interactions or bridging anions as discussed earlier. Coumarin derivatives exhibit excellent photoluminescence property which can be switched off upon dimerization. Hence, this could be touted as a potential material for optical storage if this can be fine-tuned to function in the solid state.

5.6 Switching of Magnetic Properties due to [4+4] Cycloaddition Reaction in a Cu(II) Complex of Anthracene Derivatives

Although [4+4] photodimerization of anthracene derivatives has been extensively studied in solution,[83–87] this has not been incorporated as a ligand in metal complexes for this purpose in the solid state. Here is an example reporting an aligned pair of anthracene groups after

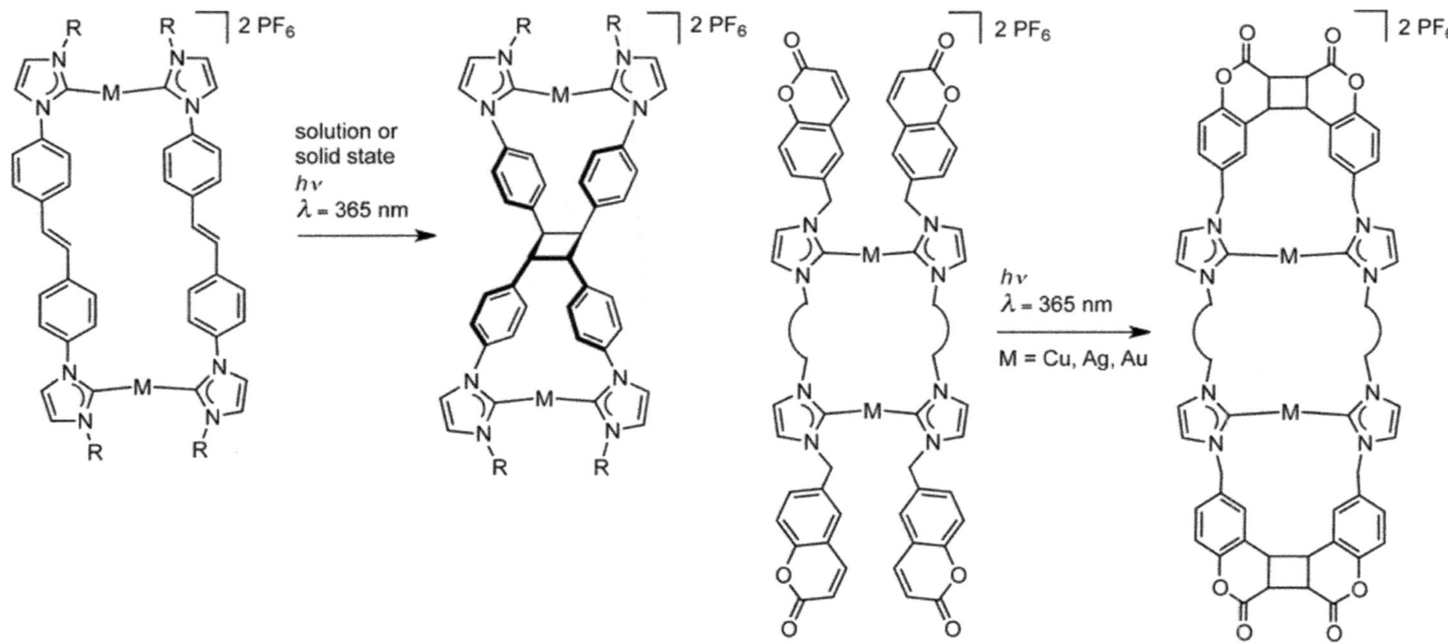

Figure 5.16 A scheme showing the synthesis, photodimerization and cycloreversion reaction involving a coumarin derivative of metal–carbene complexes.
(Reproduced with permission from Y. F. Han, G. X. Jin, C. G. Daniliuc and F. E. Hahn, Reversible Photochemical Modifications in Dicarbene-Derived Metallacycles with Coumarin Pendants, *Angew. Chem., Int. Ed.*, 2015, **54**, 4958–4962. Copyright © 2015 WILEY-VCH Verlag GmbH & Co. KGaA, Weinheim.)

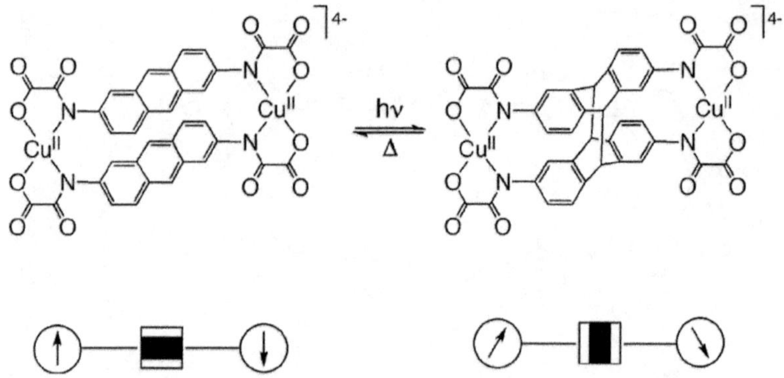

Figure 5.17 Illustration of the photomagnetic switching in dinuclear Cu(II) complex. (Reproduced from ref. 88 with permission from the Royal Society of Chemistry.)

derivatizing the backbone of a ligand in a Cu(II) complex.[88] In the complex $(nBu_4N)_4[Cu_2(anba)_2]$ (H_4anba = *N,N′*-2,6-anthracenebis(oxamic acid)), two Cu(II) atoms adopt square planar geometry by bonding to two anba ligands. The two anthracene groups are facing each other but displaced with an average intermolecular C···C distance of 3.678 Å that is slightly longer than the van der Waals contact (3.40 Å) as shown in Figure 5.17. Upon irradiation of this solid with UV light ($\lambda = 308$ nm) at room temperature for 24 h, the colour changed from green to brown. This green colour was partially recovered by heating it at 75 °C for 1 h. Although there is no direct structural evidence for the [4+4] cycloaddition reaction between the anthracene groups, the variable-temperature magnetic properties of the solid after irradiated under UV light at different intervals of time shows intriguing photomagnetic properties arising from the antiferromagnetically coupled to magnetically uncoupled (ON/OFF) switching by the photocycloaddition product. The green solid is moderately strongly antiferromagnetic ($J = -23.9$ cm^{-1}, $H = -J\,\mathbf{S}_1 \cdot \mathbf{S}_2$; $S_1 = S_2 = S_{cu} = 1/2$) due to coupling between the Cu(II) atoms that are separated by 12.477 Å.

5.7 Unusual Fusion of Interlocked Cages by [2+2] Cycloaddition Reaction

An interlocked Pd_6 cage $[\{(tmen)Pd\}_6(L)_3](NO_3)_{12}$ was synthesized by reacting 1,3,5-tris((*E*)-2-(pyridin-3-yl)vinyl)benzene (L) with $[(tmen)Pd(NO_3)_2]$.[89] In the solid state structure, one of the olefin double bonds of L was found to have a parallel conformation with its counterpart in the adjacent cage and separated by a distance of 3.656 Å, indicating the feasibility of [2+2] photochemical reaction. Two adjacent interlocked cages were covalently married together by intermolecular [2+2] cycloaddition in an SCSC fashion upon exposure to sunlight/UV irradiation as illustrated in Figure 5.18. The covalently bonded cyclobutane ring can be easily separated thermally in an

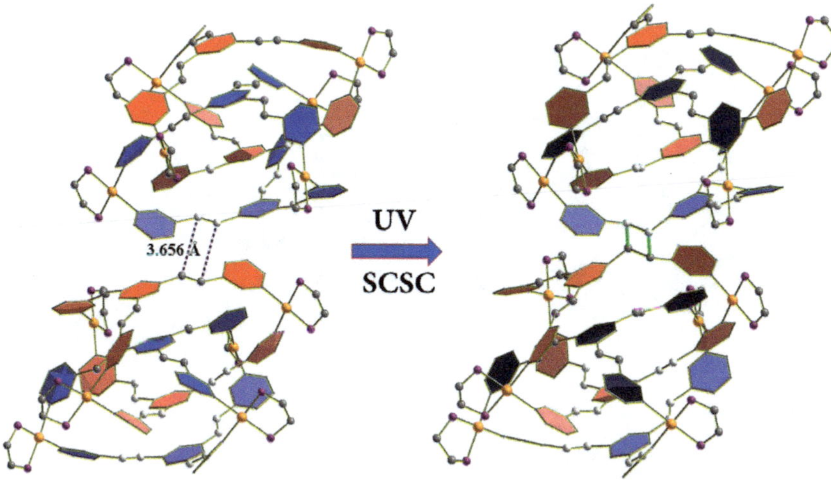

Figure 5.18 Formation of the interlocked Pd_6 cage and the product. Courtesy of Prof. P. S. Mukherjee.

aqueous medium under mild reaction conditions. This methodology could be a general route to mend cages together in the solid state through photodimerization reactions.

5.8 1D Coordination Polymers from the [2 + 2] Cycloaddition of Metal Complexes

Metal-mediated photopolymerization reactions were extensively reviewed by Georgiev and MacGillivray in 2008.[33] Hence, we present only the relevant solid-state photopolymerization *via* [2 + 2] cycloaddition reactions of metal complexes leading to the formation of coordination polymers (CPs).

5.8.1 Historical Developments

The ligand *trans,trans*-dibenzylideneacetone (dba) was synthesized as early as 1909 and its metal complexes have been investigated for their photoreactivity.[90–92] One of the olefin bonds from each dba ligand has been nicely aligned with its counterparts from the adjacent metal complexes in a slip-stacked manner in the complexes formed from tin(IV), uranyl, and other metals.[93–96] As a result, these complexes were found to undergo [2 + 2] cycloaddition reaction which connects all the monomeric metal complexes to form a chained polymer. The formation of truxillic type 1,3-bis(3-phenylprop-2-enol-oxo-2,4-bis(phenyl) cyclobutane product under UV light is evident from the packing of dba ligands in the solid-state structures of $[Sn(dba)_2Cl_4]$ and $[(UO_2)(dba)_2Cl_2]$ shown in Figure 5.19. Although the structures of the final products were not determined, they can be predicted to be polymeric from the initial way they were packed in the solid state.

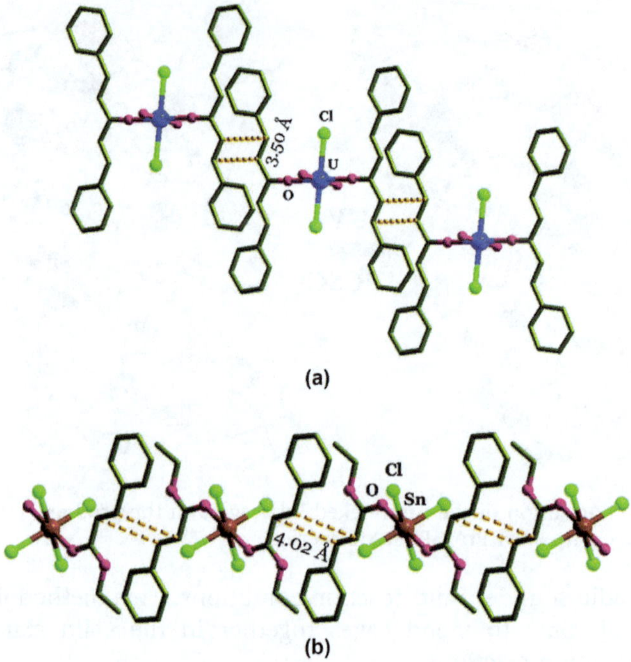

Figure 5.19 The alignment of olefin groups of dba ligands in (a) $[(UO_2)(dba)_2Cl_2]$ and (b) $[Sn(dba)_2Cl_4]$ complexes.
(Reproduced from ref. 126 with permission from the Royal Society of Chemistry.)

5.8.2 Polymerization of a Metal Complexes with bpe as a Monodentate Ligand

In the Mn(II) monomeric complex $[Mn(bpe)_2(H_2O)_4](ClO_4)_2 \cdot 2bpe$ (where bpe = 4,4′-bipyridylethelene), the bpe spacer ligand has been unusually acting as a monodentate ligand by binding the two axial positions of the octahedral Mn(II) with the equatorial sites being occupied by the aqua ligands.[97] The uncoordinated N atom of the bpe ligand is hydrogen-bonded to the aqua ligands from the neighbouring cations. These rare complementary O–H···N hydrogen bonds brought the bpe ligand pairs closer in a 'face-to-face' manner, such that the olefin bond pairs reacted under UV irradiation as shown in Figure 5.20. This leads to the formation of a hydrogen-bonded polymeric aggregate. Furthermore, the two bpe guest molecules are hydrogen-bonded to the aqua ligands to form a net and two such nets are interpenetrated in an inclined manner.

In this structure, shorter contacts were observed between the C=C bond pairs of the hydrogen-bonded guest bpe molecules (centroid-to-centroid distances of 3.95–3.98 Å) and the distances are longer for the metal coordinated bpe pairs (centroid-to-centroid distances of 4.41 Å) which is slightly above the limiting value proposed by Schmidt (<4.2 Å).[10] A concerted movement of the molecules during the photoreaction would bring the

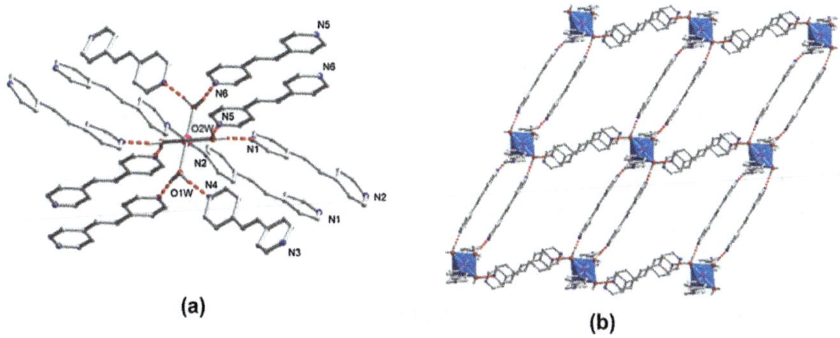

Figure 5.20 (a) A view showing the alignment of bpe bond pairs *via* O–H···N bonds and (b) 2D hydrogen-bonded layer structure.
(Reproduced from ref. 97 with permission from the Royal Society of Chemistry.)

metal-bound bpe pairs to achieve photocycloaddition. A series of photoreactive Mn(II) complexes containing *trans*-[Mn(bpe)$_2$] moieties similar to the one described here have been synthesized using both coordinating and non-coordinating anions. All these complexes have been found to be photoreactive under UV light, but do not maintain single crystallinity at the end of the reaction.[97] However, the HT alignment of the bpe pairs is expected to yield 1D CP as before.

5.8.3 Pedal Motion of Olefin Bonds by Grinding in a Zn(II) Complex

A structurally-similar Zn(II) complex $[Zn(bpe)_2(H_2O)_4](NO_3)_2 \cdot 8/3H_2O \cdot 2/3bpe$ (where bpe = 4,4′-bipyridylethelene) has also been found to be photoreactive. In this complex, the equatorial sites of Zn(II) are coordinated by water molecules and the axial sites are coordinated by the bpe ligands in a rare monodentate fashion for a spacer ligand. The uncoordinated N atom of the bpe ligand is hydrogen-bonded to the aqua ligands from the adjacent molecules in a complementary manner to form a hydrogen-bonded polymeric aggregate just like the manganese one. These complementary O–H···N hydrogen bonds brought the bpe ligands closer in a 'face-to-face' configuration congenial for photoreaction.[98] Although two-thirds of the bpe bond pairs are crisscrossed, UV irradiation of the powdered crystals still proceeded with 100% photoconversion in 25 h giving a 1D CP. Obviously, the crisscrossed olefin bond pairs undergo pedal motion[99,100] to align parallel before cyclization under UV light to form *rctt*-tpcb as before.

5.8.4 Photoreactivity in Similar Mn(II) Complexes Containing Unsymmetrical Ligands

Briceño *et al.* have utilized the *trans*-1-(2-pyridyl)-2-(4-pyridyl)ethylene (2,4-bpe) ligand to design similar photoreactive metal complexes.[101]

In $[Mn(2,4\text{-bpe})_2(OH_2)_4](ClO_4)_2 \cdot 2(2,4\text{-bpe}) \cdot 2H_2O$ and in $[Mn(2,4\text{-bpe})_2(NCS)_2(OH_2)_2]$, the two 2,4-bpe ligands are *trans* to each other and packed in an HT manner for the formation of 1D CPs. Indeed, both Mn(II) complexes have been found to undergo [2 + 2] cycloaddition reaction quantitatively as monitored by ^{1}H-NMR spectroscopy.[101] It is to be noted at this point, even though many of the reported coordination complexes in the literature have the olefin bonds nicely aligned for photodimerization, they are yet to be investigated.[102–104]

5.9 Photoreactive Polymorphs of Co-crystals Containing a Metal Complex and an Organic Molecule Yielding Two Different Cyclobutane Isomers

Here two forms of photoreactive Mn(II) complex described by Briceño have the same molecular composition $[Mn(NCS)_2(OH_2)_4] \cdot 4(2,2'\text{-bpe})$ (2,2′-bpe: *trans*-1,2-bis(2-pyridyl)ethylene), but crystallized in two different monoclinic space groups, namely $P2/n$ (polymorph **1**) and $P2_1/n$ (polymorph **2**).[105] Two different ratios of Mn(II) : 2,2′-bpe, namely 1 : 2 and 1 : 1 yielded **1** and **2** respectively. Concomitant crystallization occurred when a 1 : 1.5 ratio was employed. A Co(II) analogue similar to **2** was obtained irrespective of the metal : 2,2′-bpe ratio. These two can be described as co-crystal polymorphs rather than supramolecular isomers as these are metal complexes. In **1**, $[Mn(NCS)_2(OH_2)_4]$ are hydrogen-bonded to 2,2′-bpe through O–H···N interactions to form 2D aggregates. Two more 2,2′-bpe are aligned in a crisscrossed manner through two-fold crystallographic symmetry with the shortest centroid-to-centroid distance of 3.23 Å shown in Figure 5.21. The N atoms are hydrogen-bonded to the hydrogen atoms of an aqua ligand above and below in the cavity.

This 2,2′-bpe pairs underwent photodimerization when exposed to UV light for 7 days. The ^{1}H-NMR spectrum of the photoproduct has a singlet at 4.41 ppm, characteristic of the C–H protons of a cyclobutane ring with *rtct*-configuration.[105] This stereochemistry was also confirmed by SXRD of the isolated sample. Usually pedal motion occurs for such crisscrossed alignments in the solid state to form *rctt*-isomer. Here, since the *anti*-nitrogen atoms of the 2,2′-bpe ligands are bonded to two different Mn(II) atoms in the cavity, the pedal motions are prevented and the [2 + 2] cycloaddition afforded the *rtct* dimer.

In polymorph **2** and the Co(II) analogue, the way $[Mn(NCS)_2(OH_2)_4]$ and 2,2′-bpe hydrogen-bonded is different from polymorph **1**. One of the four 2,2′-bpe molecules hydrogen-bonded through O–H···N interactions to form a linear chain along the *a*-axis and three more 2,2′-bpe molecules are linking this chain along the *b*-axis to form a 2D aggregates. These 2,2′-bpe molecules aggregate in such a way that only one olefin pair is aligned face-to-face and separated by 3.39 Å in Mn(II) and 3.44 Å in Co(II) respectively as shown in Figure 5.22. They readily undergo photodimerization to form the *rctt*-dimer as confirmed by the singlet at 5.13 ppm in the ^{1}H-NMR spectra.

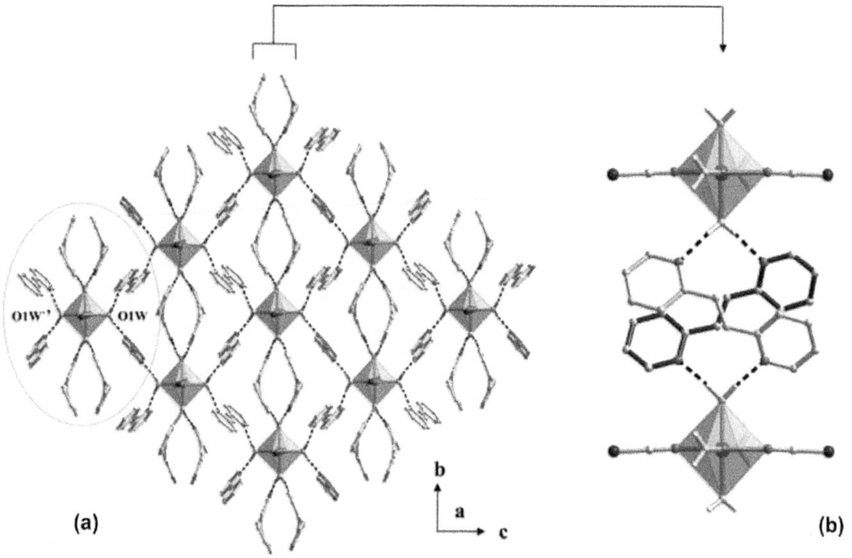

Figure 5.21 (a) Packing of $[Mn(NCS)_2(OH_2)_4]\cdot 4(2,2'\text{-bpe})$ in polymorph **1** and (b) a view showing the crisscross alignment of olefin bonds.
(Reproduced from ref. 105 with permission from the Royal Society of Chemistry.)

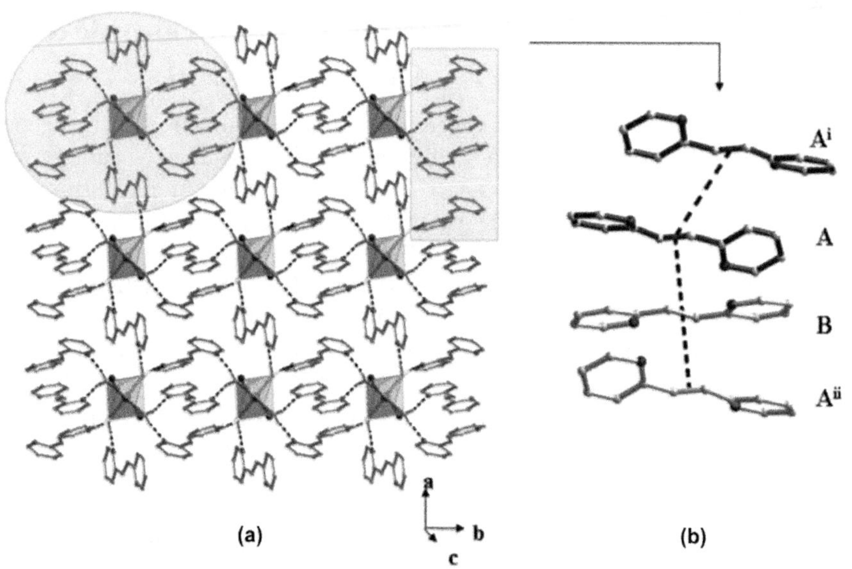

Figure 5.22 (a) Packing of $[Mn(NCS)_2(OH_2)_4]\cdot 4(2,2'\text{-bpe})$ in polymorph **2** and (b) alignment of the 2,2′-bpe.
(Reproduced from ref. 105 with permission from the Royal Society of Chemistry.)

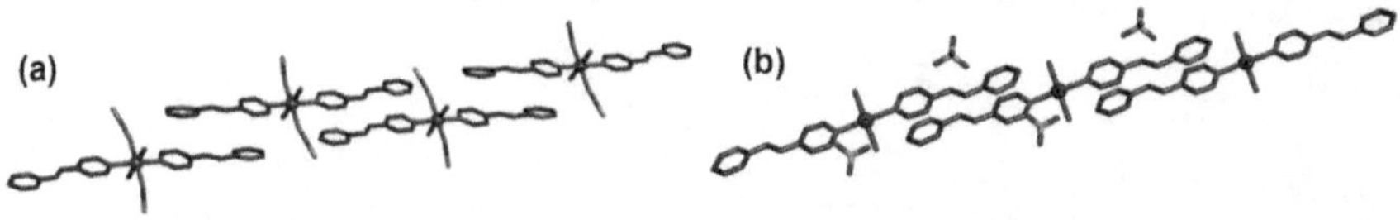

Figure 5.23 Packing in (a) $[Mn(2,4\text{-bpe})_2(OH_2)_4](ClO_4)_2 \cdot 2(2,4\text{-bpe}) \cdot 2H_2O$ and (b) $[Mn(2,4\text{-bpe})_2(NCS)_2(OH_2)_2]$ showing the alignments of 2,4-bpe ligands.

5.9.1 Photoreactivity in Similar Metal Complexes with Asymmetrical Bipyridyl Spacer Ligand

In the two photoreactive metal complexes $[Mn(2,4\text{-bpe})_2(OH_2)_4](ClO_4)_2 \cdot 2(2,4\text{-bpe}) \cdot 2H_2O$ and $[Mn(2,4\text{-bpe})_2(NCS)_2(OH_2)_2]$, the two 2,4′-bpe (2,4′-bpe = *trans*-1-(2-pyridyl)-2-(4-pyridyl)ethylene) spawned in *trans* fashion through 4-pyridyl groups of the ligands.[106] The adjacent 2,4′-bpe ligands are disposed in HT manner nicely and aligned face-to-face for [2 + 2] cycloaddition reaction as shown in Figure 5.23. The two 2,4′-bpe guest molecules in the former complex are hydrogen-bonded to the aqua ligand and nicely aligned in a HT fashion but facing each other in parallel. Both metal complexes undergo 100% photodimerization expected to yield a 1D CP. The NCS bound metal complex can be heated under reflux in a H_2O–MeOH mixture to produce the *rttt*-isomer. Even though such isomerization in solution is not something unexpected.[107]

5.10 Photoreactive Zwitter Ionic Metal Complexes Containing bpe-H$^+$ Cationic Ligand

One nitrogen atom of the bpe spacer ligand can be protonated to become like a monodentate ligand as described before.[107] Two bpe-H$^+$ cations were shown to coordinate to $[Pb(TFA)_4]^{2-}$ anion resulting in a zwitter ionic complex, $[Pb(TFA)_4(bpe\text{-}H)_2]$ (TFA = trifluoroacetate). The bpe-H$^+$ cationic ligands allowed complementary N–H···O hydrogen bonds to the oxygen atoms of the coordinated TFA ligands from the neighbouring molecule, resulting in a hydrogen-bonded 1D polymeric aggregate as shown in Figure 5.24. As a result, the bpe-H$^+$ bond pairs are aligned in parallel but in HT fashion supported by the donor–acceptor interactions between the pyridine and pyridinium groups. The lead(II) complex has been found to be photoreactive as anticipated. Interestingly, the photoreaction is accompanied by the change of crystalline solid into gel within five min before resulting in the formation of a powdered compound after 25 h. During this reaction, TFA acid has been eliminated by the transfer of proton from bpe-H$^+$ cation to TFA anion, which slowly evaporated from the photoproduct. The photoproduct was recrystallized from MeOH to confirm the formation of 1D polymer $[Pb(rctt\text{-tpcb})(TFA)_2]$ which shows slight change in the connectivity.

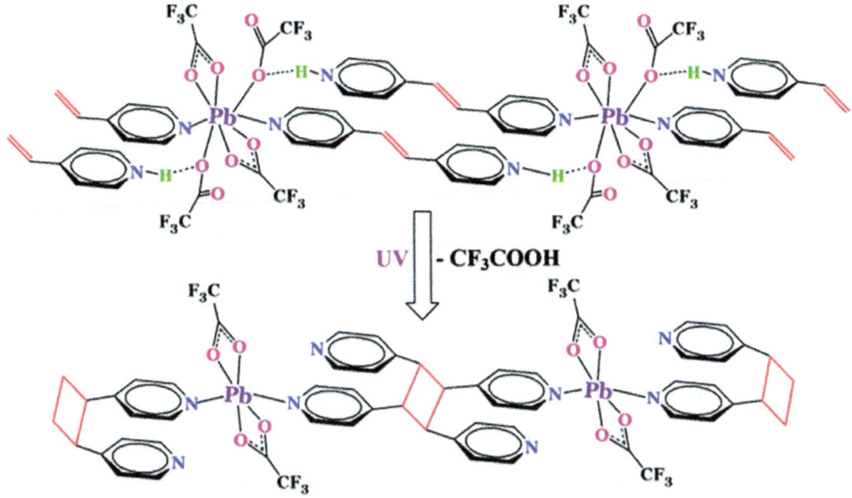

Figure 5.24 A schematic diagram showing the zigzag hydrogen-bonded zwitter ionic Pb(II) complex undergoes polymerization reaction under UV light. (Reproduced from ref. 126 with permission from the Royal Society of Chemistry.)

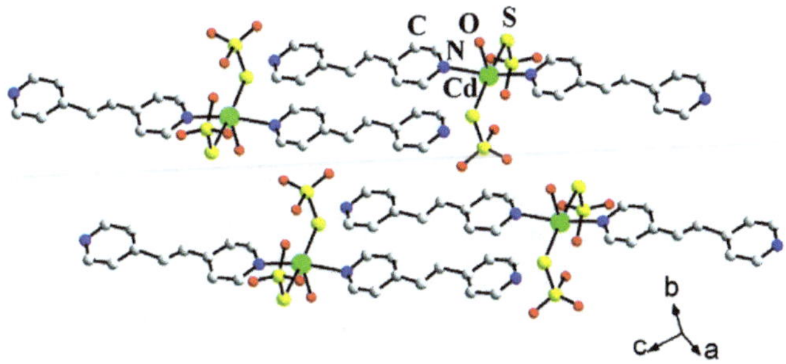

Figure 5.25 A schematic diagram depicting the alignment of monodentate bpe-H$^+$ ligands in a Cd(II) complex for the [2 + 2] polymerization. (Reproduced with permission from A. K. Paul, R. Karthik and S. Natarajan, Synthesis, Structure, Photochemical [2 + 2] Cycloaddition, Transformation, and Photocatalytic Studies in a Family of Inorganic-Organic Hybrid Cadmium Thiosulfate Compounds, *Cryst. Growth. Des.*, 2011, **11**, 5741–5749. Copyright 2011 American Chemical Society.)

Another zwitter ionic complex, [Cd(H_2O)(S_2O_3)$_2$(bpe-H)$_2$], has also been shown to have similar two monodentate bpe-H$^+$ cations coordinated to the metal centre in a *trans* fashion.[108] Further, the thiosulfate anions are hydrogen-bonded to the bpe-H$^+$ cations in a complementary fashion to achieve a 1D hydrogen-bonded aggregate as shown in Figure 5.25. Accordingly, the HT arrangement of the bpe-H$^+$ cationic pairs are separated by a distance of 3.72 Å. Quantitative solid-state [2 + 2] cycloaddition reaction between the olefin bond pairs has resulted in a 1D CP photoproduct.

5.11 Making the Terminal Ligands of the Metal Complexes Photoreactive

Occasionally, the bidentate bpe has been serendipitously monochelated to metal ions forming photoreactive metal complexes as described previously. Instead, a suitable monodentate ligand containing olefin bonds at the backbone can be incorporated in the design of photoreactive metal complexes. This idea is exemplified by using 4-styrylpyridine and its derivatives to generate an array of photoreactive metal complexes in the following section.

5.11.1 Photoreactivity of 4-Strylpyridine Ligands in Metal Complexes

In silver(I) dimer, $[Ag_2(Cl\text{-}spy)_4(ClO_3)_2]$ (Cl-spy = *trans*-1-(4-Cl-3-pyridyl)-2-(phenyl)ethylene), two chlorate anions bridge the $Ag(Cl\text{-}spy)_2$ groups forming a HT arrangements as shown in Figure 5.26.[109] The complexes are stacked in a one-dimensional array with the olefin bond pairs in parallel orientation and separated by 3.77 Å. Despite the phenyl and pyridyl groups not aligned

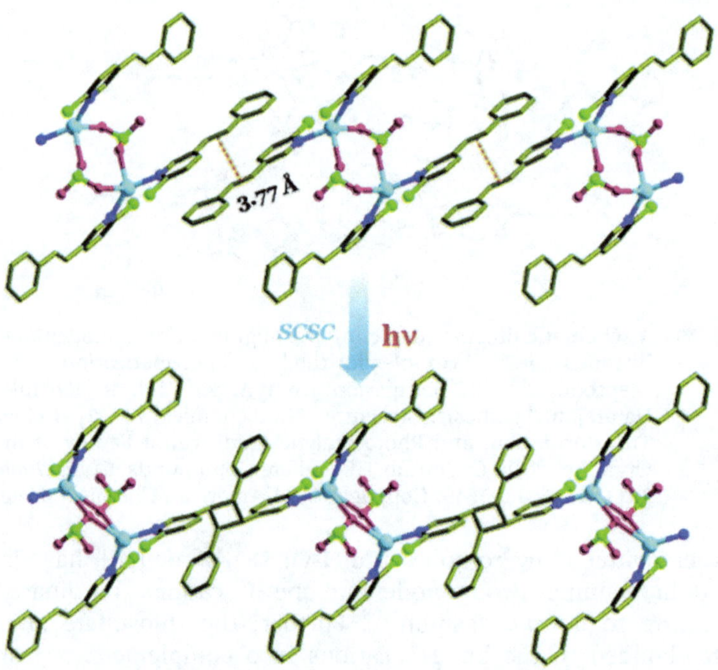

Figure 5.26 A schematic diagram showing the polymerization of Ag(I) dimer to a 1D CP through [2 + 2] cycloaddition reaction.
(Reproduced from ref. 126 with permission from the Royal Society of Chemistry.)

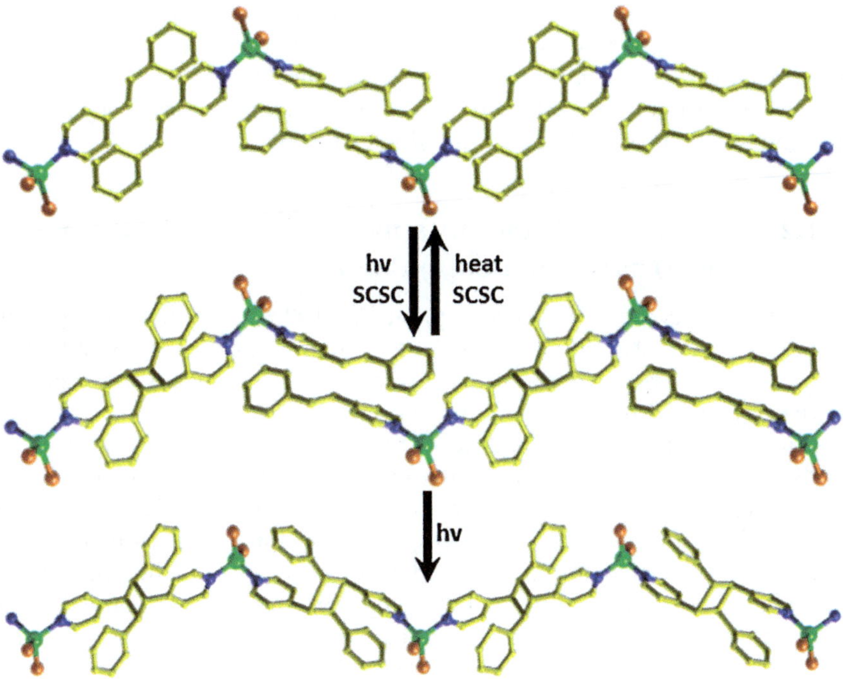

Figure 5.27 A schematic diagram showing polymerization through the dimer intermediate under UV light.
(Reproduced from ref. 110 with permission from the Royal Society of Chemistry.)

face-to-face, [2 + 2] cycloaddition reaction still occurred in an SCSC fashion to provide a 1D CP.

Usually, the 1D CPs are formed from the metal complexes in a single step reaction and no intermediates has been observed before. However, an unusual photopolymerization of Zn(II) complexes has been accomplished in a two-step reaction through the formation of cyclobutane rings.[110] Two bromide and two 4spy ligands provide a tetrahedral Zn(II) complex in the mononuclear $[Zn(L)_2Br_2]$ (where L = 4spy, 2F-4spy). These monomers are packed in a one-dimensional array such that one of these two 4spy ligands is nicely aligned in a HT way with its counterpart in the neighbouring molecule through π–π interactions between 4spy ligands. In contrast, the second 4-spy ligand is twisted with a torsional angle of 35.4° between the two rings, but stabilized *via* intermolecular C–H···π interactions in the crystal structure. Under UV light, [2 + 2] cycloaddition reaction occurs in an SCSC manner between the well-aligned 4-spy pairs yielding a dimer as shown in Figure 5.27. On prolonged UV irradiation of the single crystals, or when the crystals were ground to powder using pestle and mortar, complete photoconversion can be accomplished yielding a 1D CP, $[Zn_2($*rctt*-ppcb$)(4spy)_2Br_4]$ where *rctt*-ppcb is *rctt*-1,3-bis(4′-pyridyl)-2,4-bis(phenyl)cyclobutane.

Molecular movements are presumed to occur during grinding or photoreaction for photopolymerization to occur. The structure of the 1D CP was confirmed by recrystallization of the photoproduct. The cyclobutane ring can be cleaved in the photoproduct and converted back to monomer by heating the sample at 220 °C without loss of single crystallinity.

5.11.2 Unusual Solid-state Intermolecular Phenyl–Olefin Dimerization in a Zn(II) Complex

Intermolecular cyclization between phenyl and olefin groups is ubiquitous in solution.[111–113] Although intermolecular alignment between phenyl and olefin groups is very commonly observed in the solid state, the reaction between them was reported only for one case before.[114]

An unexpected [2 + 2] cycloaddition reaction between phenyl and olefin bonds in the crystals of a Zn(II) complex was serendipitously discovered in a metal complex.[115] Figure 5.28 shows an unusual unsymmetrical dimer core found in the Zn(II) complex, $[Zn_2(2F\text{-}4spy)_2(ptol)_4]$ (ptol = *para*-toluate). Each Zn(II) atom is chelated by a ptol ligand and further bridged by two ptol ligands. One of the oxygen atoms of a bridging ptol ligand is also bonded to the second Zn(II) atom to provide the unsymmetrical dimer. The 2F-4spy ligands

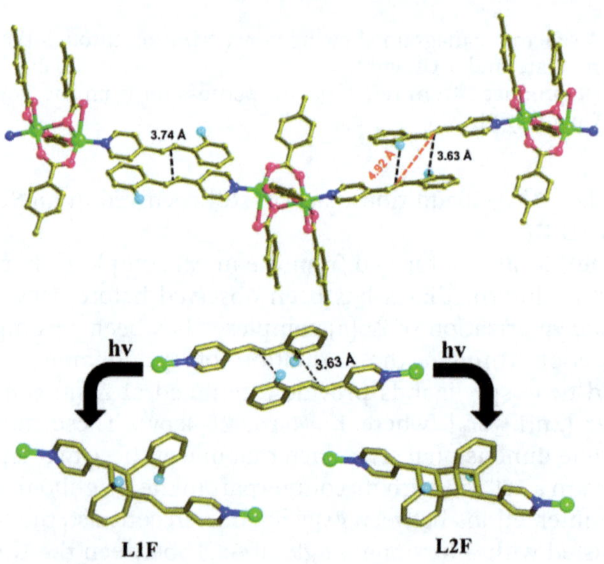

Figure 5.28 A structural diagram showing the formation of a coordination copolymer from a Zn(II) complex through [2 + 2] photoreaction between *olefin–olefin* and *phenyl–olefin* groups.
(Reproduced with permission from R. Medishetty, Z. Bai, H. Yang, M. W. Wong and J. J. Vittal, Influence of Fluorine Substitution on the Unusual Solid-State [2 + 2] Photo-Cycloaddition Reaction between an Olefin and an Aromatic Ring, *Cryst. Growth. Des.*, 2015, **15**, 4055–4061. Copyright 2015 American Chemical Society.)

are disposed in the axial sites and the dimers are packed linearly such that the neighbouring 2F-4spy ligands are aligned in a HT manner. Of these, one olefin bond pair is aligned in parallel ready for [2 + 2] cycloaddition reaction, while in the second HT pair of 2F-4spy ligands are slightly slip-stacked with the distance between the olefin centres at 4.92 Å. However, each olefin bond is closer to the C–C bond of the 2-fluorophenyl ring with a distance of separation of 3.63 Å in a complementary manner. Such bonding pairs will either be inert or photoreactive, if sliding of the ligands occurs to yield the expected cyclobutane ring. Unexpectedly, photoreaction between a phenyl group and an olefin bond occurs in an SCSC manner to form bicyclo[4.2.0]octa-2,4-diene derivative. Therefore, the product is a 1D coordination copolymer containing two different alternating repeating units as shown in Figure 5.28. Usually, for organic polymers, copolymerization requires two different monomeric units, but in this reaction, the same Zn(II) complex monomer is able to afford a hetero 1D CP.

When 2F-4spy has been replaced by 4-spy, the isotypical Zn(II) complex [Zn_2(4spy)$_2$(ptol)$_4$] has also been found to be photoreactive though no phenyl–olefin reaction was observed. When the single crystals were ground, or subjected to prolonged irradiation, normal cycloaddition reactions between olefin bond pairs occurs on both side of the 4-spy pairs quantitatively. This could be due to the sliding of molecules during the grinding process resulting in the reorientation of the misaligned olefin pairs. Further, the powder X-ray diffraction pattern of [Zn_2(4spy)$_2$(ptol)$_4$] was found to be different from that of the simulated pattern derived from single crystal data, lending a strong support to this observation. According to density functional theory (DFT) calculations, 2F-4spy is more susceptible to cycloaddition reactions due to the greater reduction of aromaticity of 2F-4spy compared to the 4-spy ligand.[115] The importance of fluorine substitution on the dynamics and reactivity in the solid state is well-documented in isomerization reactions.[116,117]

5.12 Photosalient Behaviour of Photoreactive Metal Complexes

Recently, mechanically responsive artificial materials are gaining interest as they mimic muscle movements in our body. These materials not only change their shape and size but also move from their original position in many interesting ways under external stimuli or catalysts. Understanding these phenomenon, as well as making new compounds exhibiting such behaviours, will help to design advanced actuators of molecular origin. Mechanical responses of single crystals like curling, crawling, wiggling, jumping, leaping, hopping, popping, splitting or exploding under light or heat single crystals are visually striking demonstrations for the conversion of energy to mechanical work.[118] In this context, thermosalient properties have been observed for the first time in a metal complex by Siedle and Etter.[119] Naumov, Boldyreva and co-workers reported[120] an interesting photosalient complex, $[Co(NH_3)_5(ONO)]Cl(NO_3)$ that jumps violently during photo-isomerization of a nitrite ligand from –ONO to $–NO_2$. These crystals hopped off the surface rapidly and travelled up to $10^2–10^5$ times their own lengths by the sudden release of collective energy through concerted action of intermolecular interactions at the molecular level.

On this note, solid-state [2 + 2] cycloaddition reaction is known to occur in thousands of solids since it was reported by Liberman in 1889.[121] Surprisingly, no attention was paid to photosalient behaviour. A visually appealing popping of single crystals during solid-state [2 + 2] cycloaddition reaction of three isostructural Zn(II) complexes under UV light have been reported with complexes $[Zn_2(L)_2(benzoate)_4]$ (L = 4-styrylpyridine, 2′-fluoro-4-styrylpyridine and 3′-fluoro-4-styrylpyridine).[122] All three complexes adopted paddle wheel structures with the ligands occupying the axial positions. They are stacked to form a 1D aggregate with the spy ligands disposed in a face-to-face manner for photodimerization reaction as shown in Figure 5.29.

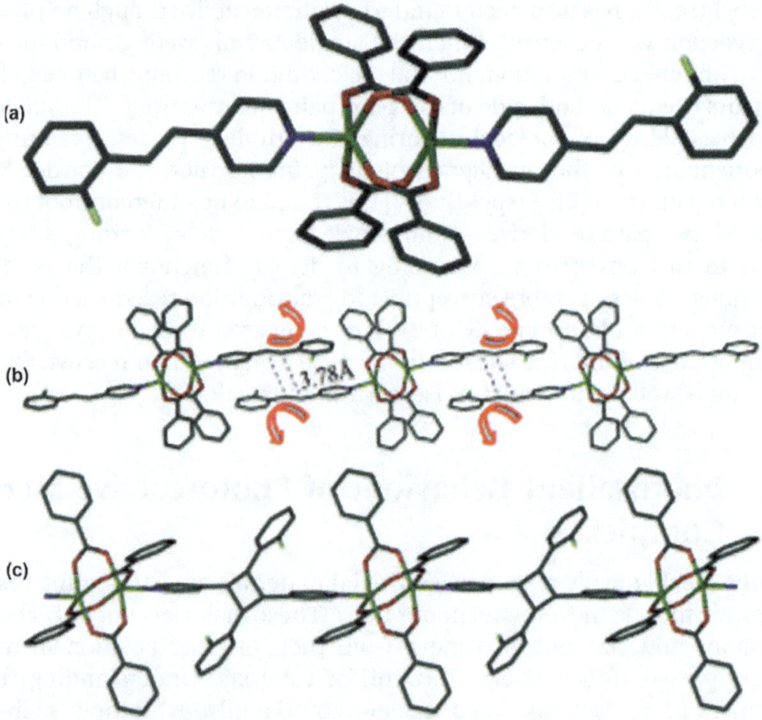

Figure 5.29 (a) Building unit of the Zn(II) complex. (b) Alignment of the olefins for photodimerization. (c) Photopolymerization of a photosalient Zn(II) complex.
(Reproduced with permission from R. Medishetty, A. Husain, Z. Bai, T. Runčevski, R. E. Dinnebier, P. Naumov and J. J. Vittal, Single Crystals Popping Under UV Light: A Photosalient Effect Triggered by a [2 + 2] Cycloaddition Reaction, *Angew. Chem., Int. Ed.*, 2014, 53, 5907–5911. Copyright 2014 WILEY-VCH Verlag GmbH & Co. KGaA, Weinheim.)

The HT arrangement of the spy ligand pairs readily undergoes [2 + 2] cycloaddition reaction and it was carefully observed that the single crystals were disintegrated under the UV light. The final photoproduct is a 1D CP which could be predicted from the crystal packing. It was further confirmed by SXRD after recrystallization of the photoproduct. Similarly, the photosalient phenomenon[120,122] is a consequence of stress accumulation at the molecular level accompanied by sudden strain released created by phase heterometry *via* an anisotropic crystal volume expansion. Understanding the origin of this effect could help in harnessing light energy and converting it into mechanical energy.

In order to probe deeper into the photosalient behaviour of photoreactive metal complexes, the study was extended to nine Ag(I) complexes $[Ag(L)_2(X)_2]$ (L = 4-styrylpyridine and 2′-fluoro-4-styrylpyridine, X = NO_3^-, BF_4^- and ClO_4^-).[123] They are all found to be photoreactive in the solid state but only six of them exhibit photosalient property (Table 1). Indeed, the photosalient activity is accompanied by a volume expansion of the crystal during photodimerization. The spy ligands in four Ag(I) metal complexes have been found to pack in a HT fashion as illustrated in Figure 5.30. Interestingly, all these Ag(I) complexes are photoreactive and two Ag(I) complexes containing 2F-4spy ligands are found to exhibit photosalient behaviour.[123] In all these complexes, the weakly coordinating anions exert little control on the geometry and hence Ag(I) exhibits linear coordination geometry by bonding to two pyridyl ligands. The HT alignment between the organic ligands is likely to produce 1D CP as depicted in Figure 5.30.

Table 5.1 Photoreactivity and photosalient properties of Ag(I) complexes.

No	Compound	C=C alignment	Photo-product^a	Rate^b/ min^{-1}	Photo-salient	Density change^c	%Vol. change^d
1	$[Ag(4spy)_2]BF_4$	HT	1D CP	0.0015	No	1.614/1.55(2)	3.97
2	$[Ag(4spy)_2]ClO_4$	HT	1D CP	0.0155	No	1.614/1.53(2)	5.20
3	$[Ag(2F\text{-}4spy)_2]BF_4$	HT	1D CP	0.0177	Yes	1.703/1.38(2)	18.97
4	$[Ag(2F\text{-}4spy)_2]ClO_4$	HT	1D CP	0.0383	Yes	1.706/1.45(2)	15.01
5	$[Ag(3F\text{-}4spy)2]BF_4$	HH	Dimer	0.1267	Yes	1.654/1.43(1)	13.54
6	$[Ag(3F\text{-}4spy)_2]ClO_4$	HH	Dimer	0.2482	Yes	1.673/1.43(3)	14.52
7	$[Ag(4spy)2]NO_3 \cdot 1/3H_2O$	HH	Dimer	0.0009	Yes	1.577/1.37(1)	13.13
8	$[AgNO_3(3F\text{-}4spy)_2]$	HH	Dimer	0.1103	Yes	1.669/1.54(4)	7.73
9	$[AgNO_3(2F\text{-}4spy)_2] \cdot MeOH$	HH	1D CP	0.0135	No	1.673/1.55(1)	7.35

^{a}The product was predicted based on the crystal packing.
^{b}The rate was determined based on first-order kinetics. In case of no. 7 the rate constant is reported only for the first step.
^{c}The density of the photoproduct was determined experimentally by the flotation method ($g\,cm^{-3}$).
^{d}Change in volume of the crystalline solid calculated from the experimentally determined density change.

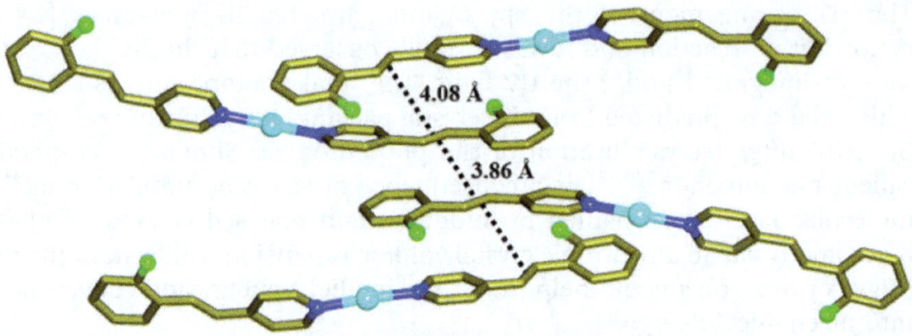

Figure 5.30 A schematic diagram showing the HT alignment of (2F-4spy)-Ag-(2F-4spy). (Reproduced with permission from R. Medishetty, S. C. Sahoo, C. E. Mulijanto, P. Naumov and J. J. Vittal, Photosalient Behaviour of Photoreactive Crystals, *Chem. Mater.*, 2015, **27**, 1821–1829. Copyright 2015 American Chemical Society.)

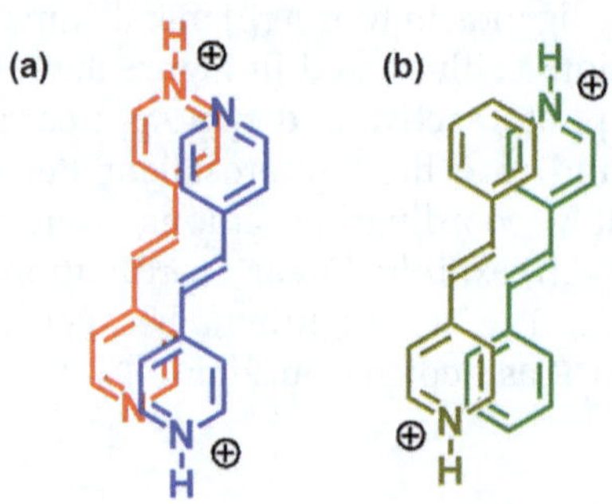

Figure 5.31 HT alignment of two pyridinium ligands, (a) 4,4′-bpe-H pairs, (b) 4-spy pairs.

5.13 Alignment of Pyridinium Cations in Metal Salts

A simple way to align a pair of dipyridyl spacer ligands like 2,2′-bpe and 4,4′-bpe is to protonate one of the pyridyl nitrogen atoms. This will force them to form donor–acceptor dimers through electrostatic interactions as shown in Figure 5.31. Protonated 4-spy ligands are also expected to align in pairs due to the donor–acceptor interactions. Briceño and Hill have cleverly used this strategy and made a number of photoreactive salts of these cations.[124]

In a number of metal complex salts, the protonated 2,2′-bpe, 4,4′-bpe and 4spy ligands have been found to align in HT manner with the pyridinium cation facing the pyridine groups similar to the perflurophenyl–phenyl interaction discussed earlier. Of the seven compounds investigated, (2,2′-bpe-H)(2,2′-bpe-H_2)$_{0.5}$[$Zn(NCS)_4$]·2(H_2O), (2,2′-bpe-H_2)[$Zn(NCS)_4$]·2(H_2O), (4-spy-H)$_2$[$Zn(NCS)_4$], (4-spy-H)$_2$[$Co(NCS)_4$], (4-spy-H)$_2$[$ZnCl_4$], (4,4′-bpe-H_2)[$Zn(NCS)_4$]$_2$ and (2,2′-bpe-H)$_4$[$Fe(NCS)_6$]·4H_2O, the first four salts have been found to be photoreactive in the solid state upon UV irradiation, leading to the formation of *rctt*-cyclobutane derivatives from moderate to high yield.

5.14 Conclusions and Summary

In this chapter, we have surveyed the current status of the solid state [2 + 2] photoreactivity of metal complexes. Various supramolecular interactions such as argentophilic, aurophilic, π–π, C–H···π, N–H···O, O–H···O, O–H···N interactions can be harnessed to design photoreactive metal complexes. Confining ligands in mostly molecular rectangles or organometallic cages, ligand design and clipping agents are alternative strategies to obtain photoreactive solids. Many of these molecular rectangles are able to maintain their structural integrity and reactivity in solution. During the formation of cyclobutane rings, various metal complexes have shown to form dimers and polymers depending on the way the ligand pairs are aligned. The cleavage of cyclobutane rings back to *trans*-olefins are gaining importance recently as such cycloreversion has not been investigated in most cases. Most often, photodimerization that did not occur quantitatively have neither been well-understood nor fully investigated. In some rare cases, the expected pedal motion did not occur for crisscrossed olefin pairs, but photodimerization yielded unexpected *rtct*-isomers in the solid state. These examples provide better understanding to control the reactivity of the olefin pairs for synthesizing the desired cyclobutane isomers. Recently, photoreactive solids were observed to undergo mechanical motions during the photochemical reactions. Such properties are usually accompanied by a sudden release in strain and an expansion in crystal volume during the photodimerization process. Although the discovery suggested the possibility of many overlooked photoreactive solids that have been documented to date, it is an encouragement that the few mentioned photosalient solids are only the tip of an iceberg waiting to be uncovered in years to come. Bending of crystals has been observed during photodimerization reaction of organic crystals,[125] though such property is yet to be discovered in metal complexes. It would be of interest to study the optical and magnetic properties accompanied by bending of crystals of metal complexes.

Acknowledgements

We thank the Ministry of Education, Singapore for funding various research projects in the past several years through NUS Tier 1 grants especially R-143-000-604-112 and R-143-000-678-112. We thank Prof. P. S. Mukherjee for providing Figure 5.18.

References

1. V. V. Boldyrev, *Reactivity of Solids: Past, Present, and Future*, Blackwell Science, 1996.

2. F. Toda, *Organic Solid State Reactions*, Springer, 2005.
3. K. Tanaka and F. Toda, *Chem. Rev.*, 2000, **100**, 1025–1074.
4. F. Toda, *Acc. Chem. Res.*, 1995, **28**, 480–486.
5. A. E. Keating and M. A. Garcia-Garibay, in *Organic and Inorganic Photochemistry*, ed. V. Ramamurthy and K. S. Schanze, CRC Press, Marcel Dekker, New York, 1998, pp. 195–248.
6. G. S. Papaefstathiou and L. R. MacGillivray, *Top. Curr. Chem.*, 2004, **248**, 201–221.
7. G. R. Desiraju, J. J. Vittal and A. Ramanan, *Crystal Engineering. A Textbook*, World Scientific, 2011.
8. G. R. Desiraju, *Angew. Chem., Int. Ed. Engl.*, 1995, **34**, 2311–2327.
9. M. D. Cohen, G. M. J. Schmidt and F. I. Sonntag, *J. Chem. Soc.*, 1964, 2000–2013.
10. G. M. J. Schmidt, *Pure Appl. Chem.*, 1971, **27**, 647–678.
11. V. Ramamurthy and K. Venkatesan, *Chem. Rev.*, 1987, **87**, 433–481.
12. C. R. Theocharis and W. Jones, in *Organic Solid State Chemistry*, ed. G. R. Desiraju, Elsevier, 1987, pp. 47–68.
13. B. S. Green, M. Lahav and D. Rabinovich, *Acc. Chem. Res.*, 1979, **12**, 191–197.
14. W. Jones, *Organic Molecular Solids: Properties and Applications*, CRC Press, Boca Raton, Florida, 1997.
15. K. Tanaka, *Molecules*, 2012, **17**, 1408–1418.
16. E. Lee-Ruff and G. Mladenova, *Chem. Rev.*, 2003, **103**, 1449–1484.
17. J. C. Namyslo and D. E. Kaufmann, *Chem. Rev.*, 2003, **103**, 1485–1538.
18. J. Nishimura, Y. Nakamura, Y. Hayashida and T. Kudo, *Acc. Chem. Res.*, 2000, **33**, 679–686.
19. E. Elacqua, T. Friščić and L. R. MacGillivray, *Isr. J. Chem.*, 2012, **52**, 53–59.
20. L. R. MacGillivray, G. S. Papaefstathiou, T. Friščić, T. D. Hamilton, D.-K. Bučar, Q. Chu, D. B. Varshney and I. G. Georgiev, *Acc. Chem. Res.*, 2008, **41**, 280–291.
21. K. Biradha and R. Santra, *Chem. Soc. Rev.*, 2013, **42**, 950–967.
22. M. Nagarathinam and J. J. Vittal, *Macromol. Rapid Commun.*, 2006, **27**, 1091–1099.
23. M. Nagarathinam, A. M. P. Peedikakkal and J. J. Vittal, *Chem. Commun.*, 2008, **42**, 5277–5288.
24. J. J. Vittal, *Coord. Chem. Rev.*, 2007, **251**, 1781–1795.
25. G. K. Kole and J. J. Vittal, *Chem. Soc. Rev.*, 2013, **42**, 1755–1775.
26. W. L. Leong and J. J. Vittal, *Chem. Rev.*, 2011, **111**, 688–764.
27. R. Medishetty and J. J. Vittal, Metal-organic frameworks for photochemical reactions, in *Metal-Organic Frameworks for Photonics Applications*, ed. B. Chen and G. Qian, Springer, 2014, pp. 105–144.
28. A. Chanthapally and J. J. Vittal, Metal-Organic Frameworks: Photoreactive Frameworks, in *Metal-Organic Framework Materials*, ed. L. R. MacGillivray and C. M. Lukehart, John Wiley & Sons, Ltd, Chichester, UK, 2014, pp. 135–157.

29. C. P. Li, J. Chen, C. S. Liu and M. Du, *Chem. Commun.*, 2015, **51**, 2768–2781.
30. A. Schneemann, V. Bon, I. Schwedler, I. Senkovska, S. Kaskel and R. A. Rischer, *Chem. Soc. Rev.*, 2014, **43**, 6062–6096.
31. C. P. Li and M. Du, *Chem. Commun.*, 2011, **47**, 5958–5972.
32. T. Friščiicì and L. R. MacGillivray, *Z. Kristallogr.*, 2005, **220**, 351–363.
33. I. G. Georgiev and L. R. MacGillivray, *Chem. Soc. Rev.*, 2007, **36**, 1239–1248.
34. H. Schmidbaur and A. Schier, *Angew. Chem., Int. Ed.*, 2015, **54**, 746–784.
35. P. Pyykko, *Chem. Rev.*, 1997, **97**, 597–636.
36. A. N. Khlobystov, A. J. Blake, N. R. Champness, D. A. Lemenovskii, A. G. Majouga, N. V. Zyk and M. Schroder, *Coord. Chem. Rev.*, 2001, **222**, 155–192.
37. M. J. Katz, K. Sakai and D. B. Leznoff, *Chem. Soc. Rev.*, 2008, **37**, 1884–1895.
38. I. G. Georgiev, D. K. Bučar and L. R. MacGillivray, *Chem. Commun.*, 2010, **46**, 4956–4958.
39. D. Chen, X. Han, W. Jin, Y. Du and P. Xu, *Chem. Commun.*, 2014, **50**, 15631–15633.
40. Q. Chu, D. C. Swenson and L. R. MacGillivray, *Angew. Chem., Int. Ed.*, 2005, **44**, 3569–3572.
41. J. Harada and K. Ogawa in *Topics in Stereochemistry*, ed. S. E. Denmark and J. Siegel, Wiley, Chichester, 2006, vol. 2, pp. 31–47.
42. K. M. Hutchins, T. P. Rupasinghe, L. R. Ditzler, D. C. Swenson, J. R. G. Sander, J. Baltrusaitis, A. V. Tivanski and L. R. MacGillivray, *J. Am. Chem. Soc.*, 2014, **136**, 6778–6781.
43. R. Santra and K. Biradha, *Cryst. Growth Des.*, 0, **10**, 3315–3320.
44. G. K. Kole, G. K. Tan and J. J. Vittal, *Cryst. Growth Des.*, 2012, **12**, 326–332.
45. R. C. Laird, M. A. Sinnwell, N. P. Nguyen, D. C. Swenson, S. V. S. Mariappan and L. R. MacGillivery, *Org. Lett.*, 2015, **17**, 3233–3235.
46. G. W. Coates, A. R. Dunn, L. M. Henling, J. W. Ziller, E. B. Lobkovsky and R. H. Grubbs, *J. Am. Chem. Soc.*, 1998, **120**, 3641–3649.
47. M. A. Sinnwell, J. Baltrusaitis and L. R. MacGillivray, *Cryst. Growth Des.*, 2015, **15**, 538–541.
48. Y. Sonoda, M. Goto, S. Tsuzuki, H. Akiyama and N. Tamaoki, *J. Fluorine Chem.*, 2009, **130**, 151–157.
49. Y. Sonoda, *Molecules*, 2011, **16**, 119–148.
50. P. A. Deck, M. J. Lane, J. L. Montgomery, C. Slebodnick and F. R. Fronczek, *Organometallics*, 2000, **19**, 1013–1024.
51. K. Autumn, Y. A. Liang, S. T. Hsieh, W. Zesch, W. P. Chan, T. W. Kenny, R. Fearing and R. J. Full, *Nature*, 2000, **405**, 681–685.
52. S. Sponberg, T. W. Kenny, R. Fearing, J. N. Israelachvili and R. J. Full, *Proc. Natl. Acad. Sci. U. S. A.*, 2002, **99**, 12252–12256.
53. J. L. Atwood, L. J. Barbour and A. Jerga, *Science*, 2002, **296**, 2367–2369.
54. J. L. Atwood, L. J. Barbour and A. Jerga, *Angew. Chem., Int. Ed.*, 2004, **43**, 2948–2950.

55. L. J. Barbour, M. R. Caira and L. R. Nassimbeni, *J. Chem. Soc., Perkin Trans. 2*, 1993, 1413–1414.
56. L. Tian and J. J. Vittal, *Cryst. Growth Des.*, 2006, **6**, 822–824.
57. X. D. Yang, D. Wu, J. D. Ranford and J. J. Vittal, *Cryst. Growth Des.*, 2005, **5**, 41–43.
58. F. P. A. Fabbiani and C. R. Pulham, *Chem. Soc. Rev.*, 2006, **35**, 932–942.
59. G. K. Kole, R. Medishetty, L. L. Koh and J. J. Vittal, *Chem. Commun.*, 2013, **49**, 6298–6300.
60. J. J. Vittal and X. D. Yang, *Cryst. Growth Des.*, 2002, **2**, 259–262.
61. M. D. Fryzuk, *Inorg. Chem.*, 2015, **54**, 96710–99674.
62. R. Robson, *Inorg. Nucl. Chem. Lett.*, 1970, **6**, 125–128.
63. G. S. Papaefstathiou, Z. Zhong, L. Geng and L. R. MacGillivray, *J. Am. Chem. Soc.*, 2004, **126**, 9158–9159.
64. G. S. Papaefstathiou, I. G. Georgiev, T. Friščič and L. R. MacGillivray, *Chem. Commun.*, 2005, **31**, 3974–3976.
65. R. J. Puddephatt, *Chem. Soc. Rev.*, 1983, **12**, 99–127.
66. M. J. Irwin, J. J. Vittal, G. P. A. Yap and R. J. Puddephatt, *J. Am. Chem. Soc.*, 1996, **118**, 13101–13102.
67. M. C. Brandys, M. C. Jennings and R. J. Puddephatt, *J. Chem. Soc., Dalton Trans.*, 2000, 4601–4606.
68. A. Deak, T. Tunyogi, G. Tarkanyi, P. Kiraly and G. Palinkas, *CrystEngComm*, 2007, **9**, 640–643.
69. M. G. Brandys and R. J. Puddephatt, *Chem. Commun.*, 2001, **16**, 1508–1509.
70. M. H. Mir, J. X. Ong, G. K. Kole, G. K. Tan, M. J. McGlinchey, Y. Wu and J. J. Vittal, *Chem. Commun.*, 2011, **47**, 11633–11635.
71. M. Lahav and G. M. J. Schmidt, *J. Chem. Soc. B*, 1967, 312–317.
72. T. Odani, S. Okada, C. Kabuto, T. Kimura, S. Shimada, H. Matsuda, H. Oikawa, A. Matsumoto and H. Nakanishi, *Cryst. Growth Des.*, 2009, **9**, 3481–3487.
73. A. Michaelides, S. Shoulika and M. G. Siskos, *Chem. Commun.*, 2011, **47**, 7140–7142.
74. Y. F. Han and G. X. Jin, *Acc. Chem. Res.*, 2014, **47**, 3571–3579.
75. Z. Z. Lu, C. C. Lee, M. Velayudham, L. W. Lee, J. Y. Wu, T. S. Kuo and K. L. Lu, *Chem. – Eur. J.*, 2012, **18**, 15714–15721.
76. B. Manimaran, R. Rajendran, Y. L. Lu, G. H. Lee, S. M. Peng and K. L. Lu, *J. Chem. Soc., Dalton Trans.*, 2001, 515–517.
77. Y. F. Han, Y. J. Lin, W. G. Jia, G. L. Wang and G. X. Jin, *Chem. Commun.*, 2008, **15**, 1807–1809.
78. W. Z. Zhang, Y. F. Han, Y. J. Lin, W. G. Jia, G. L. Wang and G. X. Jin, *Organometallics*, 2010, **29**, 2842–2849.
79. A. Michaelides, S. Skoulika and M. G. Siskos, *Chem. Commun.*, 2004, **21**, 2418–2419.
80. W. B. Yu, Y. F. Han, Y. J. Lin and G. X. Jin, *Chem. – Eur. J.*, 2011, **17**, 1863–1871.
81. Y. F. Han, G. X. Jin and F. E. Hahn, *J. Am. Chem. Soc.*, 2013, **135**, 9263–9266.

82. Y. F. Han, G. X. Jin, C. G. Daniliuc and F. E. Hahn, *Angew. Chem., Int. Ed.*, 2015, **54**, 4958–4962.
83. H. Bouas-Laurent, A. Castellan and J. P. Desvergne, *Pure Appl. Chem.*, 980, **52**, 2633–2648.
84. H. D. Becker, *Chem. Rev.*, 1993, **93**, 145–172.
85. H. Bouas-Laurent, A. Castellan, J. P. Desvergne and R. Lapouyade, *Chem. Soc. Rev.*, 2000, **29**, 43–55.
86. H. Bouas-Laurent, A. Castellan, J. P. Desvergne and R. Lapouyade, *Chem. Soc. Rev.*, 2001, **30**, 248–263.
87. H. Meier and D. Cao, *Chem. Soc. Rev.*, 2013, **42**, 143–155.
88. M. Castellano, J. Ferrando-Soria, E. Pardo, M. Julve, F. Lloret, C. Mathonière, J. Pasán, C. Ruiz-Pèrez, L. Cañadillas-Delgado, R. Ruiz-Garcia and J. Cano, *Chem. Commun.*, 2011, **47**, 11035–11037.
89. D. Samanta and P. S. Mukherjee, *J. Am. Chem. Soc.*, 2014, **136**, 17006–17009.
90. P. Praetorius, F. H. Stobbe and E. Färber, *Ber. Dtsch. Chem. Ges.*, 1925, **58**, 1548–1553.
91. P. Praetorius, Thesis, Halle, 1909.
92. P. Praetorius and F. Korn, *Ber. Dtsch. Chem. Ges.*, 1910, **43**, 2744–2746.
93. C. R. Theocharis, *J. Chem. Soc., Chem. Commun.*, 1975, **19**, 785–786.
94. N. W. Alcock, N. Herron and T. J. Kemp, *J. Chem. Soc., Chem. Commun.*, 1987, 80–81.
95. N. W. Alcock, P. de Meester and T. J. Kemp, *J. Chem. Soc., Perkin Trans. 2*, 1979, 921–926.
96. N. W. Alcock and P. de Meester, *Acta Crystallogr., Sect. B: Struct. Crystallogr.*, 1979, **35**, 609–611.
97. A. Briceño, Y. Hill, T. Gonzalez and G. Diaz de Delgado, *Dalton Trans.*, 2009, **9**, 1602–1610.
98. A. M. P. Peedikakkal and J. J. Vittal, *Chem. – Eur. J.*, 2008, **14**, 5329–5334.
99. J. Harada and K. Ogawa, *Chem. Soc. Rev.*, 2009, **38**, 2244–2252.
100. J. Harada and K. Ogawa, *Cryst. Growth Des.*, 14, **14**, 5182–5188.
101. Y. Hill and A. Briceño, *Chem. Commun.*, 2007, **38**, 3930–3932.
102. G. De Munno, D. Armentano, T. Poerio, M. Julve and J. A. Real, *J. Chem. Soc., Dalton Trans.*, 1999, 1813–1818.
103. G. De Munno, F. Cipriani, D. Armentano, M. Julve and J. A. Real, *New J. Chem.*, 2001, **25**, 1031–1036.
104. M. Du, C. P. Li and X. J. Zhao, *CrystEngComm*, 2006, **8**, 552–562.
105. A. Briceño, D. Leal, R. Atencioa and G. D. Delgado, *Chem. Commun.*, 2006, **33**, 3534–3536.
106. Y. Hill and A. Briceño, *Chem. Commun.*, 2007, **38**, 3930–3932.
107. A. M. P. Peedikakkal, L. L. Koh and J. J. Vittal, *Chem. Commun.*, 2008, **4**, 441–443.
108. A. K. Paul, R. Karthik and S. Natarajan, *Cryst. Growth Des.*, 2011, **11**, 5741–5749.
109. S. Dutta, D. K. Bučar, E. Elacqua and L. R. MacGillivray, *Chem. Commun.*, 2013, **49**, 1064–1066.

110. R. Medishetty, T. T. S. Yap, L. L. Koh and J. J. Vittal, *Chem. Commun.*, 2013, **49**, 9567–9569.
111. R. Alonso and T. Bach, *Angew. Chem., Int. Ed.*, 2014, **53**, 4368–4371.
112. N. Hoffmann, *Synthesis*, 2004, **4**, 481–495.
113. T. Bach and J. P. Hehn, *Angew. Chem., Int. Ed.*, 2011, **50**, 1000–1045.
114. Y. Ito, S. Horie and Y. Shindo, *Org. Lett.*, 2001, **3**, 2411–2413.
115. R. Medishetty, Z. Bai, H. Yang, M. W. Wong and J. J. Vittal, *Cryst. Growth Des.*, 2015, **15**, 4055–4061.
116. D. Bléger, J. Schwarz, A. M. Brouwer and S. Hecht, *J. Am. Chem. Soc.*, 2012, **134**, 20597–20600.
117. O. S. Bushuyev, A. Tomberg, T. Friščić and C. J. Barrett, *J. Am. Chem. Soc.*, 2013, **135**, 12556–12559.
118. N. K. Nath, M. K. Panda, S. C. Sahoo and P. Naumov, *CrystEngComm*, 2014, **16**, 1850–1858.
119. M. C. Etter and A. R. Siedle, *J. Am. Chem. Soc.*, 1983, **105**, 641–642.
120. P. Naumov, S. C. Sahoo, B. A. Zakharov and E. V. Boldyreva, *Angew. Chem., Int. Ed.*, 2013, **52**, 9990–9995.
121. C. Liebermann and O. Bergami, *Ber. Dtsch. Chem. Ges.*, 1889, **22**, 782–786.
122. R. Medishetty, A. Husain, Z. Bai, T. Runčevski, R. E. Dinnebier, P. Naumov and J. J. Vittal, *Angew. Chem., Int. Ed.*, 2014, **53**, 5907–5911.
123. R. Medishetty, S. C. Sahoo, C. E. Mulijanto, P. Naumov and J. J. Vittal, *Chem. Mater.*, 2015, **27**, 1821–1829.
124. A. Briceño and Y. Hill, *CrystEngComm*, 2012, **14**, 6121–6125.
125. P. Naumov, S. Chizhik, M. K. Panda, N. K. Nath and E. Boldyreva, *Chem. Rev.*, 2015, **115**, 12440–12490.
126. R. Medishetty, I.-H. Park, S. S. Lee and J. J. Vittal, *Chem. Commun.*, 2016, **52**, 3989–4001.

CHAPTER 6

Metal-organic Frameworks with Tunable Electrical and Optical Properties

NAGAPRADEEP NIDAMANURI AND SOURAV SAHA*

Department of Chemistry, Clemson University, Clemson, SC 29634, USA
*Email: souravs19@gmail.com; souravs@clemson.edu

6.1 Introduction

Looming large in front of mankind is a challenge to find sustainable ways to power and advance technological developments while limiting anthropogenic strains on our environment. To address these issues, needed are the materials that can (i) produce and store energy, (ii) improve the energy efficiency of devices, (iii) catalyse reactions through low-energy pathways, and (iv) detect and remove toxic agents from the environment. While silicon and other inorganic semiconductors remain the cornerstones of the electronics industry, high production costs associated with their high temperature synthesis and demanding top-down lithographic processing methods leave room for new materials that are accessible through facile bottom-up assembly. While conductive organic polymers[1] are easier to prepare and process compared to inorganic semiconductors, their structural disorders and modest thermal stability put them at a significant disadvantage to crystalline inorganic materials.[2]

Self-assembled from metal ion clusters and organic ligands under solvothermal conditions, highly ordered metal-organic frameworks (MOFs) offer unparalleled synthetic simplicity, processability, and control over their

Monographs in Supramolecular Chemistry No. 22
Functional Supramolecular Materials: From Surfaces to MOFs
Edited by Rahul Banerjee

Published by the Royal Society of Chemistry, www.rsc.org

structures, properties and functions that render them an attractive platform for a range of modern applications. Although their innate abilities to capture and concentrate guests inside their cavities have been exploited widely for a myriad of applications, the evolution of MOFs as powerful electronic and photonic materials has been sluggish for various reasons.[2,3]

Chief among the reasons that continue to limit our access to stimuli-responsive MOFs is the scarcity of suitable redox- and optically-active building blocks, in particular organic ligands.[2,3] Such functional building blocks are vital for molecular recognition and signal transduction capabilities of these porous materials, and to expand their utility in sensing,[3,4] electrical conduction,[2] light-harvesting[5,6] and emitting,[7–9] among other applications. Although luminescent MOFs are not so uncommon,[7–9] in most cases, rigidification of the network-bound ligands[9]—not the ligands themselves—is responsible for such emission. Some lanthanide MOFs are inherently luminescent, but they too benefit from energy transfer from the excited ligands since the *f*–*f* transitions in Ln^{3+} are forbidden,[3,9] showing the importance of optically and redox-active ligands.

The introduction of redox- and optically-active ligands[10–17] into MOFs bestows on them the desired electronic and optical properties, enabling them to communicate with and respond to various chemical and physical stimuli. The signals emanated from these interactions can propagate through the networks eliciting emerging properties that are not native to the molecular building blocks. Although at an early stage, the exploration of these potentials is gaining steam rapidly due to a burgeoning interest in MOF-based molecular electronics and photonic devices. Several research groups across the world are actively working on building stimuli-responsive MOFs using various redox- and optically active ligands, such as naphthalenediimides (NDIs),[11,12] core-substituted NDIs (cNDIs),[13] pyrene,[14] porphyrin,[5,15] tetrathiafulvalene (TTF),[16] and tetracyanoquinodimethane (TCNQ),[17] among others.

The deployment of stimuli-responsive MOFs in optoelectronic devices and their real-life applications necessitate surface-bound films. Despite an increasing attention to this aspect, making suitable MOF films remains a challenging task. The spin-coated and drop-casted films prepared from the suspensions of pre-synthesized MOFs[18] are not mechanically stable enough for device fabrication and testing. The MOF films grown on self-assembled monolayer (SAM)-coated substrates are typically more uniform,[19] but the modest thermal stability of SAMs often limits the scope of this method.[20] Moreover, the insulating nature of SAMs impairs electrical contact between MOFs and underlying electrodes, making them unfit for certain electronic devices.[2] The layer-by-layer growth method[21] yields oriented films suitable for electronic devices, but many cycles of sequential dipping of the substrates into precursor metal ion and ligand solutions are needed to reach useful film-thickness, making this method too complicated for large-scale manufacturing. Recently, MOF films have been grown *via* anodic dissolution of metal electrodes[22] and electrophoretic deposition on anodes,[23] but the question remains as to whether these methods can be adopted to develop stimuli-responsive MOF films containing redox-active ligands that could be sensitive to these conditions. A much more straightforward and practical approach involves the use of metal oxide-coated substrates on which the ligands can anchor firmly *via* covalent attachment, allowing subsequent growth of MOF films in a bottom-up fashion under simple solvothermal conditions. In recent years, this strategy has been implemented to grow films of different networks, including popular HKUST-1, IRMOF, ZIF-8, and pillared-paddle wheel (PPW) architectures. Although important, the film growth aspect remains beyond the scope of this chapter and the readers are directed to excellent review articles on this topic.[24]

In this chapter, we will focus our discussions on the advent and advancement of stimuli-responsive MOFs over the past few years, highlighting the challenges and remedies for the development of electrically and optically-tunable MOFs and their integration and advantages in modern optoelectronic devices.

6.2 Electrically Conducting MOFs

Imparting electrical conductivity to porous networks remains one of the most challenging tasks in the MOF field. Several factors contribute to this challenge. For instance, even when electroactive ligands are incorporated into porous networks, they are separated by σ-bonded metal clusters that prevent through-bond electron movement. Furthermore, large spatial gaps between the ligands prevent π-stacking, limiting through-space charge delocalization as well. To address these issues, several columnar[16] and planar (2D)[25] architectures have been developed using redox-active building blocks that show high intrinsic conductivity owing to through-space and in-plane charge mobilization. These materials will be highlighted in the first part of this chapter. In addition, guest-induced oxidation of the redox centres embedded in MOFs[26] and cross-linking of nodes with coordinating guests[27] also enhance the conductivity of intrinsically non-conducting MOFs. The success of these strategies notwithstanding, the intrinsically conducting planar and columnar architectures require specific ligands and metal ions that have certain geometry and topology.[16,25] On the other hand, the node-cross-linking strategy[27] is feasible when the nodes are coordinatively unsaturated, otherwise there is a risk of guest-induced ligand displacement that could change the structures and compositions of the networks.[28] To circumvent these issues, MOFs composed of redox-active ligands can be infiltrated with complementary guest π-systems that can form extended π-donor/acceptor (D/A) stacks within these networks. As a result, the doped MOFs can promote long-range electron delocalization through the π-stacks, displaying enhanced electrical conductivity than the undoped materials.[29]

6.2.1 MOFs with Intrinsic Electrical Conductivity

One way to develop electrically conducting MOFs is to introduce redox-active building blocks into 2D planar or 3D columnar networks that can support long-range transport of free or loosely bound charges through bonds, *i.e. via* resonance or through-space, *i.e. via* π-stacking.

6.2.1.1 Electrically Conducting 2D MOFs

In order to develop MOFs that can support facile trough-bond charge delocalization and display intrinsic electrical conductivity, Nishihara[25b] *et al.* constructed a 2D hexagonal kagomé lattice architecture composed of square-planar of Ni(II) centres coordinated with benzenehexathiol ligands (Figure 6.1a). The bulk of this material acts an insulator, but the edges display metallic character. The oxidation state of each Ni–dithiolene unit in the as-prepared material (ap-1) is −3/4, and the treatment of this material with tris(4-bromophenyl)aminiumhexachloroantimonate yields stacked oxidized nanosheets (ox-1) having zero oxidation state (Figure 6.1b). Under the van der Pauw method, microflakes of the oxidized material display an exceptionally high electrical conductivity value of 1.6×10^2 S cm^{-1}, which was attributed to facile through-bond charge transport in this graphitic hybrid

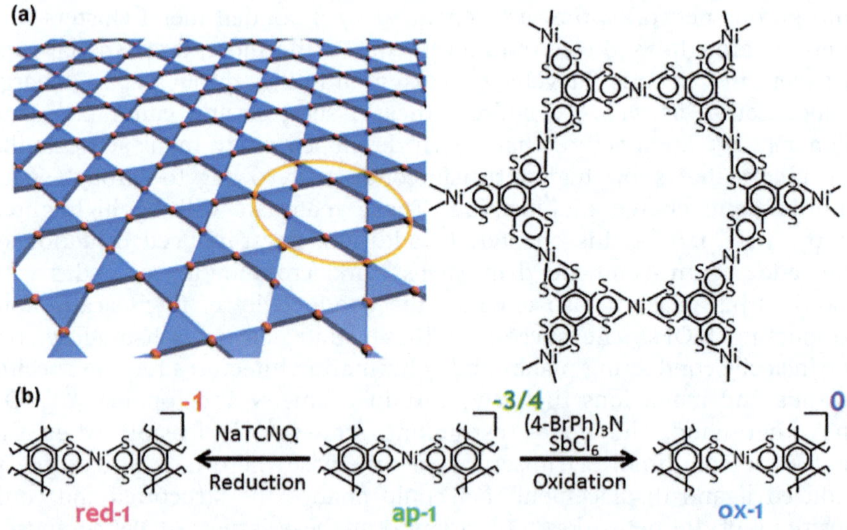

Figure 6.1 (a) The 2D structure and (b) redox-control in Ni–dithiolene-based graphitic nanosheet.
Reprinted with permission from T. Kambe, R. Sakamoto, T. Kusamoto, T. Pal, N. Fukui, T. Shimojima, Z. Wang, T. Hirahara, K. Ishizaka, S. Hasegawa, F. Liu and H. Nishihara, Redox Control and High Conductivity of Nickel Bis(dithiolene) Complex π-Nanosheet: A Potential Organic Two-Dimensional Topological Insulator, *J. Am. Chem. Soc.*, 2014, **136**, 14357. Copyright 2014 American Chemical Society.

network. Photoelectron spectroscopy and first-principles calculations further corroborated the metallic nature of these stacked nanosheets.

Another family of graphite-like 2D MOFs, $Ni_3(HITP)_2$ and $Cu_3(HITP)_2$ (HITP = 2,3,6,7,10,11-hexaiminotriphenylene) was developed by Dincă and his coworkers.[30] Concerted experimental and computational studies revealed that these isostructural materials adopted a hexagonal 2D structure with a slipped-parallel stacking of the 2D sheets resulting in 1D pores with diameters of ~1.5 nm (Figure 6.2a). Two-point probe electrical measurements of bulk powders of the Ni(II) and Cu(II) analogues revealed their conductivities as 2 $S\,cm^{-1}$ and 0.2 $S\,cm^{-1}$, respectively. Furthermore, $Cu_3(HITP)_2$ functions as a reversible chemiresistive ammonia sensor with a detection limit of 0.5 ppm (Figure 6.2b), as the devices fabricated by drop-casting $Cu_3(HITP)_2$/acetone suspensions onto interdigitated gold electrodes display a sharp increase in current upon exposure to dilute ammonia vapour under an inert (N_2) and humid conditions. The electrical response was not only reversible over several (>10) cycles, but also linearly proportional to the ammonia concentrations in the measured concentration regime.

The Dincă group constructed[31] another structurally analogous conductive MOF, namely $Cu_3(HHTP)_2$ (HHTP = 2,3,6,7,10,11-hexahydroxytriphenylene), which they incorporated into sensing arrays for the chemiresistive detection of volatile organic compounds (VOCs). These sensor arrays displayed distinct electrical signals upon exposure (30 s, 200 ppm) to vapours of different VOCs (Figure 6.3a). All three isostructural MOFs, namely $Ni_3(HITP)_2$, $Cu_3(HITP)_2$ and $Cu_3(HHTP)_2$, were responsive to polar VOCs, but not to nonpolar aliphatic hydrocarbons. Interestingly, the Cu(II) and Ni(II)-based

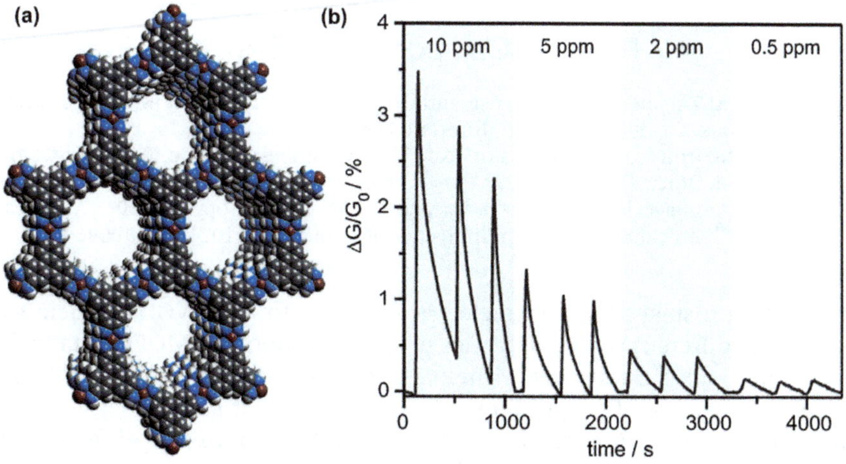

Figure 6.2 (a) The structure and (b) electrical response towards NH_3 of $Cu_3(HITP)_2$ MOF.
Reprinted with permission from M. G. Campbell, D. Sheberla, S. F. Liu, T. M. Swager and M. Dincă, Cu3(hexaiminotriphenylene)2: An Electrically Conductive 2D Metal–Organic Framework for Chemiresistive Sensing, *Angew. Chem., Int. Ed.*, 2015, **54**, 4349. Copyright 2015 WILEY-VCH Verlag GmbH & Co. KGaA, Weinheim.

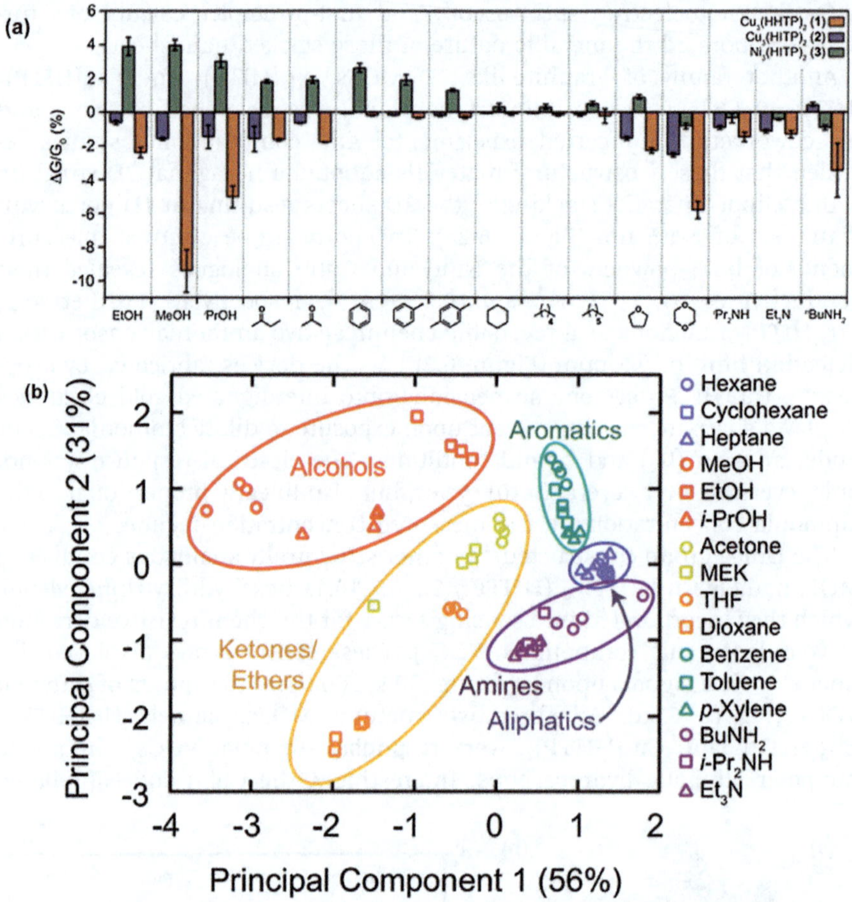

Figure 6.3 (a) The electrical response and (b) principal component analysis of MOF-based sensor arrays to different VOCs.
Reprinted with permission from M. G. Campbell, S. F. Liu, T. M. Swager and M. Dincă, Chemiresistive Sensor Arrays from Conductive 2D Metal–Organic Frameworks, *J. Am. Chem. Soc.*, 2015, **137**, 13780. Copyright 2015 American Chemical Society. (http://pubs.acs.org/doi/abs/10.1021/jacs.5b09600).

MOFs often displayed opposite electrical signals to polar VOCs, which was ascribed to different charge densities of the corresponding MOFs dictated by the d^8Ni(II) and d^9Cu(II) ions. Principal component analysis of electrical response of these MOFs to different VOCs (Figure 6.3b) showed that these MOF-based sensors are capable of classifying VOCs into five different categories with high precision and reliability.

6.2.1.2 3D MOFs with Intrinsic Electrical Conductivity and Charge Mobility

The Dincă group introduced an electron-rich ligand, tetrathiafulvalenetetrabenzoate (TTFTB) to construct a new helical columnar architecture,

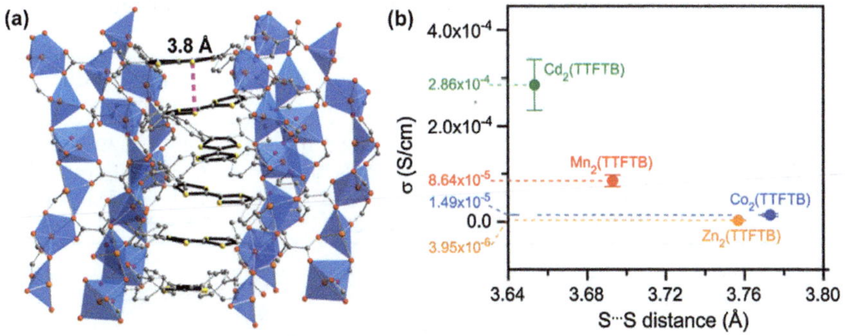

Figure 6.4 (a) The columnar structure of M_2(TTFTB), where M = Zn^{2+}, Cd^{2+}, Co^{2+} and Mn^{2+}. (b) The Correlation between the S···S distance and the electrical conductivity of isostructural MOFs.
Reprinted with permission from T. C. Narayan, T. Miyakai, S. Seki and M. Dincă, High Charge Mobillity in a Tetrathiafulvalene-Based Microporous Metal-Organic Framework, *J. Am. Chem. Soc.*, 2012, **134**, 12932. Copyright 2012 American Chemical Society. Reprinted with permission from S. S. Park, E. R. Hontz, L. Sun, C. H. Hendon, A. Walsh, T. Van Voorhis and M. Dincă, Cation-Dependent Intrinsic Electrical Conductivity in Isostructural Tetrathiafulvalene-Based Microporous Metal–Organic Frameworks, *J. Am. Chem. Soc.*, 2015, **137**, 1774. Copyright 2015 American Chemical Society. (http://pubs.acs.org/doi/10.1021/ja512437u).

namely Zn_2(TTFTB), in which the TTF units are located ~3.8 Å apart from each other and stacked along the crystallographic *c*-axis forming 1D pores (diameter ~6 Å) lined with the benzoate linkers (Figure 6.4a).[16a] Flash photolysis-time-resolved microwave conductivity measurements of single crystals revealed an exceptionally high intrinsic charge mobility (~0.2 $cm^2\ V^{-1}\ s^{-1}$), which is two orders of magnitude greater than of the free ligand, H_4TTFTB. Density functional theory (DFT) calculations confirmed that in this columnar MOF, the overlapping sulfur $3p_z$ orbitals of the stacked TTF units played a pivotal role in the through-space charge-transport pathway.

In order to fine-tune the through-space charge-transport pathway in porous 3D MOFs, the same group synthesized a library of isostructural MOFs by replacing Zn^{2+} with Co^{2+}, Mn^{2+} and Cd^{2+}.[16b] Single crystal conductivity measurements disclosed that the electrical conductivities of these isostructural MOFs are directly proportional to the ionic radii of the metal ions. The reason for this trend is that larger metal ions elongate the metal–carboxylate chairs, relaxing the MOF structures in such a way that the intermolecular S···S distances are shortened concomitantly, which in turn improves the overlap between the sulfur $3p_z$ orbitals. Thus, owing to a significantly larger ionic radius of Cd(II), which leads to the shortest S···S distances among the four analogous MOFs (Table 6.1, Figure 6.4b), Cd_2(TTFTB) displays ~72 times higher average conductivity than Zn_2(TTFTB). Nevertheless, these conductivity values are still low, making these MOFs poor semiconductors.

Table 6.1 Closest interatomic S···S distances and conductivities of four isostructural MOFs.

MOF	Ionic radius of metal (pm)	S···S (Å)	Average conductivity ($S\,cm^{-1}$)
Co_2(TTFTB)	88.5	3.7732(26)	$1.49(\pm 0.29)\times 10^{-5}$
Zn_2(TTFTB)	88	3.7568(13)	$3.95(\pm 0.56)\times 10^{-6}$
Mn_2(TTFTB)	97	3.6929(6)	$8.64(\pm 1.21)\times 10^{-5}$
Cd_2(TTFTB)	109	3.6538(23)	$2.86(\pm 0.53)\times 10^{-4}$

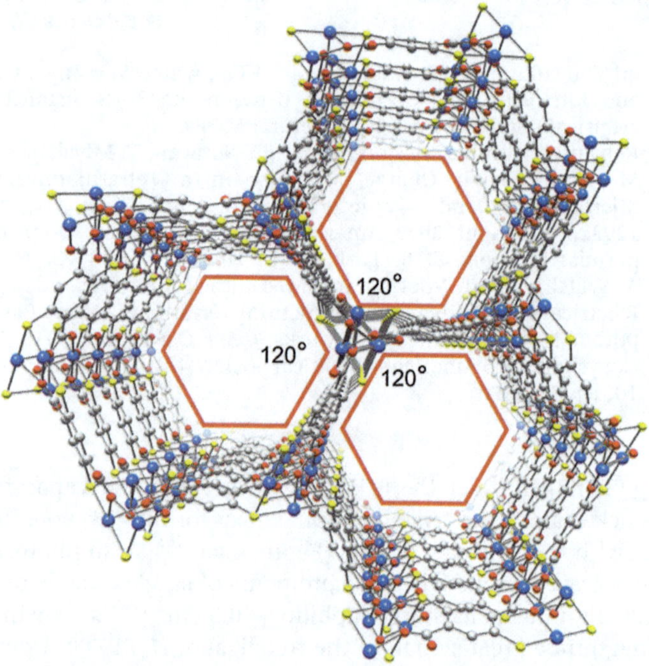

Figure 6.5 The general structure of M_2(DEBDC)$(DMF)_2 \cdot x$(DMF) MOFs (where M = Fe, Mn; E = S, O), which are analogous to canonical MOF-74. Reprinted with permission from L. Sun, C. H. Hendon, M. A. Minier, A. Walsh and M. Dincă, Million-Fold Electrical Conductivity Enhancement in Fe2(DEBDC) versus Mn2(DEBDC) (E = S, O), *J. Am. Chem. Soc.*, 2015, **137**, 6164. Copyright 2015 American Chemical Society. (http://pubs.acs.org/doi/abs/10.1021/jacs.5b02897).

Another set of columnar MOFs were synthesized from thiol and alcohol based ligands (H_4DSBDC = 2,5-disulfhydrylbenzene-1,4-dicarboxylic acid; H_4DOBDC = 2,5-dihydroxybenzene-1,4-dicarboxylic acid) and Fe^{2+} and Mn^{2+} ions and their electronic properties were explored by the Dincă group.[32] These M_2(DEBDC)$(DMF)_2 \cdot x$(DMF) MOFs (where M = Fe or Mn, and E = S or O) possessed infinite 1D pores, analogous to MOF-74 series (Figure 6.5). The electrical conductivity values of Fe_2(DSBDC) (3.9×10^{-6} $S\,cm^{-1}$) and Fe_2(DOBDC) (3.2×10^{-7} $S\,cm^{-1}$) were found to be ~6 orders of magnitude greater than those of $Mn_2(DSBDC)$ (2.5×10^{-12} S cm^{-1}) and $Mn_2(DOBDC)$ (3.9×10^{-13} S cm^{-1}). DFT calculations attributed such marked higher conductivity of the Fe(II)-based MOFs to lower binding energy of the filled β-spin *d*-band of Fe^{2+}. Nevertheless, in comparison with the TTF-based columnar MOFs, these MOF-74 analogues display even lower conductivity values, which can be attributed to the lack of redox-active ligands and a less facile through-space charge delocalization in the latter.

While the efforts to build intrinsically conducting MOFs has been gaining steam since the beginning of this decade, the long-term success of this strategy would depend on the availability of redox-active metal ions and ligands with certain topologies that can not only provide free charges, but also mobilize them over long distance through these porous materials. These strict prerequisites, however, have so far restricted the number of intrinsically conducting MOFs, provoking researchers to search for alternative strategies to impart and manipulate electrical conductivity into a wide range of hybrid networks that are already available to us.

6.2.2 Guest-induced Electrical Conductivity in Porous MOFs

To engineer electrical conductivity in a wide variety of MOFs, a second strategy was developed that involved introduction of redox-active guests into highly ordered porous networks. The encapsulated guest entities can (i) serve as redox mediators to provide mobile charges, (ii) crosslink spatially separated metallic nodes to activate through-bond charge transport, and (iii) form extended π-donor/acceptor stacks with the complementary redox-active ligands to promote through-space charge delocalization to enhance the electrical conductivity of various MOF architectures. Simple exposure of MOFs to solutions or vapours of guest entities leads to size-selective encapsulation of qualified guests, making this strategy straightforward and widely adoptable one.

Long and his team[26a] created a permanent porous MOF, $Cu[Ni(PDT)_2]$, that showed a poor intrinsic conductivity in the order of 10^{-8} S cm^{-1} at room temperature (Figure 6.6a). Oxidative doping of this material with I_2 enhanced its conductivity to 10^{-4} S cm^{-1} (Figure 6.6b,c), making it a respectable *p*-type semiconductor, and the conduction was attributed to charge transport through the framework instead of the encapsulated I_2 molecules, as the amount of the remaining encapsulated I_2 was marginal.

Zeng *et al.*[26b] demonstrated an example of I_2-induced electrical conductivity enhancement in a double-walled Zn(II)-lacticacid-4-(pyridin-4-yl)benzoic acid MOF. Upon infiltration of I_2 into the square-shaped channels of this MOF, it displayed anisotropic electrical conductivity with the parallel conductivity ($\sigma_{\parallel} = 3.42\times10^{-3}$ S cm^{-1}) being 21 times greater than the normal conductivity ($\sigma_{\perp} = 1.65\times10^{-4}$ S cm^{-1}). It was hypothesized that the encapsulated I_2 molecules are aligned in four close parallel chains inside the channels surrounded by the aromatic ligands (Figure 6.7), which support a more facile electrical conduction in a certain direction over the other. In a

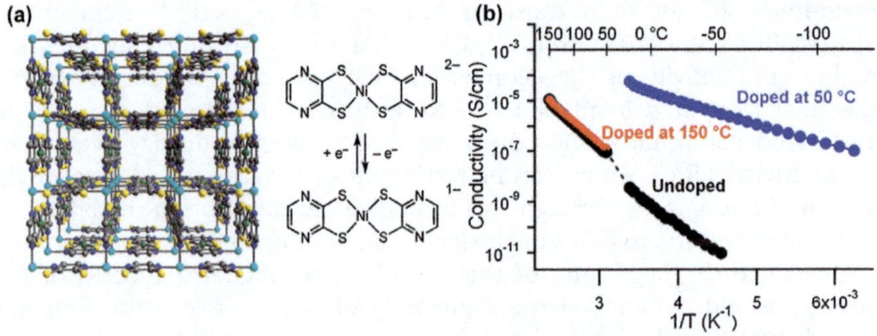

Figure 6.6 (a) The structure and (b) I_2-induced conductivity changes in $Cu[Ni(PDT)_2]$. Reprinted with permission from Y. Kobayashi, B. Jacobs, M. D. Allendorf and J. R. Long, Conductivity, Doping, and Redox Chemistry of a Microporous Dithiolene-Based Metal–Organic Framework, *Chem. Mater.*, 2010, **22**, 4120. Copyright 2010 American Chemical Society.

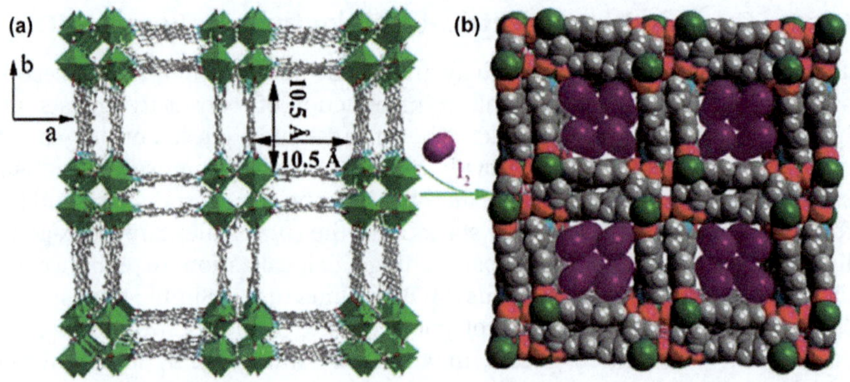

Figure 6.7 The Structure of desolvated Zn-DL-lacticacid-4-(pyridin-4yl)benzoic acid MOF (a) before (b) after I_2-infiltration.
Reprinted with permission from M.-H. Zeng, Q.-X. Wang, Y.-X. Tan, S. Hu, H.-X. Zhao, L.-S. Long and M. Kurmoo, Rigid Pillars and Double Walls in a Porous Metal-Organic Framework: Single-Crystal to Single-Crystal, Controlled Uptake and Release of Iodine and Electrical Conductivity, *J. Am. Chem. Soc.*, 2010, **132**, 2561. Copyright 2010 American Chemical Society.

subsequent paper,[33] I_2 was introduced during the synthesis of a Cu(II)-based cationic MOF, in which the polyiodide anions (I_5^- and I_7^-) were segregated inside two types of zigzag channels (Figure 6.8). The electrical conductivity of the iodine-doped cationic MOF (8.11×10^{-7} S cm^{-1}) was found to be ten times higher than that of solid I_2 itself (7.69×10^{-8} S cm^{-1}), whereas the I_2-free neutral MOF was significantly less conductive (8.04×10^{-9} S cm^{-1}).

Zhang *et al.*[34] also doped a europium and biphenyl-tricarboxylate (L) based MOF (EuL) with I_2, and studied its electrical conductivity in a

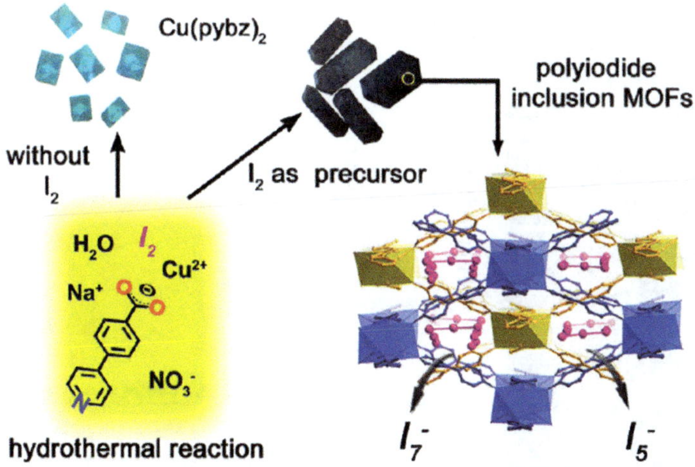

Figure 6.8 I_2-templated synthesis of the copper based MOF-{[Cu_6(4-(pyridin-4yl)-benzoic acid)$_8$(OH)$_2$] · I_5^- · I_7^-}$_n$.
Reprinted with permission from Z. Yin, Q. X. Wang and M.-H. Zeng, Iodine Release and Recovery, Influence of Polyiodide Anions on Electrical Conductivity and Nonlinear Optical Activity in an Interdigitated and Interpenetrated Bipillared-Bilayer Metal–Organic Framework, *J. Am. Chem. Soc.*, 2012, **134**, 4857. Copyright 2012 American Chemical Society.

temperature range of 25–80 °C under anhydrous conditions through impedance spectroscopy. The electrical conductivity values of the I_2-doped EuL increased linearly from 8.27×10^{-7} S cm^{-1} at 25 °C to a maximum of 2.71×10^{-5} S cm^{-1} at 80 °C. It is noteworthy that extended iodine exposure may oxidize the metal ions in MOFs. One such example was described by Yaghi[26c] where the I_2-induced oxidation of Fe^{II} to Fe^{III} in a triazole based MOF enhanced its electrical conductivity.

Through Hall Effect measurements Han *et al.*[35] demonstrated that the $Co_3(NDC)_3$ (NDC = 1,5-naphthalaene dicarboxylic acid) thin films doped with I_2 (Figure 6.9a,b) behave as a p-type semiconductor (1.88×10^{-6} S cm^{-1}) due to the formation of NDC · I_2 donor/acceptor charge transfer complex within the doped MOF films. The photoluminescence (PL) spectra of the I_2-doped MOF thin films on ITO substrates revealed the charge-transfer interactions (Figure 6.9c), confirming that the HOMO–LUMO electronic interactions between the ligands and dopants that facilitated the charge conduction in these materials.

In addition to I_2, other redox-active electron-rich or electron-deficient organic guests have also been incorporated into porous MOFs to enhance their electrical conductivities. For example, Wöll and his colleagues[36] introduced electron-rich ferrocene (Fc) molecules into HKUST-1 thin films grown on gold substrates (Figure 6.10) to improve their charge transport capability, which are otherwise insulators. While the Fc-doped HKUST-1 films show substantial electrical currents exceeding 100 μA cm^{-2} in cyclic

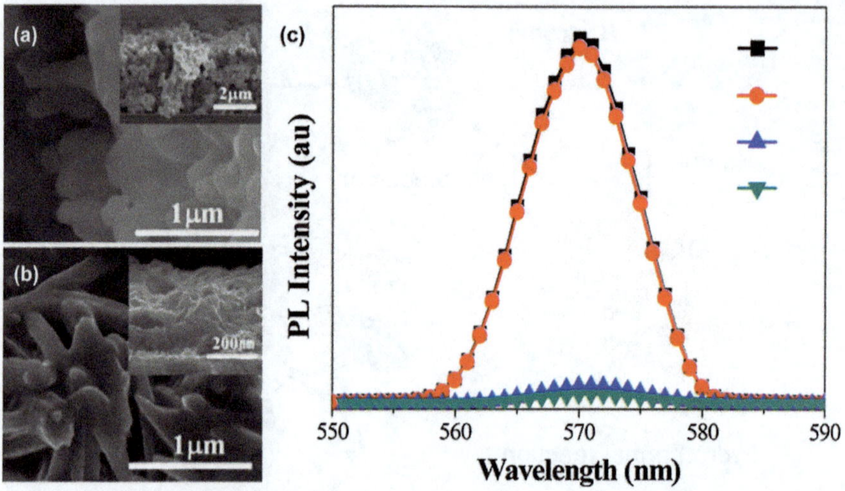

Figure 6.9 The SEM images of $Co_3(NDC)_3$ thin films on glass substrates developed from (a) LbL and (b) DB techniques. (c) Photoluminescence spectra $Co_3(NDC)_3$ thin films (■, ●) before and (▲, ▼) after doping with I_2. Reprinted with permission from D. Y. Lee, E.-K. Kim, N. K. Shrestha, D. W. Boukhvalov, J. K. Lee and S.-H. Han, Charge Transfer-Induced Molecular Hole Doping into Thin Film of Metal–Organic Frameworks, *ACS Appl. Mater. Interfaces*, 2015, 7, 18501. Copyright 2015 American Chemical Society.

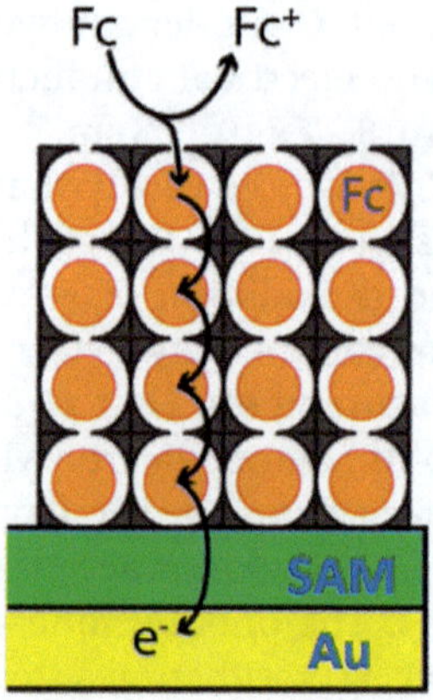

Figure 6.10 Schematic representation of electron transport in a ferrocene (Fc)-doped HKUST-1 thin films on SAM-coated gold substrate.
Reproduced from ref. 36 with permission from the Royal Society of Chemistry.

voltagrammogram, undoped HKUST-1 films show negligible currents as the latter behaves as a mere insulator.

Likewise, Allendorf *et al.*[28] introduced electron-deficient tetracyanoquinodimethane (TCNQ) into HKUST-1 by immersing activated MOF films deposited on silicon wafers into a saturated TCNQ/CH_2Cl_2 solution.

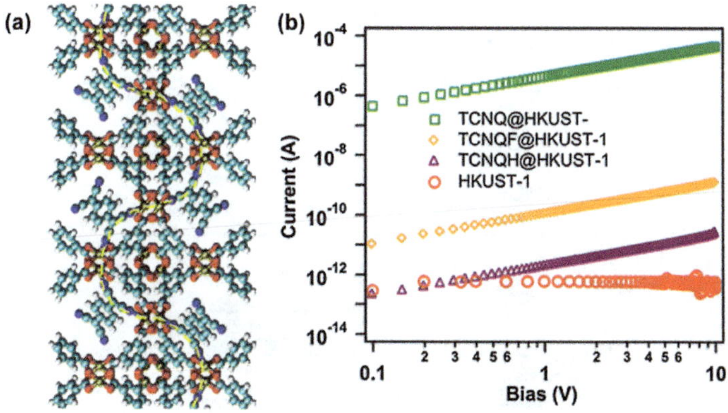

Figure 6.11 (a) HKUST-1 structure containing copper paddle wheel nodes cross-linked by TCNQ guests. (b) I/V curve of HKUST-1 films (red) doped with TCNQ (green), F_4-TCNQ (gold) and H_4-TCNQ (purple).
Reproduced from ref. 2a with permission from the Royal Society of Chemistry.

The infiltrated TCNQ molecules coordinated at the open axial sites of the unsaturated Cu_2-paddle wheel clusters cross-linking the nodes and opening the door for through-bond or resonance charge delocalization in the doped films (Figure 6.11a). As a result, in ambient conditions, TCNQ-doped HKUST-1 films exhibited an electrical conductivity of 0.07 $S\,cm^{-1}$, a value that was eight orders of magnitude greater than that of the undoped activated framework, which behaved as an insulator. This dramatic increase in conductivity was rationalized by extensive spectroscopic and modelling studies, which revealed that the cross-linking TCNQ molecules enabled a strong electronic communication between the neighbouring Cu(II) nodes through forming a continuous conductive channel in MOF unit cell *via* bridging two copper paddle wheels. The conductivity of TCNQ@HKUST-1 could be fine-tuned by controlling the exposure time of the films to guest solutions, and the efficacy of guest-mediated charge delocalization was probed systematically by introducing different analogous guests, such as [(cyclohexane-1,4-diylidene)dimalononitrile](H_4-TCNQ) and 2,3,5,6,-tetrafluoro-7,7,8,8,-tetracyanoquinodimethane (F_4-TCNQ). The introduction of H_4-TCNQ, which lacks a fully conjugated π-system, failed to increase the conductivity of HKUST-1, as the cross-linked Cu(II)-nodes were unable to communicate through resonance. On the other hand, F_4-TCNQ@HKUST-1 films displayed a modest increase in conductivity due to the high electron affinity of F_4-TCNQ, which hindered electron mobility (Figure 6.11b). These studies demonstrated that appropriate guests that can promote through-bond charge delocalization by cross-linking the redox-active metal ions are capable of dramatically improving the electrical conductivity of even the insulating MOFs.

Recently, Saha and his coworkers[37] developed a powerful new strategy to enhance and manipulate the electrical conductivity of MOFs by doping

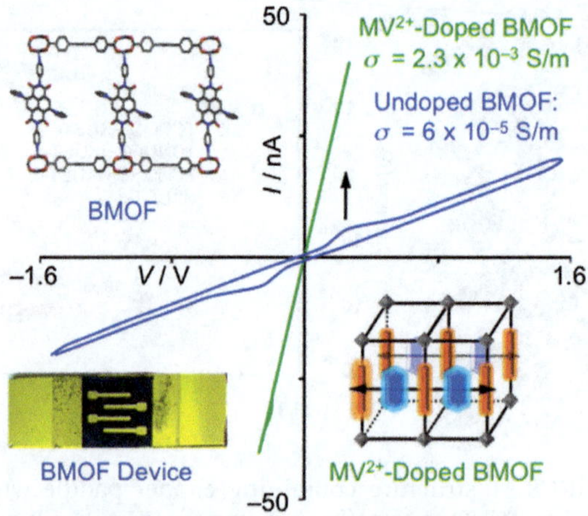

Figure 6.12 The I/V-relationship of $[Zn_2(TCPB)(BPDPNDI)]$ MOF before (blue) and after (green) MV^{2+} doping. The insets show the crystal structure (top-left), a four-probe device (bottom-left) and a graphical representation of the MV^{2+} doped material (bottom-right).
Reproduced from ref. 37 with permission from the Royal Society of Chemistry.

them with complementary guest π-systems that can form extended π-donor/acceptor stacks with redox-active ligands and thereby facilitate long-range charge delocalization (Figure 6.12). A new redox-active pillared paddle wheel architecture $[Zn_2(TCPB)(BPDPNDI)]$ composed of 1,2,4,5-tetrakis-(4-carboxyphenyl)benzene (TCPB) struts and electron-rich BPDPNDI pillars (BPDPNDI = *N*,*N*′-bis(4-pyridyl)-2,6-dipyrrolidyl naphthalenediimide) was constructed to demonstrate this concept. For electrical measurements, uniform films of this MOF were grown in a bottom-up fashion on ZnO coated substrates equipped with four gold electrodes, and their electrical conductivity was measured *via* four-probe method before and after doping with various guest entities. Upon infiltration of a strong electron-deficient guest, methyl viologen (MV^{2+}), the electrical conductivity of these films surged from 6×10^{-5} to 2.3×10^{-3} S cm^{-1}, accounting for a 40-fold improvement, but to lesser extent with a less electron-deficient guest, dinitrotoluene. On the other hand, large C_{60} molecules failed to enhance the conductivity of the MOF films, as they were size-excluded by the MOF and therefore were unable to form π-donor/acceptor interactions with the electron-rich BPDPNDI pillars. The MV^{2+}-induced conductivity enhancement was attributed to the formation of extended π-donor/acceptor stacks between the BPDPNDI pillars and intercalated MV^{2+} guests that activated the all-important long-range electron delocalization through the material, which was not possible in the undoped films because of the large gaps (≥11 Å) between the redox-active pillars in the MOF. Together, these comprehensive studies delivered a new blueprint for engineering the electrical conductivity of MOFs *via* intercalation of appropriate guest π-systems between the ligands that have complementary size, shape, and electronic properties.

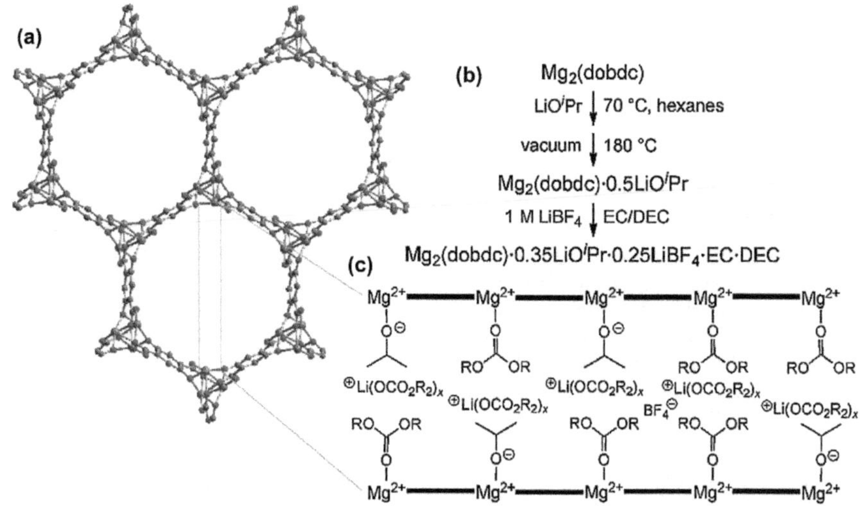

Figure 6.13 (a) The structure and (b) Li^+-doping protocol and (c) the location of mobile Li^+ ions in Mg_2(DOBDC)-based solid electrolyte.
Reprinted with permission from B. M. Wiers, M.-L. Foo, N. P. Balsara and J. R. Long, A Solid Lithium Electrolyte via Addition of Lithium Isopropoxide to a Metal–Organic Framework with Open Metal Sites, *J. Am. Chem. Soc.*, 2011, **133**, 14522. Copyright 2011 American Chemical Society.

Long *et al.*[38] impregnated another MOF, namely Mg_2(DOBDC) (Figure 6.13a,b), with LiO^iPr salts to prepare a solid lithium electrolyte that displayed higher conductivity than the Li-free material. Two-point electrical impedance measurements of the Li-doped material showed a fair conductivity of 3.1×10^{-4} S cm^{-1} at 300 K. The authors hypothesized that the high conductivity was caused by the binding of alkoxide anions with the unsaturated Mg^{2+} nodes of the framework, pinning them in place so that the Li^+ ions can move relatively freely along the channels (Figure 6.13c). In addition, this material showed a low activation barrier of 0.15 eV for conduction, which is typically dominated by intra-particle processes.

Instead of employing guest molecules, in an interesting paper, Thallapally and his colleagues[17] simply applied electrical potential to make a Cu(TCNQ) MOF more conductive. Two polymorphs of Cu(TCNQ)—Phase-I & II—were synthesized as thin films by exposing copper substrates to a hot saturated TCNQ/CH_3CN solution. Conductivity measurements indicated that Phase-II film was an insulator (5.8×10^{-7} S cm^{-1}) whereas Phase-I film was a conductor (4.8×10^{-3} S cm^{-1}) due to the presence of stacked TCNQ molecules and the interactions of the d-orbitals of Cu(II) with the p-orbitals of the cyanoligands. To determine electrical switching between two films, a cell having metal/Phase-II film (10 μm)/metal sandwich structure was constructed and *in situ* powder X-ray diffraction spectrometry was used to confirm the phase. Upon applying a voltage of 6 V, the insulating Phase-II film converted into the conductive Phase-I film through an intermediate new

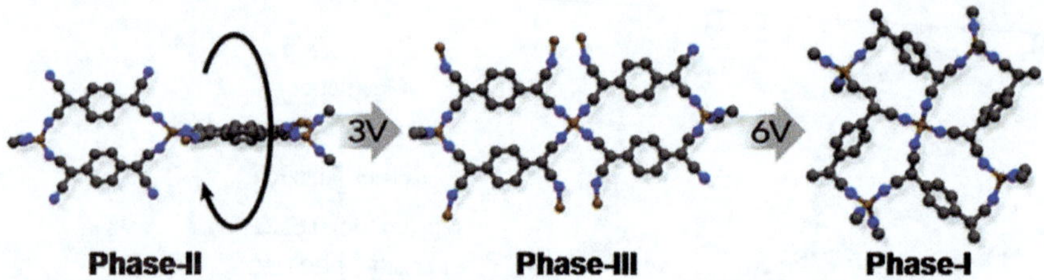

Figure 6.14 Electrical field-induced phase transition in Cu(TCNQ) MOF, going from insulating Phase-II to conductive Phase-I *via* an intermediate Phase-III. Reprinted by permission from Macmillan Publishers Ltd: Scientific Reports (C. A. Farnandez, P. C. Martin, T. Schaef, M. E. Bowden, P. K. Thallapally, L. Dang, W. Xu, X. Chen and B. P. McGrail, An Electrically Switchable Metal-Organic Framework, *Sci. Rep.*, 2014, **4**, 6114), copyright (2014).

conductive Phase-III (Figure 6.14). The actual mechanism for this reversible electrical switching remains to be elucidated.

6.3 Optically-tunable Metal-organic Frameworks

6.3.1 Electrochromic MOFs

Optically and redox-tunable organic ligands are being introduced with an increasing frequency to develop stimuli-responsive MOFs that can change their colour in response to guest entities, applied electric field, and light. Depending on the presence of electron donating and withdrawing groups on the aromatic core, the redox-active naphthalenediimide derivatives display discrete colours and their absorption and emission spectra change in response to aforesaid external stimuli. To exploit these fascinating properties in MOFs, Dincă *et al.*[13] introduced three different pyrazole-functionalized NDI derivatives—one with an unsubstituted NDI core (NDI-H), the second containing an amine-tagged NDI core (NDI-NHEt), and the third having a thioether-tagged NDI core (NDI-SEt)—to construct three isostructural MOFs, namely Zn(NDI-H), Zn(NDI-NHEt), and Zn(NDI-SEt), that contain ~16 Å wide channels lined with the NDI ligands. To demonstrate their electrochromism, thin films of these MOFs deposited on FTO-coated glass electrodes (Figure 6.15a) were introduced as working electrodes in three-electrode electrochemical cells containing a Pt-mesh counter electrode and a $Ag/Ag(cryptand)^+$ reference electrode. Cyclic voltammographs (Table 6.2, Figure 6.15b) of Zn(NDI-H), Zn(NDI-SEt), and Zn(NDI-NHEt) thin films revealed that the reduction potentials become increasingly more negative as the electron density of the NDI cores increased gradually, depending on the substituents: $X = H < SEt < NHEt$. Zn(NDI-SEt) and Zn(NDI-NHEt) MOFs showed poorer stability towards electrochemical cycling than Zn(NDI-H) because of crowding of the MOF channels by bulkier core-substituted NDI ligands that reduced the pore diameter and impeded electrolyte diffusion

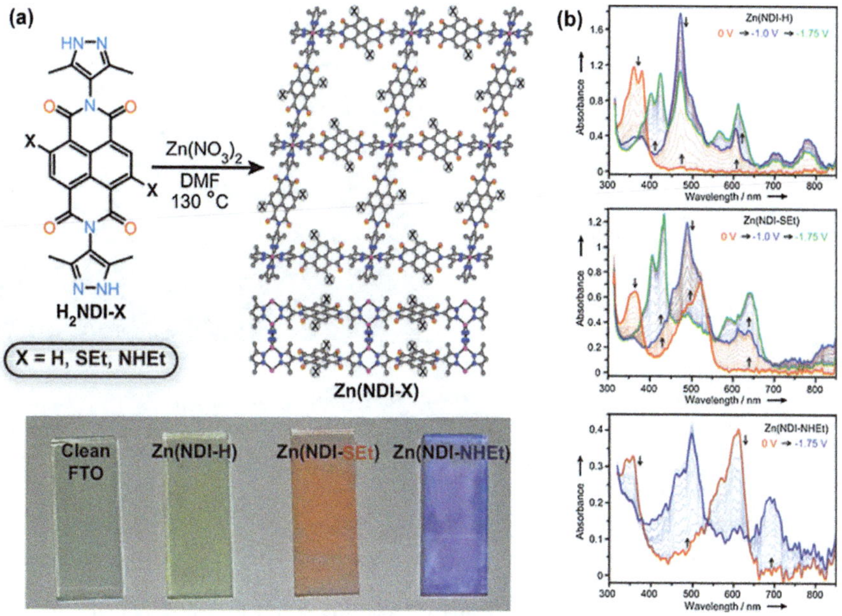

Figure 6.15 (a) The synthesis, simulated structures and photographs of Zn(NDI-H), Zn(NDI-SEt) and Zn(NDI-NHEt) films. (b) Electrochromic response of Zn(NDI-H), Zn(NDI-SEt) and Zn(NDI-NHEt) films.
Reprinted with permission from C. R. Wade, M. Li and M. Dincă, Facile Deposition of Multicolored Electrochromic Metal-Organic Framework Thin Films, *Angew. Chem., Int. Ed.*, 2013, **52**, 13377. Copyright 2013 WILEY-VCH Verlag GmbH & Co. KGaA, Weinheim.

Table 6.2 Redox potentials of three NDI-based isostructural electrochromic MOFs.

MOF	$[NDI\text{-}X]^{0/\bullet-}$ $E_{1/2}$ (V)	$[NDI\text{-}X]^{\bullet-/2-}$ $E_{1/2}$ (V)
Zn(NDI-H)	−0.56	−1.05
Zn(NDI-SEt)	−0.57	−1.08
Zn(NDI-NHEt)	−0.90	

into the structure. These NDI-based MOFs represented the first electrochromic MOFs displaying striking colour changes with significant coloration efficiency in films upon electrochemical reductions. The redox-induced colour changes in these MOFs correlate beautifully with the radical anion spectra of the corresponding NDI ligands, confirming that the reducing NDI ligands are indeed the source of the electrochromic behaviour of the MOFs.

Hupp and coworkers[14a] introduced a pyrene ligand to develop a Zr-based electrochromic MOF (Zr-PY), and deposited its thin films onto FTO-coated glass substrates (Figure 6.16a,b). The uniform Zr-PY film composed of

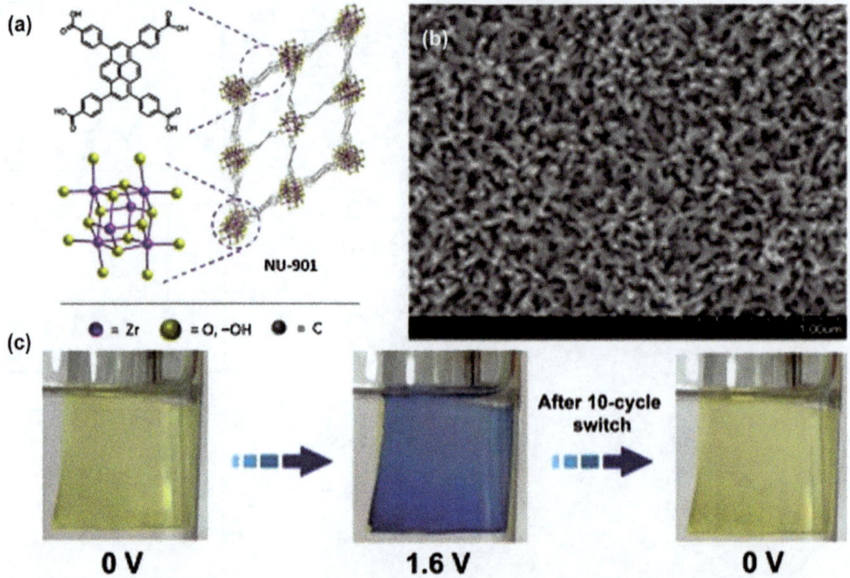

Figure 6.16 (a) The crystal structure, (b) SEM image of films and (c) electrochromic response of Zr-PY MOF.
Reprinted with permission from C.-W. Kung, T. C. Wang, J. E. Mondloch, D. Fairen-Jimenez, D. M. Gardner, W. Bury, J. M. Klingsporn, J. C. Barnes, R. V. Duyne, J. F.Stoddart, M. R. Wasielewski, O. K. Farha and J. T. Hupp, Metal-Organic Framework Thin Films Composed of Free-Standing Acicular Nanorods Exhibiting Reversible Electrochromism, *Chem. Mater.*, 2013, **25**, 5012. Copyright 2013 American Chemical Society.

free-standing acicular nanorods of width 10–30 nm. CV studies revealed a reversible oxidation of the pyrene ligand [1.45 V *vs.* Ag/AgCl/KCl], which led to reversible colour switching between yellow (neutral) and deep blue (radical cation), as shown in Figure 6.16c. The correlation between the colour switching and a pyrene radical formation was further verified by EPR spectroscopy.

By replacing pyrene with porphyrin, the Hupp team recently developed[39] another redox-active MOF (Zr-PHY), capable of electrochemically detecting nitrite in aqueous solution. MOF films were developed on conductive glass substrates (Figure 6.17a), and CV measurements in $NaNO_2$ solutions (Figure 6.17b) revealed irreversible catalytic oxidation of nitrite by the porphyrin radical cation. In addition, the amperometric experiments disclosed a wide linear range of detection (20–800 μM) with a detection limit of 2.1 μM for this MOF film.

6.3.2 Photoluminescent MOFs

Using the MOF $[Zn_2(ZnTCPP)(BPMTC)_{0.85}(DEF)_{1.15}]\cdot(DEF)_{5.15}(H_2O)_{7.25}$ (Figure 6.18a) composed of two photochromic ligands, namely bis(5-pyridyl-2-methyl-3-thienyl)cyclopentene (BPMTC) and tetrakis(4-carboxyphenyl)-porphyrin

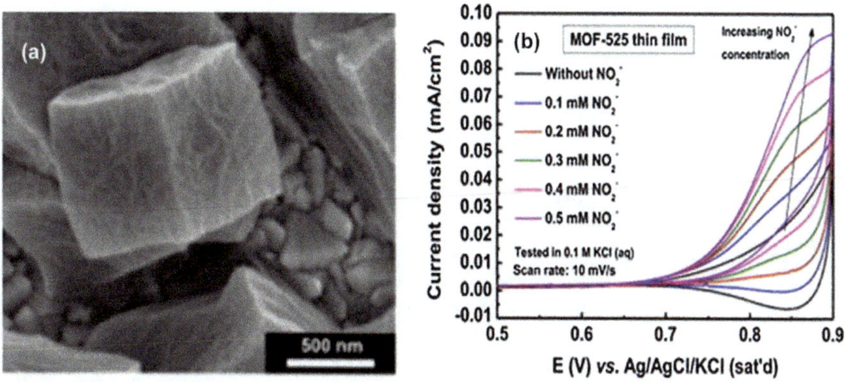

Figure 6.17 (a) SEM image and (b) CV response to NO_2^- of Zr-PHY film on glass substrate.
Reprinted from *Electrochemistry Communications*, vol. 58, C.-W. Kung, T.-H. Chang, L.-Y. Chou, J. T. Hupp, O. K. Farha and K.-C. Ho, Porphyrin-based metal-organic framework thin films for electrochemical nitrite detection, p. 51, Copyright (2015), with permission from Elsevier.

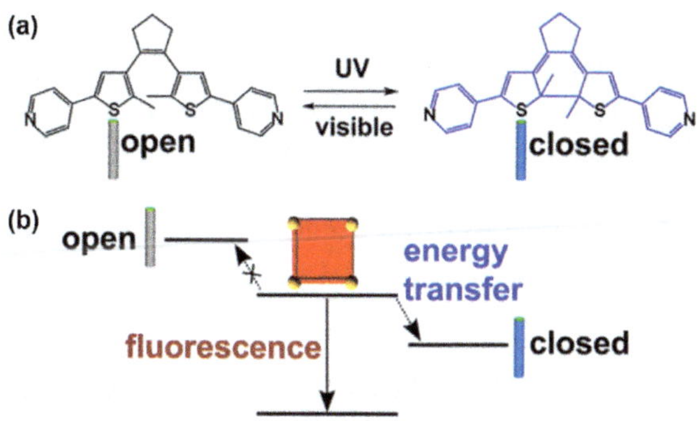

Figure 6.18 (a) Reversible photoisomerization of BPMTC ligand and (b) energy transfer mechanism in $[Zn_2(ZnTCPP)(BPMTC)_{0.85}(DEF)_{1.15}]\cdot(DEF)_{5.15}(H_2O)_{7.25}$ MOF.
Reprinted with permission from D. E. Williams, J. A. Rietman, J. M. Maier, R. Tan, A. B. Greytak, M. D. Smith, J. A. Krause and N. B. Shustova, Energy Transfer on Demand: Photoswitch-Directed Behaviour of Metal-Porphyrin Frameworks, *J. Am. Chem. Soc.*, 2014, **136**, 11886. Copyright 2014 American Chemical Society.

(TCPP), the Shustova group[40] demonstrated an instant energy transfer between the two ligands. Upon excitation at 365 nm, energy transferred from the TCPP donor to the BPMTC acceptor, converting the latter from an open to a closed form and triggering photoluminescence quenching as a consequence (Figure 6.18b). Subsequent excitation at 590 nm reversed the BPMTC isomerization from the closed to open form, leading to the recovery of photoluminescence property of the material.

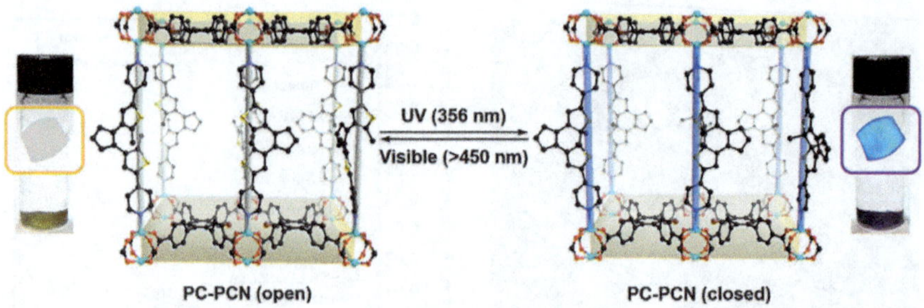

Figure 6.19 Reversible photochromism of PC-PCN *via* photoisomerization of the pillar ligand.
Reprinted with permission from J. Park, D. Feng, S. Yuan and H.-C. Zhou, Photochromic Metal–Organic Frameworks: Reversible Control of Singlet Oxygen Generation, *Angew. Chem., Int. Ed.*, 2015, **54**, 430. Copyright 2015 WILEY-VCH Verlag GmbH & Co. KGaA, Weinheim.

Zhou *et al.*[15a] exploited photoisomerization of BPMTC ligand in the pillared paddle wheel architecture (Zn_2[(Zn-TCPP)(BPMTC)]) to generate singlet oxygen, which, in turn, photo oxidized 1,5-dihydroxynaphthalene to juglone *via* heterogeneous catalysis (Figure 6.19). Upon irradiation, the BPMTC pillars undergo reversible photoisomerization between open and closed forms, enabling the MOF to display photochromism [yellow$_{(open\ form)}$$\leftrightarrow$deep–purple$_{(closed\ form)}$]. The photoinduced ^{3}[Zn-TCPP]*$\rightarrow$$^3O_2$ energy transfer in SO-PCN$_{(open)}$ facilitates $^1O_2$ generation, whereas ^{3}[Zn–TCPP]*$\rightarrow$BPDTE$_{(closed)}$ energy transfer in SO-PCN$_{(closed)}$ quenches $^1O_2$ generation, enabling SO-PCN$_{(open)}$ to serve as a heterogeneous oxidizing catalyst for 1,5-dihydroxynaphthalene to yield juglone.

6.3.3 Photovoltaic Application of Light-harvesting MOFs

Morris *et al.*[41] developed a novel photoactive MOF for solar cell applications. Two UiO-67 analogues were constructed from Ru-DCBPY-X ligands (X = OP or PS). Thin films of these MOFs grown on TiO_2-coated FTO glass substrates were introduced into photovoltaic cells (Figure 6.20), which converted light to electrical current more efficiently than the RuDCBPY/TiO_2 based devices under the same conditions. In addition to electron migration between the RuDCBPY centres in MOF film, electron injection from the MOFS into TiO_2 surface was attributed to photocurrent generation.

6.3.4 Photochromic MOFs

Han *et al.*[42] demonstrated reversible photochromic behaviour of Ca-NDITA MOF made of a *N,N′*-di(3,5-dicarboxyphenyl)-1,4,5,8-naphthalenediimide (NDITA) ligand. This doubly-interpenetrated architecture with a unique 7-connected $\{3^6\cdot4^9\cdot5^6\}$ net exhibited reversible photochromic response by changing its colour from yellowish to dark green. UV/Vis, PXRD and ESR

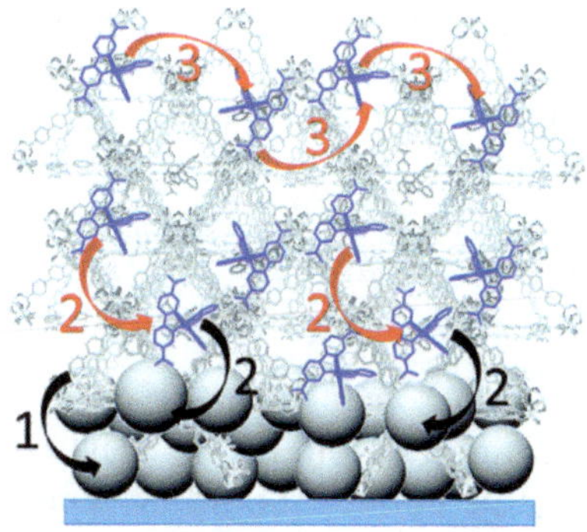

Figure 6.20 Photoinduced electron transfer mechanism in a light-harvesting MOF-based photovoltaic device.
Reproduced from ref. 41 with permission from the Royal Society of Chemistry.

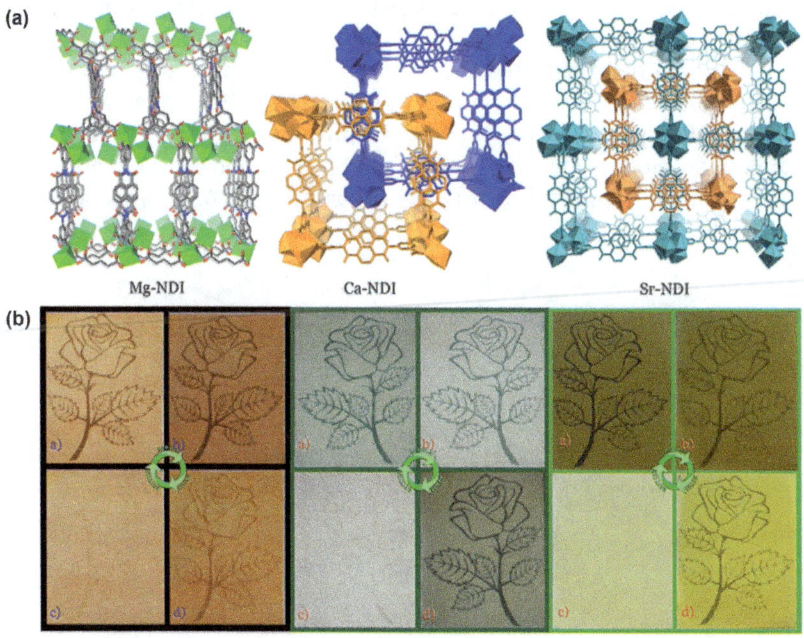

Figure 6.21 (a) Crystal structures of Mg-NDITA, Ca-NDITA and Sr-NDITA respectively. (b) Reversible printing technique based on the NDI-based photochromic MOFs.
Reproduced from ref. 43 with permission from Royal Society of Chemistry.

studies affirmed that this photoresponsive behaviour stemmed from π–π^* electron transfer leading to NDITA$^{\bullet -}$ radical anion generation.

Banerjee and coworkers[43] synthesized two more reversible photochromic MOFs, namely Mg-NDITA and Sr-NDITA using the same NDI ligand (Figure 6.21a,b). Sr-NDITA is iso-structural to Ca-NDITA and Mg-NDITA has a non-interpenetrated 3D extended structure with rectangular shaped channels (10.9×7.1 Å^2). These MOFs displayed photochromism, turning brownish black from light yellow in the presence of light. Interestingly, all three MOFs could be employed as inkless and erasable printing media by depositing them on cellulose filter papers. Upon a quick exposure to sunlight the stencil-covered MOF-coated substrates changed colour (Figure 6.21c).

6.3.5 Solvatochromic MOFs

In addition to photochromic behaviour, Mg-NDITA MOF showed solvatochromic response to organic amines (liquid or vapour).[12f] The exposure of Mg-NDITA to various amines, such as hydrazine, ethylenediamine, aniline, and triethylamine not only changed its colour from yellow to black (Figure 6.22), but also quenched the fluorescence of the MOF. These phenomena were ascribed to electron transfer from the amines to NDITA. Fluorescence quenching was linearly proportional to the amine concentration, and the amines could be detected at low concentrations (μM) *via* colorimetric changes in the MOF.

Sikdar *et al.*[29b] demonstrated the photochromic behaviour of another two-fold interpenetrated MOF, namely $\{[Cd(BPDC)(DPNDI)]\cdot 4.5H_2O\cdot DMF\}_n$ composed of *N*,*N*′-di(4-pyridyl)-1,4,5,8-naphthalenediimide) (DPNDI) and 4,4′-biphenyl-diacrboxylic acid (BPDC) ligands. Upon photoirradiation, it changed its colour from yellow to dark brown (Figure 6.23). Instead of π–π*

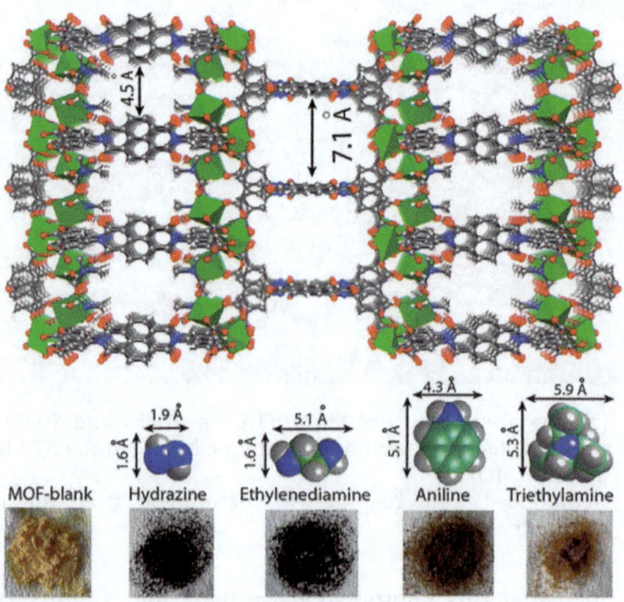

Figure 6.22 Colorimetric response of Mg-NDITA to different amines. Reproduced from ref. 12*f* with permission from the Royal Society of Chemistry.

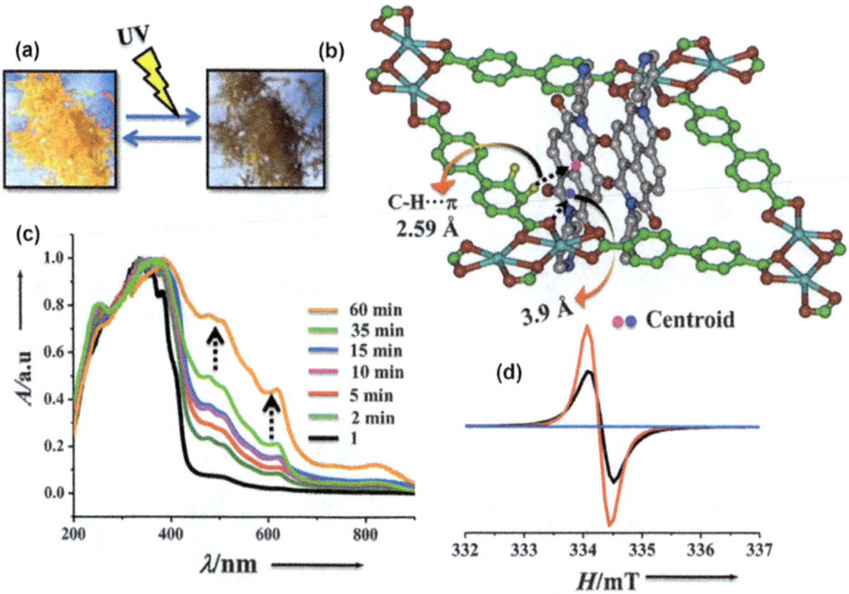

Figure 6.23 Reversible photochromism of $\{[Cd(BPDC)(DPNDI)]\cdot 4.5H_2O\cdot DMF\}_n$ caused by inter-ligand charge-transfer from BPDC to DPNDI leading to the formation of a $DPNDI^{\bullet -}$ radical anion as observed by UV/Vis and EPR spectroscopy.
Reprinted with permission from N. Sikdar, K. Jayaramulu, V. Kiran, K. V. Rao, S. Sampath, S. J. George and T. K. Maji, Redox-Active Metal-Organic Frameworks: Highly Stable Charge-Separated States through Strut/Guest-to-Strut Electron Transfer, *Chem. – Eur. J.*, 2015, **21**, 11701. Copyright 2015 WILEY-VCH Verlag GmbH & Co. KGaA, Weinheim.

electron transfer, the photochromic behaviour resulted from instant electron transfer from BPDC to DPNDI linkers. Furthermore, the guest-induced optical response of this MOF was studied with three electron-rich guests, 4,*N*,*N*-trimethylaniline, 1,5-dinaphthol, and 2,6-dinaphthol, which caused instant colour change to dark brown, green, and dark green respectively. EPR and UV/Vis studies confirmed electron transfer from the encapsulated electron-rich guests to the π-acidic DPNDI ligands. The resulting charge-separated states were found to be fairly stable over time, retaining their colours even after 3–4 months.

6.3.6 Chemical Sensors Based on Luminescent MOFs

The luminescent MOFs (LMOFs) can detect analytes by displaying guest-induced fluorescent quenching or amplification. Typically, the guest-induced fluorescence enhancement is caused by network rigidification, however, the electron-deficient and electron-rich analytes can either enhance or quench the luminescence intensity of MOFs through donor–acceptor electron transfer mechanisms. The Li group[8c] demonstrated this concept through a single MOF

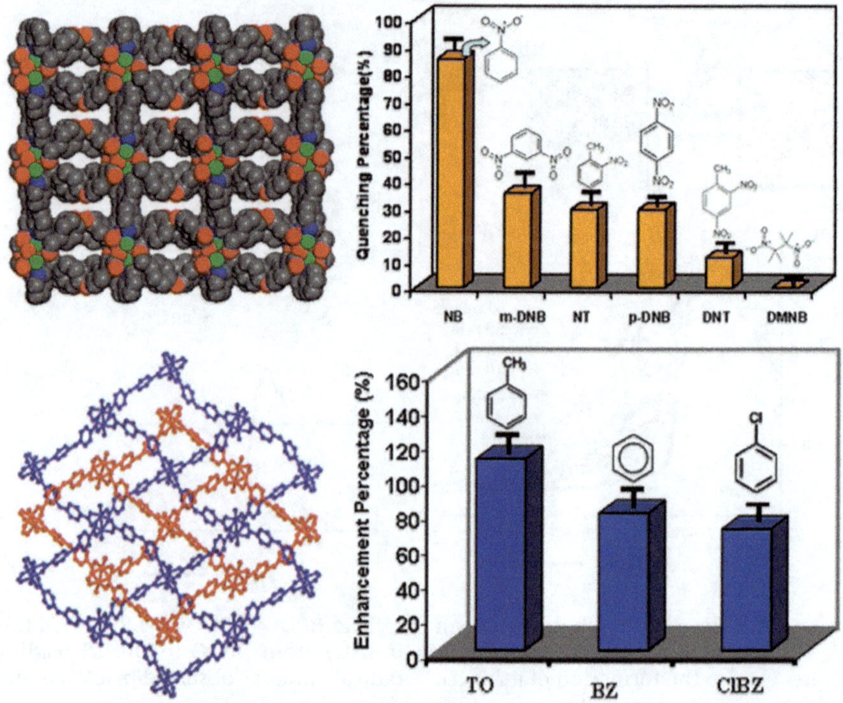

Figure 6.24 The structure and fluorescence response of LMOF-1 to different electron poor and rich guests.
Reprinted with permission from S. Pramanik, C. Zheng, X. Zhang, T. J. Emge and J. Li., New Microporous Metal–Organic Framework Demonstrating Unique Selectivity for Detection of High Explosives and Aromatic Compounds, *J. Am. Chem. Soc.*, 2011, **133**, 4153. Copyright 2011 American Chemical Society.

(LMOF-1, Figure 6.24a), which can detect both nitroaromatics and methylated aromatics. This Zn-paddle-wheel based interpenetrated layered MOF contains 1D pores of dimensions ~5.8×8.3 Å. Excitation of activated LMOF-1 at 280 nm produces a strong emission at 420 nm which is substantially quenched by nitroaromatics with varying degrees: nitrobenzene > *m*-dinitrobenzene > nitrotoluene ≈ *p*-dinitrobenzene > dinitrotoluene. On the other hand, the electron-rich analytes enhanced the fluorescent intensity in the following order: toluene > benzene > chlorobenzene (Figure 6.24b,c). These trends were dictated by the electron-withdrawing/donating nature as well as the vapour pressures of analytes. Band structure calculations explained the highly selective detection ability of LMOF-1 for nitroaromatic explosives over nitroaliphatic explosives.

When different analytes exhibit similar fluorescent behaviour, it becomes difficult to unambiguously distinguish them solely based on the change in fluorescence intensities. To overcome this difficulty, a two-variable strategy was adopted recently by the Li team where the emission frequency was

Figure 6.25 Crystal structures and guest-induced fluorescent changes of $Zn_2(BDC)_2(DPNDI) \cdot 4DMF$.
Reprinted by permission from Macmillan Publishers Ltd: Nature Communications (Y. Takashima, V. M. Martínez, S. Furukawa, M. Kondo, S. Shimomura, H. Uehara, M. Nakahama, K. Sugimoto and S. Kitagawa, Molecular decoding using luminescence from an entangled porous framework, *Nat. Commun.*, 2011, **2**, 168), copyright (2011).

evaluated in conjunction with fluorescence intensity change.[44] The validity of this strategy was exemplified by using two highly fluorescent and isostructural LMOFs namely LMOF-2 and LMOF-3. Both MOFs displayed dual response, as their fluorescence intensity as well as emission frequency changed in the presence of guests. The change in emission frequency resulted from the newly formed guest–ligand exciplex, which was further supported by computational studies. Using this strategy, different analytes can be placed precisely on a 2D Cartesian coordinate system.

Kitagawa *et al.*[12b] demonstrated an elegant molecular decoding strategy for simultaneous differentiation of analytes using a weak fluorescent two-fold interpenetrated pillared paddle wheel MOF, $Zn_2(BDC)_2(DPNDI) \cdot 4DMF$ that possesses two different void environments, namely site A and site B (Figure 6.25a,b). Upon infiltration of different aromatic VOCs, the evacuated MOF not only underwent a dynamic structural change, but also exhibited intense optical emission in the visible wavelength region depending on the electron donating ability of the encapsulated guests (Figure 6.25c). The emergence of new absorption peak between 420–500 nm affirmed the ground-state charge transfer between the π-acidic DPNDI ligands and electron-rich guest molecules. In addition, upon excitation at 370 nm, the emission of the guest-infiltrated MOF shifted to longer wavelengths displaying Mulliken dependence, *i.e.*, infiltration of stronger π-donors led to emissions at longer wavelengths caused by radiative decay of the lower-energy exciplexes.

6.4 Concluding Remarks

Opening with a brief account on how the structural features of porous MOFs rendered them attractive nanoscale containers but the scarcity of electronically and optically-tunable building blocks has limited their applications in molecular electronics and photonic devices until lately, this chapter highlights the advent and advancement of stimuli-responsive MOFs and their rapidly growing application in a variety of modern electronic and optical devices. The introduction of redox-active building blocks into 2D planar and 3D columnar architectures has opened the door for through-bond and through-space charge delocalization in porous networks, endowing them with electrical conductivity and high charge mobility. Furthermore, the guest-induced oxidation and reduction of networks, cross-linking of nodes, and the formation of extended π-donor/acceptor opened the possibility of engineering and manipulating the electrical conductivity of different MOF architectures, making them tunable semiconductors and chemiresistive sensors. The abilities of the redox-active ligands to interact with external stimuli, such as guest entities, applied electrical field, and light have gave birth to stimuli-responsive MOFs that show interesting electrochromic, photochromic, light-harvesting, and light-emitting properties, expanding the utility of such MOFs in the realms of sensing, energy conversion and storage, imaging, among other advanced applications. While increasing interest have triggered rapid expansion of stimuli-responsive MOFs, a sustained effort is needed to fully realize the potential of these materials in future optoelectronic devices.

Acknowledgements

The authors acknowledge financial support from the National Science Foundation (Awards CHE-1507529 and CHE-1660329).

References

1. (a) D. Gopalakrishnan and D. W. R. Dichtel, *J. Am. Chem. Soc.*, 2013, **135**, 8357; (b) J. L. Novotney and W. R. Dichtel, *ACS Macro Lett.*, 2013, **2**, 423; (c) T. K. Das and S. Prusty, *Polym. - Plast. Technol. Eng.*, 2012, **51**, 1487; (d) S. W. Thomas, G. D. Joly and T. M. Swager, *Chem. Rev.*, 2007,

107, 1339; (e) G. V. Zyryanov, M. A. Palacios and P. Anzenbacher Jr., *Org. Lett.*, 2008, **10**, 3681; (f) J. Janata and M. Josowicz, *Nat. Mater.*, 2003, **2**, 19.

2. (a) V. Stavila, A. A. Talin and M. D. Allendorf, *Chem. Soc. Rev.*, 2014, **43**, 5994; (b) G. Givaja, P. Amo-Ochoa, J. G. Gomez Garcia and F. Zamora, *Chem. Soc. Rev.*, 2012, **41**, 115; (c) C. H. Hendon, A. Tiana and A. Walsh, *Phys. Chem. Chem. Phys.*, 2012, **14**, 13120; (d) M. D. Allendorf, A. Scwartzberg, V. Stavila and A. A. Talin, *Chem. - Eur. J.*, 2011, **17**, 11372.
3. L. E. Kreno, K. Leong, O. K. Farha, M. Allendorf, R. P. Van Duyne and J. T. Hupp, *Chem. Rev.*, 2012, **112**, 1105.
4. (a) Z. Hu, B. J. Deibert and J. Li, *Chem. Soc. Rev.*, 2014, **43**, 5815; (b) E. Barea, C. Montoro and J. A. R. Navarro, *Chem. Soc. Rev.*, 2014, **43**, 5419; (c) M. Zhang, G. Feng, Z. Song, Y.-P. Zhou, H.-Y. Chao, D. Yuan, T. T. Y. Tan, Z. Guo, Z. Hu, B. Z. Tang, B. Liu and D. Zhao, *J. Am. Chem. Soc.*, 2014, **136**, 7241; (d) J. J. Gassensmith, J. Y. Kim, J. M. Holcroft, O. K. Farha, J. F. Stoddart, J. T. Hupp and N. C. Jeong, *J. Am. Chem. Soc.*, 2014, **136**, 8277; (e) N. B. Shustova, A. F. Cozzolino, S. Reineke, M. Baldo and M. Dincă, *J. Am. Chem. Soc.*, 2013, **135**, 13326.
5. (a) M. C. So, G. P. Weiderrecht, J. E. Mondloch, J. T. Hupp and O. K. Farha, *Chem. Commun.*, 2015, **51**, 3501; (b) T. Zhang and W. Lin, *Chem. Soc. Rev.*, 2014, **43**, 5982; (c) H.-J. Son, S. Jin, S. Patwardhan, S. J. Wezenberg, N. C. Jeong, M. So, C. E. Wilmer, A. A. Sarjeant, G. C. Schatz, R. Q. Snurr, O. K. Farha, G. P. Wiederrecht and J. T. Hupp, *J. Am. Chem. Soc.*, 2013, **135**, 852; (d) A. Fateeva, P. A. Chater, C. P. Ireland, A. A. Tahir, Y. Z. Khimyak, P. V. Wiper, J. R. Darwent and M. J. Rosseinsky, *Angew. Chem., Int. Ed.*, 2012, **51**, 7440; (e) C. Y. Lee, O. K. Farha, B. J. Hong, A. A. Sarjeant, S. T. Nguyen and J. T. Hupp, *J. Am. Chem. Soc.*, 2011, **133**, 15858.
6. (a) H. S. Quah, W. Chen, M. K. Schreyer, H. Yang, M. W. Wong, W. Ji and J. J. Vittal, *Nat. Commun.*, 2015, **6**, 7954; (b) Q. Gong, Z. Hu, B. J. Deibert, T. J. Emge, S. J. Teat, D. Banerjee, B. Mussman, N. D. Rudd and J. Li, *J. Am. Chem. Soc.*, 2014, **136**, 16724; (c) C. Wang, T. Zhang and W. Lin, *Chem. Rev.*, 2012, **112**, 1084; (d) D. F. Sava, L. E. S. Rohwer, M. A. Rodriguez and T. M. Nenoff, *J. Am. Chem. Soc.*, 2012, **134**, 3983.
7. (a) Y. Cui, Y. Yue, G. Qian and B. Chen, *Chem. Rev.*, 2012, **112**, 1126; (b) M. D. Allendorf, C. A. Bauer, R. K. Bhatka and R. J. T. Houk, *Chem. Soc. Rev.*, 2009, **38**, 1330; (c) C. A. Bauer, T. V. Timofeeva, T. B. Settersten, B. D. Patterson, V. H. Liu, B. A. Simmons and M. D. Allendorf, *J. Am. Chem. Soc.*, 2007, **129**, 7136.
8. (a) S. Pramanik, Z. Hu, X. Zhang, C. Zheng, S. Kelly and J. Li, *Chem. - Eur. J.*, 2013, **19**, 15964; (b) N. B. Shustova, B. D. McCarthy and M. Dincă, *J. Am. Chem. Soc.*, 2011, **133**, 20126; (c) S. Pramanik, C. Zheng, T. J. Emge and J. Li, *J. Am. Chem. Soc.*, 2011, **133**, 4153; (d) A. Lan, K. Li, H. Wu, D. H. Olson, T. J. Emge, W. Ki, M. Hong and J. Li, *Angew. Chem., Int. Ed.*, 2009, **48**, 2334.
9. (a) D. T. de Lill, A. de Bettencourt-Dias and C. L. Cahill, *Inorg. Chem.*, 2007, **46**, 3960; (b) B. Chen, L. Wang, F. Zapata, G. Qian and E. B. Lobkovsky, *J. Am. Chem. Soc.*, 2008, **130**, 6718; (c) J. An, C. M. Shade,

D. A. Chengelis-Czegan, S. Petoud and N. L. Rosi, *J. Am. Chem. Soc.*, 2011, **133**, 1220; (d) Y. Liu, M. Pan, Q.-Y. Yang, L. Fu, K. Li, S.-C. Wei and C.-Y. Su, *Chem. Mater.*, 2012, **24**, 1954; (e) J. Cao, Y. Gao, Y. Wang, C. Du and Z. Liu, *Chem. Commun.*, 2013, **49**, 6897.

10. (a) T.-H. Chen, I. Popov, W. Kaveevivitchai and O. S. Miljanic, *Chem. Mater.*, 2014, **26**, 4322; (b) F. A. A. Paz, J. Klinowski, S. M. F. Viela, J. P. C. Tomé, J. A. S. Cavaleiro and J. Rocha, *Chem. Soc. Rev.*, 2012, **41**, 1088; (c) W. Lu, Z. Wei, Z.-Y. Gu, T.-F. Liu, J. Park, J. Tian, M. Zhang, Q. Zhang, T. Gentle III, M. Bosch and H.-C. Zhou, *Chem. Soc. Rev.*, 2014, **43**, 5561; (d) P. Peng, F.-F. Li, V. S. P. K. Neti, A. J. Metta-Magana and L. Echegoyen, *Angew. Chem., Int. Ed.*, 2014, **53**, 160.
11. (a) S. Guha, F. S. Goodson, R. J. Clark and S. Saha, *CrystEngComm*, 2012, **14**, 1213; (b) A. Mitra, C. T. Hubley, D. K. Panda, R. J. Clark and S. Saha, *Chem. Commun.*, 2013, **49**, 6629.
12. (a) B.-Q. Ma, K. L. Mulfort and J. T. Hupp, *Inorg. Chem.*, 2005, **44**, 4912; (b) Y. Takashima, V. M. Martínez, S. Furukawa, M. Kondo, S. Shimomura, H. Uehara, M. Nakahama, K. Sugimoto and S. Kitagawa, *Nat. Commun.*, 2011, **2**, 168; (c) M. Pan, X.-M. Lin, G.-B. Li and C.-Y. Su, *Coord. Chem. Rev.*, 2011, **255**, 1921; (d) B. D. McCarthy, E. R. Hontz, S. R. Yost, T. Van Voorhis and M. Dincă, *J. Phys. Chem. Lett.*, 2013, **4**, 453; (e) C. F. Leong, B. Chan, T. B. Faust and D. M. D'Alessandro, *Chem. Sci.*, 2014, **5**, 4724; (f) A. Mallick, B. Garai, M. A. Addicoat, P. S. Petkov, T. Heine and R. Banerjee, *Chem. Sci.*, 2015, **6**, 1420.
13. C. R. Wade, M. Li and M. Dincă, *Angew. Chem., Int. Ed.*, 2013, **52**, 13377.
14. (a) C.-W. Kung, T. C. Wang, J. E. Mondloch, D. Fairen-Jimenez, D. M. Gardner, W. Bury, J. M. Klingsporn, J. C. Barnes, R. V. Duyne, J. F. Stoddart, M. R. Wasielewski, O. K. Farha and J. T. Hupp, *Chem. Mater.*, 2013, **25**, 5012; (b) K. Stylianou, R. Heck, S. Y. Chong, J. Bacsa, J. T. A. Jones, Y. Z. Khimyak, D. Bradshaw and M. J. Rosseinsky, *J. Am. Chem. Soc.*, 2010, **132**, 4119.
15. (a) J. Park, D. Feng, S. Yuan and H.-C. Zhou, *Angew. Chem., Int. Ed.*, 2015, **54**, 430; (b) W.-Y. Gao, M. Chrzanowski and S. Ma, *Chem. Soc. Rev.*, 2014, **43**, 5841.
16. (a) T. C. Narayan, T. Miyakai, S. Seki and M. Dincă, *J. Am. Chem. Soc.*, 2012, **134**, 12932; (b) S. S. Park, E. R. Hontz, L. Sun, C. H. Hendon, A. Walsh, T. Van Voorhis and M. Dincă, *J. Am. Chem. Soc.*, 2015, **137**, 1774.
17. C. A. Farnandez, P. C. Martin, T. Schaef, M. E. Bowden, P. K. Thallapally, L. Dang, W. Xu, X. Chen and B. P. McGrail, *Sci. Rep.*, 2014, **4**, 6114.
18. P. Horcajada, C. Serre, D. Grosso, C. Boissiere, S. Perruchas, C. Sanchez and G. Ferey, *Adv. Mater.*, 2009, **21**, 1931.
19. (a) N. Sakai, J. Mareda, E. Vauthey and S. Matile, *Chem. Commun.*, 2010, **46**, 4225; (b) S. Chopin, F. Chaignon, E. Blart and F. Odobel, *J. Mater. Chem.*, 2007, **17**, 4139; (c) C. Röger and F. Würthner, *J. Org. Chem.*, 2007, **72**, 8070; (d) A. Blaszczyka, M. Fischera, C. V. Hanisch and M. Mayor, *Helv. Chim. Acta*, 2006, **89**, 1986; (e) C. Thalacker, C. Röger and

F. Würthner, *J. Org. Chem.*, 2006, **71**, 8098; (f) F. Würthner, S. Ahmed, C. Thalacker and T. Debaerdemaeker, *Chem. – Eur. J.*, 2002, **8**, 4742.
20. (a) S. L. Suraru and F. Würthner, *Angew. Chem., Int. Ed.*, 2014, **53**, 7428; (b) S. V. Bhosale, C. H. Janiab and S. J. Langford, *Chem. Soc. Rev.*, 2008, **37**, 331.
21. (a) O. Shekhah, H. Wang, S. Kowarik, F. Schreiber, M. Paulus, M. Tolan, C. Sternemann, F. Evers, D. Zacher, R. A. Fischer and C. Wöll, *J. Am. Chem. Soc.*, 2007, **129**, 15118; (b) O. Shekhah, *Materials*, 2010, **3**, 1302; (c) R. Makiura, S. Motoyama, Y. Umemura, H. Yamanaka, O. Sakata and H. Kitagawa, *Nat. Mater.*, 2010, **9**, 565; (d) D. Zacher, K. Yusenko, A. Betard, M. Meilikhov, T. Ladnorg, O. Shekhah, A. Terfort, C. Wöll and R. A. Fischer, *Chem. – Eur. J.*, 2011, **17**, 1448; (e) M. C. So, S. Jin, H.-J. Son, G. P. Wiederrecht, O. K. Farha and J. T. Hupp, *J. Am. Chem. Soc.*, 2013, **135**, 15698; (f) J. Zhao, B. Gong, W. T. Nunn, P. C. Lemaire, E. C. Stevens, F. I. Sidi, P. S. Williams, C. J. Oldham, H. J. Walls, S. D. Shepherd, M. A. Browe, G. W. Peterson, M. D. Losego and G. N. Parsons, *J. Mater. Chem. A*, 2015, **3**, 1458.
22. (a) R. Ameloot, L. Stappers, J. Fransaer, L. Alaerts, B. F. Sels and D. E. De Vos, *Chem. Mater.*, 2009, **21**, 2580; (b) B. K. Keitz, C. J. Yu, J. R. Long and R. Ameloot, *Angew. Chem., Int. Ed.*, 2014, **53**, 5561.
23. I. Hod, W. Bury, D. M. Karlin, P. Deria, C.-W. Kung, M. J. Katz, M. So, B. Klahr, D. Jin, Y.-W. Chung, T. W. Odom, O. K. Farha and J. T. Hupp, *Adv. Mater.*, 2014, **26**, 6295.
24. (a) P. Falcaro, R. Ricco, C. M. Doherty and K. Liang, *Chem. Soc. Rev.*, 2014, **43**, 5513; (b) A. Bétard and R. A. Fischer, *Chem. Rev.*, 2012, **112**, 1055; (c) D. Zhacher, S. Rochus, C. Wöll and R. A. Fischer, *Angew. Chem., Int. Ed.*, 2011, **50**, 176; (d) O. Shekhah, J. Liu, R. A. Fischer and C. Wöll, *Chem. Soc. Rev.*, 2011, **40**, 1081; (e) D. Zacher, O. Shekhah, C. Wöll and R. A. Fischer, *Chem. Soc. Rev.*, 2009, **38**, 1418.
25. (a) S. Takaishi, M. Hosoda, T. Kajiwara, H. Miyasaka, M. Yamashita, Y. Nakanishi, Y. Kitagawa, K. Yamaguchi, A. Kobayashi and H. Kitagawa, *Inorg. Chem.*, 2009, **48**, 9048; (b) T. Kambe, R. Sakamoto, T. Kasumato, T. Pal, N. Fukui, K. Hoshiko, T. Shimojima, Z. Wang, T. Hirahara, K. Ishizaka, S. Hasegawa, F. Liu and H. Nishihara, *J. Am. Chem. Soc.*, 2014, **136**, 14357; (c) D. Sheberla, L. Sun, M. A. Blood-Forsythe, S. Er, C. R. Wade, C. K. Brozek, A. Aspuru-Guzik and M. Dincă, *J. Am. Chem. Soc.*, 2014, **136**, 8859.
26. (a) Y. Kobayashi, B. Jacobs, M. D. Allendorf and J. R. Long, *Chem. Mater.*, 2010, **22**, 4120; (b) M.-H. Zeng, Q.-X. Wang, Y.-X. Tan, S. Hu, H.-X. Zhao, L.-S. Long and M. Kurmoo, *J. Am. Chem. Soc.*, 2010, **132**, 2561; (c) F. Gandara, F. J. Uribe-Romo, D. K. Britt, H. Furukawa, L. Lei, R. Cheng, X. Duan, M. O'Keeffe and O. M. Yaghi, *Chem. – Eur. J.*, 2012, **18**, 10595.
27. (a) O. Karagiaridi, W. Bury, J. E. Mondloch, J. T. Hupp and O. K. Farha, *Angew. Chem., Int. Ed.*, 2014, **53**, 4530; (b) *P.* Deria, J. E. Mundloch, O. Karagiaridi, W. Bury, J. T. Hupp and O. K. Farha, *Chem. Soc. Rev.*,

2014, **43**, 5896; (c) W. Bury, D. Fairen-Jimenez, M. B. Lalonde, R. Q. Snurr, O. K. Farha and J. T. Hupp, *Chem. Mater.*, 2013, **25**, 739; (d) O. Karagiaridi, W. Bury, E. Tylianakis, A. A. Sarjeant, J. T. Hupp and O. K. Farha, *Chem. Mater.*, 2013, **25**, 3499; (e) B. J. Burnett, P. M. Barron, C. Hu and W. Choe, *J. Am. Chem. Soc.*, 2011, **133**, 9984.
28. A. A. Talin, A. Centrone, A. C. Ford, M. E. Foster, V. Stavila, P. Haney, R. A. Kinney, V. Szalai, F. E. Gabaly, H. P. Yoon, F. Léonard and M. D. Allendorf, *Science*, 2014, **343**, 66.
29. (a) S. Fujii, T. Tada, Y. Komoto, T. Osuga, T. Murase, M. Fujita and M. Kiguchi, *J. Am. Chem. Soc.*, 2015, **137**, 5939; (b) N. Sikdar, K. Jayaramulu, V. Kiran, K. V. Rao, S. Sampath, S. J. George and T. K. Maji, *Chem. - Eur. J.*, 2015, **21**, 11701.
30. M. G. Campbell, D. Sheberla, S. F. Liu, T. M. Swager and M. Dincă, *Angew. Chem., Int. Ed.*, 2015, **54**, 4349.
31. M. G. Campbell, S. F. Liu, T. M. Swager and M. Dincă, *J. Am. Chem. Soc.*, 2015, **137**, 13780.
32. L. Sun, C. H. Hendon, M. A. Minier, A. Walsh and M. Dincă, *J. Am. Chem. Soc.*, 2015, **137**, 6164.
33. Z. Yin, Q. X. Wang and M.-H. Zeng, *J. Am. Chem. Soc.*, 2012, **134**, 4857.
34. Z. Hao, G. Yang, X. Song, M. Zhu, X. Meng, S. Zhao and H. Zhang, *J. Mater. Chem. A*, 2014, **2**, 237.
35. D. Y. Lee, E.-K. Kim, N. K. Shrestha, D. W. Boukhvalov, J. K. Lee and S.-H. Han, *ACS Appl. Mater. Interfaces*, 2015, **7**, 18501.
36. A. Dragasser, O. Shekhah, O. Zybaylo, C. Shen, M. Buck, C. Wöll and D. Schlettwein, *Chem. Commun.*, 2012, **48**, 663.
37. Z. Guo, D. K. Panda, K. Maity, D. Lindsey, T. G. Parker, T. E. Albrecht-Schmitt, J. L. Barreda-Esparza, P. Xiong, W. Zhou and S. Saha, *J. Mater. Chem. C*, 2016, **4**, 894.
38. B. M. Wiers, M.-L. Foo, N. P. Balsara and J. R. Long, *J. Am. Chem. Soc.*, 2011, **133**, 14522.
39. C.-W. Kung, T.-H. Chang, L.-Y. Chou, J. T. Hupp, O. K. Farha and K.-C. Ho, *Electrochem. Commun.*, 2015, **58**, 51.
40. D. E. Williams, J. A. Rietman, J. M. Maier, R. Tan, A. B. Greytak, M. D. Smith, J. A. Krause and N. B. Shustova, *J. Am. Chem. Soc.*, 2014, **136**, 11886.
41. W. A. Maza, A. J. Haring, S. R. Ahrenholtz, C. C. Epley, S. Y. Lin and A. J. Morris, *Chem. Sci.*, 2016, **7**, 719.
42. L. Han, L. Qin, L. Xu, Y. Zhou, J. Sun and X. Zou, *Chem. Commun.*, 2013, **49**, 406.
43. B. Garai, A. Mallick and R. Banerjee, *Chem. Sci.*, 2016, **7**, 2195.
44. Z. Hu, S. Pramanik, K. Tan, C. Zheng, W. Liu, X. Zhang, Y. J. Chabal and J. Li, *Cryst. Growth Des.*, 2013, **13**, 4204.

CHAPTER 7

Conducting Framework Materials

PAVEL M. USOV,[a,b] CHANEL F. LEONG[a] AND DEANNA M. D'ALESSANDRO*[a]

[a] School of Chemistry, The University of Sydney, Sydney, New South Wales 2006, Australia; [b] Department of Chemistry, Virginia Tech, Blacksburg, Virginia 24061, USA
*Email: deanna.dalessandro@sydney.edu.au

7.1 Introduction

Over the past two decades, framework materials, including coordination polymers, metal-organic frameworks and covalent organic frameworks have emerged as a highly versatile family of materials with wide ranging potential applications. One such application which has generated considerable attention recently is that of nanoporous conductors, which, if realised, provide enormous potential technological impacts across the areas of electrocatalysis, chemiresistive sensing, thermoelectric materials, solar-electricity conversion and battery materials, amongst others. It is interesting to note that the origins of framework materials can be traced back more than 300 years to the compound known as Prussian blue, $Fe^{III}_4[Fe^{II}(CN)_6]_3 \cdot xH_2O$, which is itself a semiconducting framework solid.[1,2]

Herein, we focus on the extensive classes of framework materials known as coordination polymers and metal-organic frameworks (MOFs), which consist of metal ions or clusters (nodes) bridged by organic ligands (linkers) forming highly ordered 2D or 3D arrays.[3–6] The overall connectivity of these materials and their pore geometries depend on the coordination spheres of

Monographs in Supramolecular Chemistry No. 22
Functional Supramolecular Materials: From Surfaces to MOFs
Edited by Rahul Banerjee

Published by the Royal Society of Chemistry, www.rsc.org

the nodes and linkers.[7] While the early discoveries of framework materials were largely serendipitous and their properties were not systematically scrutinised, the seminal papers by Robson and Hoskins initiated a paradigm shift in this field.[3–5] Their early discoveries helped to establish the design principles for the construction of coordination frameworks, and opened up a new field of crystal engineering.[8,9] Frameworks also contain internal voids which are usually occupied by guests such as solvent molecules or counter-ions. The judicious selection of appropriate components for framework construction provides an exquisite level of control over the chemical and physical properties, and manipulation of host–guest chemistry offers further potential for tuning the materials.[10] A unique aspect of frameworks in comparison to their discrete molecular counterparts is the emergent properties that arise due to the synergism between multiple functionalities.[11–17]

The enormous potential of MOFs in many different applications was recognised early on.[3–5] Owing to their high internal surface areas (>7000 m^2 g^{-1}), frameworks have been extensively investigated for their potential applications in gas sorption and separation.[18,19] The interactions between guests and the framework can vary significantly depending on the characteristics of the internal void space such as the pore size, shape and electronic environment. As such, considerable research efforts have concentrated on the incorporation of reactive groups such as heteroatoms with exposed lone pairs and coordinatively unsaturated metal centres to optimise host–guest interactions. A number of frameworks have therefore been synthesised which exhibit record-breaking adsorption capacities and selectivities for important industrial gases such as H_2, CO_2 and CH_4.[58] As the field has matured over the past decade, many additional applications have been proposed and investigated which include, heterogeneous catalysis,[20] molecular sensing,[24] magnetism,[21] light harvesting,[22–24] energy storage,[25] and biomedicine.[26,27] The unparalleled chemical tunability and high structural precision of frameworks compared to amorphous carbons and other porous materials have been the main driving forces behind the fast development of this research area. In addition, considerable progress has been made in the manipulation and fabrication of these materials into nanoparticles, membranes and patterned surfaces for applications in devices.[28,29]

Despite enormous strides in the applications of framework materials over the past two decades, electronically-conducting frameworks have been relatively slow to emerge. This is due, in major part, to the fact that the metal centres typically used are redox inactive (*e.g.*, Zn^{II}) and the organic linkers (*e.g.*, carboxylates) often do not facilitate strong electron transfer between the metal centres or clusters. As a result, the majority of these materials are insulators or semiconductors. Significant opportunities at both the fundamental and applied levels can be envisioned if such conductive properties can be realised. The regular monodisperse nature of the pores, in addition to the capacity to engineer the architectures and topologies of frameworks provides an unprecedented opportunity to explore, for the first time, deeply fundamental structure–conductivity relationships in 3D coordination space. These characteristics distinguish MOFs from other types of solid-state conductors such as 'organic metals', mesoporous carbons and metal oxides. At an applied level, the realisation of conducting frameworks will spur the development of devices for a plethora of applications where signal transduction *via* conductivity changes can be exploited.

7.1.1 Scope of this Chapter

A number of recent reviews in the field of conducting coordination polymers and MOFs have outlined structural and electronic parameters which underpin the construction of these materials.[30–35] In this chapter, we provide an historical context to the emerging area of conducting frameworks, before discussing examples of conducting materials from the literature to date. While many of the early examples of coordination polymers were not permanently porous (*i.e.*, they did not retain their structures upon removal of guest molecules from the pores), they are integral to the development of the field, and shed important insights into structural design parameters that are conducive to long-range charge delocalisation. It is also worthy to note that in the historical context, conductivity measurements were often performed relatively routinely on new materials as a characterisation tool. In this regard, the development of highly conductive materials was not necessarily the sole focus of a study, however, important structure–function relationships were deduced.

As the search for conducting frameworks intensifies, an appreciation for the long and rich historical foundations of the field is timely. In moving forward, the emerging applications of framework materials in electrocatalysis, solar-electricity conversion and chemiresistive sensing, to name a few, will rely on the ability to exploit the extraordinarily high surface areas of these materials. It is also important to note that many of the MOFs investigated for their conductivities contain redox-active components (metal ions and/or ligands); modulation of the redox states of these materials can lead to markedly different conductivities – the full potential of which has yet to be exploited.[32]

This chapter opens with a discussion of the underlying mechanisms of charge transport which have been identified, to date, in coordination polymers and MOFs, followed by experimental examples from the literature. The discussion concludes with consideration of new and emerging measurement techniques, in addition to perspectives on the future development of this exciting research field.

7.2 Mechanisms of Charge Transport

Two distinct modes of charge transport have been observed in frameworks, namely ionic conduction which is facilitated by the movement of ions through the pores,[36] and electronic conduction involving migration of electrons or holes through the framework backbone.[35] Ionic conduction is typically mediated by protons due to their small radii and high mobility, and requires a hydrogen bonded network as a conduction pathway. From this point on, the chapter will primarily describe examples of electronic conduction in frameworks and outline different mechanisms by which it is facilitated.

One of the prime requirements for electronic conduction in frameworks is strong orbital overlap between the components. While the understanding of specific electronic conduction mechanisms in frameworks is in its infancy, some general design parameters have been identified. In general, it is recognised that either through-bond or through-space mechanisms can facilitate long-range charge transport; these include π-interactions (either π-conjugation or π-stacking) or redox matching which can be achieved, for example, *via* donor–acceptor or mixed-valence interactions (Figure 7.1). In the latter cases, redox activity is an important prerequisite for conductivity because it can facilitate charge migration between the components of the framework.[30] For this reason, highly conjugated ligands whose frontier orbitals can strongly overlap with metal orbitals are best suited for the design of conductive MOFs.

While the as-synthesised state of a framework material may not be conductive, application of an external stimulus can instigate charge transfer. As shown in Figure 7.1, the stimulus can take a number of forms, including the inclusion of guest molecules which may induce π-interactions such as π-stacking (through-space), or donor–acceptor interactions (through-space or through-bond). Manipulation of the redox state of a framework has also

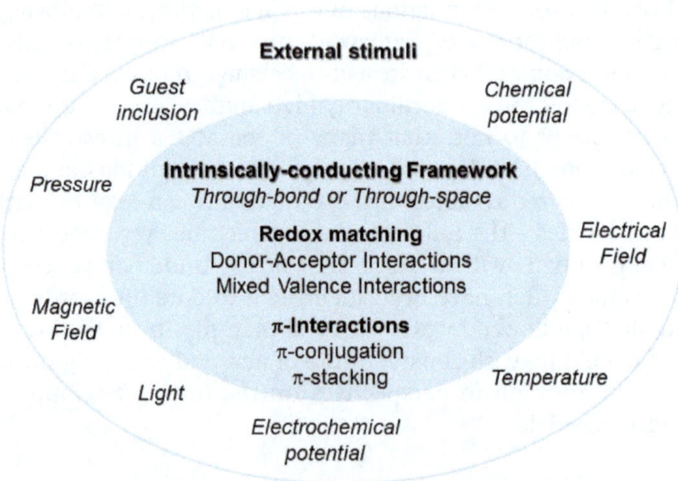

Figure 7.1 Schematic diagram of the different mechanisms identified for conducting frameworks. For intrinsically conducting frameworks, both through-bond and through-space mechanisms have been proposed. While an as-synthesised material may not be intrinsically conducting, application of an external stimulus may instigate conductivity *e.g.*, guest inclusion, a redox state change by chemical or electrochemical means, *etc.* Some, but not all of the listed stimuli have been used to date.

been shown to vary conductive properties, and in this regard, either chemical or electrochemical reduction/oxidation can be employed.[32] A number of other stimuli could potentially be used, but have not yet been demonstrated. It is important to note that different structural phases of a given framework material may exhibit different conductivities due to the different spatial arrangements of the components.

Two main charge transport regimes have been identified for MOFs, namely charge hopping and band transport.[31,37] It is generally believed that charge hopping between components is the predominant mechanism for charge transport in semiconducting frameworks. The main requirement for this mode is good communication between the components, either through a highly-delocalised connector or by virtue of their close proximity in space. Electron hopping has been identified as the main conduction mechanism in some mixed-valence frameworks such as Prussian blue[38,39] and has also been used to explain electron transfer in MOF films during electrochemical experiments.[40,41] The Variable-Range Hopping (VRH) model, originally developed for highly disordered systems with localised charge carriers, has been used to explain charge hopping conductivity in frameworks.[42] According to this model, temperature-dependent conductivity follows from eqn (7.1), where T is the temperature and d is the electronic dimensionality of the material.

$$\sigma(T) = \sigma_0 \exp\left[-\left(\frac{T_0}{T}\right)^{\frac{1}{1+d}}\right] \tag{7.1}$$

The parameters σ_0 and T_0 are determined by the electronic density of the states at the Fermi level, the localisation length and the phonon frequency. Due to the potential difficulties of fitting the electronic dimensionality factor, the VRH model has had mixed success in the interpretation of MOF conductivity data.[43] As electronic delocalisation in a framework increases, the distinction between redox components diminishes since the thermal barrier to the electron transfer between them ceases to exist.

In the high delocalisation limit, the material approaches a metallic state with essentially "free" electrons, and the conductivity is perhaps best described using band transport.[31] In this case, charge carriers (electrons or holes) are delocalised, and charge mobility is dependent upon effective orbital overlap, and minimal charge-scattering (*e.g.*, defects, impurities or grain boundaries).[31] While both hopping and band transport can give rise to long-range conductivity, the latter is believed to be more effective in achieving the highest values of conductivity.[31]

7.3 Conducting Coordination Solids: Historical Foundations

The conductive properties of coordination solids have been investigated for over 20 years. While many of these systems were semiconducting, and did not exhibit the permanent porosity that distinguishes them as MOFs, they demonstrate a number of important design principles which are pertinent to the discussion on conducting frameworks.

One of the earliest examples of conductive coordination solids was a series of salts containing the organic electron acceptor 7,7,8,8-tetracyanoquinodimethane (TCNQ).[44,45] In these studies several transition metal analogues were investigated, namely [Ag(TCNQ)], [Cu(TCNQ)], $[Fe(TCNQ)_2]\cdot 3H_2O$ and $[Mn(TCNQ)_2]\cdot 3H_2O$, which were found to exhibit moderate conductivities in the range of 1×10^{-2} to 2×10^{-5} S cm^{-1}. The first of these transition metal salts to be structurally characterised was [Ag(TCNQ)] using electron diffraction methods in 1980,[46] while its full crystal structure was obtained in 1985.[47] The system consisted of TCNQ moieties coordinated to Ag^I centres through each nitrile group which were arranged into a 3D coordination framework. The ligands in this material existed in a radical state forming an infinite π–π stack which served as a conduction pathway.

The early investigations of conductive behaviour in [M(TCNQ)] ($M = Cu^I$ and Ag^I) were marked by inconsistencies in the literature. In particular, conductivity switching measurements in thin films of these coordination solids were found to depend on the preparation conditions.[48–52] Dunbar and co-workers discovered the existence of multiple polymorphs of TCNQ frameworks which explained the variable electronic behaviour (Figure 7.2).[52,53] The structures of two polymorphs of the Cu^I analogue were determined crystallographically, hereafter denoted as Phases I and II. The structure of the former framework was reminiscent of [Ag(TCNQ)] and consists of radical TCNQ anions arranged into infinite π–π stacks. In Phase II, the TCNQ ligands were arranged further apart, with the nearest ligand–ligand distance of 6.8 Å. This structural difference has pronounced consequences on the charge transport properties of the two polymorphs. The room-temperature conductivity of Phase I measured on a pressed pellet was found to be 2.5×10^{-1} S cm^{-1}, whereas Phase II was 1.3×10^{-5} S cm^{-1} due to the lack of favourable conduction pathways in the latter case. This example illustrates the importance of the relative arrangement of structural components in determining the overall conductive properties of a material, and provides a means to control the behaviour of the framework without changing the chemical nature of the components.

Nitrile-based organic electron acceptors, such as TCNQ, are particularly well suited to generating coordination polymers because they are capable of coordinating to multiple metal centres and supporting a negative charge (often as stable radial anions or dianions) to counter-balance the positive charges of cationic metal ions. Furthermore, they have a tendency to form π–π stacks within framework structures which, as illustrated above for the example of Phase I [Cu(TCNQ)], is beneficial for charge transport. Apart from TCNQ, other nitrile-based ligands have been employed in the synthesis of conductive coordination polymers, including 11,11,12,12-tetracyano-2,6-naphthoquinodimethane (TNAP),[54] and *N,N'*-dicyanoquinonediimine (DCNQI) (Figure 7.3).[55–57] Modifications to the ligand shape and size as well as relative arrangement of

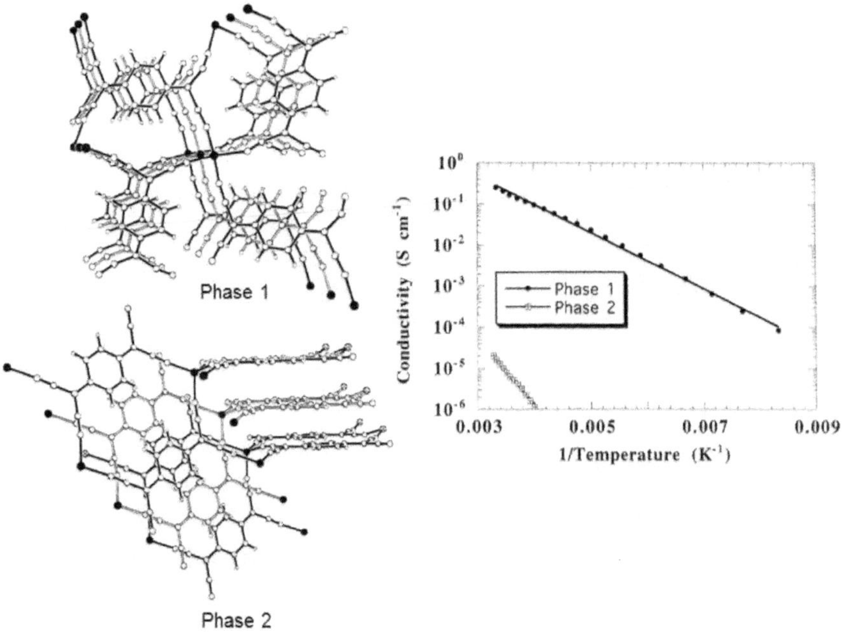

Figure 7.2 The two interpenetrating networks in Phases I and II of [Cu(TCNQ)] (*left*) where solid bonds depict one network and open bonds depict the second. A plot of the conductivity for pressed pellets of Phases I and II is shown on the *right*.
Reproduced with permission from R. A. Heintz, H. Zhao, X. Ouyang, G. Grandinetti, J. Cowen and K. R. Dunbar, New Insight into the Nature of Cu(TCNQ): Solution Routes to Two Distinct Polymorphs and Their Relationship to Crystalline Films That Display Bistable Switching Behavior, *Inorg. Chem.*, 1999, **38**, 144–156. Copyright 1999 American Chemical Society.

coordinating nitrile groups can directly affect the structure which in turn influences the physical properties. In particular, a series of Cu^{I} frameworks containing analogues of DCNQI, [Cu(R^1,R^2-DCNQI)$_2$] (R^1, $R^2 = $ H, Cl, Br, I, Me and OMe) feature highly conductive ligand π–π stacks. The conductivity of the dimethylated analogue reached 5×10^{5} S cm^{-1} at 3.5 K which is comparable to metallic copper and, to date, remains the highest conductivity value achieved in a 3D coordination polymer.

Another family of nitrile-based coordination frameworks investigated for their conductive properties feature dimeric ruthenium or rhodium secondary building units, also known as paddlewheel clusters, connected by TCNQ-derived ligands. These systems can adopt a wide range of structures including 1D ladders, 2D sheets and 3D networks and have been the subject of seminal studies by the groups of Miyasaka and Dunbar.[58–60] Conductivity studies revealed that these frameworks typically behave as semiconducting materials, the origin of which lies in donor–acceptor interactions. Since diruthenium paddlewheels and TCNQ (and its derivatives) are redox-active,

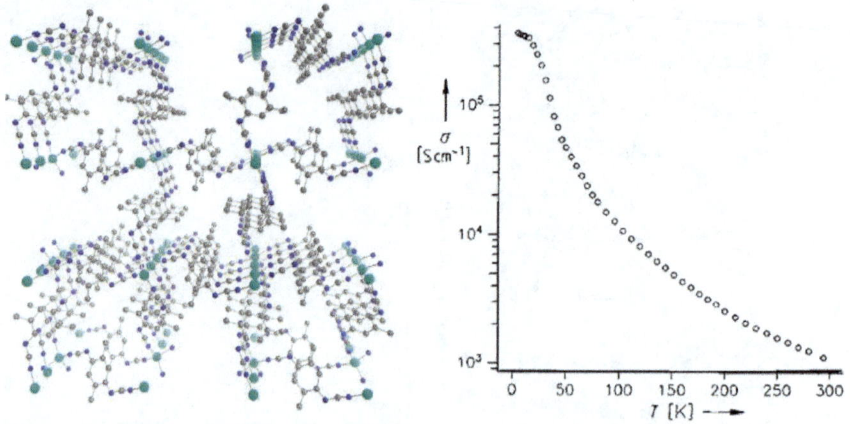

Figure 7.3 Perspective view of $[Cu(Me_2\text{-}DCNQI)_2]$ down the *c* axis (*left*). Cu, C and N are represented as cyan, grey and blue spheres, respectively (H atoms are omitted for clarity). The single crystal conductivity profile is shown on the *right*.
Reproduced with permission from A. Aumüller, P. Erk, G. Klebe, S. Hünig, J. U. von Schütz and H.-P. Werner, A Radical Anion Salt of 2,5-Dimethyl-N,N′-dicyanoquinonediimine with Extremely High Electrical Conductivity, *Angew. Chem., Int. Ed.*, 1986, **25**, 740–741. Copyright Wiley-VCH Verlag GmbH & Co. KGaA.

there is a driving force for charge transfer between these components when they are bound inside a framework. In this case the metal cluster typically acts as an electron donor while the ligand acts as the acceptor.

For the two 3D frameworks $[(Ru_2(3\text{-}FPhCO_2)_4)_2(BTDA\text{-}TCNQ)]\cdot 3.4CH_2Cl_2\cdot 1.6C_7H_7Cl$ ($3\text{-}FPhCO_2$ = 3-fluorobenzoate, BTDA-TCNQ = bis(1,2,5-thiadiazolo)tetracyanoquinodimethane and C_7H_7Cl = 4-chlorotoluene) and $[(Ru_2(3\text{-}CH_3PhCO_2)_4)_2(BTDA\text{-}TCNQ)]\cdot 1.4CH_2Cl_2\cdot 2.3C_8H_{10}$ ($3\text{-}CH_3PhCO_2$ = 3-methylbenzoate, C_8H_{10} = 1,4-dimethylbenzene), room-temperature conductivities of 9.09×10^{-5} and 1.25×10^{-5} S cm^{-1}, respectively, were found (Figure 7.4).[61] Interestingly, the latter compound underwent an anomalous increase in conductivity to 1.28×10^{-3} S cm^{-1} at 277 K due to thermally-induced charge fluctuations.

In addition to the relative spatial arrangement of the components in donor–acceptor type materials, the ionisation potential (*IP*) and the electron affinity (*EA*) of the donor (*D*) and acceptor (*A*), respectively, largely determine the degree of charge transfer (ρ), which in turn influences the conductive behaviour.[62] Here, the degree of charge transfer between *D* and *A* can be considered by $D^0A^0 \leftrightarrow D^{\delta+}A^{\delta-}$ where $0<\delta<1$. The energy of the charge transfer (CT) transition can be described by equation (7.2):

$$h\upsilon_{CT} = \Delta E_{IP-EA} - M \tag{7.2}$$

where $h\upsilon_{CT}$ is the energy of CT transition and M is the Madelung energy (defined as the energy gained if the lattice is ionic).[63] This relationship can

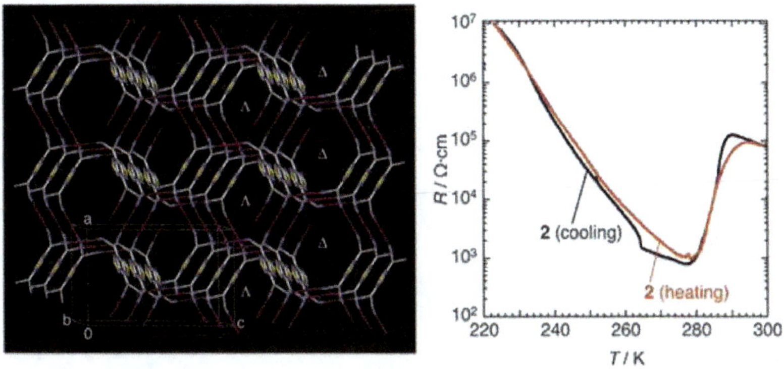

Figure 7.4 Schematic diagram (*left*) of $[(Ru_2(3\text{-}CH_3PhCO_2)_4)_2(\text{BTDA-TCNQ})]\cdot 1.4CH_2Cl_2\cdot 2.3C_8H_{10}$ showing the $[Ru_2{}^{II,III}]^+$ centres (brown), the BTDA-TCNQ$^{\bullet-}$ ligands and the helical topologies of the chains. The resistivity (R) of a single crystal (*right*) showing the disparity due to thermally-induced charge fluctuations.
Reproduced from ref. 61 with permission from the Royal Society of Chemistry.

be simplified into three regimes:[62] a neutral system where $\rho=0$ in which $\Delta E_{IP-EA}>M$; an ionic system where $\rho=1$ in which $\Delta E_{IP-EA}<M$; and a mixed-valence system where $0<\rho<1$ in which $\Delta E_{IP-EA}\approx M$. The mixed-valence regime is particularly interesting, and it is systems that exhibit this partial degree of charge transfer where conductivity values are often highest (Figure 7.5). The family of materials known as 'organic metals' best exemplify this criterion. The archetypal CT material tetrathiafulvalene-tetracyanoquinodimethane (TTF-TCNQ) discovered in 1973, was found to exhibit metallic conductivity comparable to copper (1.47×10^4 S cm^{-1} *vs.* 6×10^5 S cm^{-1} in copper).[64] Importantly, the mixed-valence regime where $0.5<\rho<0.75$ is the region in which CT complexes exhibit metallic behaviour, which is known to be dependent on the geometrical stacking arrangement of donor and acceptor units. Mixed stack materials (*i.e.*, $\cdots$*DADADA*$\cdots$) typically exhibit insulator or semiconducting behaviour, whereas segregated stack materials ($\cdots$*DDDDD*$\cdots$ and $\cdots$*AAAAA*$\cdots$) such as TTF-TCNQ tend towards metallic conductivity (Figure 7.5).

With regard to the aforementioned materials pioneered by Dunbar and Miyasaka based on dimeric ruthenium secondary building units (donors) and TCNQ-derived ligands (acceptors), partial CT from the metal node to TCNQ has been found in the systems $[(M_2(O_2CCF_3)_4)_2(\mu_4\text{-TCNQ})\cdot(C_7H_8)]$ (where M = Ru, Rh, O_2CCF_3 = trifluoroacetate, and C_7H_8 = toluene) which was attributed to significant metal-ligand π-backbonding.[60] As a result of this CT, the materials are mixed-valence and exhibit a phenomenon known as Intervalence Charge Transfer (IVCT), as shown in Figure 7.6. Appending different functional groups on the Ru paddlewheel clusters and TCNQ ligands has been shown to modulate the charge transfer behaviour through

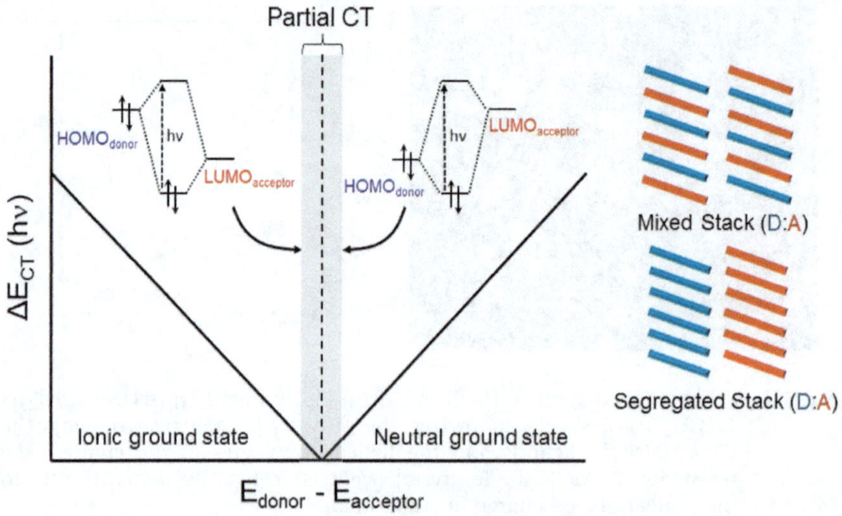

Figure 7.5 *Left:* Torrance's "V-shaped" diagram showing the relationship between the CT energy, ΔE_{CT}, and the difference between donor and acceptor *IP* and *EA*, respectively. The shaded region shows the range in which partial CT, and metallic conductivity is observed. *Right:* A general diagram representing the crystalline 1 : 1 mixed and segregated stacking arrangements of *D* (blue) and *A* (red) units that are commonly found in charge-transfer salts. Systems with a segregated stack arrangement (*e.g.*, TTF-TCNQ) typically exhibit relatively higher conductivities that fall in the shaded region of the Torrance diagram.
Left figure adapted from J. B. Torrance, An Overview of Organic Charge-Transfer Solids: Insulators, Metals and the Netural-Ionic Transition, *Mol. Cryst. Liq. Cryst.*, 1985, **126**, 55–67. Copyright 1985 Taylor & Francis. Right figure adapted from ref. 66 with permission from the Royal Society of Chemistry.

modification of the redox potentials,[67,68] which provides a means to control the physical properties of the frameworks.

Where the donor and acceptor have the same chemical identity, but different formal charges, mixed valency can result, as seen in Prussian blue, $Fe^{III}{}_4[Fe^{II}(CN)_6]_3 \cdot xH_2O$, which was first discovered to be an intrinsic semiconductor in 1954 by Fielding and Mellor.[69] As the name suggests, this material possesses a vivid blue colour and has therefore been extensively utilised as a dye and ink. Despite its long history, the structure of Prussian blue was only conclusively determined in 1936 by Keggin and Miles using X-ray diffraction methods.[7] It was found to be a cubic framework consisting of alternating Fe^{II} and Fe^{III} centres connected by cyanide linkers and containing internal voids filled with water molecules. The intense colour of Prussian blue could not be explained from the properties of its components[8] but rather was attributed to IVCT.[9] One of the complicating issues in these materials is the potential presence of counter-ions inside the framework voids which could facilitate ionic conductance. As a result,

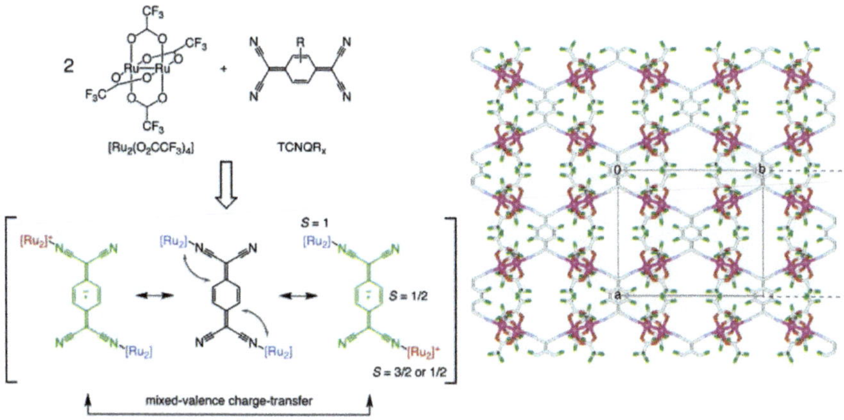

Figure 7.6 General scheme for the construction of $[(Ru_2(O_2CCF_3)_4)_2(\mu_4\text{-TCNQ})\cdot(C_7H_8)]$ frameworks showing the donor $[Ru_2(O_2CCF_3)_4]$ and acceptor $TCNQR_x$ components (*top right*), the mixed-valence charge transfer interactions leading to mixed valency and IVCT (*bottom left*) and a packing diagram of the 2D network projected along the *c* axis (*right*). Solvent molecules were omitted for clarity.
Reproduced with permission from H. Miyasaka, N. Motokawa, S. Matsunaga, M. Yamashita, K. Sugimoto, T. Mori, N. Toyota and K. R. Dunbar, Control of Charge Transfer in a Series of $Ru_2^{II,II}$/TCNQ Two-Dimensional Networks by Tuning the Electron Affinity of TCNQ Units: A Route to Synergistic Magnetic/Conducting Materials, *J. Am. Chem. Soc.*, 2010, **132**, 1532–1544. Copyright 2010 American Chemical Society.

differentiation between electronic and ionic conduction modes is a challenging task.

A more detailed study on the conduction mechanism of Prussian blue was carried out by Rosseinsky and co-workers who investigated the effect of the redox state on the magnitude of conductivity.[39] Prussian blue has two other commonly encountered redox states, fully oxidised Prussian yellow and fully reduced Prussian white, both of which were found to be significantly more conductive than the mixed-valence analogue. Berlin green, the partially oxidised form of Prussian blue, is also known to be relatively more conductive.[70] The origins of the differing conductivities have been considered semiquantitatively, and have been attributed to the fact that the ground state of Prussian blue is $Fe^{III}(t_{2g}{}^3e_g{}^2)$ $Fe^{II}(t_{2g}{}^6e_g{}^0)$: conduction through the C-coordinated Fe^{2+} is prohibited by the fully filled $t_{2g}{}^6$ character, and while e_g states are available, the energy gap is greater than the thermal energy at room-temperature, such that they are inaccessible for conduction (Figure 7.7).[70] In Berlin green, for example, the C-coordinated Fe^{3+} has unoccupied t_{2g} states which provide a viable conduction channel which permits charge hopping. One caveat to note here is that defects, hydration and counter-ions (generally K^+ ions) in different preparations of Prussian blue and its analogues have been shown to significantly influence the conductivities (in some cases by orders of magnitude).

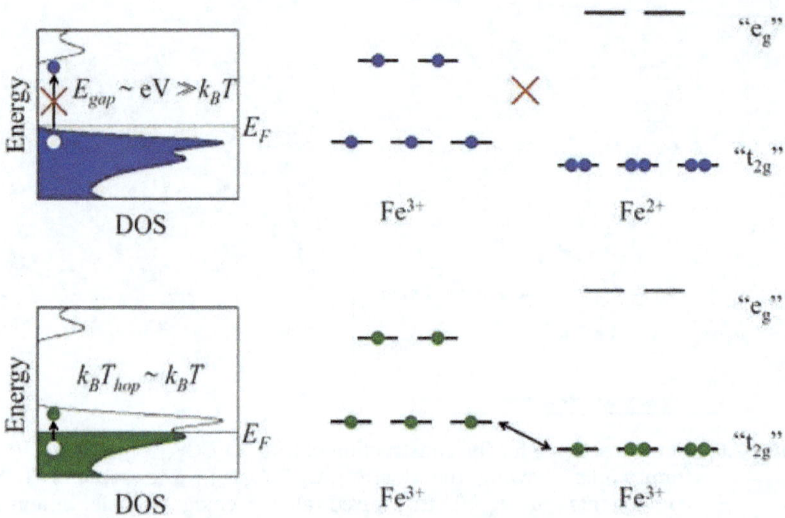

Figure 7.7 Scheme of the semiconducting nature of Prussian blue (*top*) and Berlin green (*bottom*), respectively, showing the framework and molecular formulations (*left* and *right*, respectively). In Prussian blue, the large energy gap (E_{gap}) between the valence and conduction levels precludes conduction; in Berlin green nearest-neighbour charge hopping is viable. Reproduced with permission from D. M. Pajerowski, T. Watanabe, T. Yamamoto and Y. Einaga, Electronic conductivity in Berlin green and Prussian blue, *Phys. Rev. B*, 2011, **83**, 153202. Copyright 2011 American Physical Society.

The conductivities of a number of Prussian blue analogues containing different first row transition metals as well as interstitial ions have been investigated.[71–73] Most of them exhibited similar semiconductor-type behaviour over a wide temperature range indicating that charge hopping was taking place. In an attempt to improve the conductive properties of Prussian blue, Long and co-workers substituted the second-row transition metal ruthenium into the structure.[74] The room-temperature conductivity $Fe^{III}_4[Ru^{II}(CN)_6]_3 \cdot 18H_2O$ was found to be 5.05×10^{-6} S cm^{-1} compared to 5.57×10^{-5} S cm^{-1} for $Fe^{III}_4[Fe^{II}(CN)_6]_3 \cdot 14H_2O$. This behaviour was attributed to the increase in asymmetry due to the presence of two different metal ions in the former compound which raises the barrier for electron transfer. The $K_{1.2}Ru^{III}_{3.6}[Ru^{II}(CN)_6]_3 \cdot 16H_2O$ framework, on the other hand, exhibited a markedly improved conductivity of 5.69×10^{-3} Scm^{-1} which was attributed to the more diffuse frontier orbitals of the ruthenium centres which creates better overlap with the ligand orbitals. It is important to note however, that the presence of interstitial potassium ions could have also contributed to the increased conductivity. Interestingly, the spectroscopic properties of these frameworks, particularly the energy of the IVCT band, correlated with the conductivity: the IVCT band observed in the ruthenium Prussian blue analogue occurred at lower energies compared with the other frameworks

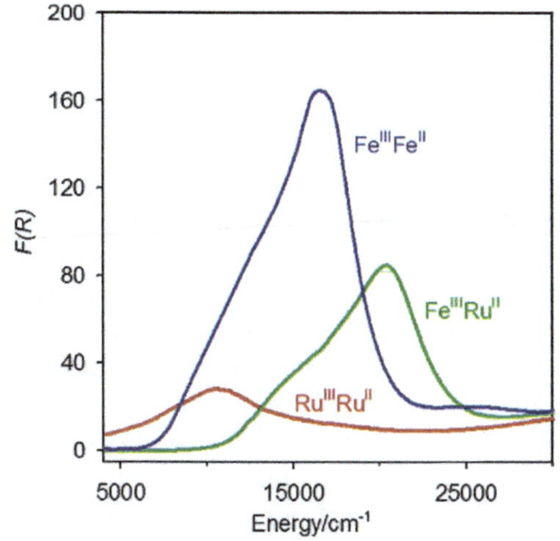

Figure 7.8 IVCT bands for $Fe^{III}_4[Ru^{II}(CN)_6]_3 \cdot 18H_2O$ ($Fe^{III}Ru^{II}$), $Fe^{III}_4[Fe^{II}(CN)_6]_3 \cdot 14H_2O$ ($Fe^{III}Fe^{II}$) and $K_{1.2}Ru^{III}_{3.6}[Ru^{II}(CN)_6]_3 \cdot 16H_2O$ ($Ru^{III}Ru^{II}$), showing the shift to lower energies as the conductivity increases due to the combined effects of electronic delocalisation and the presence of potassium ions. Reproduced with permission from J. N. Behera, D. M. D'Alessandro, N. Soheilnia and J. R. Long., Synthesis and Characterization of Ruthenium and Iron-Ruthenium Prussian Blue Analogues, *Chem. Mat.*, 2009, **21**, 1922–1926. Copyright 2009 American Chemical Society.

which is consistent with higher electronic delocalisation (Figure 7.8). Overall, this study suggested that incorporation of second and third row transition metals could improve through-bond conductivities.

7.4 Conducting Framework Materials: Recent Experimental Milestones

As the framework field has developed, many of the principles established from semiconducting and conducting coordination polymers have been extended, and the past 5 years has seen enormous strides in the advent of MOFs displaying either semiconducting or conducting properties. In this section, we discuss recent examples predominantly reported over this period. While few investigations of conduction mechanisms have appeared, examples of both through-bond and through-space charge transfer have been proposed. As shown in Figure 7.1, a number of broad design parameters have been identified, including redox matching (which can be achieved *via* either donor–acceptor or mixed-valence interactions) and π-interactions. External stimuli may also promote long-range CT, and it is important to note that the different redox states of framework materials (where accessible), can, and often do, exhibit markedly different conductive behaviours.[32]

7.4.1 Exploiting Redox Matching: Donor–Acceptor and Mixed Valency

One of the key design principles for synthesising conductive frameworks has involved utilisation of "softer" sulfur and nitrogen based ligands instead of much more commonly used carboxylate-based linkers. The notion here is that a more optimal redox match of components can be achieved in the former case, particularly with regards to the overlap of transition metal and ligand orbitals. The MOF-74 series, $[M_2(DOBDC)]$ (DOBDC = 2,5-dihydroxibenzene-1,4-dicarboxylate, M = Mg^{II}, Mn^{II}, Fe^{II}, Co^{II}, Ni^{II}, Zn^{II} and Cd^{II}) has been shown to feature both high electronic conductance and permanent porosity. Dincă and co-workers reported a Mn analogue where they replaced DOBDC with DSBDC (2,5-disulfhydrilbenzene-1,4-dicarboxylate) which exhibited a relatively high charge mobility of 0.01 $cm^2\ V^{-1}\ s^{-1}$ which is comparable to organic semiconductors (Figure 7.9).[75] Here, it was proposed that infinite Mn–S chains found within the framework serve as conduction pathways. This explanation was supported when comparing charge mobilities in $[M_2(DXBDC)]$ frameworks (M = Mn^{II} and Fe^{II}, X = O and S). Substitution of O for S resulted in an increase in conductivity from 3.0×10^{-13} to 1.2×10^{-12} S cm^{-1} for the Mn^{II} analogue and 4.6×10^{-8} to 5.8×10^{-7} S cm^{-1} for Fe^{II}.[76] This behaviour was attributed to the more diffuse orbitals of S atoms which facilitate electron hopping between the metal centres.

The notion of redox matching is inherent in mixed-valence systems, where components of the same identity, differing only in their formal redox states are involved. Mixed valency was implicated as the origin of conductivity in the framework $[Cu(Cu(PDT)_2)]$ (PDT = 2,3-pyrazinedithiolate) first reported by Takaishi and co-workers.[77] Here, the highly-delocalised redox-active Cu dithiolene complex was a structural motif which facilitated strong electronic communication between the metal nodes and metalloligands leading to a room-temperature conductivity of 6×10^{-4} S cm^{-1}. In this system, the donor

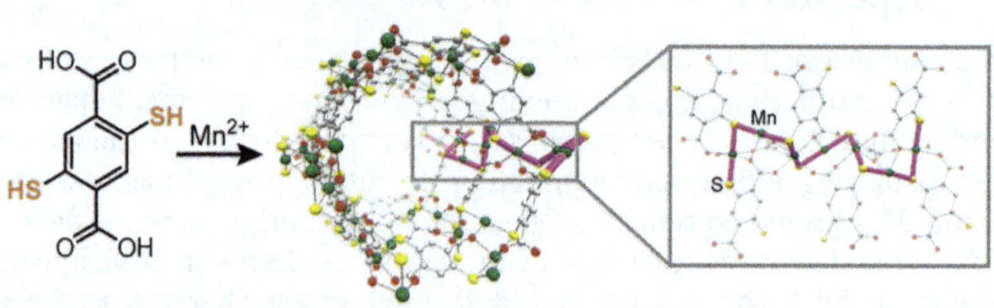

Figure 7.9 Schematic diagram of the DSBDC ligand (*left*) and the $[Mn_2(DSBDC)]$ framework showing the charge transport pathway through the material (*right*).
Reproduced with permission from L. Sun, T. Miyakai, S. Seki and M. Dincă, Mn2(2,5-disulfhydrylbenzene-1,4-dicarboxylate): A Microporous Metal–Organic Framework with Infinite (–Mn–S–)∞ Chains and High Intrinsic Charge Mobility, *J. Am. Chem. Soc.*, 2013, **135**, 8185–8188. Copyright 2013 the American Chemical Society.

Cu^{I} and acceptor $[Cu^{III}(PDT)_2]^-$ components are believed to facilitate a mixed-valence interaction that creates bistable $[Cu^{I}(Cu^{III}(PDT)_2)]$ and $[Cu^{II}(Cu^{II}(PDT)_2)]$ states. The presence of a possible IVCT band in the IR spectrum of the material also indicated the significance of a relationship between low energy CT transitions, and conductivity.

This work was expanded upon by Long and co-workers who synthesised a Ni analogue, $Cu[(Ni(PDT)_2]$.[78] One of the advantages of this framework was its permanent porosity (BET surface area of 385 $m^2\ g^{-1}$) compared to its predecessor which was found to collapse upon desolvation. The room-temperature conductivity of $Cu[(Ni(PDT)_2]$, however, was measured at 1×10^{-8} S cm^{-1} – significantly lower than that reported for the Cu analogue[77] due to the larger optical bandgap. Upon intercalation of I_2 into the MOF structure, the conductivity increased by four orders of magnitude to 1×10^{-4} S cm^{-1} due to oxidation of half of the nickel dithiolate units generating a mixed-valence state (Figure 7.10). The improved charge transport was attributed to electron hopping between the Ni centres. This work demonstrated that the conductivities of frameworks could be modulated by redox state changes, paving the way for the further development of porous conductors.

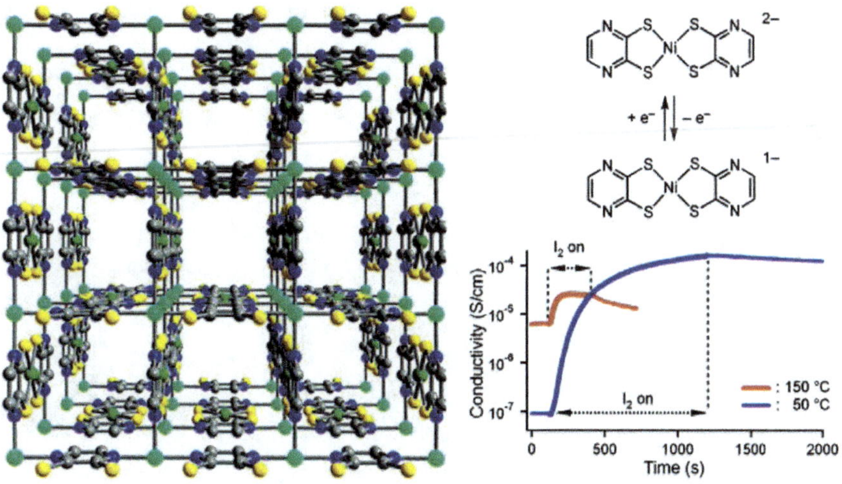

Figure 7.10 The redox-active and semiconducting framework $Cu[(Ni(PDT)_2]$ (*left*) in which the conductivity can be increased with oxidative doping with various halide guests. Intercalation of I_2, for example, induces a partial oxidation of the redox-active nickel dithiolate units in the framework (*upper right*) leading to a four order of magnitude increase in the conductivity (*lower right*). Light blue, green, yellow, dark blue, and grey spheres represent Cu, Ni, S, N, and C atoms, respectively.

The frameworks incorporating dithiolene ligands represent cases where the mixed-valence interaction is believed to originate between metal centres with different formal oxidation states; however, redox-active ligands can likewise mediate these interactions. The first example of this was reported by Long and co-workers who discovered high conductivity in a quinoid-based $(NBu_4)_2[Fe_2(DHBQ)_3]$ framework (DHBQ = 2,5-dioxidobenzoquinone)[79] originally reported by Abrahams and Robson (Figure 7.11).[80] This material exhibited room-temperature conductivity of 0.16 S cm^{-1} – the highest reported to date in a 3D MOF. Furthermore, the existence of a low energy IVCT band in the UV-Vis-NIR spectrum indicated strong electronic communication between the components of the framework which places this MOF into Class II/III of the Robin and Day classification scheme – a first such example amongst reported frameworks. Reducing $(NBu_4)_2[Fe_2(DHBQ)_3]$ with sodium naphthalenide decreased the intensity of the IVCT band and subsequently lowered the conductivity to 6.2×10^{-3} S cm^{-1} which indicates

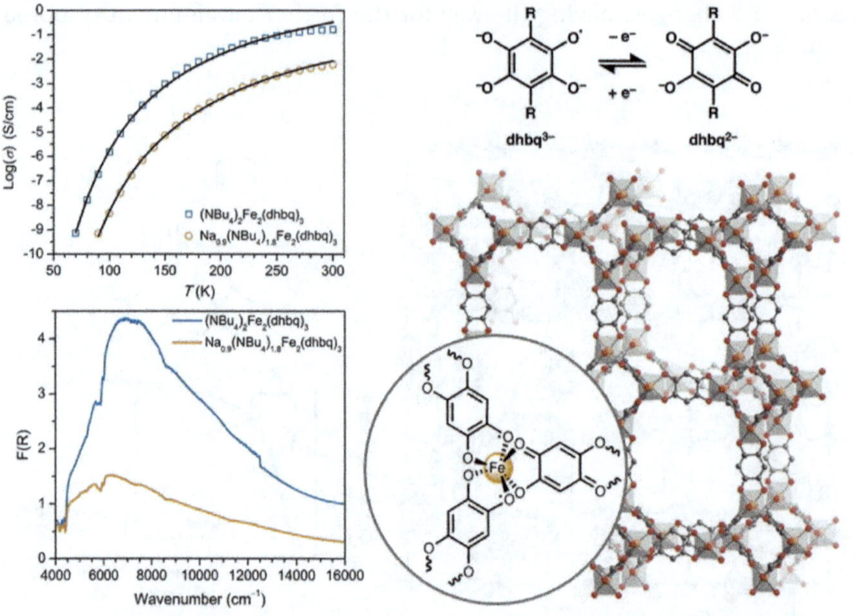

Figure 7.11 Schematic diagram of the framework $(NBu_4)_2[Fe^{III}{}_2(DHBQ)_3]$ showing the redox-active unit (*bottom right*). Chemical reduction generates $(Na)_{0.9}(NBu_4)_{1.8}[Fe^{III}{}_2(DHBQ)_3]$ which exhibits a decrease in both the conductivity (*upper left*) and the intensity of the IVCT band (*lower left*). The observations are consistent with the reduction of the mixed-valence $DHBQ^{2-/3-}$ ligands (*top right*).
Adapted with permission from L. E. Darago, M. L. Aubrey, C. J. Yu, M. I. Gonzalez and J. R. Long, Electronic Conductivity, Ferrimagnetic Ordering, and Reductive Insertion Mediated by Organic Mixed-Valence in a Ferric Semiquinoid Metal-Organic Framework, *J. Am. Chem. Soc.*, 2015, **137**, 15703–15711. Copyright 2015 the American Chemical Society.

the dependence of electronic delocalisation on the redox state of the framework. This behaviour suggested that the material may be useful in redox sensor applications.

While the concept of redox matching donor and acceptor components has been utilised to achieve charge transfer in a number of frameworks reported to date, the principle can also be applied in cases where the donor and acceptor interactions occur between the host framework and guest molecules. Talin, Allendorf and co-workers demonstrated enhancement of MOF conductivity *via* post-synthetic infiltration of [$Cu_3(BTC)_2$] (where BTC = 1,3,5-benzenetricarboxylate) with the acceptor TCNQ.[33,81] This resulted in a conductivity of 0.07 S cm^{-1} – six orders of magnitude higher than the desolvated MOF. Such an enhancement in conductivity was attributed to coordination of TCNQ at the unsaturated Cu^{II} sites, produced *via* the heat-activated removal of interstitial and coordinated water molecules, forming a new conduction pathway (Figure 7.12). The authors postulated that there was a partial charge transfer from the framework to TCNQ moieties, and provided some spectroscopic evidence. Even though the most likely source of electrons was the Cu^{II} centres, X-ray photoelectron spectroscopy (XPS) showed that they did not undergo a redox state change. Interestingly, infiltration of [$Cu_3(BTC)_2$] with the fluorinated derivative $TCNQF_4$, which has a higher *EA* than TCNQ, resulted in comparatively lower conductivity, demonstrating the importance of redox matching donor and acceptor units in tuning CT and conductivity. Furthermore, the introduction of the fully hydrogenated, π-poor $TCNQH_4$ produced an almost insulating material. Computational calculations suggested strong electronic coupling between TCNQ and the Cu^{II} $d(z^2)$ orbital which leads to fully-mixed ground and charge separated states as described by Class III of the Robin-Day Classification Scheme.[33,81] This stands as one of the few studies which applies Marcus-Hush theory to describe CT in a framework.[82]

7.4.2 Exploiting π-Interactions

A large number of the most conductive framework materials known to date exploit the strategies of through-bond π-conjugation and through-space π-stacking interactions which have been established from the extensive literature on 'organic metals'[83,84] such as TTF-TCNQ[85] which forms segregated stacks of TTF (*D*) and TCNQ (*A*) molecules (Figure 7.5). A prime example of a MOF exhibiting π-stacking interactions is [Zn_2(TTFTB)] (TTFTB = tetrathiafulvalenetetrabenzoate) which incorporates a TTF-based ligand arranged such that 1D channels and infinite TTF stacks run along the *c* axis (Figure 7.13).[86,87] As a result, the framework exhibits high electron mobility along infinite π–π stacks in the structure as well as permanent porosity making it a potential candidate for guest-dependent conductivity investigations. The conductivity was found to be strongly dependent on the separation between the TTF layers which in turn was dictated by the size of the metal ions.

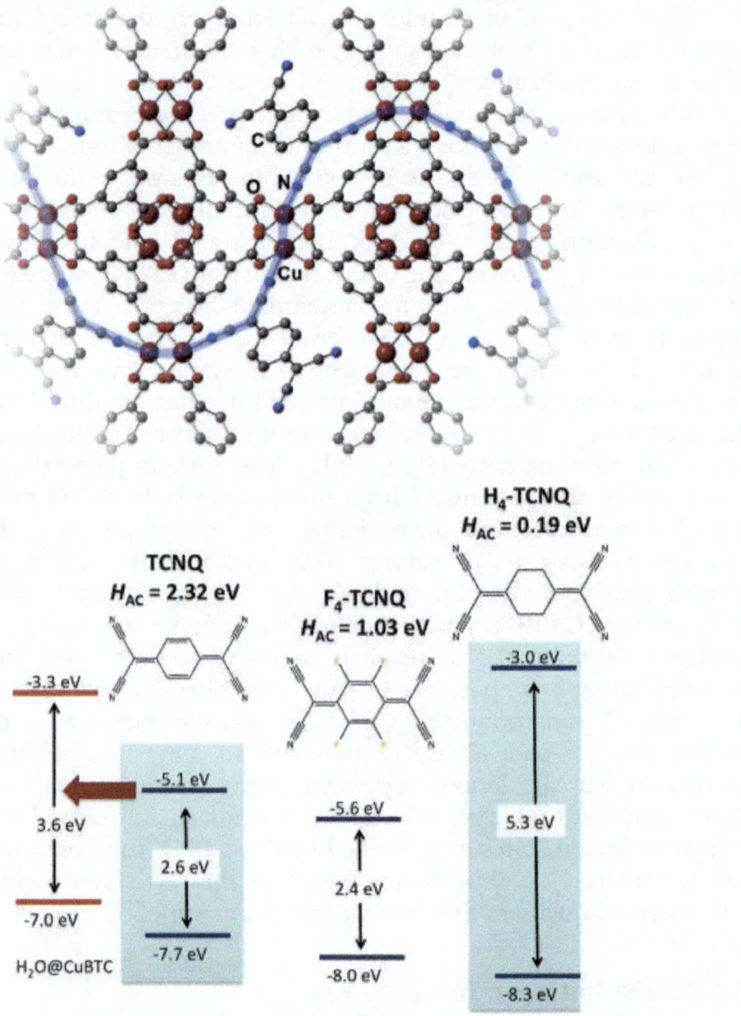

Figure 7.12 *Top:* Proposed structure of TCNQ-infiltrated $[Cu_3(BTC)_2]$ showing the bridging of neighbouring Cu_2 nodes by TCNQ molecules. The new conduction pathway is highlighted in blue. *Bottom:* Diagram of HOMO/LUMO energy levels of $[Cu_3(BTC)_2]$ (H_2O@CuBTC) and the guests TCNQ, $TCNQF_4$, and $TCNQH_4$ showing the more favourable redox match between $[Cu_3(BTC)_2]$ and TCNQ compared to the other guest molecules.

Top figure reproduced from L. Sun, M. G. Campbell and M. Dincă., Electrically Conductive Porous Metal–Organic Frameworks, *Angew. Chem., Int. Ed.*, **55**, 3566–3579. Copyright 2016 WILEY-VCH Verlag GmbH & Co. KGaA, Weinheim. Bottom figure adapted with permission from M. D. Allendorf, M. E. Foster, F. Leonard, V. Stavila, P. L. Feng, F. P. Doty, K. Leong, E. Y. Ma, S. R. Johnston and A. A. Talin, Guest-Induced Emergent Properties in Metal-Organic Frameworks, *J. Phys. Chem. Lett.*, 2015, **6**, 1182–1195. Copyright the American Chemical Society.

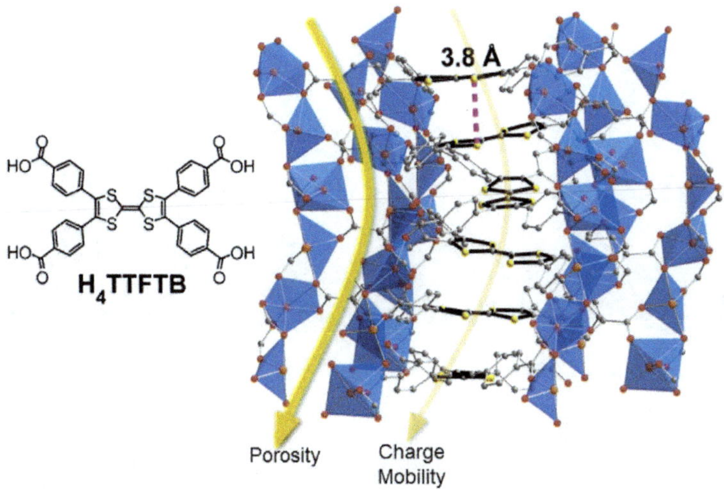

Figure 7.13 *Left:* Chemical structure of the H_4 TTFTB ligand. *Right:* A helical TTF stack in the crystal structure of [Zn_2(TTFTB)] showing the shortest intermolecular S···S contact. Orange, yellow, red and grey spheres represent Zn, S, O and C atoms, respectively (H atoms and water molecules omitted for clarity).
Adapted with permission from T. C. Narayan, T. Miyakai, S. Seki and M. Dincă, High Charge Mobility in a Tetrathiafulvalene-Based Microporous Metal-Organic Framework, *J. Am. Chem. Soc.*, 2012, **134**, 12932–12935. Copyright 2012 the American Chemical Society.

π-Stacking has also been identified as the predominant conduction pathway in the 3D framework [$ZnNa_3(L)_2(DEF)_2$] · DEF (L = 4,4′-(anthracene-9,10-diylbis(ethyne-2,1-diyl)dibenzoate, DEF = *N,N′*-diethylformamide).[88] Here, the conductivity of 1.3×10^{-3} S cm^{-1} arises from the long-range π-stacking of the ligand in the form of a zigzag chain, which was likened to double-walled carbon nanotubes. Interestingly, electroluminescence was also observed when a voltage was applied to a single crystal device in which the framework was fabricated between two ITO electrodes.

MOFs showing a structural resemblance to the well-known 2D conductor graphene have revealed some of the highest conductivities reported to date in frameworks. The common structural motif in these materials is that of a stacked honeycomb lattice which often incorporates square planar metal centres coordinated by benzene or triphenylene-derived ligands. The relative arrangement of the 2D sheets can be eclipsed, stepped-parallel or staggered, such that 1D pores propagate through the material. The source of the high conductivities is believed to be the π-conjugation through the 2D sheets, which originates from good donor–acceptor overlap, in addition to enhanced charge density as a result of ligand oxidation (to achieve charge balance with the M^{II} centres).

A number of relatively highly conductive 2D frameworks exhibiting graphitic-type topologies have been reported. For example, a family

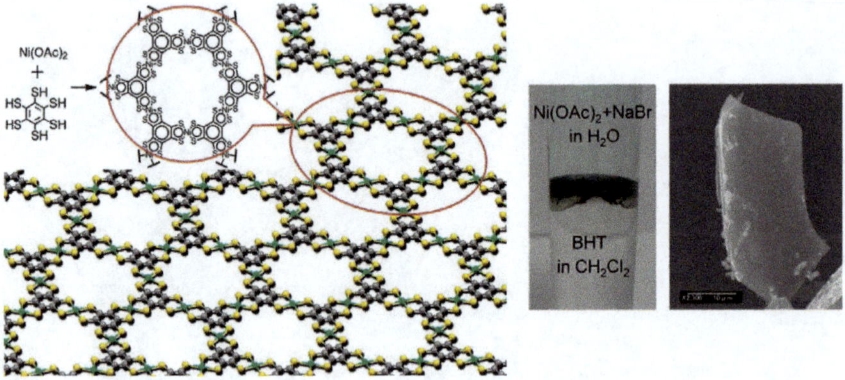

Figure 7.14 Schematic diagram (*left*) of the chemical structure of a monolayer of $[Ni_3(BHT)_2]$ with counter-ions omitted for clarity (Grey, yellow and green spheres represent C, S and Ni, respectively). The synthesis of $[Ni_3(BHT)_2]$ nanosheets at an aqueous-organic interface is shown on the *right* with a photograph of the material.
Reproduced with permission from T. Kambe, R. Sakamoto, K. Hoshiko, K. Takada, M. Miyachi, J.-H. Ryu, S. Sasaki, J. Kim, K. Nakazato, M. Takata and H. Nishihara, π-Conjugated Nickel Bis(dithiolene) Complex Nanosheet, *J. Am. Chem. Soc.*, 2013, **135**, 2462–2465. Copyright 2013 the American Chemical Society.

of catecholate frameworks constructed from hexahydroxytriphenylene (H_6HHTP) consist of 2D layers exhibiting eclipsed layering and thus substantial π–π stacking interactions. A related Cu^{II} framework exhibited a conductivity of 0.2 S cm^{-1}.[89] Nishihara and Dincă reported Ni-based frameworks with graphene-like structures containing redox-active ligands, namely $[Ni_3(BHT)_2]$ (BHT = benzenehexathiol)[90] (Figure 7.14) and $[Ni_3(HITP)_2]$ (HITP = hexaminotriphenylene).[91] These frameworks exhibited strong π-conjugation within their structures with conductivity values of 0.15 and 2 S cm^{-1}, respectively. $[Ni_3(BHT)_2]$ in particular, displayed redox-modulated behaviour upon doping with I_2 when its conductivity increased to 160 S cm^{-1} in the oxidised state.[92] A number of related 2D graphene-like MOFs have subsequently been reported,[93] and their range of applications has been expanded to include electrocatalysis[94,95] and chemiresistive sensing.[96–98] In the latter case, host–guest interactions in a series of 2D frameworks including $[Cu_3(HITP)_2]$ could be modulated by hydrogen bonding interactions with a range of different analytes.

Extensive computational modelling including first principles calculations have subsequently been performed by Shojaei and co-workers on $[Ni_3(BHT)_2]$ to further elucidate their electronic properties.[99,100] A key finding was the continuous tuning of the bandgap that could be achieved as the 2D sheets were slid with respect to one another. Interestingly, calculations suggested that rather than single layers, specific pairs of interlayer covalent bonds exist between the nanosheets, thus creating bilayers. Sliding one of the layers with respect to the others decreases the band gap from 0.15 to 0 eV, suggesting possible applications in nanoelectromechanical as well as optoelectronics devices.

The effect of dimensionality in frameworks was explored by Panda and Banerjee in their work on a family of In^{III} isophthalate frameworks.[101] The frameworks adopted either 1D or 2D architectures, with one of the structures exhibiting a π–π stacking arrangement of the phenyl rings (separated by 3.4 Å), which was absent in the other systems in which the aromatic rings were separated by 9.9 Å. The presence of π–π stacking was proposed as the origin of the high charge-carrier mobility in the system.

As interest mounts in potential device applications of conductive frameworks, attention has turned to the influence of fabrication methods for interfacing materials with surfaces. A closely related analogue of the 2D frameworks mentioned above is the framework prepared from Cu^{II} and BHT, $[Cu_3(BHT)]$.[102] Nanosheets of the framework formed at an organic/aqueous interface displayed high electrical conductivities, reaching 1580 S cm^{-1} at 300 K, the highest value reported in a 2D coordination framework to date (Figure 7.15). In addition, ambipolar charge transport behaviour and extremely high electron and hole mobilities were recorded under field modulation.

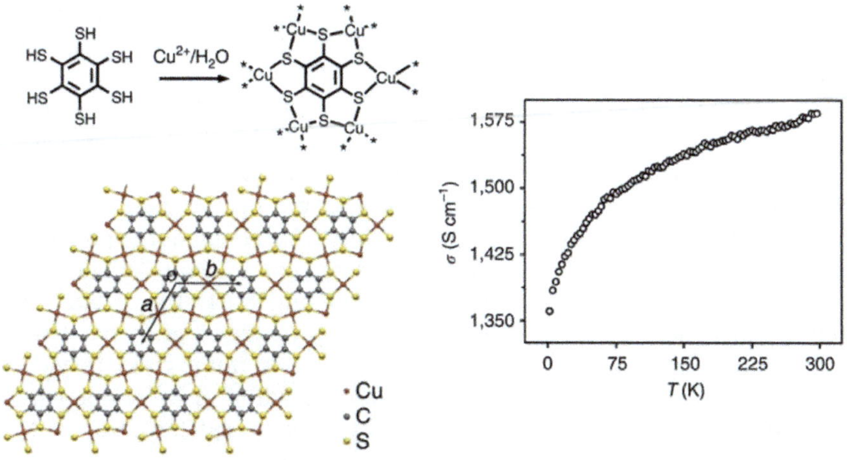

Figure 7.15 Schematic diagram of the formation of the $[Cu_3(BHT)]$ film and the 2D lattice structure (*left*) (Grey, yellow and brown spheres represent C, S and Cu, respectively). Conductivity measurements on a 150 nm thick film as a function of temperature (*right*).
Reproduced with Permission from X. Huang, P. Sheng, Z. Tu, F. Zhang, J. Wang, H. Geng, Y. Zou, C.-a. Di, Y. Yi, Y. Sun, W. Xu and D. Zhu, A two-dimensional π–d conjugated coordination polymer with extremely high electrical conductivity and ambipolar transport behaviour, *Nat. Commun.*, 2015, **6**. Published under a Creative Commons Attribution 4.0 International License (http://creativecommons.org/licenses/by/4.0/).

7.4.3 Other Examples: Conductivity Induced by External Stimuli

While a framework may possess inherent conductivity, external stimuli can be employed to induce conductive behaviour (Figure 7.1). It has also been demonstrated for 'organic metals' composed of donor and acceptor components that it is possible to modulate the degree of charge transfer using external stimuli (as depicted in Figure 7.5).[65] While not all of those stimuli shown in Figure 7.1 have been employed for studies on framework systems, the application of chemical potential (*e.g.* treatment with a chemical oxidant/reductant, or intercalation of a redox-active acceptor) has been widely used. This section considers those additional examples where a framework is not inherently conductive in its as-synthesised state, however, the material generated following application of an external stimulus (examples of guest inclusion and chemical oxidation are summarised here) does exhibit conductive properties.

In addition to guest-induced modulation of the electronic properties of framework materials, as described above for TCNQ-infiltrated $[Cu_3(BTC)_2]$ for example,[33] the judicious assembly of MOFs with regular pores can be exploited as a "template" to encapsulate guest molecules and achieve novel anisotropic properties, such as conductivity. Iodine has been used predominantly as the charge carrier medium, where an increase in conductivity is often achieved following guest inclusion. For example, Zeng and co-workers have reported a number of intrinsically semiconducting frameworks which, upon I_2 infiltration exhibited orders of magnitude increases in conductivity.[103,104] Frameworks including $[Cu_6(PYBZ)_8(OH)_2]$ and $[Zn_3(DL\text{-}LAC)_2(PYBZ)_2]$ (where DL-LAC = lactate anion and PYBZ = 4-pyridylbenzoate) have been intercalated with iodine, leading to confinement and alignment of the guests.[103,104] In the case of the zinc framework, the close host–guest proximity led to donor–acceptor interactions (I_2-π electron of the pore walls), resulting in cooperative anisotropic electrical conductivity (single crystal conductivities of 3.42×10^{-3} S cm^{-1} parallel and 1.64×10^{-4} S cm^{-1} perpendicular). These values were four orders of magnitude higher than that for solid I_2 itself (7.69×10^{-8} S cm^{-1}). In the copper framework, polyiodide ions (I^{5-} or I^{7-}) were detected in the pores, leading to a conductivity of 8.11×10^{-7} S cm^{-1}.[104] In this case, it was the well-regulated array of I_2 or polyiodide ions which facilitate long-range conductivity and render the framework semiconducting in nature. The same phenomenon was used to explain the increase in conductivity by two orders of magnitude in a europium-based framework, $[Eu_4(BPT)_4(DMF)_2(H_2O)_8]$ (where BPT = 3,4′,5-biphenyltricarboxylate) (from 8.27×10^{-7} S cm^{-1} to 2.71×10^{-5} S cm^{-1}).[105]

In an attempt to exploit the effective CT pathways in polyiodide species, the triiodide anion (I_3^-) was employed as a distinct cross-linking unit in the 2D coordination framework $[Cu(TMBP)I_3]$ (where TMBP = 3,3′,5,5′-tetramethyl-4,4′-bipyrazole).[106] The I_3^- units effectively link the Cu^I centres into infinite CuI_3 chains, such that small distortions only in the bond

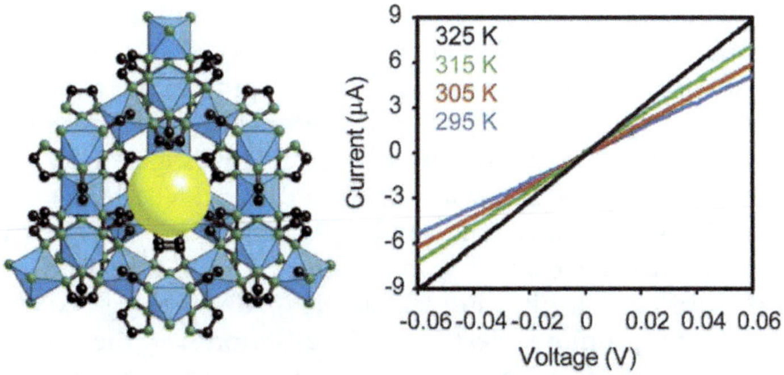

Figure 7.16 Schematic of a single pore of the 1,2,3-triazolate framework $[Fe(C_2N_3H_2)_2]$ (*left*) and the I–V curves recorded at different temperatures showing the increment in conductivity with temperature (*right*). Fe ions represented as blue spheres or polyhedra, N and C atoms are green and black spheres, respectively.
Reproduced with Permission from F. Gandara, F. J. Uribe-Romo, D. K. Britt, H. Furukawa, L. Lei, R. Cheng, X. F. Duan, M. O'Keeffe and O. M. Yaghi, Porous, Conductive Metal-Triazolates and Their Structural Elucidation by the Charge-Flipping Method, *Chem. – Eur. J.*, 2012, **18**, 10595–10601. Copyright © 2012 WILEY-VCH Verlag GmbH & Co. KGaA, Weinheim.

distances associated with the Cu–I and I–I bonds exist, giving rise to a conductivity of 9.38×10^{-4} S cm^{-1}.

The iron 1,2,3-triazolate $(C_2N_3H_2)$ framework $[Fe(C_2N_3H_2)_2]$ investigated by Yaghi and co-workers showed a room-temperature conductivity of 0.77×10^{-4} S cm^{-1} which was increased to 1.0×10^{-3} S cm^{-1} upon doping with I_2 vapour (Figure 7.16).[107] The origin of the conductivity enhancement was tentatively ascribed to oxidation of Fe^{II} to Fe^{III}, resulting in mixed-valence character akin to that found in oxides such as Fe_3O_4, however, it is also conceivable that the aforementioned formation of polyiodides within framework channels may explain the enhancement.

In the aforementioned examples, which demonstrate enhancements in framework conductivity upon oxidative doping with iodine, the origins may lie in the oxidation of metal centres; however, if this is not thermodynamically feasible, doping can arise from guest I_2-ligand π-electron interactions. In an attempt to gain insight into the underlying mechanisms for conductivity enhancements, thin films of the framework $[Co_3(NDC)_3(DMF)_4]$ (where NDC = 2,6-napthanlenedicarboxylate) were deposited onto amine-functionalised glass substrates.[108] Iodine doping induced p-type semiconducting behaviour in the material, as indicated from Hall Effect measurements. XPS measurements also indicated the presence of I_3^- in the pores, while UV/Vis measurements indicated a new donor–acceptor charge transfer process at 242 nm (in addition to the peak from iodine at 513 nm). Since oxidation of the Co^{II} ions was not thermodynamically feasible due to the unfavourable redox potentials, the origins of the enhancements were ascribed to cooperative CT upon confinement of iodine in the aromatic channels of the host framework. Density Functional Theory (DFT) calculations corroborated the findings of CT-induced hole doping in the framework.

The prospect of exploiting host–guest interactions *via* intercalation of complementary guest π-systems was recently investigated by Saha and co-workers.[109] Here, the framework [Zn_2(TCPB)(BPDPNDI)] (TCPB = 1,2,4,5-tetrakis-(4-carboxyphenyl)benzene and BPDPNDI = *N*,*N*′-bis(4-pyridyl)-2,6-dipyrrolidyl naphthalenediimide) was fabricated as a thin film on a ZnO substrate (Figure 7.17). Infiltration of π-acidic methyl viologen (MV^{2+}) guests was proposed to promote electron delocalisation *via* the formation of extended π-stacks with the framework ligands, leading to a conductivity enhancement from 6×10^{-3} to 2.3×10^{-1} S cm^{-1} and the formation of a new CT band in the electronic absorption spectrum. Infiltration of weaker π-acidic guests such as DFDNB (1,5-difluoro-2,4-dinitrobenzene) and DNT (dinitrotoluene) did not increase the conductivity as significantly as MV^{2+}, supporting the suggestion that stronger donor–acceptor interactions resulting in charge delocalisation through the π-stacks occur for more π-acidic guests. Interestingly, the conductivity was unaffected by C_{60} molecules which were

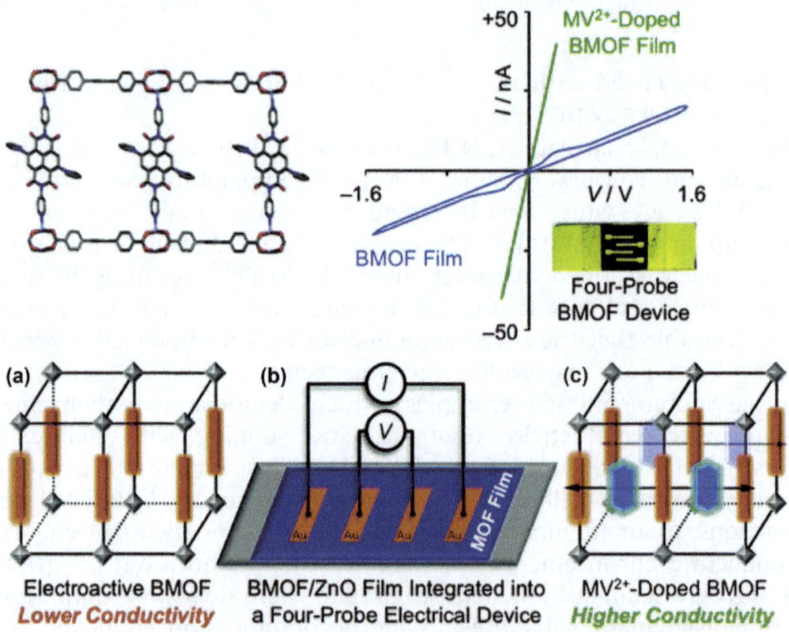

Figure 7.17 Structure of "BMOF" [Zn_2(TCPB)(BPDPNDI)] (*top left*) and the I–V relationships for the BMOF/ZnO film before and after doping with MV^{2+} (*top right*). *Bottom:* Illustrations of BMOF (a), an electrical device composed of BMOF on ZnO (b), and BMOF containing intercalated MV^{2+} leading to π-donor/acceptor stacks (and higher conductivity) (c). Reproduced from ref. 109 with permission from the Royal Society of Chemistry.

excluded from the channels on the basis of their size. The prospect of modulating donor–acceptor interactions using other external stimuli such as light also provides a basis for the future development of light harvesting and photoconducting MOFs.

7.5 Methods of Assessing Conductive Properties in Frameworks

7.5.1 Contact Methods

The most commonly used method for assessing conductivity in solid samples is the four-point probe technique in which four contacts are attached to the sample (pressed pellet, single crystal or a thin film) at equal distances (Figure 7.18a).[35] The two external contacts are used to apply current at the same time the internal contacts measure the voltage drop across the sample. The advantage of this method is that it reduces contributions from the wires and the contact-sample interfaces while improving the overall sensitivity. Currently, a major limitation with conductivity measurements is the sample preparation. Single crystals provide the best option because measurement of the conduction anisotropy becomes possible, however, crystals of sufficiently large size are necessary (>0.2 mm). For powdered samples, the material must be pressed into a pellet. Here, grain boundaries can significantly increase pellet resistance, while in the case of single crystals, problems can

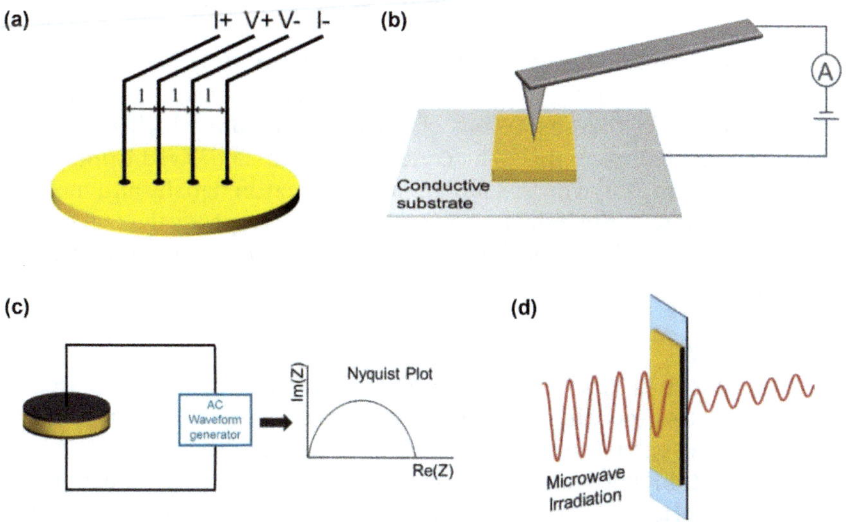

Figure 7.18 Schematic diagram of the commonly used methods to assess conductivities in framework materials: (a) Four-point probe method; (b) AFM conductivity; (c) Impedance spectroscopy and (d) Microwave conductivity.

arise from crystallographic defects. Furthermore, conductivity can be highly variable (more than an order of magnitude) even within the same sample batch,[87] which accounts for differing conductivity values reported for some materials. Decreasing the size of the sample by combining conductivity measurements with scanning electron microscopy has been used to improve upon the four-point probe method.[92]

In recent years, several additional techniques have been employed for measuring conductivities of frameworks. Another two-point probe method for obtaining the conductivity employs Atomic Force Microscopy (AFM) (Figure 7.18b).[110,111] This technique utilises a conductive AFM tip to directly probe the nanosized crystals, thus reducing contributions from grain boundaries and defects. AFM conductivity has yet to be used on frameworks, however, it has great potential to become a powerful tool. Electrical Impedance Spectroscopy (EIS) in the two-electrode mode has been extensively used to assess ion conductive frameworks (Figure 7.18c).[112,113] The real impedance (resistance) can be obtained directly from Nyquist plots, while other important electrical parameters of the material can be obtained from modelling of the impedance spectra.

7.5.2 Contactless Methods

All the aforementioned contact methods have one fundamental limitation, namely the existence of a contact-sample interface which can introduce significant variation into the measurements. As a result, the development of non-destructive contactless techniques has received considerable attention.[114] One such technique, microwave conductivity (Figure 7.18d), has been widely used to determine the intrinsic conductivity of semiconducting[115] and superconducting[116] materials and has recently been applied to frameworks.[75,86] It relies on the oscillation of charge carriers within the sample resulting in the absorption of microwaves from which the carrier mobility can be determined.[117] The conductivity can be derived from carrier mobility using eqn (7.3) where, n and μ_e are the electron density and mobility respectively, p and μ_h the hole density and mobility, respectively, and e is the elementary charge. The carrier densities can be determined from transient absorption spectroscopy.

$$\sigma = e(n\mu_e + p\mu_h) \tag{7.3}$$

As the number of conductivity techniques available to MOF researchers continues to grow, no direct comparison between them has been made to date, which is essential for standardisation of the results.

The connection between spectroscopic properties and conductivity in molecular materials was discussed in the literature as far back as the late 19th century;[118] however, no systematic investigations have been carried out to date. The first successful attempt to explore the idea was reported by Robin and Day in their seminal work,[119] where they interrogated a vast array of redox-active materials for their optical properties and devised a classification scheme based on their spectroscopic features. The Robin and Day classes can be correlated with the degree of electronic communication between redox centres in dinuclear complexes. This concept has been expanded to accommodate electron delocalisation in multidimensional materials with the notable example being Prussian blue. Class I materials have electrons completely isolated at the redox sites with no interaction between them, such that these materials are insulating. In Class II, electrons are weakly coupled but are still localised, and many semiconducting frameworks such as Prussian blue can be assigned to this class. When a material has strongly coupled redox centres with highly delocalised electrons, it belongs to the Class III regime and is expected to be highly conductive. At around the same time as the publication of the Robin and Day scheme, Hush developed a theoretical model for CT in mixed-valence systems based on classical theory for inner sphere electron transfer and the Marcus theory for outer sphere electron transfer.[120,121] The electron oscillation between the redox centres gives rise to IVCT bands which are unique to Class II and III compounds. The IVCT band parameters have previously been used to derive the rate of electron transfer in Prussian blue using the Hush model and directly correlate it with conductivity.[38] At the present time, a number of promising spectroscopic studies on MOFs, including the Prussian blue family,[74] the guest-infiltrated $[Cu_3(BTC)_2]$ family,[33,81] and semiquinone-based frameworks,[79] have commented on the potential link between CT bands and conductivities, and this may become an important complement to conventional conductivity measurements in the future. A number of other materials also exhibit CT transitions, the conductivities of which are yet to be explored.[122] Ultimately, the development of this link between theory and experiment would provide a powerful approach to exploring conductivity in framework systems.

7.6 Summary and Future Outlook

Building upon the long and rich history of coordination polymers, the advent of semiconducting and conducting framework systems heralds a new era in advanced functional materials. While a level of serendipity accompanies the synthesis of new framework systems, the exquisite levels of control that can be achieved over their structure and function bodes well for the future. Importantly, a number of design criteria have now been established for conducting frameworks, as outlined in Figure 7.1. This provides an excellent foundation for enhancing the conducting properties of known framework materials, and generating new systems which build upon the lessons learned to date. The potential for discovery of new phenomena that underlie long-range charge transport in 3D systems is enormous, particularly considering the opportunity for synergistic properties such as simultaneous magnetic and electronic functionality.

With regards to the aforementioned experimental contact and non-contact techniques, few studies have provided a comprehensive picture on the relationship between the experimental measurables reported from the various techniques on a given system.[123] Indeed, few studies have probed the mechanisms for conducting behaviour, and the vast majority of work reports room-temperature conductivity measurements only; variable temperature measurements are crucial to extract information regarding the mechanism of charge migration. Indeed, an inherent issue within frameworks is the presence of interstitial solvent and defect sites, which can impede such variable temperature measurements (*e.g.*, by causing crystal cracking).

In concert with experimental approaches, advances in methods for computational modelling of conducting frameworks are also proving valuable for guiding experimental work. Clearly, in concert with experimental measurements, the development of computational methodologies to treat long-range CT and electronic delocalisation are crucial. In this regard, DFT has been extended to deal with transition metal-based MOFs[124] and to calculate band gaps in framework systems.[125] It is known, for example, that orbital-level matching, leading to a wide bandwidth and a comparable contribution from the metal and ligand to the bands intersecting the Fermi level is essential for obtaining a conducting framework.[126]

The realisation of conductivity in frameworks also underpins their applications across a range of technologically- and industrially-useful fields, including electrocatalysis, chemiresistive sensing, electrochromic devices and energy storage systems, amongst many others. Many of these potential applications have key relevance to problems with environmental and societal impacts. In this regard, the advent of conducting frameworks also represents an opportunity to revolutionise future industries. One key requirement here is the need to interface solid materials with external circuits. Significant efforts are currently underway in this area, and while outside the scope of the present discussion, these aspects of device engineering will be central to the deployment of conducting frameworks at the technological and industrial scales.[8,81,127] In addition to new electrophoretic deposition methods for contacting framework solids with conductive substrates, epitaxial growth methods exploit the anisotropy often present in frameworks (with non-cubic topologies) for controlled growth on surfaces and for directional charge transport.

Clearly, an exceptionally bright future is envisaged for conducting framework materials, and this will necessitate the combined efforts of the scientific and engineering communities. At the applied level, a wide range of applications are envisaged; at the fundamental level, frameworks provide an unparalleled opportunity to explore deeply fundamental aspects of CT in 3D coordination space. Taken together, these challenges provide an exciting and fundamentally important pursuit for the community, with the promise of highly significant outcomes.

References

1. J. Brown, *Philos. Trans.*, 1724, **33**, 17–24.
2. J. Keggin and F. Miles, *Nature*, 1936, **137**, 577–578.
3. B. F. Hoskins and R. Robson, *J. Am. Chem. Soc.*, 1989, **111**, 5962–5964.
4. B. F. Hoskins and R. Robson, *J. Am. Chem. Soc.*, 1990, **112**, 1546–1554.
5. O. M. Yaghi and H. L. Li, *J. Am. Chem. Soc.*, 1995, **117**, 10401–10402.
6. R. S. Batten, R. N. Champness, X.-M. Chen, J. Garcia-Martinez, S. Kitagawa, L. Öhrström, M. O'Keeffe, M. Paik Suh and J. Reedijk, *Pure Appl. Chem.*, 2013, **85**, 1715.
7. S. R. Batten, S. M. Neville and D. R. Turner, *Coordination Polymers: Design, Analysis and Application*, Royal Society of Chemistry, 2008.
8. B. Seoane, S. Castellanos, A. Dikhtiarenko, F. Kapteijn and J. Gascon, *Coord. Chem. Rev.*, 2016, **307**, 147–187.
9. M. D. Allendorf and V. Stavila, *CrystEngComm*, 2015, **17**, 229–246.
10. F. Schroder and R. A. Fischer, in *Functional Metal-Organic Frameworks: Gas Storage, Separation and Catalysis*, ed. M. Schroder, 2010, vol. 293, pp. 77–113.
11. Y. J. Cui, B. Li, H. J. He, W. Zhou, B. L. Chen and G. D. Qian, *Acc. Chem. Res.*, 2016, **49**, 483–493.
12. C. J. Kepert, *Chem. Commun.*, 2006, 695–700.
13. S. Kitagawa and R. Matsuda, *Coord. Chem. Rev.*, 2007, **251**, 2490–2509.
14. S. Qiu and G. Zhu, *Coord. Chem. Rev.*, 2009, **253**, 2891–2911.
15. R. A. Fischer and C. Wöll, *Angew. Chem., Int. Ed.*, 2008, **47**, 8164–8168.
16. G. Ferey, *Chem. Soc. Rev.*, 2008, **37**, 191–214.
17. O. Shekhah, J. Liu, R. A. Fischer and C. Wöll, *Chem. Soc. Rev.*, 2011, **40**, 1081–1106.
18. B. Y. Li, M. Chrzanowski, Y. M. Zhang and S. Q. Ma, *Coord. Chem. Rev.*, 2016, **307**, 106–129.
19. D. M. D'Alessandro, B. Smit and J. R. Long, *Angew. Chem. Int. Ed.*, 2010, **49**, 6058–6082.
20. L. Q. Ma and W. B. Lin, in *Functional Metal-Organic Frameworks: Gas Storage, Separation and Catalysis*, ed. M. Schroder, 2010, vol. 293, pp. 175–205.
21. E. Coronado and G. Minguez Espallargas, *Chem. Soc. Rev.*, 2013, **42**, 1525–1539.
22. M. C. So, G. P. Wiederrecht, J. E. Mondloch, J. T. Hupp and O. K. Farha, *Chem. Commun.*, 2015, **51**, 3501–3510.
23. A. Dhakshinamoorthy, A. M. Asiri and H. Garcia, *Angew. Chem., Int. Ed.*, 2016, **55**, 5414–5445.
24. X. J. Zhang, W. J. Wang, Z. J. Hu, G. N. Wang and K. S. Uvdal, *Coord. Chem. Rev.*, 2015, **284**, 206–235.
25. L. Wang, Y. Z. Han, X. Feng, J. W. Zhou, P. F. Qi and B. Wang, *Coord. Chem. Rev.*, 2016, **307**, 361–381.

26. P. Horcajada, R. Gref, T. Baati, P. K. Allan, G. Maurin, P. Couvreur, G. Ferey, R. E. Morris and C. Serre, *Chem. Rev.*, 2012, **112**, 1232–1268.
27. M. Gimenez-Marques, T. Hidalgo, C. Serre and P. Horcajada, *Coord. Chem. Rev.*, 2016, **307**, 342–360.
28. J. Aguilera-Sigalat and D. Bradshaw, *Coord. Chem. Rev.*, 2016, **307**, 267–291.
29. P. Falcaro, D. Buso, A. J. Hill and C. M. Doherty, *Adv. Mater.*, 2012, **24**, 3153–3168.
30. D. M. D'Alessandro, J. R. R. Kanga and J. S. Caddy, *Aust. J. Chem.*, 2011, **64**, 718–722.
31. L. Sun, M. G. Campbell and M. Dincă, *Angew. Chem., Int. Ed.*, 2016, **55**, 3566–3579.
32. D. M. D'Alessandro, *Chem. Commun.*, 2016, **52**, 8957–8971.
33. M. D. Allendorf, M. E. Foster, F. Leonard, V. Stavila, P. L. Feng, F. P. Doty, K. Leong, E. Y. Ma, S. R. Johnston and A. A. Talin, *J. Phys. Chem. Lett.*, 2015, **6**, 1182–1195.
34. C. F. Leong, P. M. Usov and D. M. D'Alessandro, *MRS Bull.*, 2016, **41**, 858–864.
35. G. Givaja, P. Amo-Ochoa, C. J. Gomez-Garcia and F. Zamora, *Chem. Soc. Rev.*, 2012, **41**, 115–147.
36. S. Horike, D. Umeyama and S. Kitagawa, *Acc. Chem. Res.*, 2013, **46**, 2376–2384.
37. P. Stallinga, *Adv. Mater.*, 2011, **23**, 3356–3362.
38. S. J. England, P. Kathirgamanathan and D. R. Rosseinsky, *J. Chem. Soc., Chem. Commun.*, 1980, 840–841.
39. D. R. Rosseinsky, J. S. Tonge, J. Berthelot and J. F. Cassidy, *J. Chem. Soc., Faraday. Trans. 1*, 1987, **83**, 231–243.
40. S. R. Ahrenholtz, C. C. Epley and A. J. Morris, *J. Am. Chem. Soc.*, 2014, **136**, 2464–2472.
41. T. Shinomiya, H. Ozawa, Y. Mutoh and M.-A. Haga, *Dalton Trans.*, 2013, **42**, 16166–16175.
42. M. Tadokoro, S. Yasuzuka, M. Nakamura, T. Shinoda, T. Tatenuma, M. Mitsumi, Y. Ozawa, K. Toriumi, H. Yoshino, D. Shiomi, K. Sato, T. Takui, T. Mori and K. Murata, *Angew. Chem., Int. Ed.*, 2006, **118**, 5268–5271.
43. P. Amo-Ochoa, L. Welte, R. Gonzalez-Prieto, P. J. Sanz Miguel, C. J. Gomez-Garcia, E. Mateo-Marti, S. Delgado, J. Gomez-Herrero and F. Zamora, *Chem. Commun.*, 2010, **46**, 3262–3264.
44. L. R. Melby, R. J. Harder, W. R. Hertler, W. Mahler, R. E. Benson and W. E. Mochel, *J. Am. Chem. Soc.*, 1962, **84**, 3374–3387.
45. W. Siemons, P. Bierstedt and E. R. Kepler, *J. Chem. Phys.*, 1963, **39**, 3523–3528.
46. N. Uyeda, T. Kobayashi, K. Ishizuka and Y. Fujiyoshi, *Nature*, 1980, **285**, 95–97.
47. L. Shields, *J. Chem. Soc., Faraday Trans. 2*, 1985, **81**, 1–9.

48. J. J. Hoagland, X. D. Wang and K. W. Hipps, *Chem. Mat.*, 1993, **5**, 54–60.
49. S. Sun, X. Xu, P. Wu and D. Zhu, *J. Mater. Sci. Lett.*, 1998, **17**, 719–721.
50. H. Wachtel, M. Ohnmacht, J. U. von Schütz and H. C. Wolf, *Nanostruct. Mater.*, 1995, **6**, 291–295.
51. Q. Zhang, W. Wang, G. Ye, X. Yan, Z. Zhang and Z. Hua, *Synth. Met.*, 2004, **144**, 285–289.
52. S. A. O'Kane, R. Clérac, H. Zhao, X. Ouyang, J. R. Galán-Mascarós, R. Heintz and K. R. Dunbar, *J. Solid State Chem.*, 2000, **152**, 159–173.
53. R. A. Heintz, H. Zhao, X. Ouyang, G. Grandinetti, J. Cowen and K. R. Dunbar, *Inorg. Chem.*, 1999, **38**, 144–156.
54. G. R. Johnson, M. G. Miles and J. D. Wilson, *Mol. Cryst. Liq. Cryst.*, 1976, **33**, 67–75.
55. S. Hünig, A. Aumüller, P. Erk, H. Meixner, J. U. von Schütz, H. J. Gross, U. Langohr, H. P. Werner, H. C. Wolf, C. Burschka, G. Klebe, K. Peters and H. G. V. Schnering, *Synth. Met.*, 1988, **27**, 181–188.
56. J. U. von Schütz, M. Bair, H. J. Gross, U. Langohr, H. P. Werner, H. C. Wolf, D. Schmeiβer, K. Graf, W. Göpel, P. Erk, H. Meixner and S. Hünig, *Synth. Met.*, 1988, **27**, 249–256.
57. A. Aumüller, P. Erk, G. Klebe, S. Hünig, J. U. von Schütz and H.-P. Werner, *Angew. Chem., Int. Ed.*, 1986, **25**, 740–741.
58. H. Miyasaka, C. S. Campos-Fernández, R. Clérac and K. R. Dunbar, *Angew. Chem., Int. Ed.*, 2000, **112**, 3989–3993.
59. N. Motokawa, T. Oyama, S. Matsunaga, H. Miyasaka, M. Yamashita and K. R. Dunbar, *CrystEngComm*, 2009, **11**, 2121–2130.
60. H. Miyasaka, N. Motokawa, S. Matsunaga, M. Yamashita, K. Sugimoto, T. Mori, N. Toyota and K. R. Dunbar, *J. Am. Chem. Soc.*, 2010, **132**, 1532–1544.
61. H. Miyasaka, T. Morita and M. Yamashita, *Chem. Commun.*, 2011, **47**, 271–273.
62. H. Miyasaka, *Acc. Chem. Res.*, 2013, **46**, 248–257.
63. J. Torrance, J. Vazquez, J. Mayerle and V. Lee, *Phys. Rev. Lett.*, 1981, **46**, 253.
64. J. Ferraris, D. O. Cowan, V. Walatka and J. H. Perlstein, *J. Am. Chem. Soc.*, 1973, **95**, 948–949.
65. J. B. Torrance, *Mol. Cryst. Liq. Cryst.*, 1985, **126**, 55–67.
66. K. P. Goetz, D. Vermeulen, M. E. Payne, C. Kloc, L. E. McNeil and O. D. Jurchescu, *J. Mater. Chem. C*, 2014, **2**, 3065–3076.
67. H. Miyasaka, N. Motokawa, R. Atsuumi, H. Kamo, Y. Asai and M. Yamashita, *Dalton Trans.*, 2011, **40**, 673–682.
68. K. Nakabayashi, M. Nishio, K. Kubo, W. Kosaka and H. Miyasaka, *Dalton Trans.*, 2012, **41**, 6072–6074.
69. P. Fielding and D. Mellor, *J. Chem. Phys.*, 1954, **22**, 1155–1156.
70. D. M. Pajerowski, T. Watanabe, T. Yamamoto and Y. Einaga, *Phys. Rev. B*, 2011, **83**, 153202.
71. M. Dadic, *Croat. Chem. Acta*, 1959, **31**, 101–106.

72. P. S. Braterman, P. B. P. Phipps and R. J. P. Williams, *J. Chem. Soc.*, 1965, 6164–6169.
73. H. Inoue and S. Yanagisawa, *Keio Eng. Rep.*, 1973, **26**, 55–66.
74. J. N. Behera, D. M. D'Alessandro, N. Soheilnia and J. R. Long, *Chem. Mat.*, 2009, **21**, 1922–1926.
75. L. Sun, T. Miyakai, S. Seki and M. Dincă, *J. Am. Chem. Soc.*, 2013, **135**, 8185–8188.
76. L. Sun, C. H. Hendon, M. A. Minier, A. Walsh and M. Dincă, *J. Am. Chem. Soc.*, 2015, **137**, 6164–6167.
77. S. Takaishi, M. Hosoda, T. Kajiwara, H. Miyasaka, M. Yamashita, Y. Nakanishi, Y. Kitagawa, K. Yamaguchi, A. Kobayashi and H. Kitagawa, *Inorg. Chem.*, 2009, **48**, 9048–9050.
78. Y. Kobayashi, B. Jacobs, M. D. Allendorf and J. R. Long, *Chem. Mat.*, 2010, **22**, 4120–4122.
79. L. E. Darago, M. L. Aubrey, C. J. Yu, M. I. Gonzalez and J. R. Long, *J. Am. Chem. Soc.*, 2015, **137**, 15703–15711.
80. B. F. Abrahams, T. A. Hudson, L. J. McCormick and R. Robson, *Cryst. Growth Des.*, 2011, **11**, 2717–2720.
81. A. A. Talin, A. Centrone, A. C. Ford, M. E. Foster, V. Stavila, P. Haney, R. A. Kinney, V. Szalai, F. El Gabaly and H. P. Yoon, *Science*, 2014, **343**, 66–69.
82. S. Patwardhan and G. C. Schatz, *J. Phys. Chem. C*, 2015, **119**, 24238–24247.
83. M. R. Bryce, *Chem. Soc. Rev.*, 1991, **20**, 355–390.
84. G. Saito and Y. Yoshida, *Bull. Chem. Soc. Jpn.*, 2007, **80**, 1–137.
85. T. J. Kistenmacher, T. E. Phillips and D. O. Cowan, *Acta Crystallogr., Sect. B: Struct. Crystallogr. Cryst. Chem.*, 1974, **30**, 763–768.
86. T. C. Narayan, T. Miyakai, S. Seki and M. Dinca, *J. Am. Chem. Soc.*, 2012, **134**, 12932–12935.
87. S. S. Park, E. R. Hontz, L. Sun, C. H. Hendon, A. Walsh, T. Van Voorhis and M. Dincă, *J. Am. Chem. Soc.*, 2015, **137**, 1774–1777.
88. D. Chen, H. Xing, Z. Su and C. Wang, *Chem. Commun.*, 2016, **52**, 2019–2022.
89. M. Hmadeh, Z. Lu, Z. Liu, F. Gandara, H. Furukawa, S. Wan, V. Augustyn, R. Chang, L. Liao, F. Zhou, E. Perre, V. Ozolins, K. Suenaga, X. F. Duan, B. Dunn, Y. Yamamto, O. Terasaki and O. M. Yaghi, *Chem. Mat.*, 2012, **24**, 3511–3513.
90. T. Kambe, R. Sakamoto, K. Hoshiko, K. Takada, M. Miyachi, J.-H. Ryu, S. Sasaki, J. Kim, K. Nakazato, M. Takata and H. Nishihara, *J. Am. Chem. Soc.*, 2013, **135**, 2462–2465.
91. D. Sheberla, L. Sun, M. A. Blood-Forsythe, S. Er, C. R. Wade, C. K. Brozek, A. Aspuru-Guzik and M. Dincă, *J. Am. Chem. Soc.*, 2014, **136**, 8859–8862.
92. T. Kambe, R. Sakamoto, T. Kusamoto, T. Pal, N. Fukui, K. Hoshiko, T. Shimojima, Z. Wang, T. Hirahara, K. Ishizaka, S. Hasegawa, F. Liu and H. Nishihara, *J. Am. Chem. Soc.*, 2014, **136**, 14357–14360.
93. J. Cui and Z. Xu, *Chem. Commun.*, 2014, **50**, 3986–3988.

94. A. J. Clough, J. W. Yoo, M. H. Mecklenburg and S. C. Marinescu, *J. Am. Chem. Soc.*, 2015, **137**, 118–121.
95. R. H. Dong, M. Pfeffermann, H. W. Liang, Z. K. Zheng, X. Zhu, J. Zhang and X. L. Feng, *Angew. Chem., Int. Ed.*, 2015, **54**, 12058–12063.
96. M. G. Campbell, D. Sheberla, S. F. Liu, T. M. Swager and M. Dinca, *Angew. Chem., Int. Ed.*, 2015, **54**, 4349–4352.
97. M. G. Campbell, S. F. Liu, T. M. Swager and M. Dinca, *J. Am. Chem. Soc.*, 2015, **137**, 13780–13783.
98. M. G. Campbell, D. Sheberla, S. F. Liu, T. M. Swager and M. Dinca, *Angew. Chem., Int. Ed.*, 2015, **54**, 4349–4352.
99. F. Shojaei and H. S. Kang, *J. Phys. Chem. C*, 2014, **118**, 25626–25632.
100. F. Shojaei, J. R. Hahn and H. S. Kang, *Chem. Mat.*, 2014, **26**, 2967–2974.
101. T. Panda and R. Banerjee, *Proc. Natl. Acad. Sci. India A*, 2014, **84**, 331–336.
102. X. Huang, P. Sheng, Z. Tu, F. Zhang, J. Wang, H. Geng, Y. Zou, C.-A. Di, Y. Yi, Y. Sun, W. Xu and D. Zhu, *Nat Commun*, 2015, **6**.
103. M.-H. Zeng, Q.-X. Wang, Y.-X. Tan, S. Hu, H.-X. Zhao, L.-S. Long and M. Kurmoo, *J. Am. Chem. Soc.*, 2010, **132**, 2561–2563.
104. Z. Yin, Q.-X. Wang and M.-H. Zeng, *J. Am. Chem. Soc.*, 2012, **134**, 4857–4863.
105. Z. Hao, G. Yang, X. Song, M. Zhu, X. Meng, S. Zhao, S. Song and H. Zhang, *J. Mater. Chem. A*, 2014, **2**, 237–244.
106. J. He, P. Cao, C. Wu, J. Huang, J. Huang, Y. He, L. Yu, M. Zeller, A. D. Hunter and Z. Xu, *Inorg. Chem.*, 2015, **54**, 6087–6089.
107. F. Gandara, F. J. Uribe-Romo, D. K. Britt, H. Furukawa, L. Lei, R. Cheng, X. F. Duan, M. O'Keeffe and O. M. Yaghi, *Chem. – Eur. J.*, 2012, **18**, 10595–10601.
108. D. Y. Lee, E.-K. Kim, N. K. Shrestha, D. W. Boukhvalov, J. K. Lee and S.-H. Han, *ACS Appl. Mater. Interfaces*, 2015, **7**, 18501–18507.
109. Z. Guo, D. K. Panda, K. Maity, D. Lindsey, T. G. Parker, T. E. Albrecht-Schmitt, J. L. Barreda-Esparza, P. Xiong, W. Zhou and S. Saha, *J. Mater. Chem. C*, 2016, **4**, 894–899.
110. K. M. Hutchins, T. P. Rupasinghe, L. R. Ditzler, D. C. Swenson, J. R. G. Sander, J. Baltrusaitis, A. V. Tivanski and L. R. MacGillivray, *J. Am. Chem. Soc.*, 2014, **136**, 6778–6781.
111. V. Palermo, A. Liscio, M. Palma, M. Surin, R. Lazzaroni and P. Samori, *Chem. Commun.*, 2007, 3326–3337.
112. J. J. Gassensmith, J. Y. Kim, J. M. Holcroft, O. K. Farha, J. F. Stoddart, J. T. Hupp and N. C. Jeong, *J. Am. Chem. Soc.*, 2014, **136**, 8277–8282.
113. B. M. Wiers, M.-L. Foo, N. P. Balsara and J. R. Long, *J. Am. Chem. Soc.*, 2011, **133**, 14522–14525.
114. J. Krupka, *Meas. Sci. Technol.*, 2013, **24**(062001), 062013.
115. P. Das and D. K. Ferry, *Solid-State Electron.*, 1976, **19**, 851–855.
116. P. J. Hirschfeld, W. O. Putikka and D. J. Scalapino, *Phys. Rev. Lett.*, 1993, **71**, 3705–3708.
117. M. Kunst and G. Beck, *J. Appl. Phys.*, 1986, **60**, 3558–3566.

118. H. L. Wells, *Am. J. Sci.*, 1922, **3**, 417–422.
119. M. Robin and P. Day, *Adv. Inorg. Chem. Radiochem.*, 1967, **10**, 247–422.
120. N. S. Hush, *Electrochim. Acta*, 1968, **13**, 1005.
121. N. S. Hush, *Prog. Inorg. Chem.*, 1967, **8**, 391–444.
122. C. F. Leong, B. Chan, T. B. Faust and D. M. D'Alessandro, *Chem. Sci.*, 2014, **5**, 4724–4728.
123. L. Sun, S. S. Park, D. Sherberla and M. Dinca, *J. Am. Chem. Soc.*, 2016, **138**(44), 14772–14782.
124. M. A. Addicoat, N. Vankova, I. F. Akter and T. Heine, *J. Chem. Theory Comput.*, 2014, **10**, 880–891.
125. J. X. Liu, W. C. Zhou, J. X. Liu, I. Howard, G. Kilibarda, S. Schlabach, D. Coupry, M. Addicoat, S. Yoneda, Y. Tsutsui, T. Sakurai, S. Seki, Z. B. Wang, P. Lindemann, E. Redel, T. Heine and C. Wöll, *Angew. Chem., Int. Ed.*, 2015, **54**, 7441–7445.
126. T. Komatsu, J. M. Taylor and H. Kitagawa, *Inorg. Chem.*, 2016, **55**, 546–548.
127. V. Stavila, A. A. Talin and M. D. Allendorf, *Chem. Soc. Rev.*, 2014, **43**, 5994–6010.

CHAPTER 8

Functional Metal-organic Frameworks for Enzyme/ Protein Immobilization

WAN-LING LIU,[a,b] STEPHEN LIRIO,[b] HSI-YA HUANG[b] AND SHENGQIAN MA*[a]

[a] Department of Chemistry, University of South Florida, 4202 East Fowler Avenue, Tampa, Florida 33620, USA; [b] Department of Chemistry at Chung Yuan Christian University, 200 Chung Pei Road, Chung-Li 320, Taiwan
*Email: sqma@usf.edu

8.1 Introduction

Enzymes are functional proteins and biological catalysts that are naturally evolved and efficient catalysts, which speed up the rate of reaction. They are important for achieving an environmentally friendly chemical industry.[1] Enzymes are generally active and selective catalysts while operating under mild conditions and exhibit high selectivity for the synthesis of desired products over by-products.[2,3] These merits prompted the persistent exploration of employing enzymes for the chemical, pharmaceutical, food industries and life science.[4,5] However, some of the disadvantages of enzymes include aggregation, auto-digestion, low thermal stability, low stability in organic solvent and expense.[6]

Enzyme immobilization has been proven as an effective strategy to enhance the stability of enzyme under harsh conditions as well as ease of

Monographs in Supramolecular Chemistry No. 22
Functional Supramolecular Materials: From Surfaces to MOFs
Edited by Rahul Banerjee

Published by the Royal Society of Chemistry, www.rsc.org

separation and reusability.[7,8] Immobilization of enzymes was first introduced in 1916 and different strategies have been developed.[9] In general, enzymes are immobilized *via* three major routes: (1) adsorption (binding to support), (2) chemical bonding (cross-linking) and (3) encapsulation.[10] Each technique has different advantages and disadvantages. For years, several materials have been developed for enzyme immobilization such as nanoparticles,[11] polymer monoliths[12] or silica mesoporous materials[13] in which different applications including protein extraction, protein digestion, protein separation, drug delivery, biocatalysis and biosensor fabrication have been applied.[14–16] However, these materials may lead to low enzyme loading capacity of nonporous materials, alteration of enzyme structure or decreased in enzyme activity during the immobilization process.[17]

Metal-organic frameworks (MOFs) are crystalline inorganic–organic hybrids that are constructed from metal ions and organic ligands *via* covalent bonds. MOFs, which belong to a class of nanoporous materials, have fascinating properties due to their high surface areas, tunable pore sizes, good thermal and chemical stabilities.[18–20] Owing to these advantages, the applications of MOFs have attracted much attention in gas storage chromatography, sensing, extraction, magnetism and catalysis.[21–24] Furthermore, their application has also been extended in biological analyses, which includes peptide extraction, biosensors, and immobilization of proteins and enzymes.[25–27] The specific interactions of enzymes in MOFs are deemed plausible with the aid of metal oxide or organic ligand effects as well as π–π interactions.[28] In this chapter, we will highlight the recent developments and methods for enzyme immobilization on different MOFs. This chapter is divided into three categories: chemical binding (cross-linking), physical adsorption on a carrier (support) and encapsulation (Figure 8.1).[10]

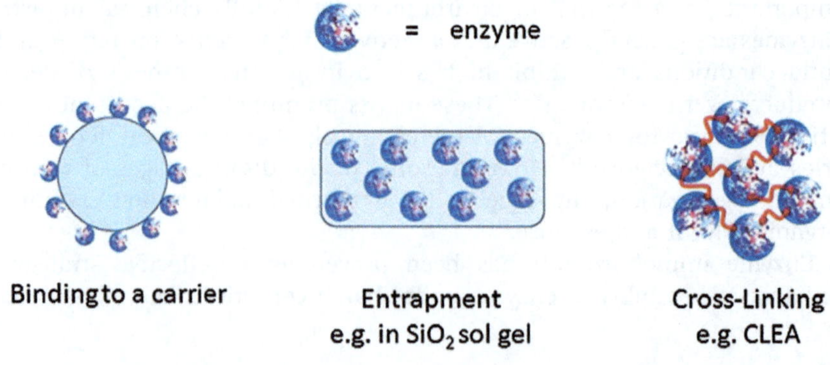

Figure 8.1 Three methods of enzyme immobilization. (Reproduced from ref. 10 with permission from the Royal Society of Chemistry.)

8.2 Enzyme Immobilization MOFs *via* Chemical Binding

Covalent coupling is one of the most typical strategies to immobilize the enzyme. The typical approach for this technique is the utilization of functional groups containing gluteraldehyde, glyoxyl, divinylsufone, or epoxy-activated support to covalently bond with the reactive functional groups of the enzyme.[29,30] Since MOFs are highly tunable, immobilization of enzymes *via* chemical bonding with MOFs has successfully been reported. In 2011, Jung *et al.* first demonstrated that functional proteins, *Candida antartica* Lipase B, can covalently bond on MOFs without losing their catalytic activity.[31] It has been known that several isoreticular metal-organic frameworks (IRMOFs) are constructed with Zn_4O nodes and dicarboxylate groups, thus, it is imperative that the presence of the latter's group on the surface of MOF could be activated and modified *via* simple organic reaction. The immobilization of enzyme was carried by activating the carboxylate group on the MOFs surface using 1-ethyl-3-(3-dimethylaminopropyl) carbodiimide (EDC) or dicyclohexyl carbodiimide (DCC) and subsequently conjugated with the target protein. This idea was first employed by using enhanced green fluorescent protein (EGFP) as a model protein and conjugated with 1D-polymer, 2D-, 3D-MOFs bearing carboxylate groups, which can easily be tracked by fluorescence microscopy (Figure 8.2).[31]

The successful application of EGFP led Jung *et al.* to the utilization of this strategy in conjugating the MOFs and CALB. The conjugated CALB-MOF was applied in transesterification of (±)-1-phenylethanol, which showed 400–1000 times higher catalytic activity than free enzyme in organic solvents and kept the enantio-selectivity. Similarly, Shih and co-workers immobilized trypsin on MIL-101(Cr), MIL-88B(Cr) and MIL-88B-NH_2(Cr) by using the abovementioned method (Figure 8.3). The carboxylate groups in MIL-101(Cr) and MIL-88(Cr) were the key target for the conjugation of the protease trypsin. Meanwhile, the presence of the amine group in MIL-88B-NH_2(Cr) further increased the affinity of trypsin, which provided a more stable covalent bond.[32] The trypsin-MOFs were applied in proteolysis and afforded higher digestion efficiency with good reusability than the free enzyme. Furthermore, this method retained the substrates structure as well as its enzymatic activity.

The application of MOFs in enzyme immobilization *via* chemical bonding further extends its application for biocatalysis. However, chemical bonding contains several parameters that should be considered prior to enzyme immobilization. This includes the density of reactive groups, steric hindrance, and stability of enzymes during immobilization process, which could affect the efficiency and performance of the immobilized enzymes.

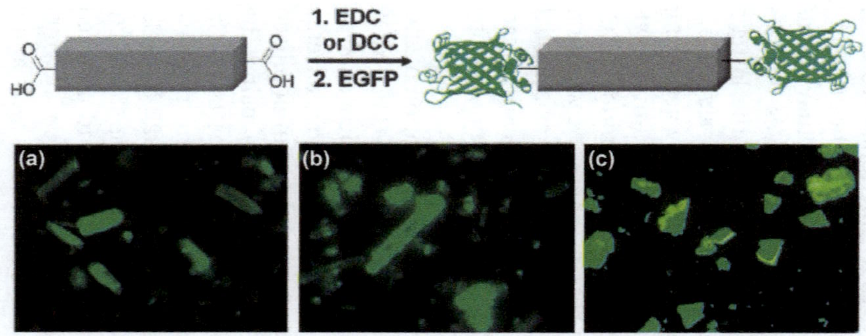

Figure 8.2 Schematic representation of the bioconjugation of the 1D-polymer with EGFP. Fluorescence microscopic images of EGFP coated MOFs. (a) 1D + EGFP. (b) 2D + EGFP. (c) 3D + EGFP.
(Reproduced from ref. 31 with permission from the Royal Society of Chemistry.)

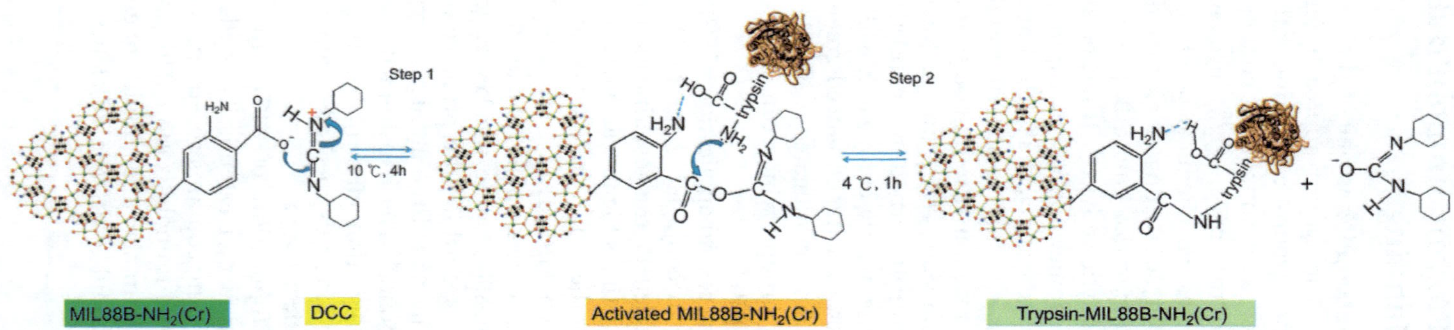

Figure 8.3 Trypsin immobilization onto activated MOFs.
Reproduced from Y.-H. Shih, S.-H. Lo, N.-S. Yang, B. Singco, Y.-J. Cheng, C.-Y. Wu, I.-H. Chang and C.-H. Lin, Trypsin-Immobilized Metal-Organic Framework as a Biocatalyst In Proteomics Analysis, *ChemPlusChem*. Copyright 2012. Wiley-VCH.

8.3 Enzyme Immobilization MOFs *via* Physical Adsorption

In some cases, modification of solid supports may not be suitable for enzyme immobilization, which could lead to a decrease in their enzyme activity.[33] Physical adsorption is a simple, fast and most widely used technique to immobilize enzyme/protein on nonporous or porous materials. In contrast to chemical bonding, an enzyme solution is commonly brought into contact with a solid support *via* Van der Waals forces, hydrogen bonding, ionic binding and hydrophilic/hydrophobic forces.[34] However, several parameters such as pH value and salt concentration should carefully be investigated for it strongly affects the electrostatic protein-support interactions.[35]

Pisklak *et al.* first reported the adsorption of proteins on MOFs.[2] A microperoxidase-11 (MP-11) was selected for immobilization in the periodic mesoporous organosilica (PMO) and MOF materials. MP-11 is the product of proteolytic degradation of cytochrome c and the active site in peroxidase involves an iron heme group that can oxidize organic molecules in the presence of hydrogen peroxide. The CuMOF was used as solid support, which was first synthesized by Seki *et al.*[36] The channel-type CuMOF is comprised of copper dimers and connected *via* 4,4′-biphenyldicarboxylate. This MOF has a pore size of 17.8 nm, which enabled the adsorption of small enzymes and proteins. The MP-11@MOF was evaluated based on the oxidation efficiency of methylene blue (Figure 8.4),[2] and the reaction was monitored by UV-Vis spectroscopy.

In Figure 8.5, good catalytic efficiency was observed when MP-11 immobilized in MOF.[2] In 2011, Ma and co-workers demonstrated a proof of concept for the immobilization of enzymes onto mesoporous MOF, Tb-TATB. The immobilization of MP-11 on Tb-TATB demonstrated superior catalytic performance than its mesoporous silica counterpart, MCM-41.[37] The Tb-TATB contains nanoscopic cages of 3.9 and 4.7 nm in diameter that

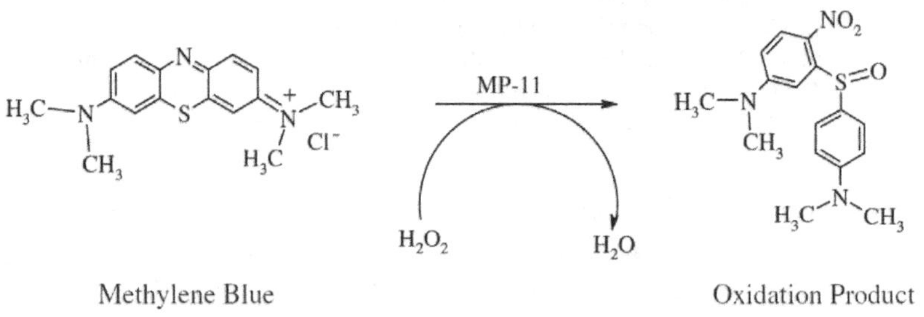

Figure 8.4 Scheme for the oxidation of methylene blue.
Reproduced from T. J. Pisklak, M. Macías, D. H. Coutinho, R. S. Huang and K. J. Balkus Jr., Hybrid materials for immobilization of MP-11 catalyst, *Top. Catal.*, 2006, **38**, 269. © Springer Science + Business Media, Inc. 2006, with permission of Springer.

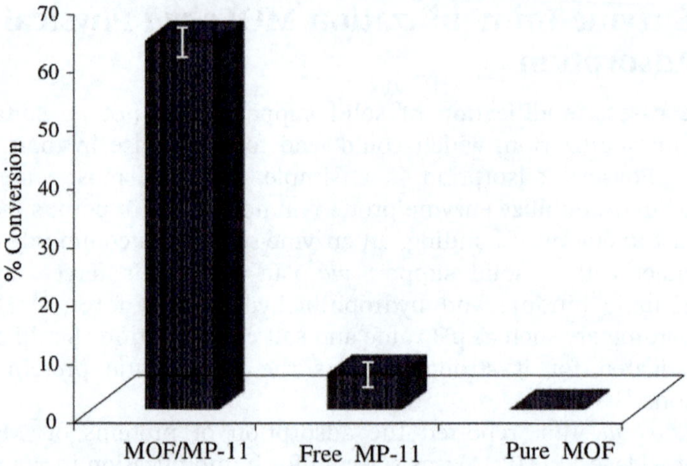

Figure 8.5 The percent conversion of methylene blue by MOF immobilized MP-11, free MP-11, and pure MOF from a solution of 2.35 μM methylene blue, and 10 μL 30% H_2O_2.
Reproduced from T. J. Pisklak, M. Macías, D. H. Coutinho, R. S. Huang and K. J. Balkus Jr., Hybrid materials for immobilization of MP-11 catalyst, *Top. Catal.*, 2006, **38**, 269. © Springer Science + Business Media, Inc. 2006, with permission of Springer.

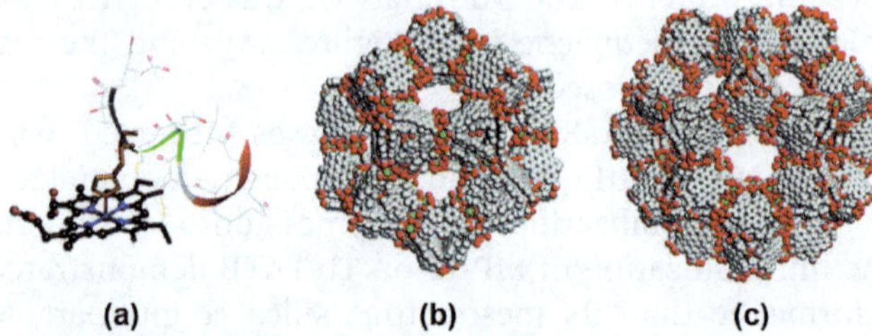

Figure 8.6 (a) Molecular structure of MP-11; (b) 3.9 nm diameter cage and (c) 4.7 nm diameter cage in Tb-MOF.
Reprinted with permission from V. Lykourinou, Y. Chen, X.-S. Wang, L. Meng, T. Hoang, L.-J. Ming, R. L. Musselman and S. Ma, Immobilization of MP-11 into a Mesoporous Metal-Organic Framework, MP-11@mesoMOF: A New Platform for Enzymatic Catalysis, *J. Am. Chem. Soc.*, 2011, **133**(27), 10382. Copyright 2011 American Chemical Society.

provided space to accommodate the MP-11 (Figure 8.6).[37] The pore sizes are dominantly distributed around 3.0 and 4.1 nm in addition to a micropore size around 0.9 nm (Figure 8.7).[37] For comparison, MCM-41 was selected to adsorb the MP-11 (hereafter denoted MP-11@MCM-41). Based on the results, the MP-11@Tb-MOF demonstrated higher loading capacity (~6-fold) than to that of the MCM counterpart. The low loading capacity for MCM-41 was due to lower surface area (BET surface area: ~1000 $m^2\ g^{-1}$) as compared to Tb-MOF (BET surface area: 1935 $m^2\ g^{-1}$).

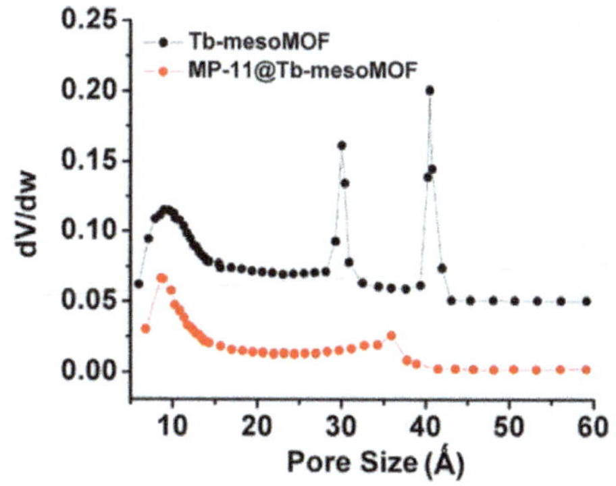

Figure 8.7 Pore size distributions of Tb-MOF and MP-11@Tb-MOF. Reprinted with permission from V. Lykourinou, Y. Chen, X.-S Wang, L. Meng, T. Hoang, L.-J. Ming, R. L. Musselman and S. Ma, Immobilization of MP-11 into a Mesoporous Metal-Organic Framework, MP-11@mesoMOF: A New Platform for Enzymatic Catalysis, *J. Am. Chem. Soc.*, 2011, **133**(27), 10382. Copyright 2011 American Chemical Society.

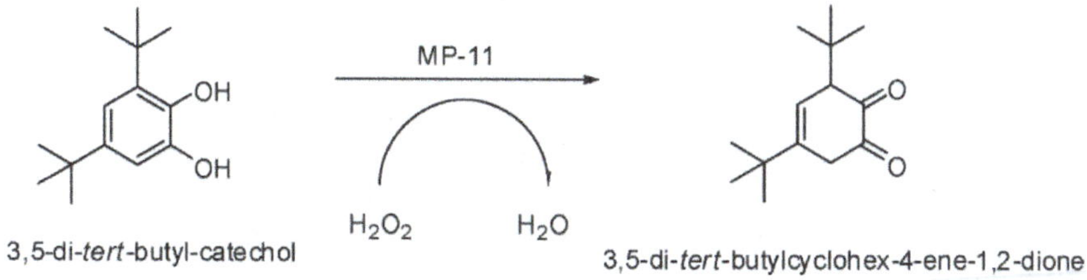

Figure 8.8 Reaction for oxidation of 3,5-di-*t*-butylcatechol to *o*-quinone. Reprinted with permission from V. Lykourinou, Y. Chen, X.-S Wang, L. Meng, T. Hoang, L.-J. Ming, R. L. Musselman and S. Ma, Immobilization of MP-11 into a Mesoporous Metal-Organic Framework, MP-11@mesoMOF: A New Platform for Enzymatic Catalysis, *J. Am. Chem. Soc.*, 2011, **133**(27), 10382. Copyright 2011 American Chemical Society.

The catalytic activities of MP-11@Tb-mesoMOF and MP-11@MCM-41 were evaluated by monitoring the oxidation of chromogenic substrate 3,5-di-*t*-butyl-catechol (DTBC) to form *o*-quinone product (Figure 8.8).[37] The reaction rate for Tb-MOF is relatively slower than the MP-11@MCM-41. Interestingly, the MP-11@Tb-MOF had faster reaction rate during the initial time period of 30 min. The catalytic activity of the MP-11@Tb-MOF also offers a good recyclability as depicted in Figure 8.9.[37]

In comparison, the activity of MP-11@MCM-41 had 28% activity during the third cycle, which could be attributed to the leaching of MP-11 in the supernatant. Overall, this new approach demonstrated for the first time the successful immobilization of MP-11 into a mesoporous MOF consisting of nanoscopic cages, which provided excellent enzymatic catalysis performances. The successful immobilization of MP-11@mesoMOF revealed that strong π–π interaction between the heme of MP-11 and the conjugated

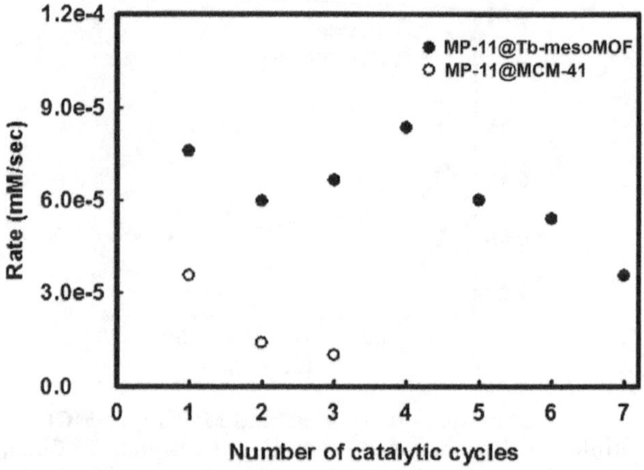

Figure 8.9 Reaction rates of MP-11@Tb-MOF and MP-11@MCM-41 at different cycles. Reprinted with permission from V. Lykourinou, Y. Chen, X.-S Wang, L. Meng, T. Hoang, L.-J. Ming, R. L. Musselman and S. Ma, Immobilization of MP-11 into a Mesoporous Metal-Organic Framework, MP-11@mesoMOF: A New Platform for Enzymatic Catalysis, *J. Am. Chem. Soc.*, 2011, **133**(27), 10382. Copyright 2011 American Chemical Society.

triazine and benzene rings in the organic ligand of Tb-mesoMOF facilitated the retention of MP-11 molecules within the MOF pores. In contrast, the MCM-41 demonstrated severe leaching of MP-11 due to the lack of specific interactions between the enzyme and the mesoporous silica material.[28]

In general, microporous MOFs are commonly used in the adsorption of proteins, which have dimensions of 3 to 8 nm. Therefore the MOF, as solid support, must contain enough pore size to easily facilitate the adsorption of the protein. This technique is rather critical because it delimits the accessibility of protein or enzyme into the MOFs, which is due to the small pore size. This drawback was addressed by Huang and co-workers, in which a dye-tagging strategy was introduced to immobilize a protein molecule into a microporous MOF.[38] In this proof of concept, Huang's group have hypothesized that conjugation of fluorescein isothiocyanate (FITC) dye with trypsin can stably anchored the FITC-trypsin onto the MOF pores. Trypsin is a macromolecule while FITC, which has a dimension of $0.9 \times 1.2 \times 1.4$ nm, is a small molecular dye that can easily access into the pore of MOFs (Figure 8.10).[38] Therefore, for trypsin to be immobilized in the MOF, FITC was used to carry the enzyme and anchored onto the MOF pores. This new strategy has successfully been applied on various MOF supports ((MIL-101(Cr) and CYCU-4(Al)) using FITC dye-tagging strategy.

The immobilization of trypsin was carried out by conjugating the FITC with trypsin *via* 2 min microwave heating and subsequently added onto the MOF. The immobilization procedure was conducted based on 30 min vortex mixing (hereafter denoted Try-FITC@CYCU-4 and Try-FITC@MIL-101). The resultant bioreactor was applied in protein digestion and the sequence coverage was in the range of 63 to 72% after 4th reuse. This novel strategy

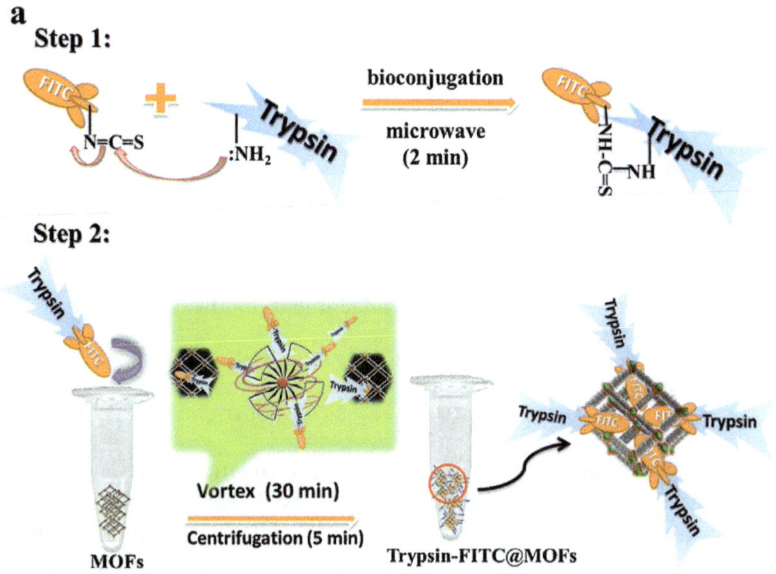

Figure 8.10 Scheme for bioreactor fabrication.
Reproduced from ref. 38 with permission from the Royal Society of Chemistry.

provided three main innovations: (1) hampered any chemical modifications on the MOF surface; (2) increased in catalytic activity of the conjugated FITC-trypsin; (3) simple, fast, cheap and green preparation of bioreactor *via* 30 min vortex-assisted host–guest interaction. The Try-FITC@MOFs exhibited excellent biocatalytic performance over the free enzyme digestion, as well as other enzyme immobilized material reactors reported so far. As part of their continuing research, Huang *et al.* also presented the formation of trypsin bioreactors using a different fluorescent dye, 4-chloro-7-nitrobenzofurazan (NBD).[39] The molecular size of NBD is smaller than FITC, suggesting that it could be more useful for enzyme immobilization in various microporous MOFs. The NBD also feature its ability to directly calculate the loading capacity of MOF due to its fluorescent emission when tagged with enzymes. This approach was successfully applied in two MOFs (UiO-66(Zr) and CYCU-4(Al)) with characteristic pore size of less than the size of the trypsin protein molecule. It is noteworthy to mention that this strategy led to the development of multipoint attachment of enzymes onto the microporous MOFs. Multipoint attachment is an important issue in enzyme bioreactors because it further enhanced the stability of enzyme onto the solid support as has been discussed in previous reports.[40,41] One of the prerequisites for traditional multipoint attachment methods includes complex surface modification of the support material, which generates more organic waste, is high cost, labour-intensive, and time consuming in preparation.[42,43] This dye-tagging strategy by means of multipoint attachment to the solid support was achieved as illustrated in Figure 8.11.[39]

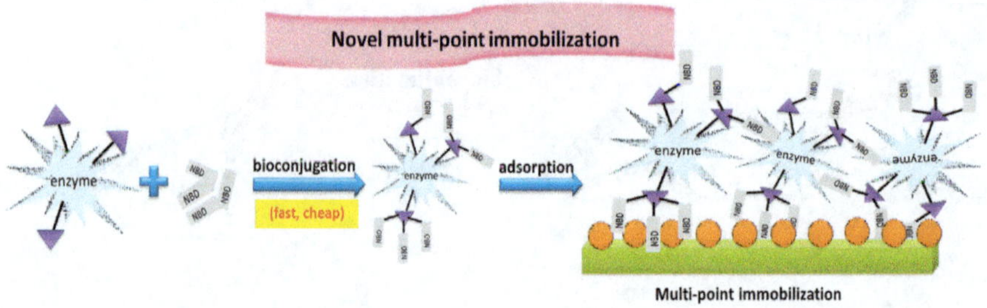

Figure 8.11 Scheme for a multipoint attachment to the solid support *via* dye-tagging strategy.
Reproduced from W.-L. Liu, C.-Y. Wu, C.-Y. Chen, B. Singco, C.-H. Lin and H.-Y. Huang, Fast Multipoint Immobilized MOF Bioreactor, *Chem. - Eur. J.*, 2014, **20**, 8923–8928. Copyright 2014. Wiley-VCH.

A rapid multipoint technique of enzyme immobilization, which was demonstrated onto a MOF solid support, without any chemical modifications bounded at least three or four NBD moieties for every trypsin molecule. The conjugation of NBD–enzyme by nucleophilic attack of the amine group on trypsin was carried out in 4 min *via* microwave heating and subsequent adsorption onto the MOFs generated Try-NBD@UiO-66 and Try-NBD@CYCU-4 materials. The hydrogen bonding and strong π–π interaction govern the host–guest attraction between the dye and the organic linkers of the MOF structure. The proposed biocatalyst, trypsin-NBD@UiO-66, possesses good performance, which is comparable to the previous study in terms of its catalysis ability, reusability and stability. Furthermore, chemical modification on the material surface prior to biomolecule capture is not necessary and the crystallinity of the MOFs is retained even after trypsin-NBD solution treatment. This strategy is a green fabrication technique due to the elimination of organic solvents usually employed in chemical modifications.

8.4 Enzyme Immobilization MOFs *via* Encapsulation

The preparation of various protein@MOF materials has greatly extended the applications of MOFs as biocatalysts. Current strategies for enzyme immobilization involve the adsorption on meso- or microporous MOFs, covalent bonding (*i.e.* surface modification), and dye-tagging strategy to anchor the enzyme or biomolecules to the MOF pores. For the adsorption of enzymes, the main challenge lies in maintaining the catalytic activity of the enzyme as it may undergo some conformational changes. Besides, a large pore of MOF is necessary to facilitate the enzyme immobilization. On the other hand, surface modification of the substrate is likely to decrease the enzyme activity when covalently bonded to the MOF structure. The abovementioned methodologies often consist of two major steps in which the MOFs structure or enzymes are post-functionalized that result in a time consuming preparation process for enzyme immobilization. An alternative strategy used to address these challenges is the introduction of a one-step immobilization technique that offers a synthesis of solid support and immobilization of enzymes, which could generate a multipoint attachment *via* covalent bonds between the enzyme molecules and the solid support. In 2012, Ge *et al.* reported a new method for the synthesis of hybrid organic–inorganic nanoflowers using copper(II) ions as the inorganic component and proteins as the organic component.[44] The proteins would form complexes with the copper ions, and these complexes become the nucleation sites for primary crystals of copper phosphate (as depicted in Figure 8.12).[44]

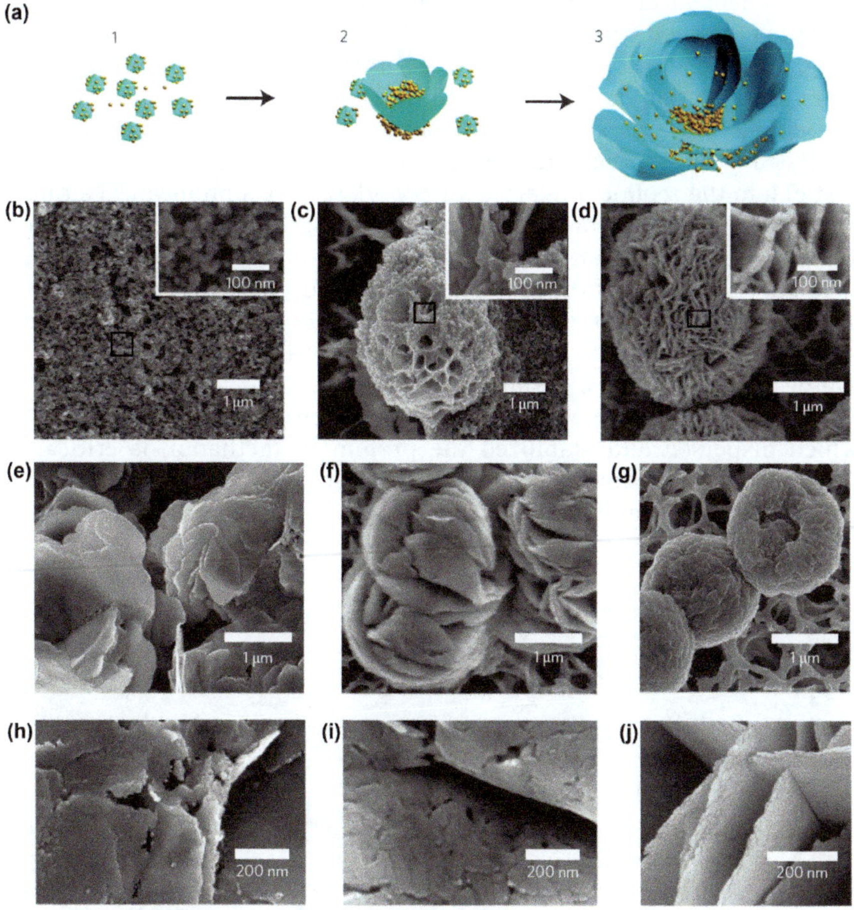

Figure 8.12 Formation of BSA-incorporated $Cu_3(PO_4)_2 \cdot 3H_2O$ nanoflowers. (a) Proposed mechanism: (1) nucleation and formation of primary crystals; (2) growth of crystals; (3) formation of nanoflowers. Yellow spheres indicate protein molecules. SEM images (b–d) at 2 h (b), 12 h (c) and 3 days (d). Insets: high-resolution images of the regions indicated by boxes. SEM images (e–g) of nanoflowers made with NF-2, treated by calcination (e), trypsin (f) and glutaraldehyde and EDTA (g). h–j, High-resolution SEM images of the petals of calcined nanoflowers in NF-2 (h), trypsin-treated NF-2 (i) and in NF-2 without any treatment (j).
Reprinted by permission from Macmillan Publishers Ltd: J. Ge, J. Lei and R. N. Zare, Protein-inorganic hybrid nanoflowers, *Nat. Nanotechnol.*, 2012, 7, 428–432. Copyright 2012. Nature Publishing Group.

The interaction of the proteins and copper ions lead to the growth of micrometre-sized particles that are shaped like flower petals. This the first example of enzyme–inorganic crystal hybrid nanoflowers based on a *de novo* synthesis process *via* a one-step precipitation in solution. In this study, laccase (a copper ion-containing oxidase), carbonic anhydrase (a zinc-containing enzyme) and lipase (a metal-free enzyme) were separately investigated. The hybrid nanoflower demonstrated higher catalytic efficiency as compared to the free enzyme. It was suggested that high surface area and the confinement of the enzymes in the nanoflowers were the key points in achieving the good enzymatic activity. The hybrid nanoflower provided a new insight in the development of enzyme immobilization. In the succeeding years, investigation on MOF as platform has led to the development of new and simple route for enzyme immobilization. In 2014, Liu and co-workers successfully introduced a new approach of embedding the Cyctochrome c (Cyt c) into the zeolitic imidazole framework (ZIF-8), a prototypical structure of zeolites.[45] Since the molecular size of Cyt c is bigger than the pore size of ZIF-8, the immobilization of enzyme was carried out using a one-pot synthesis method. A mixed solution containing Cyt c and polyvinylpyrrolidone (PVP) with a methanol, zinc nitrate hexahydrate and 2-ethylimidazole was used to successfully encapsulate the Cyt c. The complete illustration for one-pot synthesis of Cyt c@ZIF-8 is shown in Figure 8.13.

This remarkable discovery led to a conclusion that the presence of PVP, which dispersed and stabilized the protein in methanol, is critical for enzyme encapsulation. Several kinds of enzyme such as horseradish peroxidase (HRP) and lipase (LP) were also conducted *via* this strategy

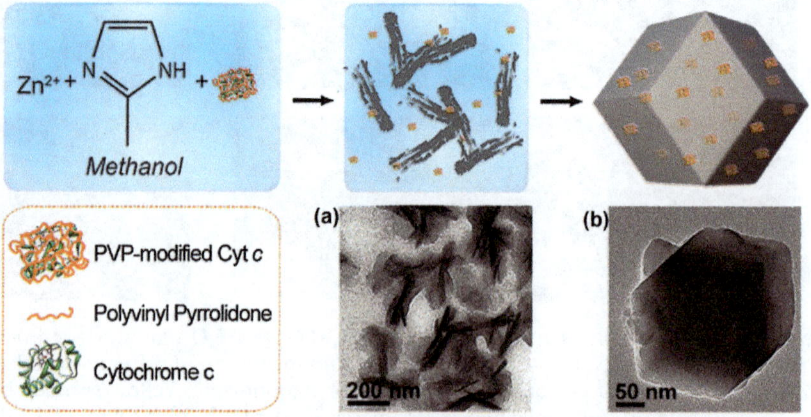

Figure 8.13 Schematic showing the growth process of Cyt c/ZIF-8 and time dependent morphology during the growth process.
Reprinted with permission from F. Lyu, Y. Zhang, R. N. Zare, J. Ge and Z. Liu, One-Pot Synthesis of Protein-Embedded Metal-Organic Frameworks with Enhanced Biological Activities, *Nano Lett.*, 2014, **14**(10), 5761–5765. Copyright 2014 American Chemical Society.

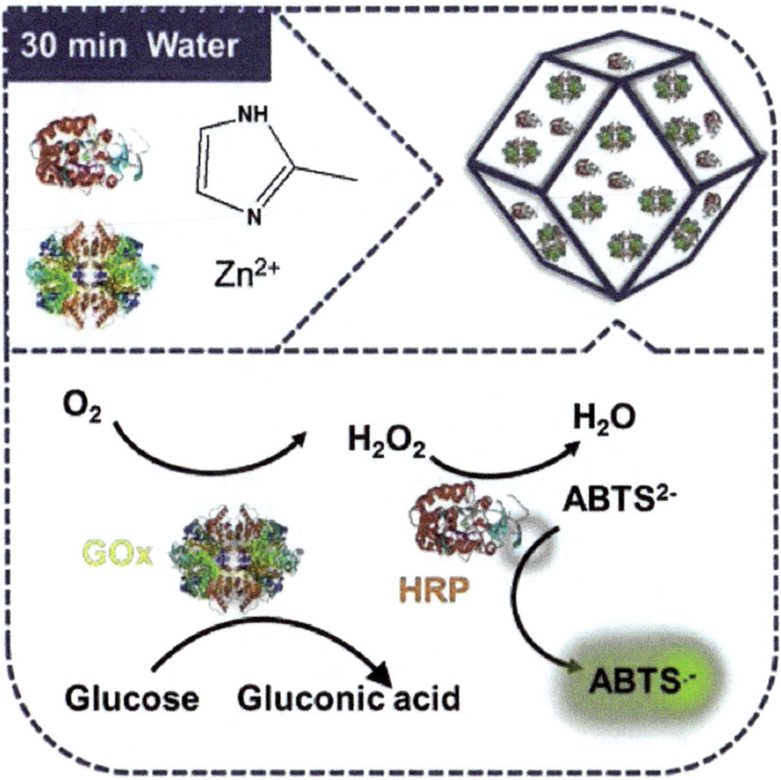

Figure 8.14 Schematic synthesis of multi-enzyme-containing ZIF-8. Reproduced from ref. 46 with permission from the Royal Society of Chemistry.

under the same research group. Taking the advantage of enzyme encapsulation, Liu and co-workers also reported the first example of a multiple enzyme-incorporated MOF.[46] The enzymes@MOF was prepared *via* a co-precipitation procedure in aqueous environment at ambient conditions (as shown in Figure 8.14).[46]

Glucose oxidase (GOx) and HRP were chosen to synthesize the multiple enzyme-embedded ZIF-8 (hereafter denoted GOx&HRP@ZIF-8), which was formed within 30 min at 25 °C. The GOx&HRP@ZIF-8 demonstrated enhanced thermal stability and prevented the enzymes from proteolysis and chelating. The GOx&HRP@ZIF-8 also exhibited high catalytic efficiency for an enzyme cascade reaction. Since the pioneering work presented by Liu and co-workers, Tsung *et al.* have also reported similar concept.[47] The catalase was embedded into uniformly sized ZIF-90 crystals *via* a *de novo* approach (hereafter denoted CAT@ZIF-90) as illustrated in Figure 8.15.[47]

Protease and catalase are two useful enzymes in the textile industry. Protease is a proteolytic enzyme used for silk degumming while the function

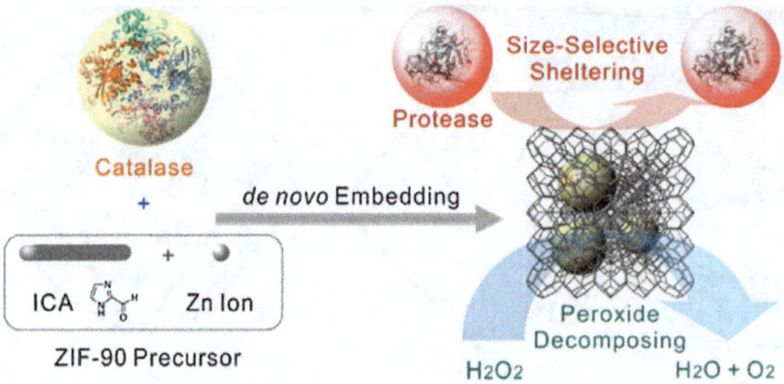

Figure 8.15 Water-based synthesis of ZIF-90 with encapsulated catalase enzyme and its functional activity.
Reprinted with permission from F.-K Shieh, S.-C. Wang, C.-I. Yen, C.-C. Wu, S. Dutta, L.-Y Chou, J. V. Morabito, P. Hu, M.-H. Hsu, K. C.-W. Wu and C.-K. Tsung, Imparting Functionality to Biocatalysts via Embedding Enzymes into Nanoporous Materials by a *de Novo* Approach: Size-Selective Sheltering of Catalase in Metal-Organic Framework Microcrystals, *J. Am. Chem. Soc.*, 2015, **137**(13), 4276–4279. Copyright 2015. American Chemical Society.

of catalase is for peroxide decomposition, which can be applied in wastewater treatment. However, it is believed that the coexistence of these enzymes in solution results in protease cleavage. Therefore, a method of sheltering the catalase from protease during industrial processes is necessary. Tsung *et al.* demonstrated the retained peroxidase activity of the catalase by directly embedding into ZIF-90. The ZIF-90 protected the enzymatic function of the catalase, which prohibits any contact with proteinase K. This strategy of imparting the catalase into the ZIF-90 offered a new function for biomolecules. Encapsulation or *de novo* approach bestows a powerful strategy of allowing the MOF materials, with pore sizes smaller than the size of the proteins/enzymes, to be useful. Furthermore, this method of encapsulating a multi-enzyme into the MOF structure is easy to operate that has a great potential in extending the application for biocatalysis.

8.5 Conclusions

Immobilization of enzymes onto MOF offers a new platform in opening a new avenue for heterogenous biocatalytic applications. This new strategy demonstrated an enhance enzyme stability, recyclability, activity as well as selectivity. This proof of concept can be beneficial for industrial, agrochemical, bio-engineering, catalysis, and biomolecular delivery applications. The current concept of enzyme/protein immobilization is rather in the early stage of development. Several issues concerning the stability of enzymes must carefully be addressed. First, the presence of metal ions from MOF may affect the enzyme activity and could lead into its denaturation during immobilization. Second, organic solvents and high temperatures in the immobilization procedure will decrease enzyme activity due to the enzyme denaturation, thus it is a limitation for immobilization or application. Finally, the relationship between the pore size of MOF and the encapsulated enzyme may cause leaching or mass transfer difficulties.

The rapid development of MOFs and the knowledge about protein/enzyme will indeed deliver a bright prospective in the advancement of catalysis and material science. With the continuous efforts, we believe that a new breakthrough for protein/enzyme immobilization using MOF will be introduced in the future.

References

1. S. J. Benkovic and S. Hammes-Schiffer, *Science*, 2003, **301**, 1196.
2. T. J. Pisklak, M. Macías, D. H. Coutinho, R. S. Huang and K. J. Balkus Jr., *Top. Catal.*, 2006, **38**, 269.
3. D. E. De Vos, B. F. Sels and P. A. Jacobs, *CATTECH*, 2002, **6**, 14.
4. O. Kirk, T. V. Borchert and C. C. Fuglsang, *Curr. Opin. Biotechnol.*, 2002, **13**, 345.
5. A. Schmid, J. S. Dordick, B. Hauer, A. Kiener, M. Wubbolts and B. Wtholt, *Nature*, 2001, **409**, 258.
6. M. Hartmann and D. Jung, *J. Mater. Chem.*, 2010, **20**, 844.
7. C. Garcia-Galan, A. Berenguer-Murcia, R. Fernández Lafuente and R. C. Rodrigues, *Adv. Synth. Catal.*, 2011, **353**, 288.
8. P. V. Iyer and L. Ananthanarayan, *Process Biochem.*, 2008, **43**, 1019.
9. J. M. Nelson and E. G. Griffin, *J. Am. Chem. Soc.*, 1916, **38**, 1109.
10. R. A. Sheldon and V. P. Sander, *Chem. Soc. Rev.*, 2013, **42**, 6223.
11. M. Kato, K. Inuzuka, K. Sakai-Kato and T. Toshimasa, *Anal. Chem.*, 2005, **77**, 1813.
12. S. Wu, L. Sun, J. Ma, K. Yang, Z. Liang, L. Zhang and Y. Zhang, *Talanta*, 2011, **83**, 1748.
13. H. H. P. Yiu and A. P. Wright, *J. Mater. Chem.*, 2005, **15**, 3690.
14. Y.-J. Han, G. D. Stucky and A. Butler, *J. Am. Chem. Soc.*, 1999, **121**, 9897.
15. I. I. Slowing, B. G. Trewyn and V. S. Y. Lin, *J. Am. Chem. Soc.*, 2007, **129**, 8845.
16. T. Osathanon, C. M. Giachelli and M. J. Somerman, *Biomaterials*, 2009, **30**, 4513.
17. A. M. Klibanov, *Science*, 1983, **219**, 722.
18. J. R. Long and O. M. Yaghi, *Chem. Soc. Rev.*, 2009, **38**, 1213.
19. M. O'Keeffe and O. M. Yaghi, *Chem. Rev.*, 2012, **112**, 675.
20. K. M. L. Taylor-Pashow, J. D. Rocca, Z. Xie, S. Tran and W. Lin, *J. Am. Chem. Soc.*, 2009, **131**, 14261.
21. A. J. Fletcher, E. J. Cussen, T. J. Prior, M. J. Rosseinsky, C. J. Kepert and K. M. Thomas, *J. Am. Chem. Soc.*, 2001, **123**, 10001.
22. Y. Li, A. Pang, C. Wang and M. Wei, *J. Mater. Chem.*, 2011, **21**, 17259.
23. R. P. Van Duyne and J. T. Hupp, *Chem. Rev.*, 2012, **112**, 1105.

24. F. Millange, N. Guillou, M. E. Medina, G. Férey, A. Carlin Sinclair, K. M. Golden and R. I. Walton, *Chem. Mater.*, 2010, **22**, 4237.
25. J. Lei, R. Qian, P. Ling, L. Cui and H. Ju, *Trends Anal. Chem.*, 2014, **58**, 71.
26. M. Zhao, X. Zhang and C. Deng, *Chem. Commun.*, 2015, **51**, 8116.
27. W. Ma, Q. Jiang, P. Yu, L. Yang and L. Mao, *Anal. Chem.*, 2013, **85**, 7550.
28. Y. Chen, S. Han, X. Li, Z. Zhang and S. Ma, *Inorg. Chem.*, 2014, **53**, 10006.
29. J. Pedroche, M. del Mar, Yust, C. Mateo, R. Fernández Lafuente, J. Girón-Calle, M. Alaiz, J. Vioque, J. M. Guisán and F. Millán, *Enzyme Microb. Technol.*, 2007, **40**, 1160.
30. L. Betancor, F. López-Gallego, A. Hidalgo, N. Alonso-Morales, G. D. O. C. Mateo, R. Fernández-Lafuente and J. M. Guisán, *Enzyme Microb. Technol.*, 2006, **39**, 877.
31. S. Jung, Y. Kim, S.-J. Kim, T.-H. Kwon, S. Huh and S. Park, *Chem. Commun.*, 2011, **47**, 2904.
32. Y.-H. Shih, S.-H. Lo, N.-H. Yang, B. Singco, Y.-J. Cheng, C.-Y. Wu, I.-H. Chang, H.-Y. Huang and C.-H. Lin, *ChemPlusChem*, 2012, **77**, 982.
33. X. Wu, M. Hou and J. Ge, *Catal. Sci. Technol.*, 2015, **5**, 5077.
34. D. S. Raja, W.-L. Liu, H.-Y. Huang and C.-H. Lin, *Comments Inorg. Chem.*, 2015, **35**, 332.
35. U. Hanefeld, L. Gardossi and E. Magner, *Chem. Soc. Rev.*, 2009, **38**, 453.
36. K. Seki and W. Mori, *J. Phys. Chem. B*, 2002, **106**, 1380.
37. V. Lykourinou, Y. Chen, X.-S. Wang, L. Meng, T. Hoang, L.-J. Ming, R. L. Musselman and S. Ma, *J. Am. Chem. Soc.*, 2011, **133**, 10382.
38. W.-L. Liu, S.-H. Lo, B. Singco, C.-C. Yang, H.-Y. Huang and C.-H. Lin, *J. Mater. Chem. B*, 2013, **1**, 928.
39. W.-L. Liu, C.-Y. Wu, C.-Y. Chen, B. Singco, C.-H. Lin and H.-Y. Huang, *Chem. – Eur. J.*, 2014, **20**, 8923.
40. R. M. Blanco, J. J. Calvete and J. M. Guisán, *Enzyme Microb. Technol.*, 1989, **11**, 353.
41. J. M. Guisán, A. Bastida, C. Cuesta, R. Fernández Lafuente and C. M. Rosell, *Biotechnol. Bioeng.*, 1991, **39**, 75.
42. J. Pedroche, M. M. Yust, J. Girón-Calle, J. Vioque, M. Alaiz, C. Mateo, J. M. Guisán and F. Millan, *Enzyme Microb. Technol.*, 2002, **31**, 711.
43. L. Betancor, F. López-Gallego, A. Hidalgo, N. Alonso-Morales, G. Dellamora-Ortiz, J. M. Guisán and R. Fernández Lafuente, *J. Biotechnol.*, 2006, **121**, 284.
44. J. Ge, J. Lei and R. N. Zare, *Nat. Nanotechnol.*, 2012, **7**, 428.
45. F. Lyu, Y. Zhang, R. N. Zare, J. Ge and Z. Liu, *Nano Lett.*, 2014, **14**, 5761.
46. X. Wu, J. Ge, C. Yang, M. Hou and Z. Liu, *Chem. Commun.*, 2015, **51**, 13408.
47. F.-K. Shieh, S.-C. Wang, C.-I. Yen, C.-C. Wu, S. Dutta, L.-Y. Chou, J. V. Morabito, P. Hu, M.-H. Hsu, K.-C. W. Wu and C.-K. Tsung, *J. Am. Chem. Soc.*, 2015, **137**, 4276.

CHAPTER 9

Porous Metal Organic Polygons and Polyhedra – Intrinsic vs. *Extrinsic Porosity*

HAYLEY R. GREEN AND GARETH O. LLOYD*

Institute of Chemical Sciences, Heriot-Watt University, Edinburgh, UK, EH14 4AS
*Email: G.O.Lloyd@hw.ac.uk

9.1 Introduction

Porous materials encompass a range of species from the crystalline materials such as zeolites, metal organic frameworks (MOFs) and covalent organic frameworks (COFs), to amorphous species such as porous organic polymers and porous carbons. These materials are rapidly growing areas of research that typically display good stability, diversity, designability and are significantly important industrial materials. Due to their high chemical stabilities they are insoluble, making them incongruous for solution processing. However, porous *molecular* solids, by their nature, are generally more soluble than the former examples, meaning solution processing is possible.[1] This chapter will focus on porous metal organic polygons and polyhedra (MOPs). We aim to provide an introduction to the history, design and applications of these varied porous materials with more in depth reading contained within the references provided.

Over the last century, since the awarding of the Nobel Prize in 1913 to the "father of coordination chemistry", Alfred Werner, coordination chemistry has grown extensively through better understanding of the coordination

Monographs in Supramolecular Chemistry No. 22
Functional Supramolecular Materials: From Surfaces to MOFs
Edited by Rahul Banerjee

Published by the Royal Society of Chemistry, www.rsc.org

geometries of metal ions and the dynamics of coordination based supramolecular assembly. This understanding established rules by which scientists could create rational synthetic methodologies for the incorporation of ligands and metals together to give predesigned structures. Within the field of coordination chemistry, the last decade has seen a divergence, resulting in two recognisable branches: MOFs and supramolecular coordination complexes (SCCs). SCCs have also been known as MOPs. Midway through the 20th century it was established that complementary small molecules could exhibit intermolecular recognition using noncovalent interactions, such as Van der Waal's, hydrogen bonding and Pi–Pi interactions. These rules were then applied to build larger more complex entities from molecular species that again are held through supramolecular interactions.[2,3]

This can manifest as crystalline materials which show solid-state host : guest chemistry. This is historically described as inclusion chemistry and/or clathrate chemistry, which has a very rich history in purely organic phases, as well as coordination complexes such as Hofmann and Werner complex clathrates.[4–6] An excellent description/summary of the potential of inclusion chemistry to show porosity was produced by Nassimbeni[7] over a decade ago. The production of a potentially porous crystalline material from a *molecular* building component requires some form of crystallisation process (Figure 9.1). As nature has an aversion to a vacuum/empty void, this often results in the "inclusion" of a guest molecular species resulting in a multicomponent material and/or a "close-packed" phase. This close-packed solid form is described as the apohost. Apo- meaning "furthest point from" (the host structure). The apohost structure has been shown through computational work and crystal structure prediction (CSP) to be isostructural to the solvated or guest included materials in some examples, *e.g.* Dianin's compound.[7–9] An apohost lattice is "close-packed" and normally exhibits the most stable lattice energy for the host building molecule and is referred to as the α phase. Recrystallisation of the apohost, from the dissolved molecules in a solution that may contain potential guest molecules, can result in an inclusion complex (β phase). Ibragimov has formulated a set of general rules for which topology, for the β phase, is likely to arise from crystallisation at different temperatures.[10] These may be summarized as follows:

(i) The guest : host ratio decreases as the crystallisation temperature increases.
(ii) The topology change from low temperature to high temperature is intercalate – tubulate – cryptate – apohost.[11]

There can be a number of β phases based upon different guests where the host framework is different. When materials are obtained with the same lattice compiled of the complexes with different guest forms, then these forms are all the same β phase. "Isoskeletal" has been suggested by Barbour to describe this phenomenon.[11,12] Partial decomplexation can occur resulting in a different host : guest ratio and expulsion of guests giving a new

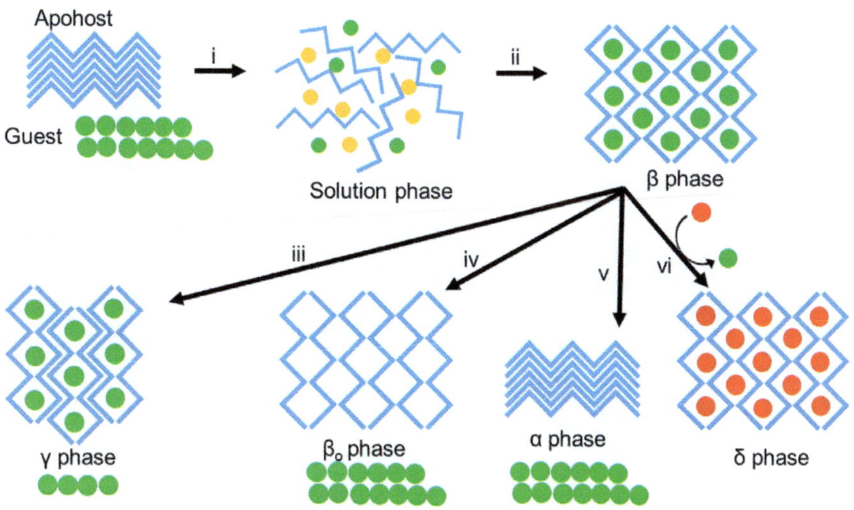

Figure 9.1 Schematic of the formation and decomposition of inclusion compounds. (i) The "close-packed" apohost lattice α phase is dissolved into a solution that may contain potential guest molecules (green dots), (ii) Recrystallisation results in an inclusion complex (β phase) with guests. (iii) Partial decomplexation can occur, resulting in a new lattice (γ phase). There can be a number of γ phases, *i.e.* stepped decomplexation. (iv) Complete decomplexation but retention of the host lattice structure ($β_o$ phase). (v) Collapse of the host lattice reforming the α phase. (vi) Post-assembly modification through guest exchange resulting in a δ phase. Reprinted (adapted) with permission from (L. R. Nassimbeni, *Acc. Chem. Res.*, 2003, **36**, 631–637). Copyright (2003) American Chemical Society.

lattice (γ phase). There can be a number of γ phases. Complete decomplexation can result in total loss of the guests but retention of the host lattice structure, resulting in a $β_o$ phase. Conversely, complete loss of the guests normally results in collapse of the host lattice, resulting in the reformation of the α phase (apohost) and occasionally a close-packed polymorph of the α phase. Post-assembly modification (also referred to as post-synthetic modification) can be made of the host : guest complex phase, *e.g.* β phase, through guest exchange resulting in a δ phase. This transformation must occur through a crystal-to-crystal mechanism and must not be able to form through a crystallisation from solution. If it forms through solution crystallisation then it is simply another β phase. A δ phase may show its own versions of the γ and $β_o$ phases but will also be able to revert to the α phase. In terms of porosity this represents the different thermodynamic and kinetic routes to getting a potentially porous phase. For example, the $β_o$ phase and $δ_o$ phases normally represent kinetically-trapped porous phases and pathway complexity arises through the use of the intermediate phases and the phases may not be accessible directly from solution. This description of inclusion solid-state chemistry allows for design of a porous material, a subset of the general field of crystal engineering.

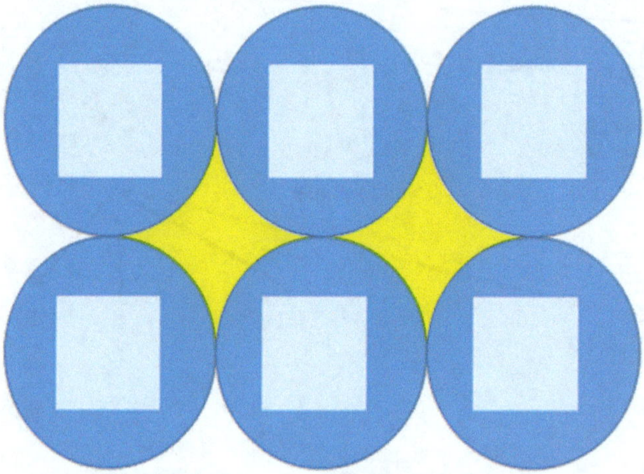

Figure 9.2 Intrinsic (light blue square) and extrinsic (yellow space between spherical molecules) pore spaces formed by molecular species in a periodic assembly.

Porous molecules can be classified as intrinsically and/or extrinsically porous (Figure 9.2). Intrinsic porosity is defined as porosity that is already inherent in the molecular structure when viewed as a discrete molecule, *i.e.* a concave cavity within the host that can fit a convex guest to form a molecular complex.[13] For example, shape-persistent voids, clefts, or cavities, traditionally found in purely organic materials such as calixarenes, cyclodextrins, molecular cages and cucurbiturils are viewed as intrinsically porous.[3,14] Extrinsic porosity is where the molecules pack together inefficiently forming a porous structure, *i.e.* the molecules form a lattice inclusion compound that has cavities, channels, or layers which can accommodate guest molecules (*i.e.* form intercalates, tabulates and cryptates). Molecules within molecular solids tend to pack together as closely as possible to maximise intermolecular contacts; this means that finding a molecular crystal with open channels or discrete lattice voids of more than 25 $Å^3$ is rare.[15] Examples of purely organic compounds that show extrinsic porosity due to inefficient packing in the solid state are 4-phenoxyphenol[16] and Dianin's compound.[17] The Werner complexes, such as $[Ni(4\text{-methylpyridine})_4(NCS)_2]$, as coordination examples, also show extrinsic porosity. Although the intrinsic–extrinsic porosity is a useful set of definitions, they should be viewed as being a grey scale, *i.e.* nearly all molecular porous materials will have some pore volume that can be viewed as intrinsic and some that is extrinsic. The definitions can aid the production of a porous material through design of the crystalline lattice's molecular building blocks by utilising intrinsic and extrinsic porosity and their connectivity, in conjugation with crystal engineering concepts.

Intrinsically porous molecules can be utilised in host–guest chemistry and are synonymous with the development of supramolecular chemistry.

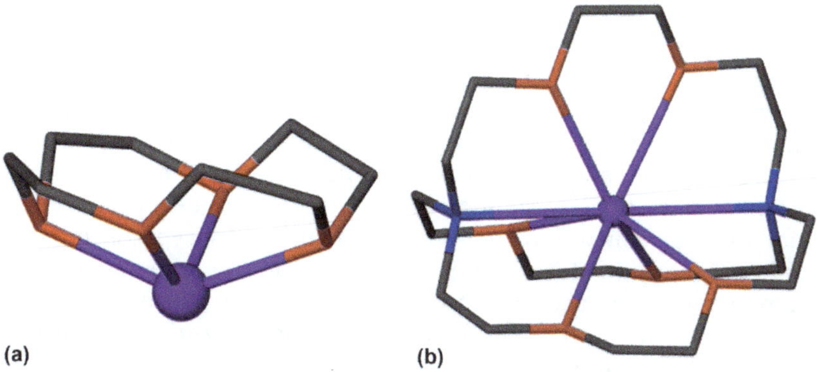

Figure 9.3 (a) Shows a metal complexed crown ether whilst (b) is a complexed cryptand. The cation is displayed in purple, O in red, N in blue and C in grey. The Figure also represents the design principle of a simple polygon-shaped molecule to a polyhedral-shaped molecule.

The 1987 Nobel Prize for Chemistry was awarded to Donald Cram, Jean-Marie Lehn and Charles Pedersen for the discovery of cryptands and crown ethers, in part. Cryptands can be simplistically described as three-dimensional crown ethers (Figure 9.3) that are suited to the selection and binding of certain cations.[18] The binding of metals within the "intrinsic pore space" of these hosts is the analogue we shall utilise for describing the development of metal organic porous molecular materials in the form of metal organic polygons (metallocycles) to metal organic polyhedra (metallocryptands). It is therefore recommended to the reader to obtain and assimilate the design principles of supramolecular hosts, begun by the excellent works of Cram, Lehn and Pedersen, and continues today utilising many complex procedures.

This completes the introduction of molecular host–guest and solid-state chemistry. The remainder of this chapter will introduce metal organic polygon design principles and their solid-state chemistry, metal organic polyhedron design and solid-state chemistry, finishing with applications and characterisation of both classes of compounds, as these two topics are closely related. This chapter is not an extensive collection of the literature, but is provided as a tool to better understand molecular metal organic porous materials and their intrinsic and extrinsic porosity.

9.1.1 Polygons

The simplest of all the intrinsically porous metal organic compounds to make are the metal organic polygons (metallocycles) (*e.g.* M_2L_2, M_3L_3 *etc.*, where M refers to the metal and L the organic ligand). The rational design of these complexes can be simple as all that is normally needed is complementarity between a convergent or divergent ligand (*e.g.* C-shaped) and a functionality match with the convergent or divergent coordination species of the metal arising in a macrocycle formation.[19] Note that when both species are divergent in a two-component system, a polymer must result. Therefore, metallocycles form with convergent–convergent, convergent–divergent and divergent–convergent combinations of metal and ligand. Further detail of rational design in MOPs can be found in the next section on polyhedra. This type of chemistry can result in a dynamic mixture of macrocycles and oligomers. Preferential crystallisation from the solution-based mixture of species through constitutional dynamic chemistry concepts is important and allows for design of preferential crystallisation from the mixture.[20] The development of metallocycles is as old as the development of coordination chemistry but it is only really in the last two decades that the porosity of the solid materials has been studied.

To understand the assembly processes of such complexes, the chemist needs to have some understanding of the mathematics of polygons. In geometry, a polygon is a plane figure that is bounded by a finite chain of straight-line segments closing in a loop to form a closed chain or circuit (*e.g.* macrocycle). These straight-line segments are called edges and the points where two edges meet are the polygon's vertices. The interior of the polygon is called its body, and often describes the intrinsic porosity of a molecular host. An *n*-gon is a polygon with *n* sides, so for example when $n=3$ you have a triangle (trigon), $n=4$ quadrilateral (tetragons like squares and rectangles) and $n=5$ is a pentagon, *etc.* We are all familiar with regular convex simple polygons such as squares and equilateral triangles. They are defined under Euclidean geometry to have equal angles between sides (equiangular) and equal lengths of sides (equilateral). Most people are familiar with the regular convex polygons, the non-convex regular polygons are the regular star polygons. A polygon is a two-dimensional example of the general polytopes which mathematically can be of any dimension. The three-dimensional versions are polyhedra (polyhedrons), and are described in the next section on metal organic polyhedra.[19]

An early example of the applicability of these intrinsically porous metal organic polygons is by Hupp and co-workers.[21,22] By utilising divergent, rigid dipyridyl ligands and coordination sites that are 90° to each other, the researchers showed the regular assembly of a series of Re squares (Figure 9.4a). The applications of these assemblies are described in the applications and characterisation Section (9.3). This particular assembly is also a useful example of the dynamics of both the conformational and configurational space available for large coordination assemblies and how the different solvates and inclusion compounds (β, δ or γ phases) can result in remarkably different properties for essentially the same complex.[23] An "ideal" square conformation of the complex (Figure 9.4b), with positions of the coordinated chloride relative to each other undetermined due to disorder in the crystal, results from the crystallisation of a 3 : 1 host-to-guest ratio material (the guest in this case is acetone). "Ideal" refers here to a regular convex simple polygon (square with equal lengths and angles). The opposing crystal form, a 2 : 1 host-to-guest ratio material, has the

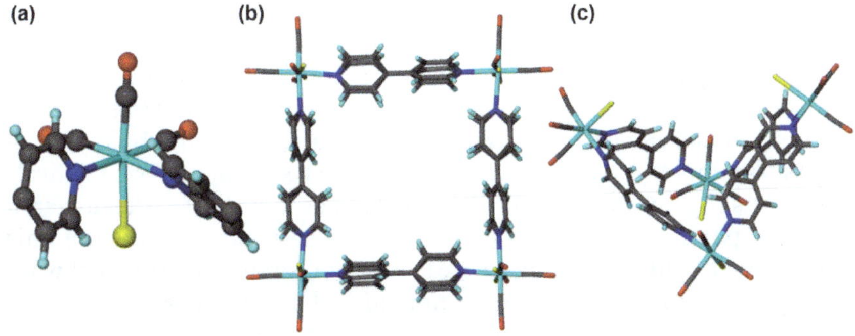

Figure 9.4 (a) Coordination sphere of the Re complex meaning connectivity is effectively 90° and planar. (b) "Ideal" square complex (regular convex simple polygon – square with equal lengths and angles). (c) Folded form of the same complex in different solvate material.

coordinated chloride ordered in an up-down alternating pattern. However, the assembly is far from "ideal" square with the conformation in a folded form (Figure 9.4c). This example of the structure illustrates how different phases can have very different packing efficiencies. The 2 : 1 ratio material is less well packed (38% void space of the unit cell volume calculated using Mercury, 1.2 Å probe over a 0.7 Å grid), and potentially more porous, than the 3 : 1 ratio material (22% void space of the unit cell volume).

As mentioned in the introduction, one issue of the crystal engineering of molecular porous materials is the connectivity of the pore spaces (the extrinsic porosity). This was highlighted through the work on a M_2L_2 metallocycle consisting of tetrahedral metal 2^+ dihalide coordination spheres with a dipyridyl diamide ligand.[24] Work originally published by Puddephatt showed that the metallocycles could be produced as a number of solvates.[25–28] However, the solvent-occupied void space (intrinsic porosity) is not connected. By changing the solvate form (ethanol or methanol inclusion phases) the intrinsic porosity is linked together into one-dimensional channels (tubulate).[24] This phase can be easily desorbed through a single-crystal-to-single-crystal transformation resulting in a porous material, as shown through I_2 sorption experiments and standard gas isotherm determination. See Section 9.3 for characterisation techniques for porous MOP materials.

Non-porous porosity refers to a crystalline material in which the static crystallographic structure shows no sign of porosity, yet the material clearly shows porous behaviour during experimental tests.[29–33] The characterisation of pore volumes and connectivity is described later in the section on characterisation and involves studying the crystal structures. The non-porous porosity behaviour is related to the gating phenomena found in some porous MOFs and zeolites. The thermally induced dynamics of the crystalline material results in transient porosity or the pressure (chemical activity) of the guests can force a transient opening of the material at a specific gate opening pressure. This behaviour often requires cooperative events between close neighbours within a porous framework.

Cooperativity between molecules within materials built from molecular entities can be utilised to control properties and functions of the materials. This was particularly well characterised in the example from Barbour and co-workers.[29] In this M_2L_2 metallocycle the intrinsic void can hold two relatively small guests (*e.g.* two MeOH molecules or two gas molecules). This material shows non-porous porosity as mapping of the static crystal structure only shows discrete void spaces (Figure 9.5). Through careful experimental work utilising gas sorption isotherms, computational and physical theory work involving statistic mechanics and *in situ* gas crystallography, they have shown how the deformation of the macrocycles, induced by the inclusion of

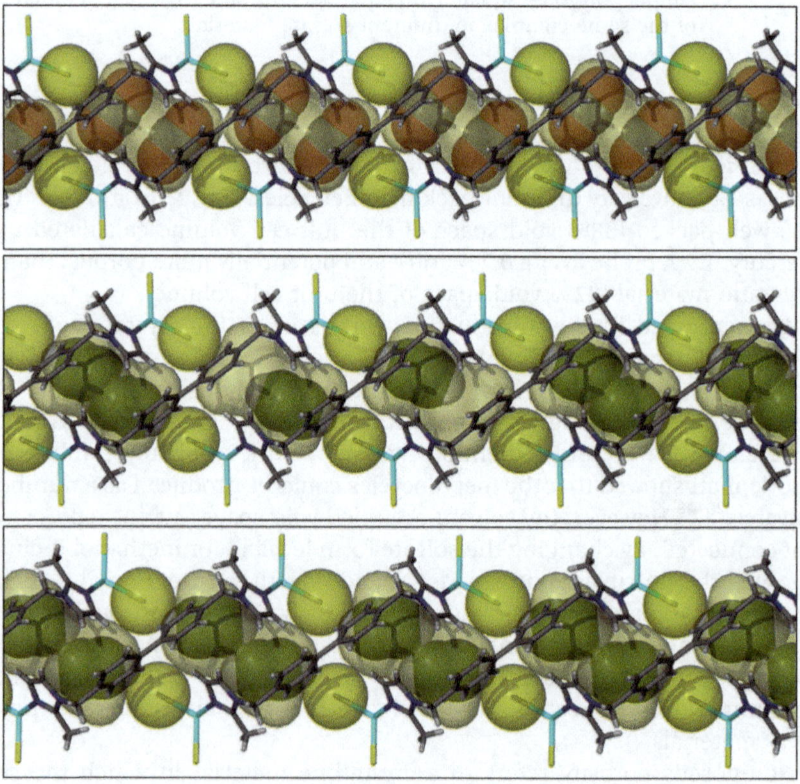
(a) (b) (c)

Figure 9.5 Crystal structures of a non-porous metal organic polygon under *in situ* gas pressure. The void spaces (semi-transparent yellow surfaces) can be occupied by two guest molecules of CO_2 or C_2H_2. (a) In the case of CO_2 there is ample room for both molecules within the void space. (b) However, in the case of C_2H_2 the void is too small for two guest molecules as shown from the crystal structure at only one occupancy. (c) The structure therefore deforms, resulting in an increase in void space and a reorientation of the interactions of the guest C_2H_2 molecules within the void. This rearrangement occurs in a cooperative manner.

two guests in the intrinsic void of the macrocycle material, enhances the probability of the neighbouring macrocycles deforming. This makes the neighbouring macrocycles become energetically more likely to absorb two guests. This particular example occurs when the gas being sorbed into the material is acetylene. This gives rise to a type VI gas isotherm (stepped isotherm) which has a noticeable inflection half way through the full saturation of the material with the guest. This saturation occurs at a 1 : 2 host-to-guest ratio, with the inflection occurring at half loading (1 : 1 ratio). This acetylene guest uptake is in strong contrast to that of CO_2, where the uptake shows a typical type I isotherm. The reasoning behind this difference can clearly been seen when the host–guest crystalline structure is determined under *in situ* crystallographic conditions (Figure 9.5). The binding of CO_2 occurs in an orientation that means that no structural rearrangement in the host framework needs to occur as the void space can easily accommodate two CO_2 guests. However, in the case of acetylene, the interaction of the acetylene with the host, and its resultant orientation, results in lack of space for two guests, with the void large enough to contain one guest with no deformation. Therefore, the framework deforms to fit the two guests. This deformation must occur through a cooperative effect resulting from the elastic coupling between host macrocycles, *i.e.* deformation of one host macrocycle "pressures" the neighbouring macrocycles to "conform", resulting in their transformation being energetically more favourable.

In summary, the metal organic polygons are the simplest of the MOP materials that can be constructed. However, this does not mean they show uncomplicated behaviour. The crystal engineering of the void space is still required and careful attention must always be made to selecting the correct host–guest system to get to a porous material. The simple building blocks and structures mean that complex behaviour relating to porosity, such as non-porous porosity and cooperativity, can be studied in great detail.

9.1.2 Polyhedra

Metal organic polyhedra are discrete molecular complexes assembled from a combination of metals coordinating to organic ligands which can be defined as having a particular solid topology. Their well-defined "cage-like" structures and confined cavities allow for guests such as solvent molecules, dissolved species or counter-ions to reside within the structure.[34] The polyhedra topologies, like most three-dimensional solids, can be mathematically categorised as Platonic solids, Archimedean solids, Johnson solids, faceted polyhedra or stellated polyhedra. This is analogous to the concept of there being only 230 space groups in crystallography. Platonic and Archimedean solids are simpler, high symmetry and familiar three-dimensional shapes, that are closely related to each other.[19,35] Platonic solids (Figure 9.6) are polyhedra created from one regular polygon, with the five possible Platonic solids being the tetrahedron, hexahedron (cube), octahedron, icosahedron and dodecahedron. Meanwhile, Archimedean solids are semi-regular solids

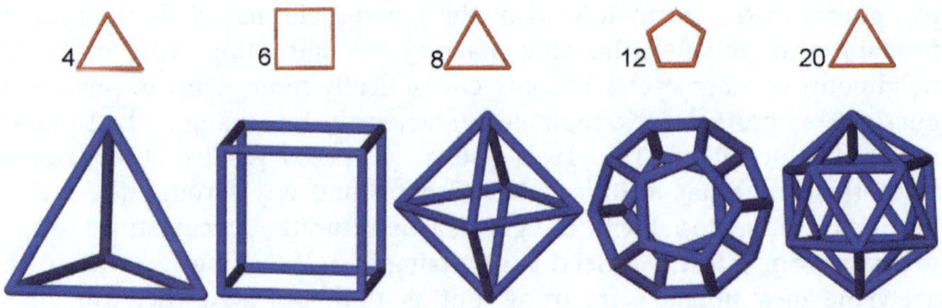

Figure 9.6 The five Platonic solids. The tetrahedron, cube, octahedron, icosahedron and dodecahedron shown, from left to right respectively. Shown above the blue solids are the number of regular polygon from which each of the Platonic solids is formed. Each of the shapes is a representation of an experimentally determined crystal structure, showing that molecular structures can exhibit these solid forms.

that contain at least two different regular polygons to create the faces. The dodecahedron, truncated tetrahedra and snubs cubes are three examples of the possible fifteen Archimedean solids.[36] The nine faceted polyhedra are derived from the Archimedean solids but share vertices as opposed to edges as seen with the Platonic and Archimedean solids. Stellated polyhedra are rarely seen in MOP structures, they are star-shaped polyhedra formed from elongating the polyhedron in three dimensions. The Johnson solids are non-uniform and convex polyhedra, where each face is a regular polygon but there are no rules regarding how the polygons join around vertices or that each face must be the same, this allows for ninety two polyhedra to be formed. It is very rare to see MOP materials that can be clearly described as Johnson solids.[37] It must be noted that whilst the majority of compounds will be Archimedean or Platonic in nature, due to the majority of assemblies being highly symmetric, some will fail to crystallise as any of these 'regular' polyhedra.

Rational design of metal organic polyhedra has been utilised by many researchers over the years to produce complexes of a predetermined shape.[36,38–40] This rational design can be split into two principles, although in truth they are similar and should always be considered in conjunction:

- the symmetry interaction model,
- the molecular library model, alternatively named the directional bonding model.

The metal organic polyhedral structures can be simplified by viewing them in terms of nodes and spacers, as described by Wells in the 1950s.[35,40] All linear ligands/connections are termed spacers, whilst nodes are viewed as anything that diverges from linear. From this, it is then possible to determine which net the structure belongs too using the (n,p) parameters, for example.[40] However, this simplistic methodology does not cover all possible polyhedra and polygons and therefore more complex methods have been developed to create uniform symbols, for example, Schläfli- and

Delaney-symbols.[41] In the 1990s the foundation stones of node and spacer were fully utilised in the development of methods that would allow for the rational design and construction of discrete self-assembled complexes. Stang coined the term "directional bonding" to describe the self-assembly process.[39,40] This is also known as the "molecular library model". This process began with formation of 2D macrocycles, in the shape of a regular polygon, by designing both the organic ligand and the metal occupying the edges and vertices of the target shape. To do this, the polygon is broken along its edges to provide sections with no more than a 2-fold symmetry axis, the angles of these pieces can then be equated to the angles that the linkers and nodes will be required to take on. For example, a square is broken into four 90° nodes (these can also be referred to as acceptors) and four 180° spacers (donors); this means four ditopic ligands are required to link four ditopic metal centres (Figure 9.7). By the same approach, a triangle can be broken into three 60° nodes and three 180° spacers. By incorporating higher symmetry tritopic linkers with the ditopic species, it becomes possible to design cages or polyhedra. The formation of higher symmetry linkers can be accomplished through adding additional binding sites on the donor, increasing the amount of labile coordination sites or the number of metals in an organometallic acceptor. For example, a cube is formed by incorporating twelve 180° ditopic linkers with eight 90° tritopic linkers.

The "molecular library model" described as "directional bonding" was proposed in 2008 by Yaghi and co-workers to aid the reticular synthesis of a family of MOPs.[34] This family of materials is described in the Section 9.2.2 – Tetrahedral MOPs. Reticular chemistry is the process of assembling rationally designed rigid molecular building blocks into predetermined ordered structures (networks). Reticular chemistry can be used in the design and synthesis of compounds built from chemically bonded secondary building units (SBU).[34] A SBU is best described as a molecular complex or cluster. They can be connected by polytopic ligands to form a closed discrete MOP or network.[34,42]

Finally, the "symmetry interaction model" is based upon controlling the bonding vector direction between the symmetry building units. More simply

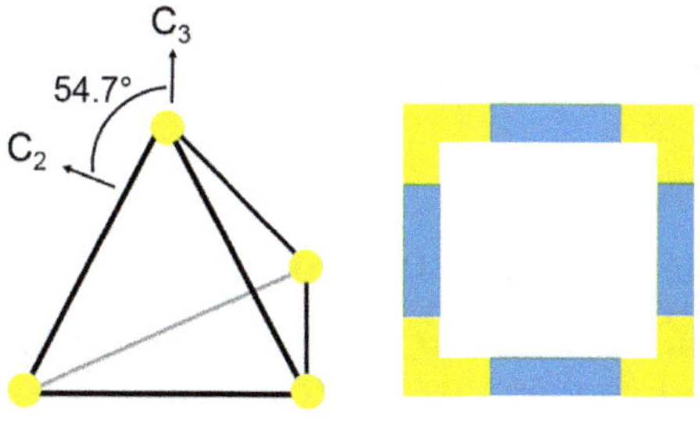

Figure 9.7 The "symmetry interaction model" shown with a tetrahedron. Whilst the molecular library model shows the formation of a square through ditopic nodes (yellow 90° sections) and spacers (blue 180° sections).

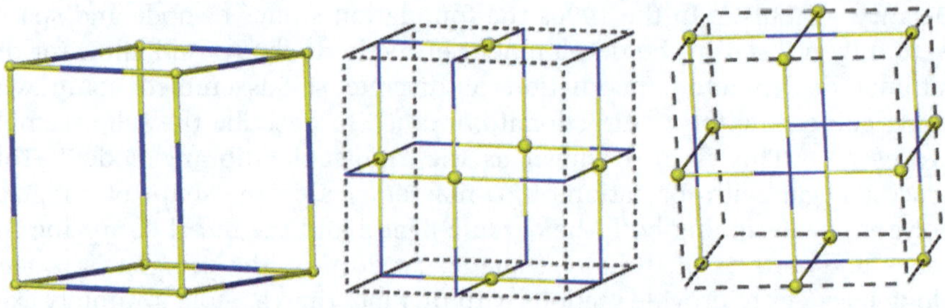

Figure 9.8 The assembly of cubic cage MOP materials can use one of three design configurations. From left to right. The cubes can be vertex points (C_3 symmetry connected metal centres, yellow spheres) connected through the edges (effectively linear ditopic ligands, blue) giving a M_8L_{12} stoichiometry. Four-coordinate metal centres at the faces connected by right-angled ditopic ligands giving a M_6L_{12}. Finally, two-coordinate metal centres of the edge of a cube connected by C_4 symmetry ligands positioned on the faces of a cube to give a $M_{12}L_6$ stoichiometry.

put, this method relies on knowledge of the ligands-favoured geometry when chelated to metals, based upon binding energies. Further understanding of the symmetry laws required to make the desired polyhedra, and the geometric relationship between the ligand and the metal, allows for the construction of the MOP.[38] Both models are represented in Figure 9.7.

There can be a number of general approaches to the formation of these MOP assemblies, and two common approaches are to utilise edge-directed and face-directed assembly of a particular solid form. As implied by the name, edge-directed assembly uses the organic linker to define the edges of the MOP/polyhedron, whilst face-directed means that the linker covers some, or all, of the faces of the MOP/polyhedron. This self-assembly process is maybe best visualised in terms of a cube (Figure 9.8). An edge-directed cube can be formed from eight 90° tritopic linkers connected to twelve linear ditopic linkers [M_8L_{12}], Whilst a face-directed cage requires twelve 90° ditopic linkers connected to six tetratopic planar 90° faces [M_6L_{12}].[43,44]

The design of polyhedra allows for applications in material porosity. The majority of the MOP materials published in the literature to date are charged or only studied in solution. We thus present in the next section examples of the polyhedra that have been successfully shown to be porous within crystalline phases.

9.2 Materials and Their Properties

Within this section, we present two case studies of two types of polyhedra, the copper nanoball and tetrahedral MOPs. These two classes of materials are presented as they represent two well-studied materials and they give good descriptions of much of the complex understanding that is required to truly make these materials applicable and useful.

9.2.1 Copper Nanoball

Within the literature there is an identifiable family of MOPs containing the paddlewheel secondary building unit.[45–55] This paddlewheel motif, commonly seen in MOF synthesis, is constructed of four carboxylate linkers coordinated to two metal atoms, typically copper, as shown in Figure 9.9. The metal atoms have an additional ligand bonded to them, providing square pyramidal coordination. These ligands are typically neutral nitrogen and oxygen donors, such as solvent, which can be subsequently removed creating open coordination metal sites.

There are two opinions when it comes to the naming of polyhedra built from this copper paddlewheel and isophthalate ligands. These are the small rhombihexahedron and the cuboctahedron. The small rhombihexahedron is an example of a faceted polyhedron; with 48 edges, 24 vertices and displays two-fold symmetry. The small rhombihexahedron is very similar to the solid rhombicuboctahedron, which is the dual of a cuboctahedron. The cuboctahedron is an Archimedean solid displaying 12 vertices, 24 edges and fourfold symmetry. The two solid forms are closely related to each other, and a copper nanoball structure can be viewed in both ways. The point of fourfold symmetry within the cuboctahedron is the paddlewheel building unit, whereas the paddlewheel building unit can be located within the centre of the coloured squares of the diagram of the small rhombihexahedron shown within Figure 9.10. Faceted polyhedra, polyhedra containing both concave and convex faces, are often excellent descriptions of the MOPs. This particular MOP discussed here, generated from reacting copper nitrate with

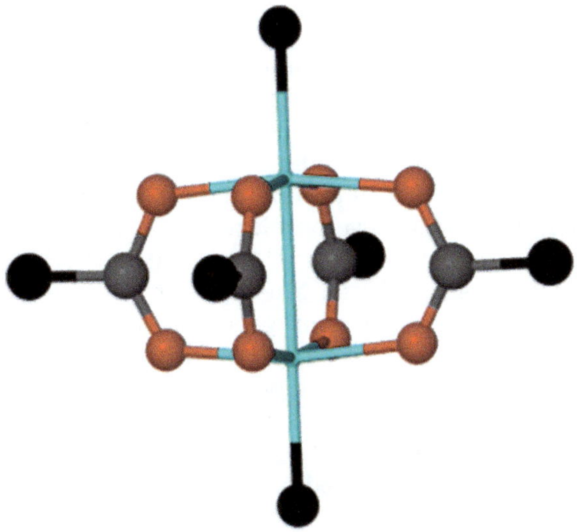

Figure 9.9 Copper paddle wheel secondary building unit where the aqua points are Cu, red spheres are O, grey spheres are C and black spheres are neutral donor groups.

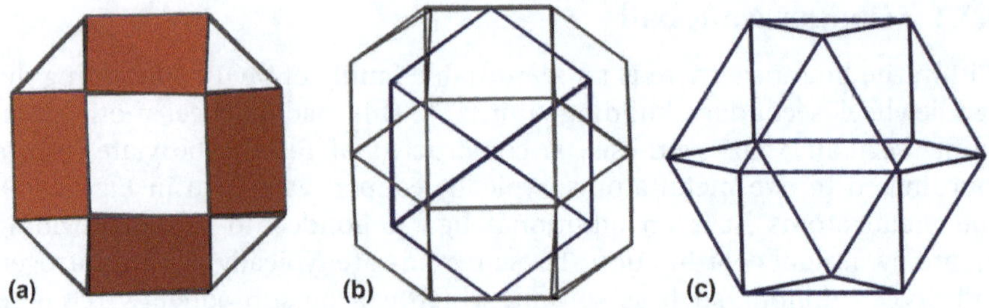

Figure 9.10 (a) A rhombihexahedron representation where the red squares are the paddlewheel SBU, the white spaces (triangles and squares) are the facets within the polyhedron. (b) Shows an overlay of the rhombihexahedron (grey) and the cuboctahedron (blue), where the blue fourfold symmetry points correspond to the red square in part a. (c) Displays the cuboctahedron where the points/vertices of fourfold symmetry are the paddle wheel SBU. These shapes are generated from a single crystal structure of the copper paddle wheel isophthalate "nanoball".

isophthalic acid in methanol in the presence of a templating agent such as nitrobenzene, is often described by both methods in the literature.[47] The intrinsic pore of one such cage is approximately 1 nm^3 in volume and is large enough to encapsulate a C_{60} molecule.[47]

The paddlewheel formation is not strictly limited to using copper as its metal source. Zhou and co-workers[49] have shown that it is possible to use molybdenum in the formation of the paddlewheel building unit. This slight difference in metal provided some interesting results in terms of the structure of the polyhedra formed. They recognised that by slightly varying the organic linker used it was possible to form both cuboctahedra and anticuboctahedra. As already mentioned, cuboctahedra are Archimedean solids that contains eight triangular faces and six square faces and displays the O_h point group symmetry, whilst the anticuboctahedra are isomers of the cuboctahedra. Whilst it still contains six square faces and eight triangular faces, it displays D_{3h} point group symmetry and is one of the Johnson solids, of which there are ninety two.[49] Zhou and co-workers give an elegant description of the assembly of the anticuboctahedra structure. The paddlewheel contains four dicarboxylic acid linkers and one Mo–Mo unit. These linkers, however, can be orientated to provided four geometric isomers, with the point group symmetries C_{4v}, C_s, C_{2h} *and* D_{2d}. Due to their energetic similarities, they are all equally likely to be formed but only the C_{4v} is capable of forming a closed polyhedra (Figure 9.11).[49]

The C_{4v} paddlewheel templates the formation of the *cis* triangles, which in turn form halves of the polyhedra, referred to as a triangular cupola. Depending upon how the two halves align, a cuboactahedron or anticuboctahedra is formed (Figure 9.12). If the triangles align to the squares, a cuboctahedron is formed (a). Meanwhile, if the triangles align with each other the anticuboctahedron (b) results.[49]

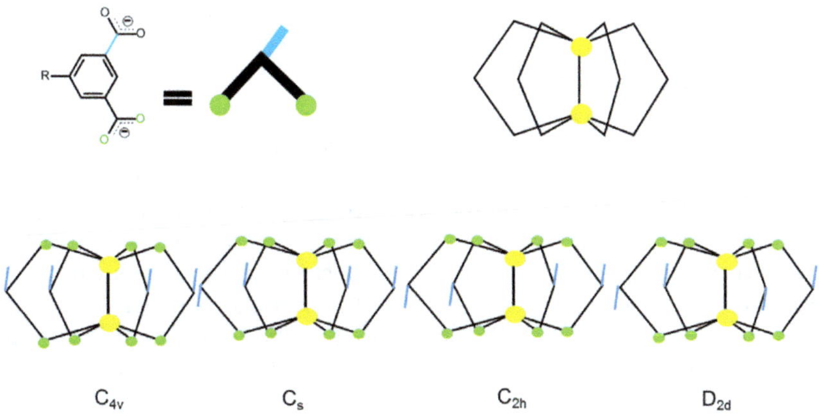

Figure 9.11 The four geometric isomers of the paddle wheel structure which orientates the functionality up or down relative to the plain of the carbon centres of the carboxylate groups bound to the metals, as shown. Yellow represents the metal centres and green the single set of oxygens of a carboxylate within an isophthalate ligand.

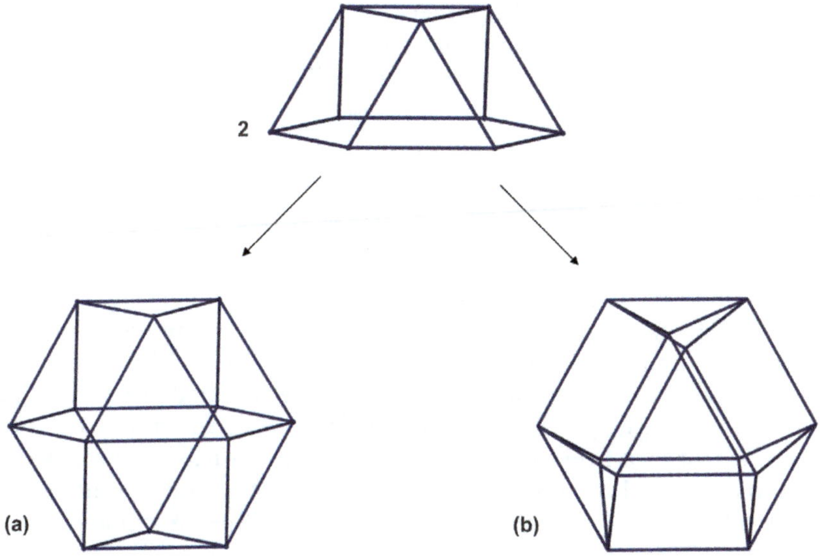

Figure 9.12 (a) The formation of a cuboctahedron from two triangular cupolas. (b) Shows the anticuboctahedron. They form dependent upon how the two half polyhedra are orientated relative to each other.

In 2010 Li and Zhou[48] reported the use of a bridging ligand substitution reaction for the synthesis and isolation of a number of MOPs containing the copper paddlewheel secondary building unit linked by a range of different carboxylate ligands. They initially created three MOPs which displayed cuboctahedral geometries. The MOPs where then soaked in solvents containing an excess of an additional carboxylate ligand. This allowed for exchange of ligands and the formation of new polyhedra. In some cases the new polyhedra contained both of the organic linkers and formed a new geometry, whilst in other cases the original linker was substituted entirely resulting in a polyhedra containing just the new ligand. This post-synthetic modification, or post-assembly modification, produced cages of different geometries (octahedral opposed to cuboctahedral) determined by the ligand species and the conditions used. Aside from the isolation of new structures, the variation in the cages alters both the packing of the solids within the crystalline materials, ultimately altering the extrinsic porosity of the materials. The variation in the ligands functionality causes changes in the intermolecular interactions experienced between the cages, which therefore alter the extrinsic porosity.

As a good example of this extrinsic porosity crystal engineering, Li and co-workers reported the formation of two cuboctahedral cages synthesised from isophthalic acid and 5-hydroxyisophthalic acid.[48] Both cages display the same cuboctahedral topology and will have similar intrinsic porosities; however, their extrinsic porosity is different. The bulkier 5-hydroxy-isopthalate actually shows the greater amount of void space per unit cell, at 66.3%, whilst the less bulky isophthalate cage shows 63.2% void space per unit cell. This shows that intermolecular forces between the cages are of significant importance, where controlling extrinsic porosity is concerned.[3,56,57]

9.2.2 Tetrahedral MOPs

Another family of porous MOPs identified from the literature are those based around the tetrahedral structure. The tetrahedra is a Platonic solid formed of four triangular faces. The tetrahedra can be produced by three different assemblies; the M_4L_6, the M_4L_4 and the M_6L_4. The former two assemblies are most commonly seen providing the edge and face form of the tetrahedra, respectively. The M_4L_6 tetrahedra is formed from six ditopic linker and four metal centres. The M_4L_4 requires four tripodal linkers connected to four metal centres whilst the M_6L_4 connects six metal centres with four tripodal ligands. Figure 9.13 displays these three stoichiometries of tetrahedra.

In 2005 Yaghi and co-workers reported a family of M_4L_6 truncated tetrahedra based on a derivative of the basic iron acetate secondary building unit.[58] The SBU is based on the well-known basic chromium acetate SBU where the chromium is substituted with iron and three of the acetates are replaced by sulphate anions. The three equivalent octahedrally bound iron atoms connect through three bridging carboxylate ligands and three bridging sulphate ligands. A triply coordinated oxide at the centre of the SBU connects the three metal centres together. Finally, a neutral donor atom, typically from a solvent atom such as water or pyridine, completes each of

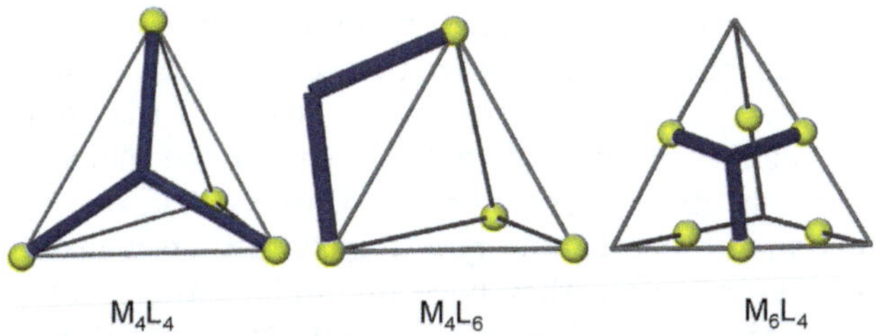

Figure 9.13 Three possible assemblies of tetrahedra through use of face connectivity with vertices (M_4L_4), edge connectivity with vertices (M_4L_6) and face connectivity with edges (M_6L_4). These are schematics of real crystal structures. The three tetrahedral assemblies are represented in grey, where the yellow spheres represent the metal atoms and the blue represents a single ligand.

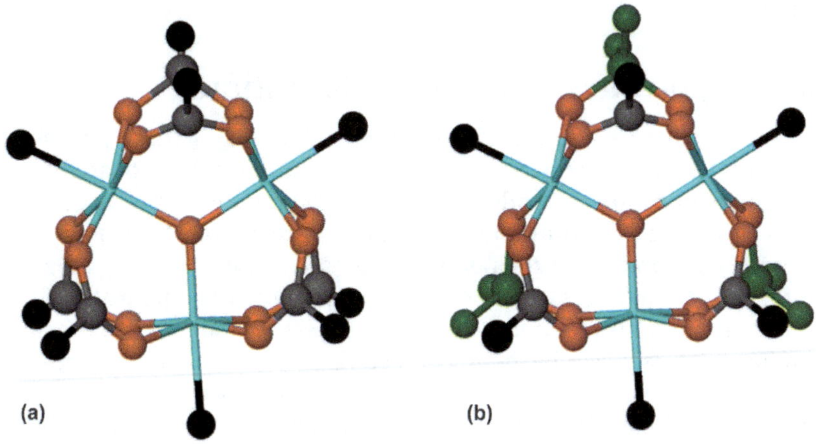

Figure 9.14 The basic iron/chromium acetate SBU is shown is shown as (a). In the design of the Yaghi tetrahedral SBU, half the carboxylates are replaced with sulphates resulting in a C_3 symmetric secondary building unit (b). Within this Figure the iron is coloured aqua, oxygen is red, carbon is grey and R groups are black.

the iron's coordination spheres. This leads to a C_3 symmetric SBU. The iron acetate SBU (Figure 9.14) is connected with a set of ditopic or tritopic carboxylate ligands which ultimately lead to the formation of the polyhedra.

By systematically increasing the length of the ligand (Reticular chemistry), the formation of a family of tetrahedra with an increasingly large pore volume was formed. The isoreticular system formed showed some interesting features in terms of its porosity. The intrinsic porosity was obviously increased as a result of the larger ligand being incorporated into the structure, increasing the intrinsic pore volume from 7.1% up to 26.7%. Whilst the extrinsic porosity is dependent on the packing of the polyhedra, this value varied from 27.2% to 62.7%.[58] Yaghi established that both the intrinsic and extrinsic pores are interconnected *via* the open faces of the tetrahedra, showing it would be possible for guest species to move through the material. Upon desolvation, the material experiences a loss of crystallinity due to the weak forces between the tetrahedra being lost and the framework order being lost.[58] Yaghi went on to establish whether the material maintained permanent porosity. By using gas sorption isotherms, it was established that the materials shows a reversible type 1 isotherm. So whilst the materials have lost long range order, they have maintained their intrinsic porosity, whilst the extrinsic porosity has been altered. This work showed that MOPs have surface areas comparable to other porous materials such as zeolites and MOFs.[58] One other example of a porous tetrahedra was reported in 2004 by Bai *et al.*[59] This neutral tetrahedron is based on the M_4L_6 structure with four iron centres connected by six ditopic hydroxamic acid ligands. The four windows of the cage contain dimethylformamide (DMF) solvent and as yet no work has been published on the gas sorption or permanent porosity of this material.

9.3 Applications and Characterisations

Due to the varied chemical and material properties of these porous MOP materials, the porosity can be targeted towards a wide variety of applications. Catalysis, separation and adsorption are all common target uses for materials of a porous nature and MOPs are no exception. Possibly the most conventional application for porous MOPs are those pertaining to gas separation and storage. This is not only due to the potential applications of this gas sorption but also as it is a key characterisation tool of the porosity. We therefore aim to describe both characterisation routes and processes required to understand the material properties and therefore their potential applications. To begin with, we present a collective set of procedures to ensure the complete characterisation of the material properties. Each step necessarily requires feedback to others, which results in a dynamic process. We break this process down to structural determination, determination of host–guest chemistry (can include gas sorption isotherms), guest removal characterisation and finally porosity testing (Figure 9.15).

Characterising molecular materials is of great importance in terms of understanding the porosity of the materials created, their properties and potential applications. Initially single crystal X-ray crystallography (SCXD) is often utilised to determine the structure of the material. Powder X-ray crystallography (PXRD) can subsequently or separately be used to determine bulk phase purity or structural information when SCXD is not possible. From this dataset it is possible to quantify some measure of the porosity of the material. By using pore mapping software (for example, as found in the Mercury, Platon, XSeed and Olex2 single crystal crystallography programmes) it is possible to calculate the free volume within the material when a clear host framework can be defined. The majority of these utilise the "rolling ball" algorithm originally developed to map enzyme surfaces and the accessible surface area or solvent-accessible area.[60–65] By using a probe, a

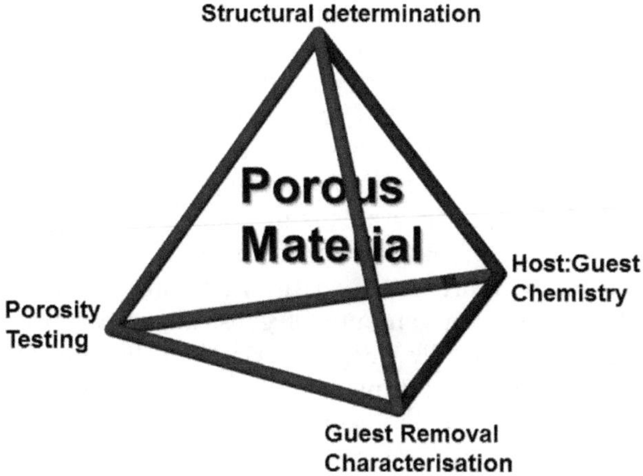

Figure 9.15 Characterisation of a porous material. The tetrahedron represents the feedback from several characterisation techniques. This allows for the determination of the structure of the material, its lattice and host–guest chemistry, how to remove the included guests, and testing of the resultant porosity.

sphere with a radius typically between 1.2–1.7 Å (selection of sphere is crucial and should always be stated in the analysis) the programme of choice maps the volume that the probe can sweep out within a molecular void or channel. The vast majority of the materials produced contain guest species within the pores of the molecules, to determine the porosity of the materials using this procedure these guests need to be removed; this can be done virtually by removing the guest species from the asymmetric unit within the crystallography software being used prior to the pore mapping regime.[15] This "virtually" empty structure is packed to show any channels or cavities formed, but it should be noted that this may not provide the most accurate representation because removal of the guest within molecular species can often cause the system to rearrange or collapse leading to less porous systems (as described in Section 9.1 and Figure 9.1). The connectivity/topology of the pore space, as well as the framework built from the molecular species, can be determined as it is sometimes important in understanding the properties but more importantly, the design of the material(s). Programmes such as Topos allows for this analysis, although experience is often required as the "connectivity" interactions are not as directional as those seen in MOFs.[15] Ideally, the host species should be removed and the crystal structure re-determined as the solvent free materials and mapping done on this structure. The process of the removal of the guests is not trivial.

Once the material has been structurally mapped and there is confidence that the system displays cavities or channels that could be potentially porous, the guest can be removed utilising heat and chemical potential.[66,67] Heating is often combined with dynamic vacuum to remove mobile guests from hypothetically porous materials. To circumvent the possible collapse of the materials to a more thermodynamically stable α phase, techniques such as freeze drying and super-critical drying have been utilised to trap intermediate, metastable states. Thermogravimetric analysis and differential scanning calorimetry (TGA and DSC, respectively) are crucial analytical steps in determining this "emptying" of materials. TGA accurately records the weight of the material as it is heated at a specific rate, or held at a constant temperature, under a specific atmosphere. Using TGA, it is possible to accurately determine the amount of included guest within the material[24,68] and gauge the temperature at which the material can be completely desolvated or dehydrated. The amount of guest shown within the crystallographic structure should tally with the TGA data, helping to confirm the material is desolvated, and thus giving definitive structural characterisation.[69] DSC works simply by recording the difference in the amount of energy required to heat the reference and the sample to the same temperature and rate. DSC can provide information on phase changes and enthalpies due to changes in the materials' heat capacity. Understanding whether the materials undergo phase transitions upon desorption of guests can be useful in determining amorphous and crystalline phases produced as part of "processing" the materials for porosity. Clearly, after characterisation and emptying of the material, PXRD and SCRD need to be repeated to confirm structural coordinates.

The standard measurement for porosity requires the determination of gas sorption isotherms for a variety of gases. There are many isotherms types and models for their interpretation. IUPAC categorise isotherms as one of six types, Types I–VI. One of the most common models utilised for quantifying porosity between materials is the Brunauer–Emmett–Teller model (BET).[70] BET values are referred to as surface areas, typically as $cm^2\,g^{-1}$ or $m^2\,g^{-1}$, however, care should be always taken to insure the theory still holds and the other representations of the isotherm data is not more appropriate. For example, the materials discussed here are discrete molecular species which will show absolute host–guest ratios. We would therefore recommend, when possible, that the host–guest ratios at specific pressures or vapour pressures be reported, with typical $mg\,g^{-1}$ given for CO_2 sorption isotherms being easily converted in molar quantities and reported. One very simple method for testing porosity is simply to absorb a guest. Iodine (I_2) represents an easy method for testing porosity as it changes the colour of the material, although some care must be taken to ensure that the I_2 is not just bound to the surface of the crystals.[24,30,71] Tests such as elemental analysis and TGA can also confirm sorption of I_2. There has been much interest in I_2 sorption into porous materials, mostly due to interest in radioactive I_2 produced from nuclear reactors.[71,72] Related to that is the interest in porous materials that absorb dyes out of solution.[71] This has been shown to occur in MOP materials as well.[71]

In situ crystallography,[73,74] where the structure of the crystals are determined in the presence of the guest in the form of liquid, vapour pressure or gas pressure under controlled experimental values, allows for determination of not only the binding positions of the guests (if possible) but also host–guest inclusion properties. This technique requires the use of a capillary or sealed vessel loaded with the gas or vapour at a desired pressure, at which the data collection is performed. This can allow for refinement of crystal structures in which the location of the gas molecules trapped within the pores of the material can be determined. This, of course, can be complemented by the many computational techniques available for determining gas sorption and the gas binding sites.[75]

Due to the control that can be exerted over these materials in terms of their pore shape, size, and chemical functionality, they are an obvious choice when dealing with gas separation and storage. The intrinsic cavity of the cage can be designed with properties idealised to capturing and trapping molecules of the desired gas. Due to the environmentally harmful effects of greenhouse gases such as CH_4, CO_2 and N_xO_x, they have become common targets for gas separation and storage across the field of porous materials, with MOPs being no exception. One such example is the selective separation of CO_2 from multi-gas systems. Zhou and co-workers[76] developed a MOP cavity to specifically capture a single CO_2 molecule. The intrinsic porosity of the cage was designed to contain one molecule of CO_2. The cavity is designed to optimise the interactions between the cavity and the CO_2 molecule, therefore increasing its selectivity towards CO_2 and limiting the adsorption of other gases within a gas mixture.[76] More specifically the angle and length of the ligands optimise the metal–metal length and orientation to allow the CO_2 to anchor between the metals within the MOP. When separation is the targeted properly there are other analyses that can be utilised to quantify the materials' efficiencies. These are the calculation of a separation factor from the isotherms of different gases and the use of break-through experiments.[77] Separation of other gaseous compounds is possible with MOPs, for example, the separation of alkane, alkene and aromatic organic compounds is important industrially. A Johnson-solid-based material has been shown to have selectivity towards larger C_2H_6 and C_3H_8 over methane.[37] Designed to produce an asymmetric MOP by utilising the organic building blocks in the form of a calixarene and 5-(pyridin-4-yl) isophthalate, this example shows how complex the MOP designs can become.

Sensing based on a colorimetric change in a material is an important methodology for development of sensors. There are a number of MOP examples shown in the literature,[21,23,78] especially based on solution sensing, but the examples of solid materials that are porous in nature are scarcer. A good example to show the potential of these materials for applications is given of a metal organic polygon based on a colorimetric sensing group and platinum-based assembly (Figure 9.16).[78] This material is constructed from a Pt-based assembly of a metallocycle with a C-shaped organic ligand, which contains a reporting colorimetric group. Upon desolvation, the empty material can phase change upon exposure to CH_2Cl_2 vapours within minutes. This new phase is not only metastable, *i.e.* can maintain its phase upon CH_2Cl_2 removal, it is also a different colour. The material converts from

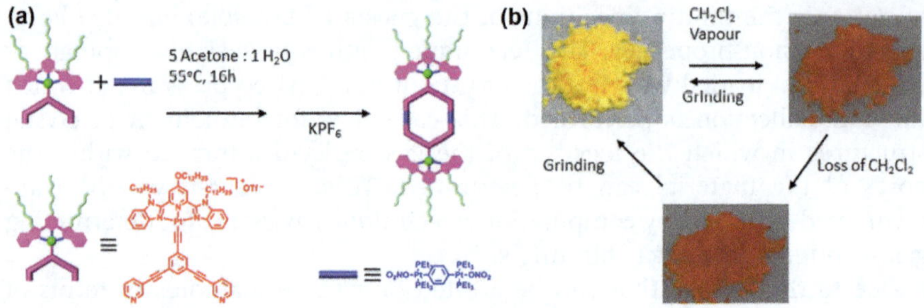

Figure 9.16 Colour hange material. (a) Assembly of the Pt-based metal organic polygon incorporating a colorimetric sensing group. (b) The empty material changes from yellow to red within minutes of being exposed to CH_2Cl_2 vapours. Even on loss of the CH_2Cl_2 the material maintains this colour. It is only through a mechanochemical induced phase change that the material can be returned to its original yellow form.
Reprinted (adapted) with permission from (B. Jiang, J. Zhang, J. Q. Ma, W. Zheng, L. J. Chen, B. Sun, C. Li, B. W. Hu, H. Tan, X. Li and H. B. Yang, *J. Am. Chem. Soc.*, 2016, **138**, 738–741). Copyright (2016) American Chemical Society.

yellow to orange. To reform the yellow phase the orange material needs to be mechanically ground.

MOPs have a clear future in membrane development as they have an advantage over many porous materials due to their processability. This processability is due to their solubility, as they are molecular species.[21,79,80] Membrane technologies are seen as a low cost, energy efficient solution to the problem of separation.[80] Within the literature there are examples of MOPs being combined with a variety of polymers to form a material suited to seperation.[80,81] The separation of methane from carbon dioxide is vital in industry due to the harmful effects of carbon dioxide and the improvement in the combustion of the methane. The incorporation of MOPs into polymer membranes can significantly improve both the permeability and the separation factors of the material.[80] The post-synthetic modification of these MOPs also plays a role in altering the separation and selectivity of the materials. Due to the high solubility's displayed by the MOPs, they display excellent adhesion and dispersion properties within the polymers. This is shown to improve permeability and the separation factors by up to 81% and 60%, respectively.[80] Conventional polymer membranes have a trade-off between selectivity and permeability so by incorporating MOPs there is potential to create a highly selective and controlled permeability in a material. Due to the high solubility of MOPs it is thought that they can be better dispersed throughout the material thus avoiding aggregation problems seen with other additives in mixed membrane systems.[80] Another example shows the separation of liquids and uses a soluble copper-based MOP with its organic linker functionalised to improve solubility and functionality when incorporated into a highly branched polymer solution.[81] It forms a material capable of separating a 1 : 1 toluene, *n*-heptane solution, with a separation factor of 19. The material can also enrich benzene from a 50% benzene–cyclohexane solution, with a separation factor of 15.4 towards benzene.[81] The final example given here is the Hupp Re square described earlier in Section 9.1.1.[21] By incorporating the square into a membrane, they were able to show a stable porous material with good selectivity.

MOP materials are also being developed for use within pharmaceutical delivery. The aim is to develop new routes for drug delivery due to the size of discrete MOP being analogous to that of enzymes.[82] MOPs can be seen to be favourable over other potential carriers as they are easily functionalised and decorated upon their surface, thus allowing for biocompatibility. They can selectively encapsulate the target drug molecule, store and transport it to the target site before controlling the release of the drug at a specific site within the body. By controlling the design of the cavity, the encapsulation of the drug can be controlled and by controlling the composition of both the ligand and the metal used within the cage, it is possible to control biocompatibility or even use metals as part of the treatment. For example, Ruthenium(II) complexes are used as anti-cancer agents and building blocks of the MOPs.[82] As a potential drug formulation, the crystalline phase of these materials will be important.

Given the importance of porous materials within industrial heterogeneous catalysis, for example the use of zeolite ZSM-5 for the cracking of long chain hydrocarbons, it is surprising that no crystalline porous molecular material has yet been utilised as a heterogeneous catalyst. A particular advantage over traditional inorganic solid material catalysts that the MOP materials could potentially have is their high processability due to their solubility. This makes their separation from products or regeneration processes much easier. Within the literature, there are many examples of homogeneous supramolecular catalysis; one such example, of many, is a Nazarov cyclisation within a catechol-based tetrahedral cage.[83] The field is continuously developing and there are many exciting opportunities for stereo- and enantio-selective synthesis mimicking biological enzymes. These catalysts are often described as being biomimetic as the intrinsic pore space of the MOPs acts as a controlling environment reminiscent of those found within enzymes. One major difference between heterogeneous and homogeneous supramolecular catalysis is that in the solid-state it is required that there be extrinsic porosity to allow for the diffusion of the products and reactants through the material and into the intrinsic pore where the catalysis takes place. This makes the crystal engineering of such materials incredibly difficult and the stability of such materials can be compromised through decreases in mechanical and chemical stabilities. The mechanical properties of the MOP crystalline family of materials has not been investigated in depth, which is in contrast to most other materials.[84]

9.4 Conclusion

Crystal engineering and material characterisation of molecular porous materials constructed from discrete MOPs represent a growing applicable class of materials for deployment in porous applications within industry and academia. Even with this progress, there are still a significant number of challenges we need to overcome. To truly design the materials and apply them to solve societal and industrial problems such as renewable and sustainable energy, and anthropometric induced climate change, the materials need to be designed. The design of cage structures is still fraught with difficulties, such as there are no easy use software and theory tools, however there will definitely be significant strides forward in the coming decades on this front. Even if you can make the cage compounds, crystallisation and the crystal structure are still nearly impossible to predict. Crystal engineering tools are available, and intrinsic porosity of the MOPs means that obtaining porous materials is likely. However, CSP is not universally accurate for all molecule types and when kinetics dominate the assembly process, even more so. The material properties need to be better characterised, especially that of the mechanical properties, as is happening in the literature at the moment with MOFs.[84] Molecular porous compounds can display porosity with BET surface areas of up to 2000 $m^2 g^{-1}$, which is comparable to nearly all other classes of materials. It is the host–guest chemistry of the intrinsic and extrinsic porosity that truly differentiates these materials from their more illustrious cousins. The well-established supramolecular chemistry of host–guest chemistry means that scientists have an excellent opportunity to design, tune and eventually make applicable materials for specific tasks. This is not the only advantage over other classes of materials. The solubility of the molecular entities means that processability of the materials should be significantly easier, cost affordable and flexible meaning that the engineering of the applications should be possible. This could be no truer than when discussing the production of membranes for separation technologies.

In conclusion MOPs are a rapidly expanding field of porous material research. Due to their rational synthesis, high selectivity, display of both intrinsic and extrinsic porosity, and high solubility they are a fascinating area of research which encompasses many different techniques in understanding their structure and potential applications. We have discussed the history of MOPs, the design principles for the synthesis of both these materials and finally the characterisation and applications of the materials alongside some real-life examples. The aim of this chapter was to provide a foundation of knowledge into the key principles of MOPs and their porosity. The references provided and those within will allow for further understanding into the ever-expanding field of porous MOPs. We challenge the reader to keep pushing our knowledge and understanding resulting in increasing impact of MOP materials in solving societal and economic issues.

References

1. A. G. Slater and A. I. Cooper, *Science*, 2015, **348**, 8075.
2. T. R. Cook, Y. R. Zheng and P. J. Stang, *Chem. Rev.*, 2013, **113**, 734–777.
3. S. Jiang, J. T. A. Jones, T. Hasell, C. E. Blythe, D. J. Adams, A. Trewin and A. I. Cooper, *Nat. Commun.*, 2011, **2**, 207.

4. J. L. Atwood, J. E. D. Davies, D. D. MacNicol and J. Lipkowski, *Inclusion Compounds Formed by Werner MX2A4 Coordination Complexes*, Academic Press, London, 1st edn, 1984.
5. J. L. Atwood and J. W. Steed, *Encyclopedia of Supramolecular Chemistry*, CRC Press, New York, 2004, vol. 2.
6. D. V. Soldatov, G. D. Enright and J. A. Ripmeester, *Cryst. Growth Des.*, 2004, **4**, 1185–1194.
7. L. R. Nassimbeni, *Acc. Chem. Res.*, 2003, **36**, 631–637.
8. A. J. Cruz-Cabeza, G. M. Day and W. Jones, *Chemistry*, 2009, **15**, 13033–13040.
9. A. J. Cruz-Cabeza, S. Karki, L. Fábián, T. Friscić, G. M. Day and W. Jones, *Chem. Commun.*, 2010, **46**, 2224–2226.
10. B. Ibragimov, *J. Inclusion Phenom. Macrocyclic Chem.*, 1999, **34**, 345–353.
11. L. J. Barbour, D. Das, T. Jacobs, G. O. Lloyd and V. J. Smith, *Supramolecular Chemistry: From Molecules to Nanomaterials*, ed. J. W. Steed and P. A. Gale, John Wiley & Sons Ltd, 2012, pp. 2869–2904.
12. G. O. Lloyd, J. Alen, M. W. Bredenkamp, E. J. C. de Vries, C. Esterhuysen and L. J. Barbour, *Angew. Chem., Int. Ed.*, 2006, **45**, 5354–5358.
13. D. J. Cram, The *Design* of *Molecular Hosts*, *Guests* and *Their Complexes*, Nobel Lecture, Nobelprize.org, 1987, 420–436.
14. M. Mastalerz and I. M. Oppel, *Angew. Chem., Int. Ed.*, 2012, **51**, 5252–5255.
15. L. J. Barbour, *Chem. Commun.*, 2006, 1163–1168.
16. T. Jacobs, V. J. Smith, L. H. Thomas and L. J. Barbour, *Chem. Commun.*, 2014, **50**, 85–87.
17. R. M. Barrer and V. H. Shanson, *J. Chem. Soc., Chem. Commun.*, 1976, 333.
18. J. P. Konopelski, F. Kotzyba-Hibert, J. M. Lehn, J. P. Desvergne, F. Fages, A. Castellan and H. Bouas-Laurent, *J. Chem. Soc., Chem. Commun.*, 1985, 433.
19. H. S. M. Coxter, *Regular Polytopes*, Dover publications Inc, New York, 3rd edn, 1973.
20. *Constitutional Dynamic Chemistry*, ed. M. Barboiu, Springer-Verlag, Berlin Heidelberg, 2012.
21. S. Bélanger, J. T. Hupp, C. L. Stern, R. V. Slone, D. F. Watson and T. G. Carrell, *J. Am. Chem. Soc.*, 1999, **121**, 557–563.
22. R. V. Slone, J. T. Hupp, C. L. Stern and T. E. Albrecht-Schmitt, *Inorg. Chem.*, 1996, **35**, 4096–4097.
23. L. Dobrzańska, G. O. Lloyd, C. Esterhuysen and L. J. Barbour, *Angew. Chem., Int. Ed.*, 2006, **45**, 5856–5859.
24. C. D. Jones, J. C. Tan and G. O. Lloyd, *Chem. Commun.*, 2012, **48**, 2110.
25. T. J. Burchell, D. J. Eisler and R. J. Puddephatt, *Inorg. Chem.*, 2004, **43**, 5550–5557.
26. T. J. Burchell, D. J. Eisler and R. J. Puddephatt, *Chem. Commun.*, 2004, **788**, 944–945.
27. T. J. Burchell, D. J. Eisler, M. C. Jennings and R. J. Puddephatt, *Chem. Commun.*, 2003, 2228–2229.

28. T. J. Burchell, D. J. Eisler and R. J. Puddephatt, *Cryst. Growth Des.*, 2006, **6**, 974–982.
29. T. Jacobs, G. O. Lloyd, J. A. Gertenbach, K. K. Müller-Nedebock, C. Esterhuysen and L. J. Barbour, *Angew. Chem., Int. Ed.*, 2012, **51**, 4913–4916.
30. L. Dobrzańska, G. O. Lloyd, H. G. Raubenheimer and L. J. Barbour, *J. Am. Chem. Soc.*, 2006, **128**, 698–699.
31. J. L. Atwood, L. J. Barbour, A. Jerga and B. L. Schottel, *Science*, 2002, **298**, 1000–1002.
32. G. M. Espallargas, M. Hippler, A. J. Florence, P. Fernandes, J. van de Streek, M. Brunelli, W. I. F. David, K. Shankland and L. Brammer, *J. Am. Chem. Soc.*, 2007, **129**, 15606–15614.
33. S. Bin Baek, D. Moon, R. Graf, W. J. Cho, S. W. Park, T. U. Yoon, S. J. Cho, I. C. Hwang, Y. S. Bae, H. W. Spiess, H. C. Lee and K. S. Kim, *Proc. Natl. Acad. Sci. U. S. A.*, 2015, **112**, 14156–14161.
34. D. J. Tranchemontagne, Z. Ni, M. O'Keeffe and O. M. Yaghi, *Angew. Chem., Int. Ed.*, 2008, **47**, 5136–5147.
35. A. F. Wells and R. R. Sharpe, *Acta Crystallogr.*, 1963, **216**, 857.
36. J. J. Perry, J. A. Perman and M. J. Zaworotko, *Chem. Soc. Rev.*, 2009, **38**, 1400–1417.
37. X. Hang, B. Liu, X. Zhu, S. Wang, H. Han, W. Liao, Y. Liu and C. Hu, *J. Am. Chem. Soc.*, 2016, **138**, 2969–2972.
38. G. Seeber, B. E. F. Tiedemann and K. N. Raymond, *Top. Curr. Chem.*, 2006, **265**, 147–183.
39. S. Leininger, B. Olenyuk and P. J. Stang, *Chem. Rev.*, 2000, **100**, 853–907.
40. T. R. Cook, Y. R. Zheng and P. J. Stang, *Chem. Rev.*, 2013, **113**, 734–777.
41. V. A. Blatov, M. O'Keeffe and D. M. Proserpio, *CrystEngComm*, 2010, **12**, 44–48.
42. M. Eddaoudi, D. B. Moler, H. Li, B. Chen, T. M. Reineke, M. O'Keeffe and O. M. Yaghi, *Acc. Chem. Res.*, 2001, **34**, 319–330.
43. S. R. Seidel and P. J. Stang, *Acc. Chem. Res.*, 2002, **35**, 972–983.
44. P. J. Stang and B. Olenyuk, *Acc. Chem. Res.*, 1997, **30**, 502–518.
45. M. Tonigold, J. Hitzbleck, S. Bahnmüller, G. Langstein and D. Volkmer, *Dalton Trans.*, 2009, **114**, 1363–1371.
46. J. Yang, M. Lutz, A. Grzech, F. M. Mulder and T. J. Dingemans, *CrystEngComm*, 2014, **16**, 5121.
47. B. Moulton, J. Lu, A. Mondal and M. J. Zaworotko, *Chem. Commun.*, 2001, **2**, 863–864.
48. J. R. Li and H. C. Zhou, *Nat. Chem.*, 2010, **2**, 893–898.
49. Y. Ke, D. J. Collins and H. C. Zhou, *Inorg. Chem.*, 2005, **44**, 4154–4156.
50. H. Kim, M. Oh, D. Kim, J. Park, J. Seong, S. K. Kwak and M. S. Lah, *Chem. Commun.*, 2015, **51**, 3678–3681.
51. C. D. Ene, A. M. Madalan, C. Maxim, B. Jurca, N. Avarvari and M. Andruh, *J. Am. Chem. Soc.*, 2009, **131**, 4586–4587.
52. A. D. Burrows, C. G. Frost, M. F. Mahon, M. Winsper, C. Richardson, J. P. Attfield and J. A. Rodgers, *Dalton Trans.*, 2008, **2**, 6788–6795.

53. H. Abourahma, B. Moulton, V. Kravtsov and M. J. Zaworotko, *J. Am. Chem. Soc.*, 2002, **124**, 9990–9991.
54. M. Eddaoudi, J. Kim, J. B. Wachter, H. K. Chae, M. O'Keeffe and O. M. Yaghi, *J. Am. Chem. Soc.*, 2001, **123**, 4368–4369.
55. Z. Ni, A. Yassar, T. Antoun and O. M. Yaghi, *J. Am. Chem. Soc.*, 2005, **127**, 12752–12753.
56. M. J. Bojdys, M. E. Briggs, J. T. A. Jones, D. J. Adams, S. Y. Chong, M. Schmidtmann and A. I. Cooper, *J. Am. Chem. Soc.*, 2011, **133**, 16566–16571.
57. J. R. Holst, A. Trewin and A. I. Cooper, *Nat. Chem.*, 2010, **2**, 915–920.
58. C. Sudik, R. Millward, N. W. Ockwig, P. Cote, J. Kim and O. M. Yaghi, *J. Am. Chem. Soc.*, 2005, **127**, 7110–7118.
59. Y. Bai, D. Guo, C. Y. Duan, D. Bin Dang, K. L. Pang and Q. J. Meng, *Chem. Commun.*, 2004, 186–187.
60. B. Lee and F. M. Richards, *J. Mol. Biol.*, 1971, **55**, 379–400.
61. A. Shrake and J. A. Rupley, *J. Mol. Biol.*, 1973, **79**, 351–371.
62. M. L. Connolly, *J. Appl. Crystallogr.*, 1983, **16**, 548–558.
63. T. J. Richmond, *J. Mol. Biol.*, 1984, **178**, 63–89.
64. M. L. Connolly, *J. Mol. Graphics*, 1993, **11**, 139–141.
65. M. L. Connolly, *Science*, 1983, **221**, 709–713.
66. L. Ma, A. Jin, Z. Xie and W. Lin, *Angew. Chem., Int. Ed.*, 2009, **48**, 9905–9908.
67. A. P. Nelson, O. K. Farha, K. L. Mulfort and J. T. Hupp, *J. Am. Chem. Soc.*, 2009, **131**, 458–460.
68. J. B. DeCoste and G. W. Peterson, *Chem. Rev.*, 2014, **114**, 5695–5727.
69. E. J. Carrington, I. J. Vitórica-Yrezábal and L. Brammer, *Acta Crystallogr., Sect. B: Struct. Sci., Cryst. Eng. Mater.*, 2014, **70**, 404–422.
70. S. Brunauer, P. H. Emmett and E. Teller, *J. Am. Chem. Soc.*, 1938, **60**, 309–319.
71. E. Amayuelas, A. Fidalgo-Marijuán, B. Bazán, M. K. Urtiaga, G. Barandika and M. I. Arriortua, *CrystEngComm*, 2016, **18**, 1709–1712.
72. B. J. Riley, J. D. Vienna, D. M. Strachan, J. S. McCloy and J. L. Jerden, *J. Nucl. Mater.*, 2015, **470**, 307–326.
73. L. J. Barbour, *Acta Crystallogr., Sect. B: Struct. Sci., Cryst. Eng. Mater.*, 2014, **70**, 403.
74. S. A. Moggach, S. Parsons and P. A. Wood, *Crystallogr. Rev.*, 2008, **14**, 143–184.
75. C. E. Wilmer, M. Leaf, C. Y. Lee, O. K. Farha, B. G. Hauser, J. T. Hupp and R. Q. Snurr, *Nat. Chem.*, 2012, **4**, 83–89.
76. J. R. Li, J. Yu, W. Lu, L. B. Sun, J. Sculley, P. B. Balbuena and H. C. Zhou, *Nat. Commun.*, 2013, **4**, 1538.
77. D. Peralta, G. Chaplais, A. Simon-Masseron, K. Barthelet, C. Chizallet, A. A. Quoineaud and G. D. Pirngruber, *J. Am. Chem. Soc.*, 2012, **134**, 8115–8126.
78. B. Jiang, J. Zhang, J. Q. Ma, W. Zheng, L. J. Chen, B. Sun, C. Li, B. W. Hu, H. Tan, X. Li and H. B. Yang, *J. Am. Chem. Soc.*, 2016, **138**, 738–741.

79. C. Zhao, N. Wang, L. Wang, H. Huang, R. Zhang, F. Yang, Y. Xie, S. Ji and J. R. Li, *Chem. Commun.*, 2014, **50**, 13921–13923.
80. J. Ma, Y. Ying, Q. Yang, Y. Ban, H. Huang, X. Guo, Y. Xiao, D. Liu, Y. Li, W. Yang and C. Zhong, *Chem. Commun.*, 2015, **51**, 4249–4251.
81. C. Zhao, N. Wang, L. Wang, H. Huang, R. Zhang, F. Yang, Y. Xie, S. Ji and J. R. Li, *Chem. Commun.*, 2014, **50**, 13921–13923.
82. N. Ahmad, H. A. Younus, A. H. Chughtai and F. Verpoort, *Chem. Soc. Rev.*, 2015, **44**, 9–25.
83. C. J. Hastings, M. D. Pluth, R. G. Bergman and K. N. Raymond, *J. Am. Chem. Soc.*, 2010, **132**, 6938–6940.
84. J. C. Tan and A. K. Cheetham, *Chem. Soc. Rev.*, 2011, **40**, 1059–1080.

CHAPTER 10

Functional Metallo-supramolecular Polyhedral Capsules and Cages

AIDAN J. BROCK, HYDAR AL-FAYAAD, MICHAEL C. PFRUNDER AND JACK K. CLEGG*

School of Chemistry and Molecular Biosciences, The University of Queensland, St Lucia, QLD 4072, Australia
*Email: j.clegg@uq.edu.au

10.1 Introduction

The first metallo-supramolecular polyhedron was reported in 1988 and the design, preparation and use of these materials has captured the imagination and attention of chemists ever since.[1] There are many reasons for the continued growth of this research area. Firstly, metallo-supramolecular materials are inherently beautiful. The combination of metals with various coordination geometries (tetrahedral, square planar, octahedral *etc.*) with organic ligands in which the donor atoms are arranged to produce binding modes with controlled directionality, a wide range of high symmetry cages and capsules have been prepared and characterised. Secondly, as the principles of metallo-supramolecular chemistry were developed it became clear that careful design of the organic components and judicious choice of metal ions could be used to control the size, shape and symmetry of the resulting multi-nuclear architectures. Accordingly, the investigation of this synthetic approach became a champion of supramolecular design, from which

Monographs in Supramolecular Chemistry No. 22
Functional Supramolecular Materials: From Surfaces to MOFs
Edited by Rahul Banerjee

Published by the Royal Society of Chemistry, www.rsc.org

fundamental understanding of self-assembly resulted and was applied to the production, not just of metallo-supramolecular capsules and cages, but also of a variety of other materials created by supramolecular chemists, including helicates, knots, MOFs, molecular machines, *etc.* Finally, metallo-supramolecular capsules and cages, as their names imply, enclose a well-defined space. This space can be accessed by smaller molecules (guests) and, therefore, the careful design of these capsules and cages can result in the creation of spaces of particular shape and size, with suitable interior surface functionality such that they can both display selectivity towards particular guests and catalyse reactions within them. This mode of functional binding and reactivity has often been compared to the way that enzymes function in biological systems. This, of course, results in an almost infinite number of potential applications for these materials, ranging from selective encapsulation and sequestration[2] to high value separations, catalysis[3] and sensing.[4]

Given the vast number of metallo-supramolecular materials so far prepared, we have limited our discussion to capsules and cages that have geometrical features that resemble polyhedra, although other functional non-polyhedral architectures have been prepared.[5] In later sections we focus on describing metallo-supramolecular polyhedra that have been demonstrated to perform in a functional manner, describing the features of the growth of these materials and the current (and future) applications for these remarkable complexes. Before we do though, it is worthwhile considering the basic design principles that are the foundations of the chemistry of functional metallo-supramolecular polyhedra.

10.2 Design Principles

The majority of metallo-supramolecular capsules and cages so far produced have geometrical features that resemble polyhedra, that is, they are three-dimensional objects composed of flat polygonal faces joined by straight edges and vertices. In simple terms, therefore, a metallo-supramolecular polyhedron is produced when the reaction of metal ions and bridging ligands produces a polymetallic complex where the arrangement of apex positions defines a polyhedron. The apices of the assembly may be defined by either metallic or organic components. Metallo-supramolecular capsules and cages can therefore be considered a sub-class of metallo-supramolecular polyhedra where the internal cavity of the polyhedron is of sufficient size to be accessible to a smaller molecule. As we have only selected capsules and cages to feature in this chapter we will also use the general term polyhedra for these materials.

The potential for the self-assembly of diverse metallo-supramolecular capsules and cages is immense. The particular geometry of any one metallo-supramolecular architecture results from the specific combination of the directional constraints of the organic bridging ligands and the angles produced at the coordinated metal ions. Due to crystal packing effects and molecular motion it is common for the geometries of metallo-supramolecular

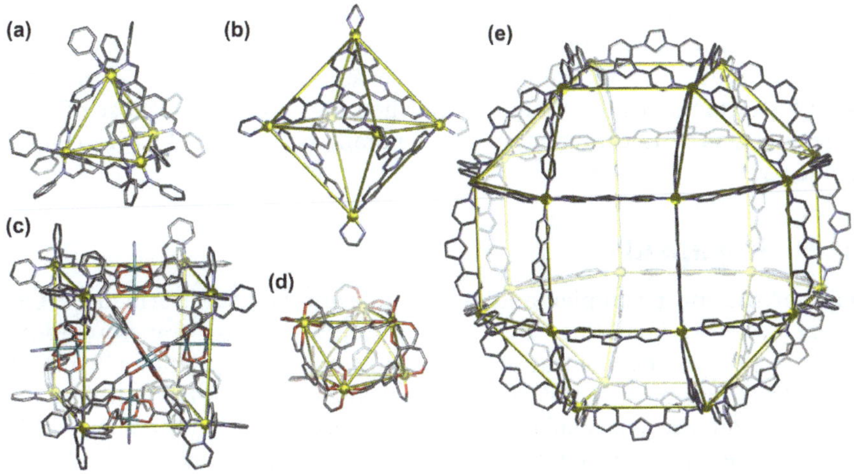

Figure 10.1 Selected examples of metallo-supramolecular polyhedra, (a) an [M_4L_6] tetrahedron, (b) an [M_6L_4] octahedron, (c) an [M_8L_6] cube, (d) an [M_6L_4] trigonal anti-prism, and (e) an [$M_{24}L_{48}$] rhombicuboctahedron.

architectures to deviate from strict mathematical definitions of regular polyhedra. It is therefore common in the literature to describe the geometry of a particular molecular architecture by the polyhedral shape that it most closely approximates. A selection of metallo-supramolecular polyhedra are presented in Figure 10.1.[6–8]

Many approaches have been employed for the design of self-assembled metallo-supramolecular capsules and cages in a systematic fashion.[9–12] These include the "metal-directed", "coordination vector", "symmetry interaction" and "metallo-ligand" approaches.[11,13] While each of these design methods have subtle differences, each can be reduced to the consideration of the summation of the inherent steric, electronic, geometric and chemical information present in each of the sub-components which come together to form the desired architecture.[9] Or, to put it in another way, molecular components (*i.e.* metals and ligands) are selected such that after combination they will confer the symmetry elements of the point symmetry of the desired polyhedron. For example, a tetrahedron can by constructed by incorporating metals and ligands that will give rise to the production of the necessary combination of C_2 and C_3 axes of tetrahedral point groups. By extending this approach, a wide variety of geometric architectural types, including tetrahedra, cuboids, cubes, spheres and prisms, have been produced. The synthesis and design features of metallo-supramolecular polyhedra have been extensively reviewed[14] and are therefore not discussed in detail here. We instead focus on the potential uses and applications of these fascinating and beautiful assemblies.

As the presence of an accessible central cavity is a defining feature of this class of materials, it is perhaps not surprising that their resulting functions arise from the encapsulation of smaller molecules within the micro-environment defined by the metallo-supramolecular superstructure. In other words, the functions that arise as a consequence of the self-assembly of metallo-supramolecular polyhedra as opposed to those of their individual synthetic components can generally be described as host–guest chemistry.

10.3 Encapsulation

Many of the most complex and remarkable functions performed by biological systems, including the highly specific catalytic transformations achieved by enzymes, the transcription and transfer of genetic information within the ribosome and the storage and transport of molecules around the body, rely on the encapsulation of one molecule (or molecules) within another. The guest molecule can be completely or partially surrounded by the host; the encapsulated molecule is contained within a volume of space that is chemically and physically distinct from the conditions outside. The isolation of the guest from the bulk environment can produce substantially different behaviour to that otherwise observed in solution, leading to new modes of reaction of the trapped guests, producing unusual and potentially useful outcomes. Changes in the chemical behaviour of the encapsulated molecules often arise from the presence of a more restricted set of interactions between the guest and the host or a small number of other guest molecules compared to the random collisions with other molecules that normally occur in bulk solution. Encapsulation within such a micro-environment can also lead to interesting changes in the physical properties of guest(s), including inducing unusual spin-states,[15] disfavourable conformations[16] and exposing the guest(s) to more elevated concentrations and/or pressures than would readily be accessible under ambient conditions.[17] Metallo-supramolecular capsules and cages are only a small subset of the types of artificial materials that act as functional host molecules in this way (which are often collectively described as "container" molecules or "molecular flasks"). Others examples include zeolites,[18] metal-organic frameworks (MOFs),[19] and purely organic capsules[20] including cavitands,[21] cyclodextrins[22] and cucurbiturils,[23] as well as nanotubes[24] and fullerenes.[25]

The observation of a guest within a host raises a number of fundamental questions related directly to the processes of host–guest chemistry which have been a focus of many investigations – how did the guest get in? Can it get out? How strongly is it bound? What about the kinetics of this process? While a detailed discussion of all of these factors is beyond the scope of this chapter, it is worth commenting on the first two questions, which can be collectively grouped together – what is the mechanism of guest exchange? The answer will depend on the properties of each metallo-supramolecular host and its guest, however, we can suggest that there are two general classes of guest exchange modes; the guest can either move in and out of a host through any open pores (or faces) within the host (termed "through-face" exchange), or the host can reversibly open and close capturing the guest normally through a de-ligation event ("de-ligation" exchange). Studies of the kinetics of these processes can give some insight into the particular mechanism that occurs for any particular host-guest system.

A detailed discussion of functional metallo-supramolecular polyhedra follows. Functions discussed include selectivity, separation and sequestration (Section 10.4) including guest binding and reaction modulation (including both catalysis and inhibition) (Section 10.5). The process of encapsulation and selective guest binding can also lead to a variety of emerging functions such as stimuli responsive behaviour which is highlighted in each relevant section.

10.4 Selectivity, Separations and Sequestration

Because the space within metallo-supramolecular capsules and cages is defined by the molecular components employed in the self-assembly formation reaction, it follows that the chemical and physical properties of the enclosed volume are directly related to those of the components that make up the structure. By changing the size or chemical nature of the organic groups (or even by changing the metal ion) the volume, shape and chirality of the encapsulated space can be varied. The chemical nature of the walls of metallo-supramolecular polyhedra can be tailored to affect the polarity or charge surrounding the internal cavity and ligands can be designed to incorporate guest-recognition or catalytically-active functional groups into the capsule or cage. Similarly, the "openness" of the structure, that is, the relative porosity of faces compared to the magnitude of encapsulated volume can be readily controlled. Varying this parameter is expected to result in a trade-off between the specificity of guest binding and rates of guest exchange. In other words, the more "open" a structure, the more exposure any bound guest molecule has to the external environment and the lower the potential for stabilising host–guest interactions. This should, in turn, allow for any guests to move more readily in between the encapsulated and bulk environment, increasing the rates of exchange, while also potentially rendering the host molecule more promiscuous towards a variety of guests. Conversely a more "closed" host would be expected to bind guests with higher degrees of specificity and slower rates of exchange.

Systematic changes in the properties of the host and careful selection of guests, in turn, can be used to investigate fundamental factors that influence the selectivity and strength of guest binding, the kinetics of guest exchange and the type (and mechanism) of reactions that can be catalysed inside the capsules' cavities. The resulting materials often display relatively high binding constants and, in many cases, exquisite selectivity for one particular guest compared to others of a similar chemical nature. A number of systems have also been prepared in which more than one guest molecule can be bound inside a cage simultaneously; some metallo-supramolecular capsules also display solvent-dependent functionality. There is an almost infinite scope to the type of guest molecules that can be encapsulated in polyhedra of this type, including neutral and charged molecules, anti-cancer drugs, small proteins, nanoparticles, gases and even molecules that would otherwise be unstable under similar conditions. In the following sections we discuss examples of the encapsulation of gases, anions, cations and other neutral small molecules within metallo-supramolecular capsules and cages.

10.4.1 Anion Recognition

Anions are omnipresent. Chloride, for example, is a significant natural component of oceans and also a key constituent in biological fluids. Phosphates and nitrates are used extensively as fertilisers and perchlorates have been used in the manufacture of explosives for many years. Inorganic anions, in particular, are also significant contributors to pollution including acid rain, agricultural run-off and dangerous products of the nuclear fuel cycle. Biomineralisation resulting in the formation of bones, teeth, shells and exo-skeletons relies on the presence of carbonates, silicates and phosphates in living systems. Indeed, anions are crucial for life; even DNA is anionic and a large proportion of cofactors and enzyme substrates are negatively charged. If anions are inefficiently transported around the body or their concentrations are ineffectively regulated through malfunctioning proteins or genetic mutations, lethal consequences can result. Cystic fibrosis is an example of such a disease, which arises from malfunctioning trans-membrane chloride channels and receptors.[26]

Not surprisingly, therefore, the design and preparation of synthetic anion receptors (sometimes called anion coordination chemistry[27]) has received a great deal of interest from the chemical community in the last 50 years. As many metallo-supramolecular polyhedra have been prepared from the combination of charge-neutral organic components (ligands) with metallic cations it follows that the resulting assemblies carry positive charges and are, by definition, associated with anions. The presence of both a defined cavity and the cationic nature of the polyhedron have led a number of groups to investigate the use of these architectures as selective anion-binding agents.[2] There are a number of other examples of metallo-supramolecular capsules that display anion-binding behaviour which do not fall into the category of polyhedral architectures, and have not been reviewed here.[2,28]

In an early and very significant study the group of Fujita combined six equivalents of the *cis*-capped ethylenediamino-palladium(II) complex **1** with four equivalents of the tritopic tripyridyltriazine ligand (tpt, **2**) in water, resulting in the self-assembly of an $[Pd_6 2_4]^{12+}$ octahedron. The palladium ions are located at the vertices of the assembly and four of the eight faces are panelled by the electron-poor tpt ligands, while the other four are open to allow guest entry. X-ray and solution ^1H NMR studies showed that the cationic polyhedron bound four adamantylcarboxylate molecules within its central cavity.[7]

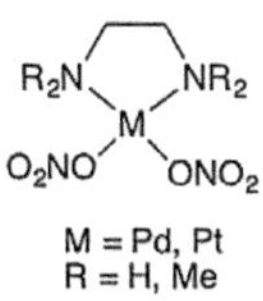

1

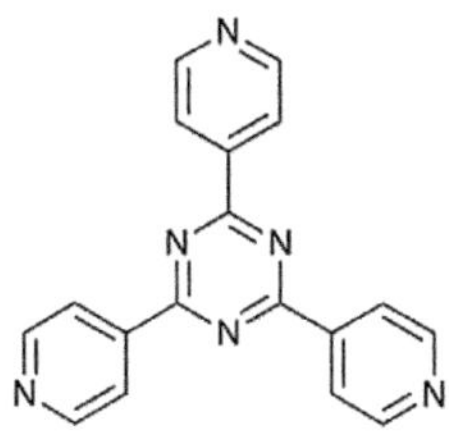

2

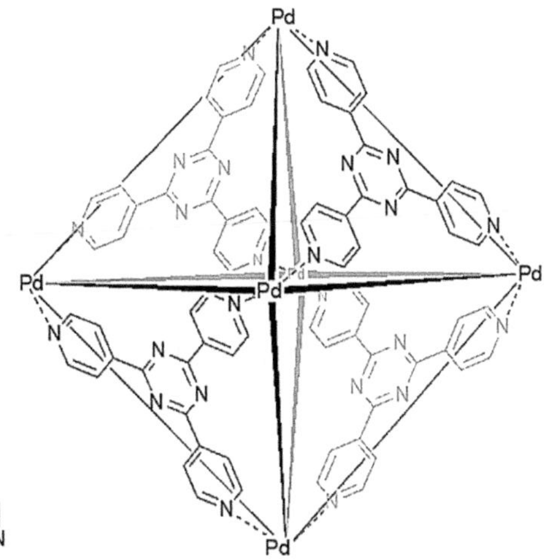

$[Pd_6\mathbf{2}_4]^{12+}$

Soon afterwards, the group of Ward reported the first example of anion-binding within an $[M_4L_6]$ metallo-supramolecular cage.[29] The reaction of six equivalents of the *bis*-pyridylpyrazole ligand **3** with four equivalents of cobalt(II) acetate followed by the addition of sodium tetrafluoroborate ($NaBF_4$) resulted in the formation of a $[Co_4\mathbf{3}_6]^{8+}$ metallo-supramolecular tetrahedron.

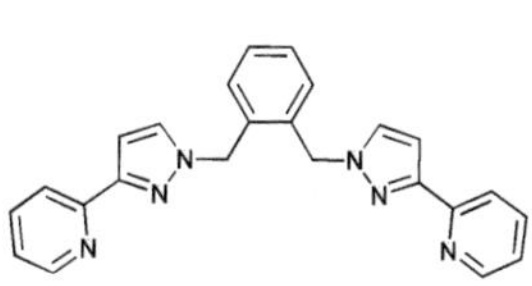

3

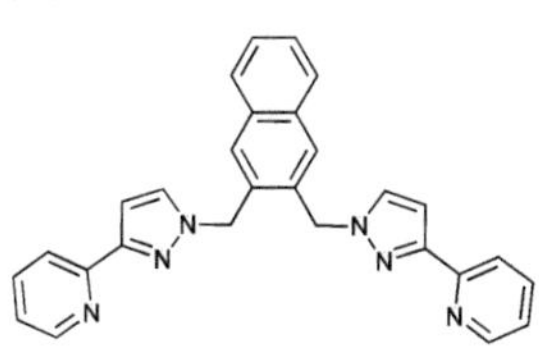

4

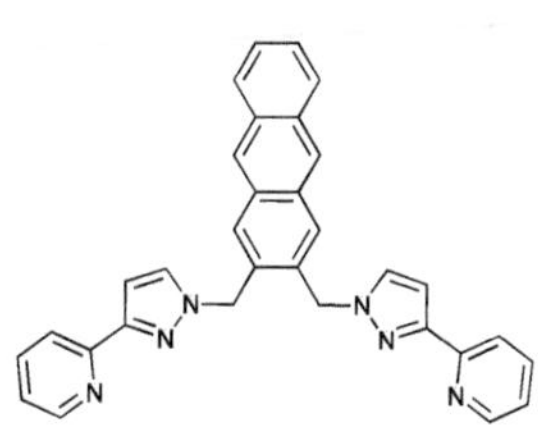

5

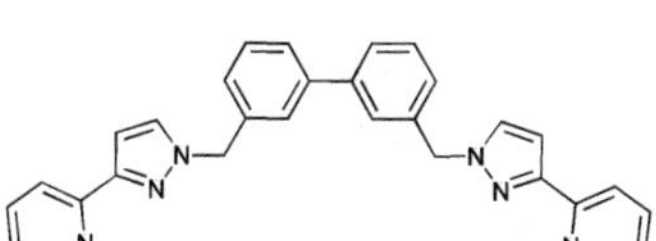

6

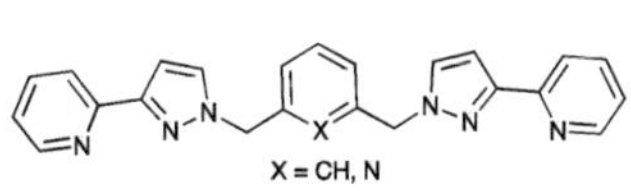

7

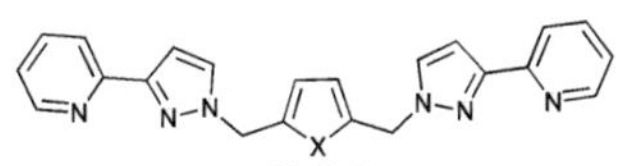

8

^{19}F and ^{11}B NMR spectra showed that the BF_4^- anions were present in two different environments in a 1 : 7 ratio that were in slow exchange on the NMR timescale, consistent with an observation of BF_4^- encapsulation previously described by Hüttner *et al.*[30] The X-ray crystal structure (Figure 10.2) demonstrated that the anion was encapsulated within the central cavity

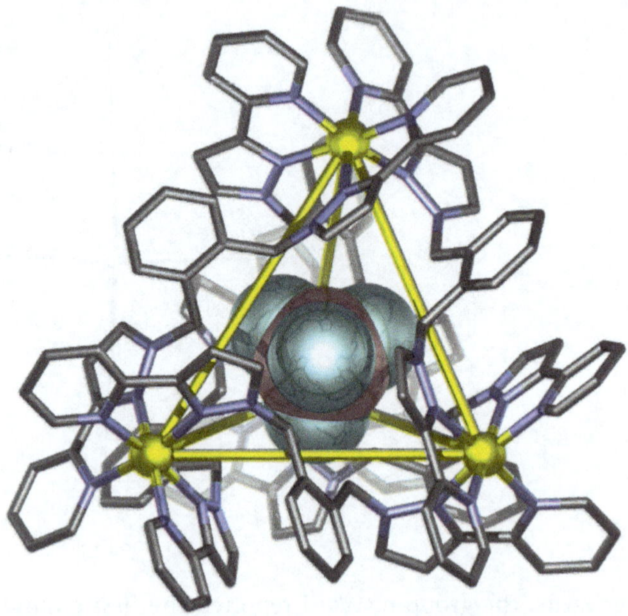

Figure 10.2 The crystal structure of $[BF_4{\subset}Co_4\mathbf{3}_6]^{7+}$. The encapsulated anionic guest is shown as a space-filling model. The fluorine atoms point towards the windows of the metallo-supramolecular tetrahedron.

defined by the metallo-supramolecular tetrahedron. The four fluorine atoms were arranged to point towards the faces of the tetrahedral assembly rather than the four metal centres, leading the authors to conclude that size and shape complementarity was the predominant driving force for the strong guest binding rather than electrostatic attraction between the cationic metals and anionic guest. It was suggested that the encapsulated anion was acting as a template for the formation of the tetrahedral cage.

Changing the ligand backbone from *o*-phenylene (**3**) to 2,3-napthalene (**4**) or 2,3-anthracene (**5**) resulted in the formation of similar $[M_4L_6]^{8+}$ tetrahedra when either cobalt(II) or zinc(II) was employed.[31] Based on solid-state studies small anions such as perchlorate (ClO_4^-) and tetrafluoroborate could be successfully bound inside these cages but the larger hexafluorophosphate (PF_6^-) anion was not encapsulated. Interestingly, however, the large dianion hexafluorosilicate (SiF_6^{2-}) was located inside the cavity of the $[Zn_4\mathbf{4}_6]^{8+}$ cage. Perhaps the higher electrostatic attraction resulting from the dianionic guest and cationic host is sufficient to overcome any of the steric effects that prevented the binding of PF_6^-. Kinetic studies suggested that the mechanism of guest exchange was a through-face process as guest diffusion rates were significantly faster than those observed for de-ligation processes.

Increasing the separation between pyridylpyrazole moieties through the use of a 3,3′-biphenyl-spacer produced a series of larger C_3-symmetric tetrahedral $[M_4\mathbf{6}_6]^{8+}$ (where M = Co(II), Zn(II), Cd(II), Cu(II) and Hg(II)) cages.[32]

The size of the enclosed cavity was found to have increased so that hexafluorophosphate could now be encapsulated, as could iodide and the smaller perchlorate and tetrafluoroborate. The larger size of the pores in the faces of the tetrahedron resulted in the guest molecules diffusing quickly on the NMR timescale.

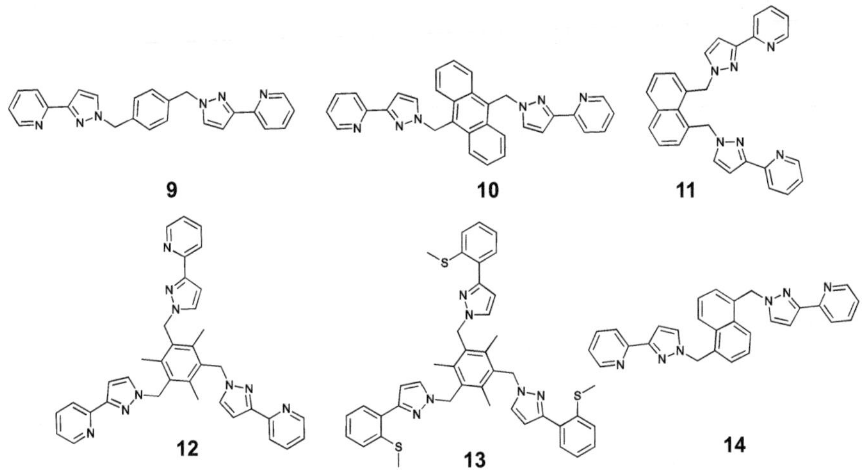

9 10 11 12 13 14

In subsequent studies, changing the directional properties of the ligands by using either a *m*-substituted 6-membered ring (**7**) or 2,5-disubstituted 5-membered ring (**8**) resulted in a remarkable shift in structure. The reaction of these ligands with cadmium(II), zinc(II) or copper(II) was found to produce $[M_8L_{12}]^{16+}$ cubic architectures. The resulting polyhedra were found to encapsulate either BF_4^- or ClO_4^-. When **7** (X=N) was used only one perchlorate was found to be encapsulated while the analogous structure formed from **7** (X=CH) bound two BF_4^- guests. Similar $[M_8L_{12}]^{16+}$ cubic structural motifs were observed when **10** and **14** were used; the cavity within $[M_8\mathbf{10}_{12}]^{16+}$ was found to be sufficiently large to bind one tetrafluoroborate and a number of solvent molecules, however, the X-ray structure of $[M_8\mathbf{14}_{12}]^{16+}$ did not reveal any anion-binding behaviour.[33]

The reaction of **9** or **11** with a variety of divalent metal ions in a 3 : 2 ratio produced higher-order polyhedra than previously observed, both of which enclosed substantial cavities. While **11** produced $[M_{12}L_{18}]^{24+}$ truncated tetrahedra that bound four BF_4^- or ClO_4^- guests, **9** yielded $[M_{16}L_{24}]^{32+}$ tetracapped truncated tetrahedral polyhedra that encapsulated eight ClO_4^- anions and six acetonitrile solvent molecules! A similar number of anionic guests (eight ClO_4^- anions and four water molecules) were found to bind within the cavity defined by an $[Cd_6L_4]^{12+}$ octahedron formed from the C_3-symmetric ligand **13**. An even more elaborate cuboctahedral polyhedron resulted when ligands **11** and **12** were combined in a reaction with cadmium(II). The resulting $[Cd_{12}\mathbf{11}_{12}\mathbf{12}_4]^{24+}$ structure encapsulated "numerous counter-ions and solvent molecules". No attempts to quantify the anion-binding of the many metallo-supramolecular polyhedra prepared by Ward *et al.* discussed above have been reported.

In 2008, the group of Lindoy first attempted to qualitatively probe anion-binding selectivity within a metallo-supramolecular tetrahedron. The reaction of the quaterpyridine ligand **15** ($n = 0$) with iron(II) in acetonitrile produced a $[Fe_4\mathbf{15}_6]^{8+}$ tetrahedron with overall T-symmetry.[34] The single crystal structure of this material (Figure 10.3) showed that each of the metal centres within each self-assembled molecule have the same chirality, although both enantiomeric forms are present in the crystal lattice. X-ray analysis and solution studies showed that the cage was capable of binding a variety of anions including BF_4^-, PF_6^- and $FeCl_4^-$.[34,35] The polyhedron was demonstrated to be selective for PF_6^- over BF_4^-; the addition of hexafluorophosphate to a solution of $[BF_4 \subset Fe_4\mathbf{15}_6] \cdot 7BF_4$ resulted in the quantitative displacement of the encapsulated tetrafluoroborate. Similarly, the addition of excess BF_4^- to a solution of $[PF_6 \subset Fe_4\mathbf{15}_6] \cdot 7PF_6$ did not displace the bound guest. Significantly, by preparing the cage with a bulky $[ZnCl_4]^{2-}$ anion it was demonstrated that the tetrahedron could be formed "empty" of an encapsulated anion – only disordered acetonitrile solvent molecules were located within the cage. This was an interesting finding, as it demonstrated that the anionic guest molecules did not play a templating role in the formation of the polyhedron, which paved the way for later quantitative studies by other groups.

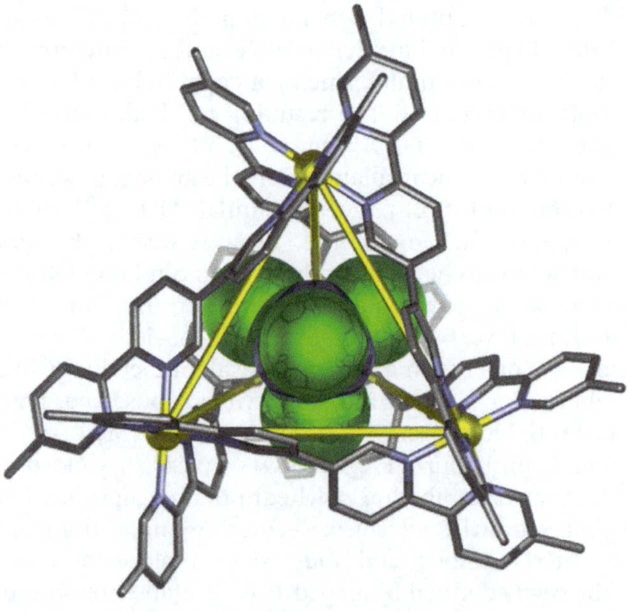

Figure 10.3 Crystal structure of the $[FeCl_4 \subset Fe_4\mathbf{15}_6]^{7+}$ (where $n = 0$) homochiral tetrahedron. The encapsulated anionic guest is shown as a space-filling model. The chlorine atoms point towards the windows of the cage.

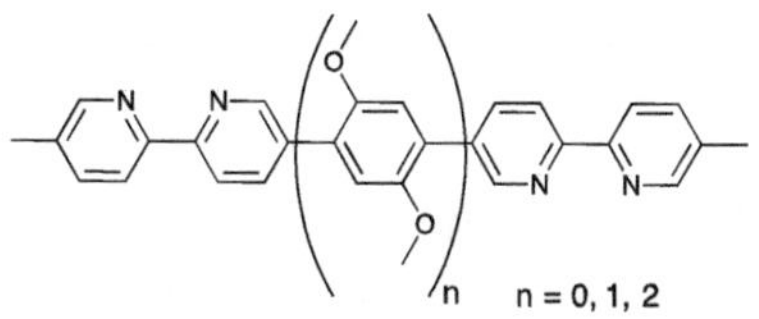

15

Increasing the length of the ligand by including 2,5-dimethyoxyphenylene spacers increased both the volume within the cavity and the area of the windows of the faces, making the structure more open.[36] The crystal structure of $[Fe_4\mathbf{15}_6]^{8+}$ ($n=1$) demonstrated that the accessible void volume within the tetrahedron increased to 227 Å^3 compared to 102 Å^3 for $[Fe_4\mathbf{15}_6]^{8+}$ ($n=0$). This volume is large enough to simultaneously bind both a PF_6^- anion and a disordered solvent molecule. Further increasing the spacer length resulted in the formation of $[Fe_4\mathbf{15}_6]^{8+}$ ($n=2$) with an encapsulated volume of 417 Å^3, while substituting the iron(II) with nickel(II) further increased the cage volume to 839 Å^3! ^{1}H NMR studies showed that this cage was sufficiently large to bind BPh_4^- anions, and the faces of the assembly are so large that these substantial guest molecules diffuse quickly through the faces of the cage on the NMR timescale. Notably, changes in the fluorescence spectrum indicate that BPh_4^- binding led to a reduction in solvent mediated quenching of the tetrahedron.

While each of the tetrahedral capsules discussed so far are formed from the combination of four octahedral metal ions and six bis-bidentate C_2-symmetric ligands resulting in a structure with $[M_4L_6]$ stoichiometry, it is also possible to form a molecular tetrahedron from four tris-bidentate ligands with idealised C_3-symmetry and four octahedral metals. In contrast to the $[M_4L_6]$ structures which have "open" faces as the ligands define the edges of the polyhedron, the resulting $[M_4L_4]$ tetrahedra are "face-capped" as the ligands are situated on the faces of the prismatic assembly. Thus, the reaction of ligands **16** or **17** in a 1 : 1 ratio with iron(II) produced two distinct $[Fe_4L_4]^{8+}$ tetrahedra, which were shown to bind anions.

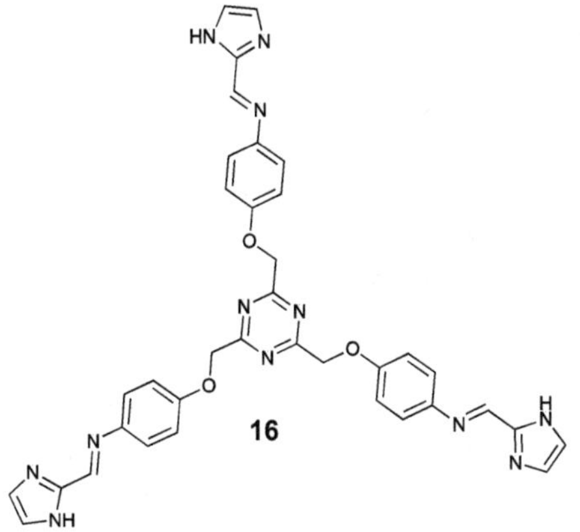

16

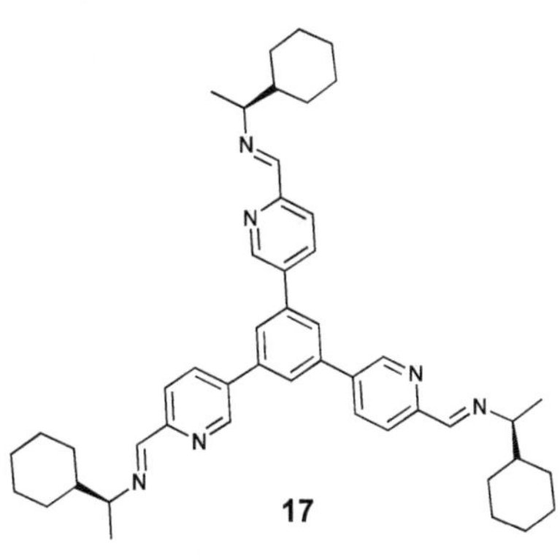

17

The $[Fe_4\mathbf{16}_4]\cdot 8BF_4$ tetrahedron, prepared by the group of Kruger,[37] was shown to have an Fe–Fe separation of 14.6 Å and enclose a volume of 106 $Å^3$. A single tetrafluoroborate anion was found to be encapsulated within the central cavity. Each of the fluorine atoms is directed towards a face of the tetrahedron, resulting in a strong interaction between each fluorine and the electron-deficient triazine rings of the ligand (Figure 10.4). The twelve imidazole groups on the periphery of the capsule act as hydrogen bond donors, binding the remaining seven tetrafluoroborate anions on the exterior of the polyhedron and forming a network of hydrogen bonds throughout the crystal lattice. The $[Fe_4\mathbf{16}_4]^{8+}$ also showed interesting spin-crossover and LIESST effects.

The ligand (**17**) used to form the $[Fe_4\mathbf{17}_4]^{8+}$ tetrahedron is more rigid than that just discussed (**16**) and the reduced flexibility increases the volume of the central cavity to 133 $Å^3$ despite the observation of a Fe–Fe separation of 11.9 Å. This larger cavity was found to house a single triflate (OTf^-) anion by both X-ray crystallography and ^{19}F NMR. The presence of the resolved stereocentres in the ligand resulted in the $[Fe_4\mathbf{17}_4]^{8+}$ polyhedron displaying *T*-symmetry, with each of the iron(II) centres displaying the same (Δ) chirality. The $[Fe_4\mathbf{16}_4]^{8+}$ capsule, in contrast, contained two Δ and two Λ stereocentres.

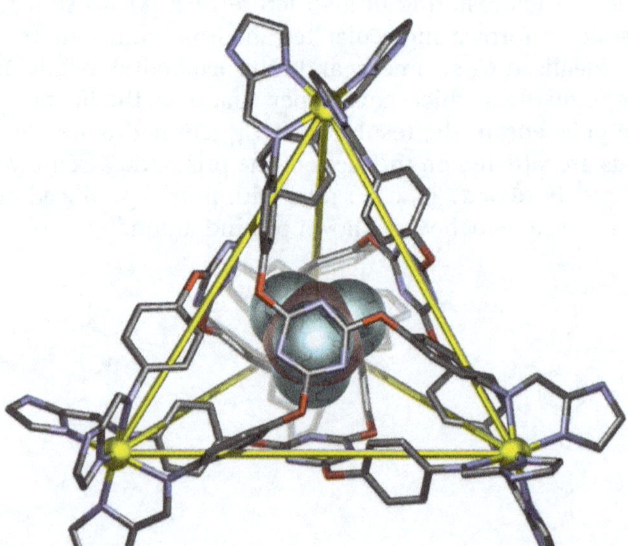

Figure 10.4 Crystal structure of the $[BF_4\subset Fe_4 7_4]^{7+}$ spin-crossover tetrahedron. The encapsulated anionic guest is shown as a space-filling model. The fluorine atoms point towards the electron-deficient walls of the cage.

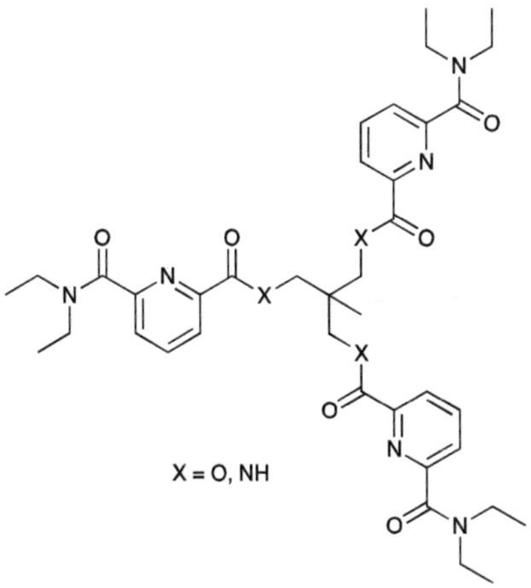

18

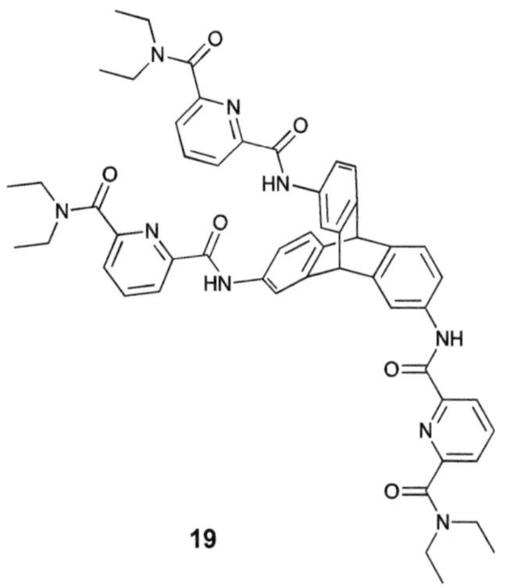

19

By employing three 1,3-dicarbonyl pyridine moieties as metal coordination groups as part of the tripodal ligands **18** and **19**, Hamacek *et al.* were able to form face-capped $[M_4L_4]^{12+}$ tetrahedra with lanthanide ions situated at the vertices.[38,39] X-ray structures of both the $[Eu_4\mathbf{18}_4]^{12+}$ and the larger $[Eu_4\mathbf{19}_4]^{12+}$ polyhedron, crystallised in the presence of perchlorate, demonstrated the encapsulation of an anion in both cases. Like $[Fe_4\mathbf{17}_4]^{8+}$ there are no specific interactions present between the walls of the cage and the anion and the arrangement is stabilised mainly by weak electrostatic interactions. ^{35}Cl NMR studies on $[Eu_4\mathbf{18}_4]^{12+}$ confirmed that the arrangement persists in solution, as two different signals in an 11 : 1 ratio, corresponding to ClO_4^- anions inside and outside the cage, were observed to change slowly on the NMR timescale. The addition of tetrafluoroborate, imidazolate (im^-) or iodide anions resulted in the displacement of the perchlorate guests with the hierarchy of binding found to be $I^- > im^- > ClO_4^- > BF_4^-$. The larger $[Eu_4\mathbf{19}_4]^{12+}$ cage[39] was found to bind perchlorate and tetrafluoroborate as well as the larger triflate anions, however, solution studies indicated fast exchange of guests due to the larger pore windows. Interestingly, guests were seen to bind inside this cage with a different specificity compared to the smaller analogue, such that $BF_4^- > ClO_4^- \sim OTf^-$.

The group of Hardie have investigated a series of coordination cages based on the pyridyl-substituted cyclotriveratrylene ligand **20**.[40,41] Of particular note is the combination of this ligand with palladium(II) which resulted in the formation of a highly symmetric stella octanuclear prism $[Pd_6\mathbf{20}_8]^{12+}$. This cage has a diameter of approximately 3.1 nm and eight windows on the surface of the prism to allow guest exchange. ^{1}H NMR studies in DMSO demonstrated that a series of detergent molecules could be bound within the cage cavity. Two molecules of each of octyl-, dodecyl-, or tetradecyl sulfate could be taken up into the cage with the smallest guest bound approximately five times more strongly than the largest.[41]

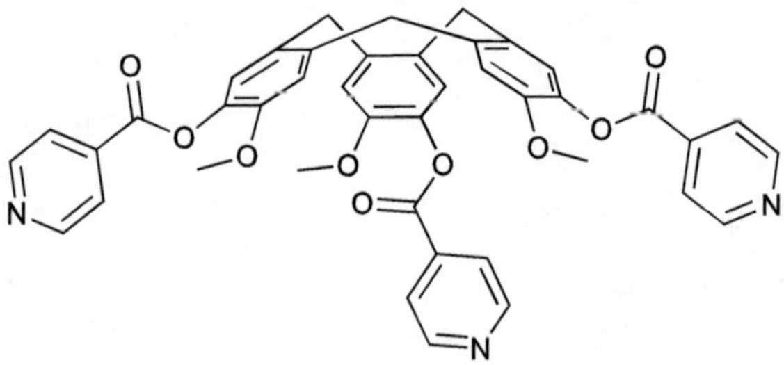

20

Each of the metallo-supramolecular polyhedra discussed above have cavities that are lined with comparatively apolar organic functional groups resulting in hydrophobic cavities that are suitable for binding anions from the lipophilic range of the Hofmeister series – that is those that are large, monovalent and have low solvation energies. Accordingly, the anion-binding investigations have been conducted in organic solvents. The binding of harder and/or multivalent anions, particularly those of biological or environmental relevance, in aqueous solutions is a more significant challenge as significant free energies of hydration need to be overcome for a binding event to occur.

To design a metallo-supramolecular cage capable of selectively binding hydrophilic tetrahedral oxoanions such as sulfate or phosphate in aqueous solutions Custelcean and Hay used a computer-aided design strategy to prepare the urea functionalised ligand **21**.[42,43] The design of the ligand is such that after reaction with a suitable octahedral metal ion such as nickel(II) or zinc(II) an $[M_4\mathbf{20}_6]^{8+}$ tetrahedron would result in which the internal cavity would be surrounded by six inward facing urea groups, allowing twelve hydrogen bond donors to be orientated to saturate the hydrogen bond requirements of an encapsulated anion.

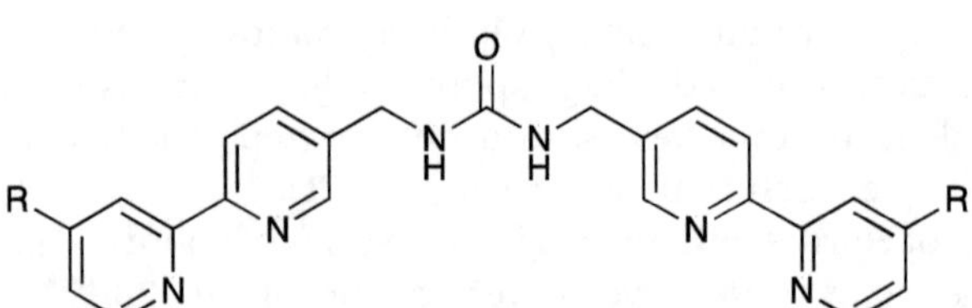

21

The X-ray structure of $[SO_4{\subset}Ni_4\mathbf{21}_6]\cdot 3SO_4$ (Figure 10.5) confirmed that the designed molecule functioned as predicted, with the bound guest saturated by hydrogen bonds. Competitive gravimetric experiments involving $BaSO_4$ in water allowed the measurement of the apparent sulfate binding constant within $[Ni_4\mathbf{21}_6]^{8+}$, which was determined to be greater

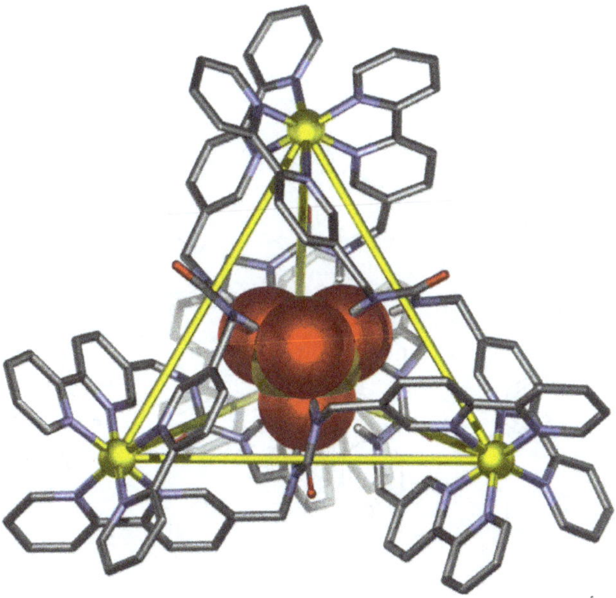

Figure 10.5 Crystal structure of the computer-designed $[SO_4 \subset Ni_4\mathbf{21}_6]^{6+}$ tetrahedron. The encapsulated anionic guest is shown as a space-filling model. The oxygen atoms point towards the faces of the capsule. The bound sulfate ion is stabilised by twelve hydrogen bond donors incorporated into the edges of the tetrahedron.

than 6×10^6 M^{-1}. So precise was the computer-aided design process that the presence of a tetrahedral oxoanion $EO_4^{\,n-}$ (E = S, Se, Cr, Mo, W, P; $n = 2, 3$) was necessary for the capsule to form; the guest molecule templated the self-assembly reaction which generated the polyhedron. No cage formation was observed when anions with different shapes were employed. Anions tested included F^-, Cl^-, Br^-, I^-, NO_3^-, BF_4^-, ClO_4^-, ReO_4^-, PF_6^-, OAc^-, $CH_3SO_3^-$, OTf^-, CO_3^{2-}, SO_3^{2-} and SeO_3^{2-}. Quantitative ^{77}Se NMR studies found that the anion's hydrogen bond accepting ability (*i.e.* charge density and Lewis basicity) rather than size determined their binding strength. The following trend was established: $PO_4^{3-} > CrO_4^{2-} > SO_4^{2-} > SeO_4^{2-} > MoO_4^{2-} > WO_4^{2-}$.[43] An analogous cage, in which the bipyridine groups were substituted with pyridyl-imine moieties, unsurprisingly produced similar results.[44]

The potential mechanism of guest exchange in the $[SO_4 \subset Ni_4\mathbf{21}_6]\cdot 3SO_4$ system was examined using molecular dynamics (MD) simulations.[43] The results suggested that guest exchange occurred *via* a "through-face" process, that is, the encapsulated sulfate could be ejected through conformational distortion of the cage and did not require the rupture of a metal–ligand bond.

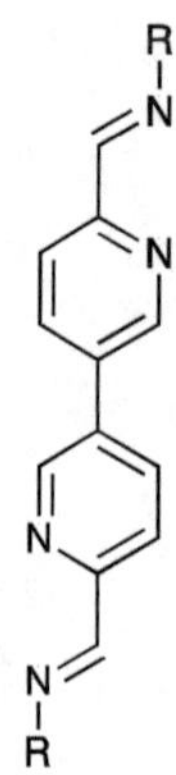

22

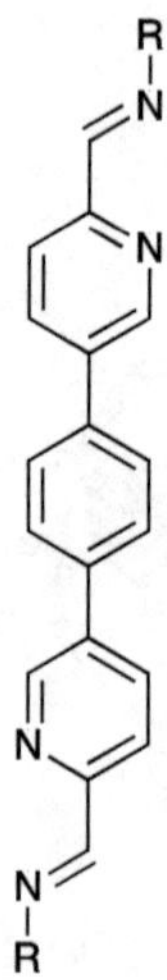

23

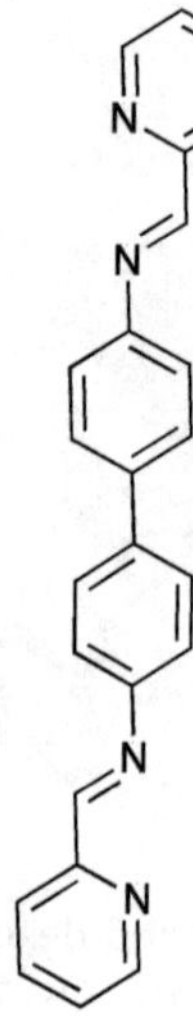

24

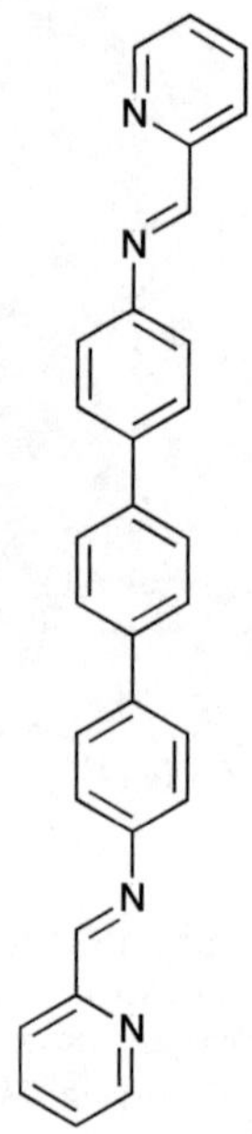

25

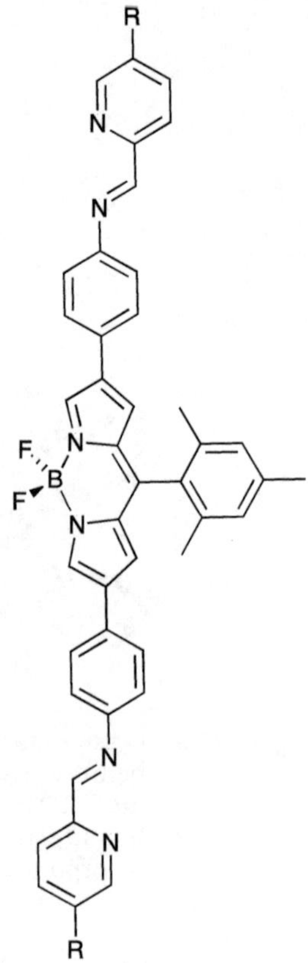

26

The group of Nitschke have extensively investigated the use of pyridylimine ligands as organic components for the preparation of metallo-supramolecular polyhedra, many of which have been shown to selectively encapsulate anions. For example, the reaction of six equivalents of **22** with iron(II) triflimide produced an $[Fe_4\mathbf{22}_6]^{8+}$ tetrahedron which was isolated without an encapsulated anionic guest.[45] It was reasoned that the triflimide anion (NTf_2^-) was too large (156 Å^3) to fit inside the cavity of the polyhedron; this result also served to demonstrate that the cage could be formed without the use of an anionic template in a similar fashion to $[Fe_4\mathbf{15}_6]^{8+}$ ($n = 0$). The isolation of an "empty" cage allowed for the absolute determination of anion-binding constants in acetonitrile (Table 10.1). Differences in selectivity were accounted for by considering the volume of the encapsulated guest; the highest affinity was observed when hexafluorophosphate was employed – larger and smaller anions bound less strongly. When the longer ligand **23** was used in a parallel set of experiments a substantially larger tetrahedron $[Fe_4\mathbf{23}_6]^{8+}$ resulted.[46] This cage was now observed to

Table 10.1 Anion volumes and anion-binding constants within capsules $[Fe_4\mathbf{22}_6]^{8+}$ and $[Fe_4\mathbf{23}_6]^{8+}$.

Anion	Anion volume (Å^3)	K_a for $[Fe_4\mathbf{22}_6]^{8+}$ (M^{-1})	K_a for $[Fe_4\mathbf{23}_6]^{8+}$ (M^{-1})
ClO_4^-	55	5.7(2)×10^5	1.4(5)×10^4
BF_4^-	53	2.3(4)×10^4	4.0(1)×10^3
OTf^-	85	5.2(8)×10^4	3.0(1)×10^3
PF_6^-	75	1.3(5)×10^6	2.8(1)×10^3
NTf_2^-	156	no binding observed	1.4(5)×10^3

encapsulate triflimide as well as the other anions that were bound within the smaller analogue (Table 10.1). It is worth noting that the same size-discrimination effects were not observed within the larger capsule and each of the guests investigated were bound with similar affinities (although perchlorate was now the most competent guest).

With these two cage molecules in hand, and the binding constants determined, a sequential anion-exchange chain reaction was developed (Figure 10.6).[46] Commencing with a mixture of the two cages in acetonitrile (Equilibrium A), triflimide anions were bound exclusively inside the larger cage ($[NTf_2 \subset Fe_4\mathbf{23}_6]^{7+}$) while the smaller cage remained "empty" ($[Fe_4\mathbf{22}_6]^{8+}$). Addition of perchlorate resulted in the formation of Equilibrium B with the ClO_4^- trapped within the smaller cage ($[ClO_4 \subset Fe_4\mathbf{22}_6]^{7+}$) while the triflimide remained inside the larger cage ($[NTf_2 \subset Fe_4\mathbf{23}_6]^{7+}$). The addition of hexafluorophosphate to this mixture produced Equilibrium C in which the perchlorate guest had been displaced from within the smaller cage by the more strongly bound PF_6^- to produce ($[PF_6 \subset Fe_4\mathbf{22}_6]^{7+}$). The free perchlorate in solution then displaced the triflimide from the larger cage to produce ($[ClO_4 \subset Fe_4\mathbf{23}_6]^{7+}$). Due to the similar binding constants the anion selectivity in this chain reaction was not perfect, however, the demonstration of the use of sequential equilibria and binding events points the way towards the development of more elaborate chemical systems that might approach the complexity of those used in biological systems.

The same group investigated the effects of anion templation using the same ligand (**22**) while changing the metal ions. As well as the iron(II) complexes just discussed, similar self-assembly reactions were explored with cobalt(II), nickel(II), zinc(II) and cadmium(II). Depending on the combination of anion and cation used in the reaction (or the addition of anions at a later time) four different anion-binding metallo-supramolecular polyhedra, two circular helicates, a triple helicate and one coordination polymer were produced and could be interconverted between.[47–50] The polyhedra produced include $[M_4\mathbf{22}_6]^{8+}$ of the type just discussed, distorted $[M_8\mathbf{22}_{12}]^{16+}$ cuboids, barrel-shaped $[M_{10}\mathbf{22}_{15}]^{20+}$ pentagonal prisms and remarkable $[Cd_{12}\mathbf{22}_{18}]^{24+}$ hexagonal prisms. The reaction of $Co(NTf_2)_2$ with ligand **22** resulted in the formation of a dynamic library of metal complexes. Addition of different

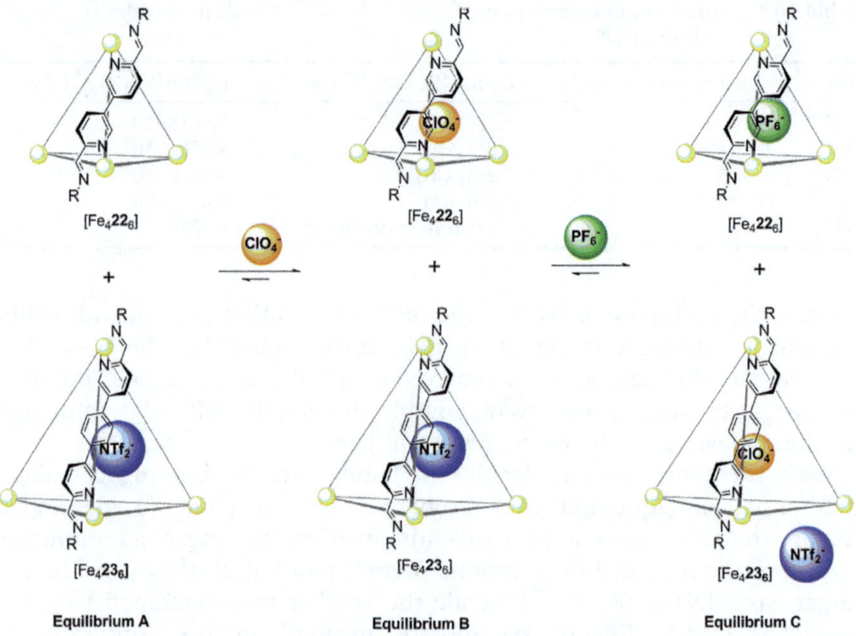

Figure 10.6 The sequence of anion-exchange reactions that occur with cages $[Fe_4\mathbf{22}_6]^{8+}$ and $[Fe_4\mathbf{23}_6]^{8+}$ upon the addition of different guests.

anions to this mixture resulted in the templation of different metallo-supramolecular architectures. For example, the addition of triflate or hexafluorophosphate produced a $[OTf \subset Co_4\mathbf{22}_6]^{7+}$ tetrahedron, while nitrate yielded a $[Co_8\mathbf{22}_{12}]^{16+}$ cuboid with four encapsulated anions and perchlorate induced a structural rearrangement to the produce a $[Co_{10}\mathbf{22}_{15}]^{20+}$ pentagonal prism. X-ray studies revealed the presence of five peripheral anion-binding pockets, each of which housed a perchlorate anion, as well as a central cavity that was shown to bind a variety of anions such as Cl^-, HF_2^-, Br^-, OCN^- and SCN^- in a tight pocket surrounded by ten CH hydrogen bond donors (Figure 10.7). The chloride ion was estimated to be bound with an association constant of greater than 6×10^5 M^{-1}. Once the pentagonal prism had formed, the templating perchlorate anions could be exchanged for hexafluorophosphate or triflate anions.[47] The larger $[Cd_{12}\mathbf{22}_{18}]^{24+}$ hexagonal prism displayed a large enough central cavity to even encapsulate triflimide (NTf_2^-).[49] Subsequent studies investigated the effects of the interconversion between different polyhedra upon the addition of different chemical signals in the form of cations or anions.[49,50]

The reaction of 4,4′-diaminobiphenyl (benzidine) with two equivalents of 2-formylpyridine produced ligand **24**, which, in the presence of iron(II) yielded an $[Fe_4\mathbf{24}_6]^{8+}$ tetrahedral cage[51] (Figure 10.8). The tetrahedral cage was found to exist as a complex mixture of interconverting diastereomers

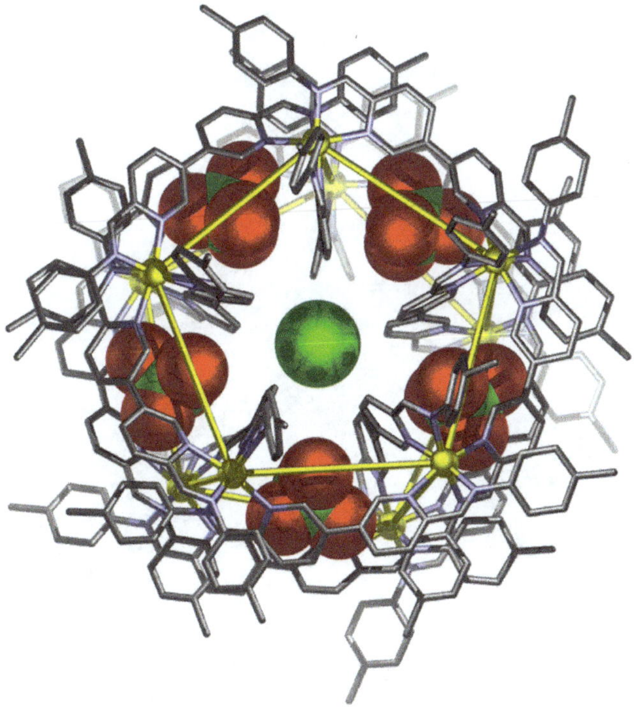

Figure 10.7 Crystal structure of the $[Cl(ClO_4)_5 \subset Co_{10}\mathbf{22}_{15}]^{14+}$ pentagonal prism. The encapsulated anionic guest is shown as space-filling models. The centrally bound chloride ion is stabilised by ten hydrogen bond donors, while the five perchlorate ions are encapsulated in distinct pockets tightly closed off from the surrounding environment.

in solution, each of which was capable of binding anions with differing affinities, with the *T*-diastereomer binding guests most strongly. Anions were stabilised within the cavity by twelve CH-hydrogen bonds and an extremely high binding constant (1.7×10^7 M^{-1}) was observed for iodide within the cage. Guest exchange was found to proceed *via* a through-face process, while diastereomer interconversion occurred *via* bond-cleavage. This cage was also found to respond to the presence of different anions in solution by changing the ratio of the three diastereomers present depending on the identity of the anion. The volume of the cage is highly flexible, changing through bond rotations to optimally accommodate different-sized anions. Thus, $[Fe_4\mathbf{24}_6]^{8+}$ was demonstrated to adapt in response to external stimuli. Responses were found both at the molecular level through conformational distortions, and at the system level through diastereomeric interconversions. Inserting an additional phenylene spacer into the backbone of the ligand (to produce **25**) resulted in the formation of a larger more open $[Fe_4\mathbf{25}_6]^{8+}$ tetrahedron, which did not display any significant anion-binding

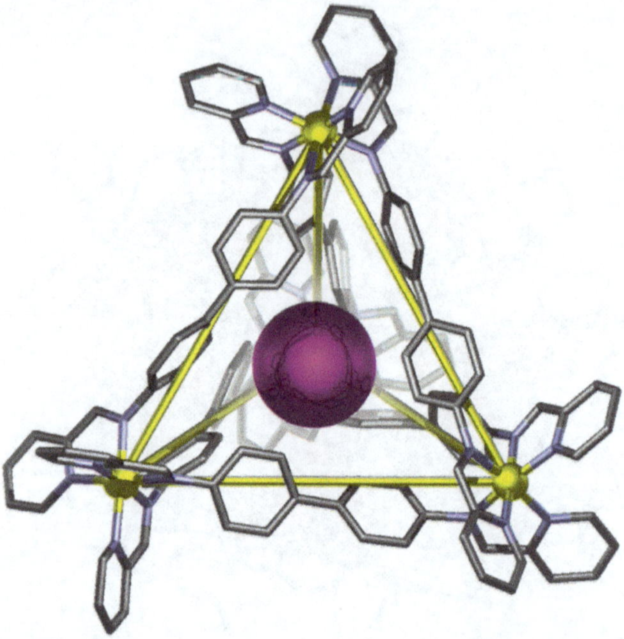

Figure 10.8 Crystal structure of the $[I\subset Fe_4\mathbf{24}_6]^{7+}$ tetrahedron. The encapsulated anionic guest is shown as a space-filling model. The iodide ion is stabilised by twelve hydrogen bond donors resulting in an extremely high binding constant. This cage respond to external symmetry by adapting both its size and the ratio of diastereomers present to provide an optimal guest-binding cavity.

behaviour. The more closed cage incorporating the BODIPY fluorophore $[M_4\mathbf{26}_6]^{8+}$, on the other hand, produced some interactions with anions such as azide and acetate that could be visualised through changes in fluorescence.[52]

When the *tetrakis*-bidentate ligand **27** was reacted with copper(I) ions in a 1:1 ratio a $[Cu_8\mathbf{27}_4]^{8+}$ rectangular prism was produced.[53] The prismatic assembly encloses a tubular shaped cavity that was found to be large enough to bind two acetonitrile solvent molecules. The addition of numerous potential guests including $[Ag(CN)_2]^-$, $[Cu(CN)_2]^-$, CN^-, OCN^-, SCN^-, SCN^- and N_3^- resulted in no observable guest binding. The addition of one equivalent of $[Au(CN)_2]^-$, however, partially decomposed the cage releasing copper(I) ions into solution. Solution and solid-state structures showed that these copper(I) ions reacted with the $Au(CN)_2^-$ yielding the complex $[Cu\{Au(CN)_2\}_2]^-$ which was encapsulated and solubilised within the $[Cu_8\mathbf{27}_4]^{8+}$ assembly. A binding constant of 1.21×10^9 M^{-3} was recorded. Once formed the encapsulated copper(I) could be exchanged for silver(I), which was better stabilised by the aromatic groups surrounding the central cavity. The encapsulated guests could be released by treatment with a more competitive solvent such as DMSO. Similar results were obtained when silver(I) ions and/or longer ligands of similar geometry were employed.[54]

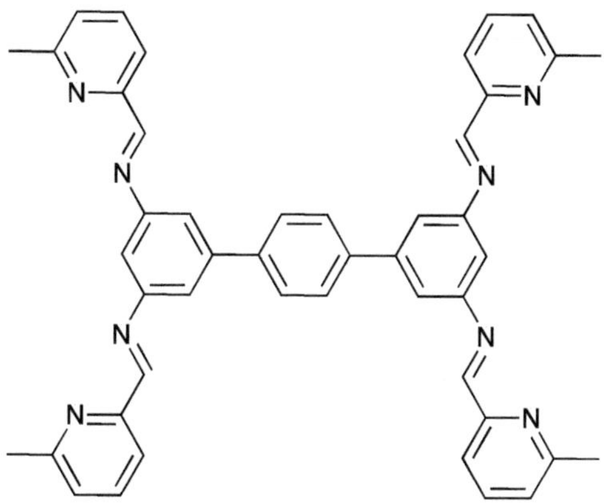

27

If a C_4-symmetric *tetrakis*-bidentate ligand is reacted with an octahedral metal ion the resulting metallo-supramolecular architecture expected to be produced is an $[M_8L_6]$ cube with the metal ions located on the corners of the assembly and the ligands arranged on the faces. For example, the reaction of 4-anilinopyridine with Pt(II) followed by the condensation of the product with 2-pyridylaldehyde produced ligand **28** which formed a metallo-supramolecular cube, $[Fe_8\mathbf{28}_6]^{28+}$, in the presence of iron(II) in 78% yield.[55] The geometry of the ligands (Figure 10.9) is such that the cage is significantly porous, with the cavity not effectively closed off from the outside environment, and although four anions were found to be encapsulated within the cube in the solid state (Figure 10.9) no evidence of anion (or other guest molecule) binding was observed in solution. In contrast, both a larger cage designed incorporating anthracene groups into the ligand (**29**) to better restrict the size of the pores of the cube[56] and a smaller cube formed from ligand **30** displayed anion-binding behaviour.[8,57] The $[Cd_8\mathbf{29}_6]^{28+}$ cube was shown to encapsulate a series of mono- and dianionic guests, which the smaller $[Fe_8\mathbf{28}_6]^{28+}$ did not. The observation that the dianionic $[Mo_6O_{19}]^{2-}$ and $[B_{12}F_{12}]^{2-}$ guests bound an order of magnitude more strongly than the monoanions BPh_4^-, $CB_{11}H_{12}^-$ and $B(C_6F_5)_4^-$ was consistent with attractive electrostatic forces with the highly-charged cube constituting the primary reason for encapsulation. The presence of the anthracene groups reduced the accessible void volume from over 7000 Å^3 to approximately 4200 Å^3, however, the largest guest only occupied 11% of the cavity within the host molecule.[58]

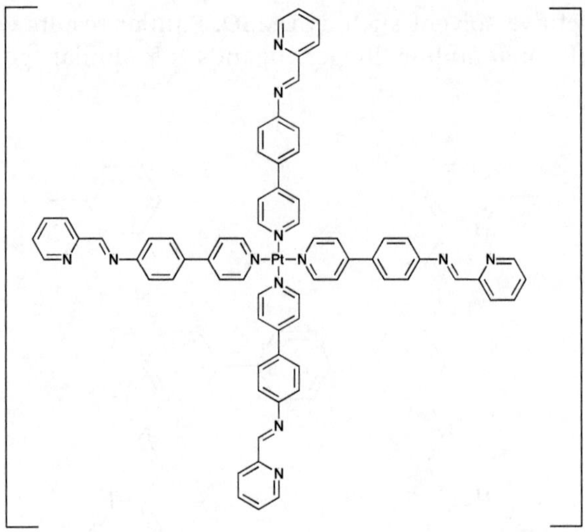

28

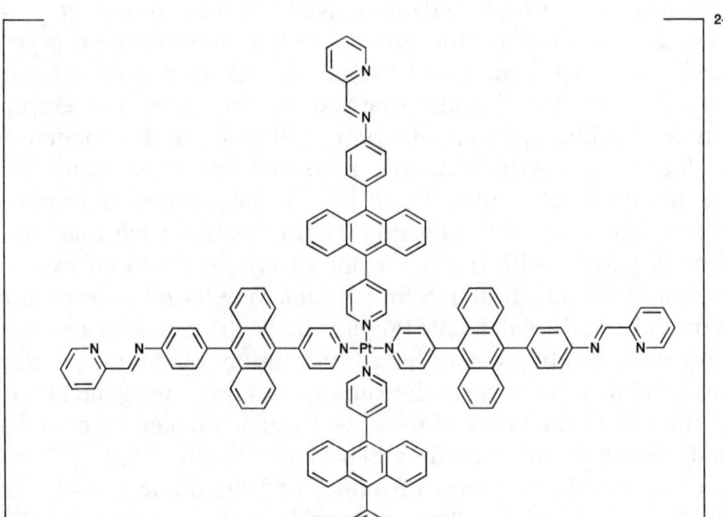

29

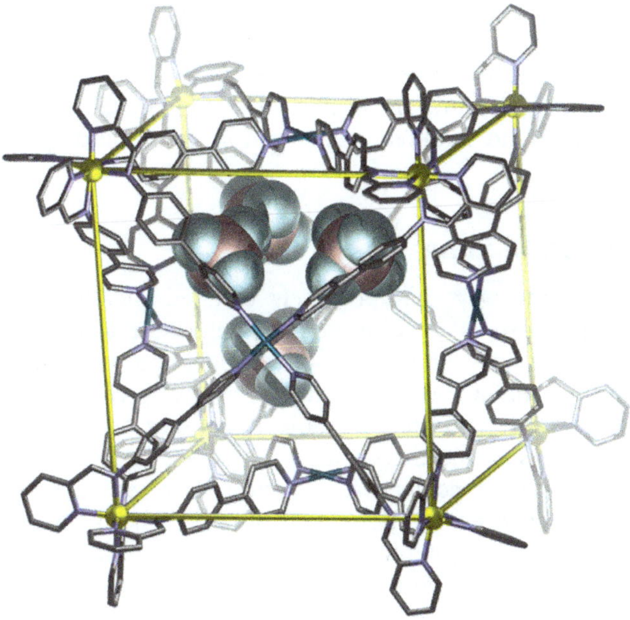

Figure 10.9 Crystal structure of the $[(BF_4)_4 \subset Fe_8\mathbf{28}_6]^{284+}$ face-capped metallo-supramolecular cube. The encapsulated anionic guests are shown as a space-filling model. No anion-binding was observed in this polyhedron in solution.

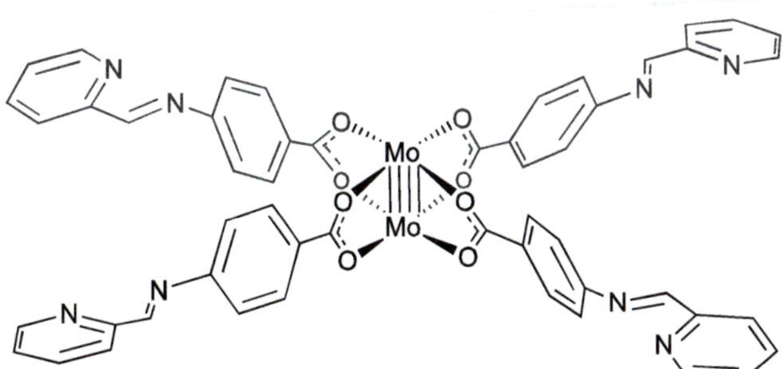

30

A smaller $[Fe_8\mathbf{30}_6]^{16+}$ that encloses 559 Å^3 was found to bind halides through electrostatic interactions inside its cavity or large tetraphenylborate anions on its exterior.[8,57] Its crystal structure is given in Figure 10.1c. The addition of small ligating groups such as ammonia, trimethylamine, trimethylphosphine oxide and trifluoroacetate resulted in their coordination to the molybdenum centres inside the cubic cavity. It was found that the halide binding affinity was significantly increased by the presence of these co-ligands, which displaced competitive triflate anions from the cage cavity thus enhancing Columbic interactions. Favourable hydrogen bonding interactions between the coordinated co-ligands and the halide ions were also found to be significant. The binding of tetraphenylborate to the outside of the cage was found to allosterically inhibit anion-binding in the cube's cavity.[57]

10.4.2 Cation Binding

In much the same way that positively charged metallo-supramolecular polyhedra display extensive anion-binding behaviour, it follows that negatively charged capsules should be effective cation binding agents. For example, the use of ligands with either negatively charged coordinating groups (such as **31–33**) or other ionisable moieties (*e.g.* **34**) can be combined with metal ions of suitable charge and geometry to form negatively charged metallo-supramolecular polyhedra.[1,59] A number of these cages have been shown to encapsulate both cations, and due to their solubility in aqueous media, a variety of hydrophobic charge-neutral guest molecules.

In an early study, Saalfrank *et al.* prepared a negatively charge mixed-valent $[Fe_4\mathbf{31}_6]^-$ (where **31** is in its doubly deprotonated form) tetrahedron from the *in situ* reaction of oxalyl chloride and the corresponding di-protected malonate in the presence of iron(II) chloride.[60] The presence of three iron(III) centres and one iron(II) centre was confirmed by electrochemistry, Mössbauer spectroscopy and X-ray analysis. The latter result also demonstrated the encapsulation of an ammonium counter-ion within the central cavity of the tetrahedron, which was presumed to act as a template for the formation of the tetra-nuclear assembly.

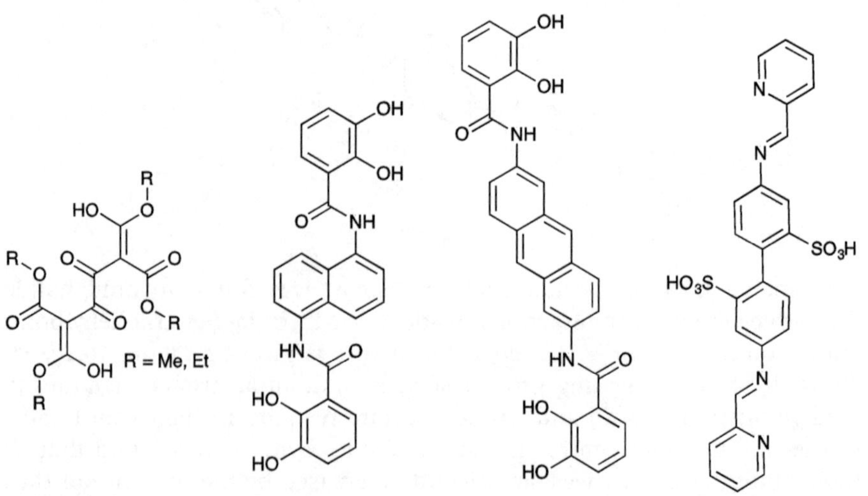

31 **32** **33** **34**

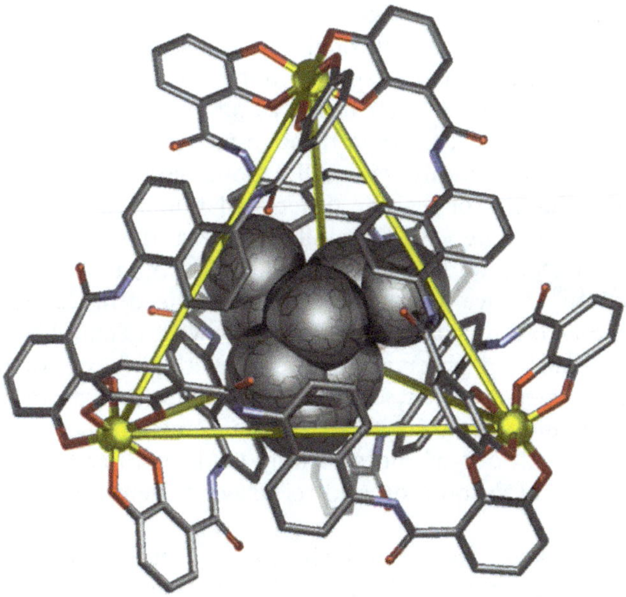

Figure 10.10 Crystal structure of the $[NEt_4{\subset}Ti_4\mathbf{32}_6]^{11-}$ homochiral tetrahedron. The guest molecule is shown as a space-filling model with hydrogens removed. The tetraethylammonium guest is bound more strongly than smaller cations such as tetramethylammonium.

Raymond *et al.* prepared a longer ligand, dicatechol ligand **32**, to synthesise a series of anionic $[M_4\mathbf{32}_6]^{n-}$ (where **32** is quadruply deprotonated and $M = Al^{3+}$, Ga^{3+}, Fe^{3+}, Ti^{4+}) *T*-symmetric metallo-supramolecular polyhedra (Figure 10.10).[61] The identity of the metal ion has no effect on the structure of the assembly, but the size and shape of the cavity was observed to change significantly depending on the encapsulated guest. The volume of the cavity is substantially larger than that in the $[Fe_4\mathbf{30}_6]^-$ assembly just discussed, varying between 253–434 Å^3.[61] While the $[M_4\mathbf{32}_6]^{n-}$ cages do not require a template to form, in the absence of Me_4N^+ a parallel reaction between **32** and titanium(IV) produced $[Ti_2\mathbf{33}_3]^{4-}$ triple helicates instead of the corresponding polyhedron.[62]

The scope of cations that can be bound inside the interior of the $[M_4\mathbf{32}_6]^{n-}$ cage is wide, ranging from simple organic cations to organometallic catalysts.[63] Guests encapsulated include quaternary ammonium and phosphonium cations, protonated secondary and tertiary amines and phosphines, including cyclic and linear species, some of which are bound as protonated homo-dimers.[61,64] Small alkali-metal crown ether complexes[65] and organometallic sandwich cations such as $[Cp_2Fe]^+$, $[Cp_2Co]^+$, $[CpRu(\eta^6\text{-}C_6H_6)]^+$ and $[CpRu(p\text{-cymene})]^+$ have also been shown to be encapsulated.[66] A series of zwitterionic molecules was found to be partially bound within the cavity – their hydrophobic cationic head groups were enclosed within the cage while their hydrophilic negatively charged sulfonate tails (attached *via* alkyl chains of various lengths) protruded through the walls of the capsule.[67] The selectivity of cation encapsulation appears to be determined by the size of the potential guest rather than their basicity and extensive NMR studies have shown guest exchange occurs via a through-face process.[68]

Notably, the strong binding of small cations has allowed for the characterisation of a number of reactive (intermediate) species in aqueous solutions that would not otherwise be possible. For example, iminium ions,[69] the phosphine – acetone adduct $[Me_2C(OH)PEt_3]^+$[70] and the organometallic $[CpRu(cod)(CO)]^+$[71] have been successfully stabilised in water, under conditions that they are not otherwise stable. As well as endohedral binding, cations have also been observed to interact with the exterior of the cage (through either π–π, cation-π, or CH–π interactions). External binding is enthalpy driven and mainly due to attractive interactions between the guests and the exterior surface of the assembly while encapsulation is entropy driven as a result of desolvation and release of solvent molecules from the host cavity.[72]

A further remarkable feature of the $[Ga_4\mathbf{32}_6]^{12-}$ tetrahedron is that each of the metal ions within the individual polyhedral assemblies adopts the same chiral configuration. Upon treatment of a racemic mixture of the $[Ga_4\mathbf{32}_6]^{12-}$ cages with the chiral cation *S*-nicotinium, the ΔΔΔΔ enantiomer of $[Ga_4\mathbf{32}_6]^{12-}$ selectively precipitated leaving the antipodean cage in solution and allowing for the separation of the two enantiomerically-pure cages.[73,74] After cation exchange the cages were found to remain enantiomerically pure after eight months in solution and display some enantioselective guest binding and catalytic properties, including the dynamic resolution of diastereomeric ruthenium complexes.[74,75] In later studies, resolved stereogenic centres were incorporated into the ligand backbone which allowed the generation of enantiomerically pure polyhedra directly through the self-assembly process,[76] allowing for the chiral cage molecules to be employed as chiral catalysts.[77]

The group of Abrahams and Robson have used cyclotricatechylene ($H_6\mathbf{35}$) as a ligand for the formation of a series of tetrahedral $[M_6\mathbf{35}_4]^{12-}$ complexes.[78] Unlike the other tetrahedral cages so far discussed in which the metal ions are located at the vertices of the polyhedron, in the case of the $[M_6\mathbf{35}_4]^{12-}$ cages, the four $\mathbf{35}^{6-}$ ligands are located at the vertices with divalent metal ions acting as linkers between them. A range of metals have been used to form these cages including VO^{2+}, Co^{2+} and Mn^{2+} with the catechol oxygens coordinating in an equatorial (or basal) plane in each case. Crystallographic studies showed that when calcium(II) was employed as a counter-ion up to four dioxane molecules could be incorporated inside the cavity of the tetrahedra. When alkali earth metals were used instead, encapsulation of these counter-ions occurred. The $[M_6\mathbf{35}_4]^{12-}$ complexes bind both a $[Na(H_2O)_6]^+$ cation at the centre of the cage and four Cs^+ or Rb^+ cations in the corners. Cs^+ and Rb^+ ions are stabilised by the π-electron density of the ligands and coordination to bridging aqua and catechol oxygens.

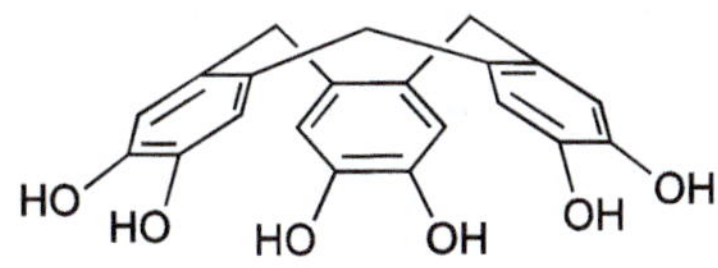

35

The reaction of ligand **34** with iron(II) and tetramethylammonium hydroxide in water produces a $[Fe_4\mathbf{34}_6]^{4-}$ (where **34** is doubly deprotonated) tetrahedron as the exclusive product.[79] The cage forms such that three sulfonate groups are located on each face of the tetrahedron surrounding the cage windows. While $[Fe_4\mathbf{34}_6]^{4-}$ was not observed to bind small organic cations within its central cavity, the arrangement of the sulfonate groups is such that one guanidinium cation can be bound to each face of the tetrahedron with an intrinsic equilibrium constant of $263 \pm 63\ M^{-1}$. The crystal structure of the guanidinium complex (Figure 10.11) shows that the presence of guanidinium effectively closes off the cage cavity from the outside environment. Indeed, it was shown that the presence of the guanidinium groups effectively reduced the size of the host apertures significantly slowing the rate of exchange of cyclopentane guest molecules compared to the tetramethylammonium analogue.[80]

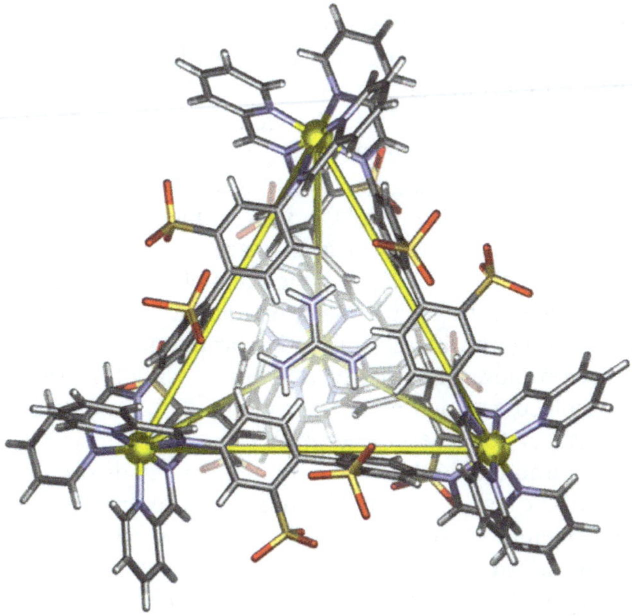

Figure 10.11 Crystal structure of the guanidinium salt of the $[Fe_4\mathbf{34}_6]^{4-}$ tetrahedron. The guanidinium cations bind to the faces of the cage through six hydrogen bond interactions with the sulfonate groups closing the windows of the tetrahedron and slowing the rate of guest exchange.

10.4.3 Charge-neutral Guest Encapsulation

The targeted and selective encapsulation of charge-neutral molecules within metallo-supramolecular polyhedra cannot simply be controlled by electrostatic interactions between host and guest, as is the driving force for many of the cation and anion-binding events already discussed. Indeed, while interactions with functional groups incorporated into the walls of the polyhedral host molecule often play a significant role in determining the type of guest enclathrated, solvent effects, particularly the hydrophobic effect, play a significant role in the binding of neutral guest molecules.[81] In fact, a large number of organic receptor molecules have relied on the hydrophobic effect to drive the self-assembly process.[82] We have therefore divided the following discussion by the conditions under which neutral guest binding is observed, whether aqueous, non-aqueous or in the solid state.

10.4.3.1 Neutral Guest Recognition in Aqueous Solution

The group of Fujita have designed and prepared an extensive range of water soluble metallo-supramolecular polyhedra through the combination of *cis*-capped square planar (palladium(II), or platinum(II)) metal ions (*e.g.* **1**, **38–40**) and various nitrogen donor ligands (*e.g.* **2**, **36–37**). As mentioned in Section 10.4 the reaction of six equivalents of **1** with four equivalents of **2** produced a metallo-supramolecular $[Pd_6\mathbf{2}_4]^{12+}$ octahedron of approximately 2 nm in diameter with four open and four closed faces. The electron-deficient triazine groups of the ligands are located on the four closed faces, creating a hydrophobic environment within the cage. Similar cages could be formed when the capping ligand was changed for any of **38–40**.[83] This design strategy has been extended to explore the use of tetradentate bridging ligands to form a series of rectangular and trigonal prisms.[84]

The $[Pd_6\mathbf{2}_4]^{12+}$ octahedron binds a wide range of hydrophobic guest molecules in water, particularly those that are sufficiently electron-rich to interact favourably with the electron-poor walls of the cage. The guest molecules bound range from small polyaromatic molecules including tetra-azaporphine,[16,85] polyfluorinated aliphatic compounds,[86] N-oxide radicals, organometallic complexes, adamantyl derivatives, small peptides as well as aggregates of molecules, such as DNA duplexes[87] or halogen bonded complexes.[88] The size of the guests determines the number and geometry in which they are bound inside the hydrophobic cavity. Four smaller molecules such as *o*-carborane can be simultaneously bound inside the host in a tetrahedral arrangement, while two medium sized guests such as diphenylmethane are enclathrated orthogonally and the largest guests such as peptides bound only with one molecule per cage.[89] The ability to encapsulate more than one guest inside the cage has been used extensively for other applications, including catalysis discussed in Section 10.5 below.

The importance of the hydrophobic and electron-deficient nature of the cage is highlighted by the observations of sequence-selective recognition of short peptides within its cavity. It was determined that the presence of residues containing an aromatic functional group such as Trp, Tyr or Phe was required for the effective encapsulation of similar hexa-peptides within the octahedron's cavity. The presence of a charge-transfer interaction between the bound peptides and the cage wall led to a colour change upon binding that could be observed with the naked eye.[90] It was further determined that peptides containing the Trp-Trp-Ala sequence in particular were bound more effectively ($K_a > 10^6\ M^{-1}$) through cooperative CH–π and π–π interactions than peptides containing other sequences (*e.g.*, Trp-Ala-Trp or Ala-Trp-Trp) of the same residues.[91] Due to the high positive change on the cage (12+) peptides with positively charged residues (*e.g.* His) were not observed to bind and reducing the pH was also found to eject incorporated guests.

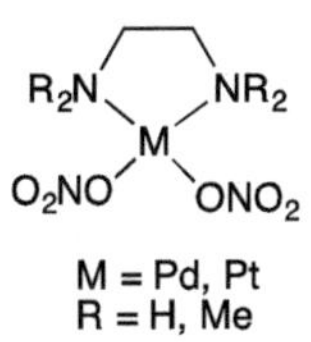

1

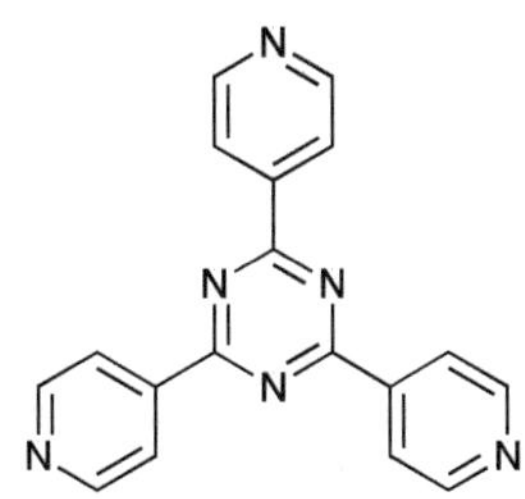

2

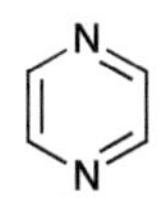

36

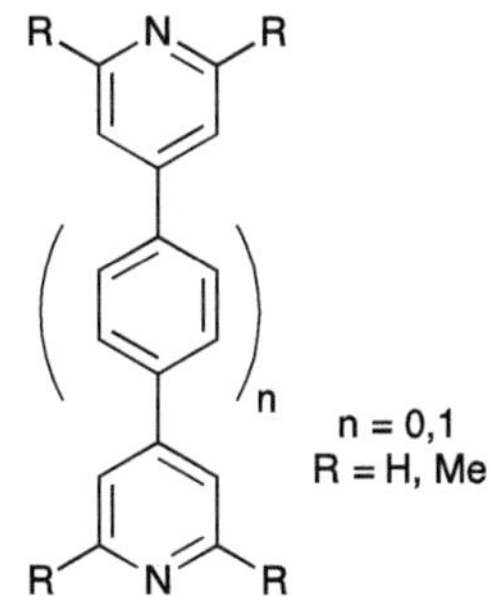

37

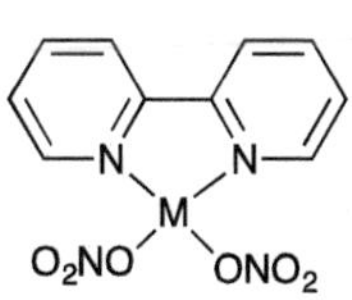

38

39

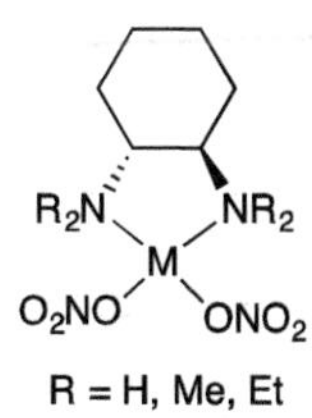

40

The combination of the *cis*-capped square planar metal centres (*e.g.* **1**, **38–40**) in water with two different ligands – the electron-deficient tpt (**2**) and a suitable difunctional ligand (**36** or **37**) produced pillared trigonal prismatic $[Pd_6L_3\mathbf{2}_2]^{12+}$ complexes. Each pyridyl group of the tpt (**2**) ligands is coordinated to three different *cis*-capped metal centres which are further are connected *via* (three) linear difunctional heterocyclic amine co-ligands (such as pyrazine (**36**)). These ligands serve as "pillars" or struts between corresponding metal centres. The trigonal prismatic products require the presence of a large planar template molecule such as hexamethoxytriphenylene to form (Figure 10.12). Once assembled, the template can be removed (by extraction with chloroform) and the empty cage readily incorporates large unsaturated guests including, for example, coronene,

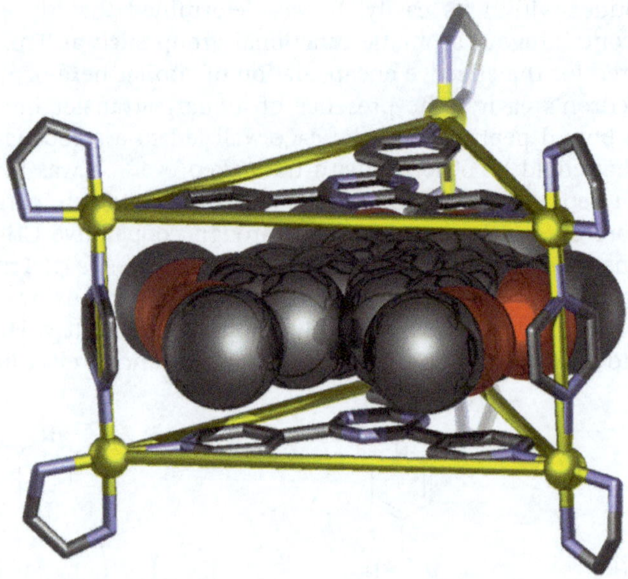

Figure 10.12 Crystal structure of the $[Pd_6\mathbf{36}_3\mathbf{2}_2]^{12+}$ trigonal prism encapsulating a hexamethoxytriphenylene guest. The guest is shown as a space-filling model and is stabilised by strong π–π interactions with the electron-deficient walls of the cage.

pyrene and a dibenzo-substituted *β*-diketone which was found to be bound exclusively in its planar enol tautomer.[92] Similar host–guest binding results were observed by another group when the *cis*-capped metal ions were replaced with cyclopentadiene-capped ruthenium centres.[93]

Replacing the pyrazine (**36**) pillar with a longer difunctional ligand such as **37** ($n=0$ or 1) under similar reaction conditions produced an analogous trigonal prismatic structure with an elongated cavity.[94] When **37** ($n=0$, R = Me) was employed the resulting cage was found to encapsulate two pyrene molecules, which were stabilised by π-stacking between both the two included guests and the two metal-coordinated ligands forming the faces of the cage. Other planar molecules[95] including porphines, and metal complexes such as $[M(acac)_2]$ (M = Pt, Pd or Cu; acac = acetylacetonato) can also be bound in a 2 : 1 guest : host ratio. The close proximity of metal centres within the cage led to unusual magnetic behaviour.

Further elongating the pillaring ligands to **37** ($n=1$, R = Me or H) led to the encapsulation of three guest molecules – two electron-rich corenene or porphine templates and an uncoordinated electron-deficient tpt (**2**) ligand which was sandwiched between the two guests forming an alternating donor–acceptor stack of five different aromatic groups.[96] Three tri-metallic gold(I) clusters (**41**) could also be assembled into a columnar 3×3 gold(I) array inside the larger $[Pd_6L_3\mathbf{37}_2]^{12+}$ ($n=1$, R = H) cage. The X-ray structure clearly shows that aurophilic Au···Au interactions are present, with the gold ions separated by only 3.2 Å; significantly shorter than the sum of the van der Waals radii of gold(I) (3.6 Å).[97] Increasing the length of the pillaring ligand even further allowed for the incorporation of up to five aromatic guest molecules.[96] Similar prismatic architectures have been formed by replacing tpt with copper(I) or silver(I) complexes similar to **41**.[98]

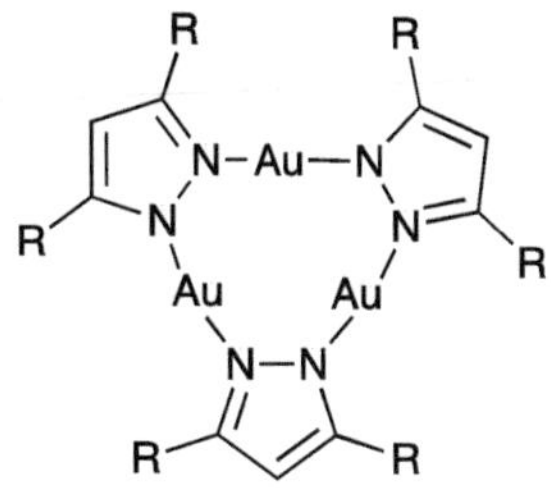

41

The significant role solvent and hydrophobic effects can play on encapsulation behaviour was determined through the examination of the guest-binding properties of Ward's $[Co_8\mathbf{14}_{12}]^{16+}$ octanuclear cubic co-ordination cage (discussed above) which did not reveal any anion-binding behaviour despite its significant positive charge. This polyhedron encloses a cavity of volume of 400 Å^3, which is accessible through 4 Å diameter portals in the cube faces. ^{1}H NMR studies in MeCN solution revealed that the cube can act as a size- and shape-selective host for small organic guests such as coumarin[99] ($K = 78$ M^{-1}) and other bicyclic molecules of comparable size, shape and functionality such as isoquinoline-*N*-oxide ($K = 2100$ M^{-1}). Binding was observed to only occur when guest molecules contained: (i) a suitable a polar oxygen atom (carbonyl or similar) which could interact with a series of convergent CH protons inside the cavity; and (ii) suitable functionality to allow a combination of aromatic and van der Waals interactions with the interior surface of the cage. The strength of the first component was found to vary linearly with the H-bond-accepting ability of the guest; the second component was found to be fixed at approximately 10 kJ mol^{-1}.[100] The preparation of an isostructural, but water soluble (due to its appended alcohol groups), analogue of the cage ($[Co_8\mathbf{42}_{12}]^{16+}$) allowed for direct comparison of the binding behaviour and quantification of the solvophobic contribution to guest binding. This was determined to be approximately 10 kJ mol^{-1} which correlates closely to the change in free energy expected for the transfer of toluene from MeCN to water.[101]

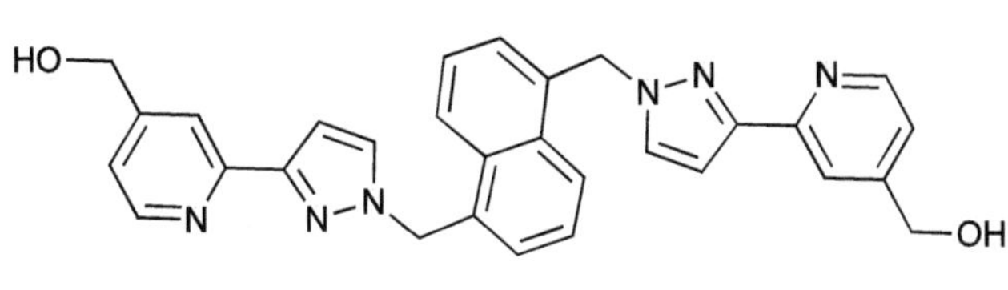

42

The solvophobic contribution is enough to result in significantly enhanced (both in terms of affinity and scope of substrates recognised) guest-binding abilities when water is the solvent. Thus, compared to the small number of guests bound by $[Co_8\mathbf{14}_{12}]^{16+}$ in MeCN, the water soluble $[Co_8\mathbf{42}_{12}]^{16+}$ cube binds a much wider range of guest molecules with association constants up to three orders of magnitude greater than the equivalent encapsulation in acetonitrile. For example, the affinity of the water soluble $[Co_8\mathbf{42}_{12}]^{16+}$ cage towards aliphatic ketones of increasing size (from C_5 to C_{11}) was investigated by a fluorescence assay.[102] It was found that linear aliphatic ketones were not encapsulated, but preorganised (cyclic) ketones were bound with association constants of 10^4–10^5 M^{-1}. The highest association constant ($K = 1.2 \times 10^6$ M^{-1}) was recorded for cycloundecanone, with the association constant progressively decreasing for larger guests. A crystal structure (Figure 10.13) of the cycloundecanone inclusion complex showed that the guest carbonyl oxygen is directed into a binding pocket defined by a convergent set of CH groups, which act as weak hydrogen bond donors, and also shows close contacts between the exterior surface guest and the interior surface of the cage cavity. Based on these promising results, the same group then employed virtual screening to identify potential guest molecules for the $[Co_8\mathbf{42}_{12}]^{16+}$ cube. Using this method it was possible to identify 15 new guests with a diverse range of functional groups and accurately predict their binding constants in aqueous solution.[103]

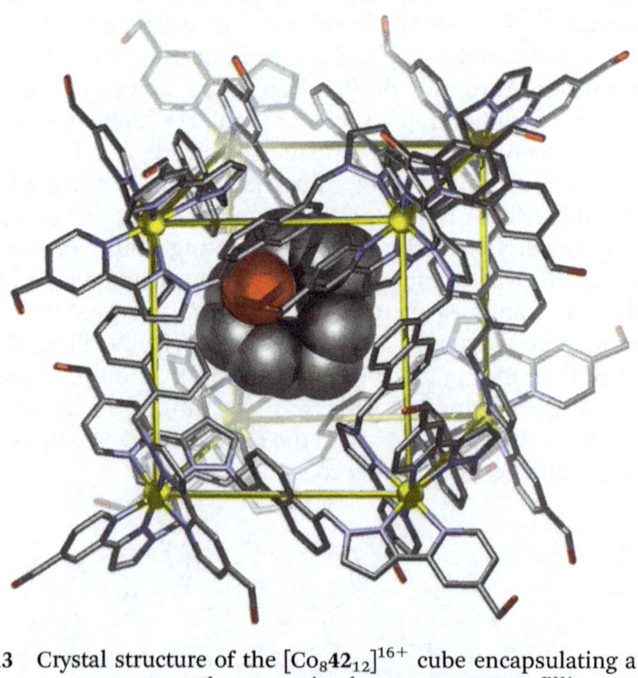

Figure 10.13 Crystal structure of the $[Co_8\mathbf{42}_{12}]^{16+}$ cube encapsulating a cycloundecanone guest. The guest is shown as a space-filling model and is stabilised by weak hydrogen bonding interactions.

Using water as a solvent also allows for the potential of pH-sensitive guest binding to be explored. Using the same $[Co_8\mathbf{42}_{12}]^{16+}$ cube, Ward *et al.* demonstrated that a wide variety of guests could be reversibly encapsulated and released through the control of pH. The functional groups that could be controlled in this way include primary and tertiary amines, pyridine, imidazole and carboxylic acids,[104] many of which are commonly found in drug molecules, pointing towards the potential use of metallo-supramolecular polyhedra as drug-delivery vehicles. The Parkinson's drug 1-amino-adamantane was found to bind strongly (10^4 M^{-1}) under basic conditions while the binding reduced by three times in the acidic conditions, effectively releasing the drug into solution. This idea could be extended to control the selection of one of three guests from aqueous solution through pH control. Switching between different guests was possible because neutral organic guests bind strongly in the cage due to the hydrophobic effect, whereas charged (protonated or deprotonated) molecules are hydrophilic and do not bind. The guests used were: adamantane-1,3-dicarboxylic acid which was bound at low pH when it is neutral but not when it is deprotonated; 1-amino-adamantane which binds at high pH when it is neutral but not when it is protonated; and cyclononanone whose binding is not pH dependent and is therefore the default guest at neutral pH. Thus an increase in pH can reversibly switch the host between the three different bound states in succession.[105]

The group of Raymond have exploited the hydrophobic effect to encapsulate a variety of small hydrophobic organic guests within their chiral $[Ga_4\mathbf{32}_6]^{12-}$ tetrahedron. The scope of guests that can be incorporated is wide including small aromatics, natural products,[106] ruthenocene[107] derivatives, as well as a number of tertiary amides.[108] In many cases, the simultaneous encapsulation of multiple species is observed. The molecular host has also been found to recognise different substitutional isomers of disubstituted benzenes, with *ortho* substitution leading to the encapsulation of two guest molecules, but *meta* or *para* substitution leading to the encapsulation of only one guest molecule. In some cases diastereoselectivity is observed,[106] and even the subtle changes of isotopic substitution can distinguished based on kinetic isotope effects.[109]

Similar hydrophobic guest binding in aqueous solution has been observed by Nitschke *et al.* in a series of $[M_4\mathbf{34}_6]^{4-}$ tetrahedra. When iron(II) was employed as the metal ion, the cage was shown to bind a series of small organic guest molecules such as dichloromethane, cycloalkanes and benzene up to a maximum size of *n*-pentane.[79,80,110–114] The scope of the guests incorporated was expanded, however, when cobalt(II) or nickel(II) was employed. The longer bond lengths of these metals led to an increase in the volume of the enclosed cavity within the polyhedron, which in turn allowed for the incorporation of larger guest molecules such as toluene and cyclooctane.[110] The same group explored the solvent-dependent host–guest chemistry of the face-capped $[Fe_4\mathbf{43}_4]^{8+}$ tetrahedron. Depending on the counter-ion employed, this capsule was found to be soluble either in water or in acetonitrile. Quantitative binding studies were undertaken, revealing that both aromatic and aliphatic guests bind in water, whereas only aliphatic guests bind in acetonitrile.[114,115]

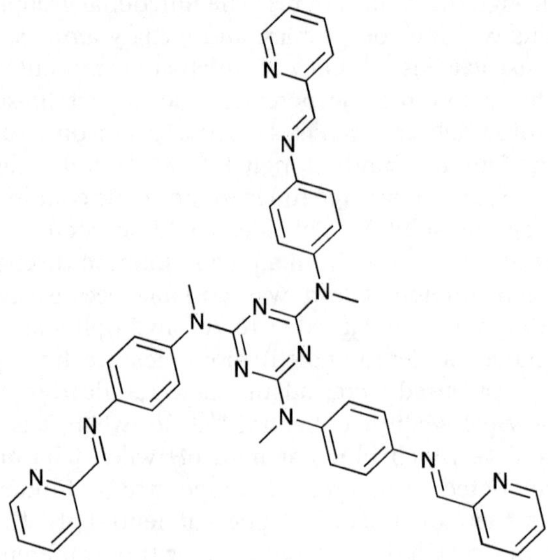

43

10.4.3.2 Non-aqueous Guest Binding

As will be clear from the discussion above, there are relatively few examples of charge-neutral metallo-supramolecular polyhedra. Lindoy *et al.* have used the bis-β-diketonato ligand **44** (with either $n=1$ or 2) in combination with either four Fe(III) or four Ga(III) centres to produce charge-neutral $[M_4L_6]$ tetrahedra.[116] The X-ray structure of the $[Fe_4\mathbf{44}_6]$ (with $n=1$) revealed that the cavity of the assembly was sufficiently large to encapsulate a tetrahydrofuran solvent molecule. Increasing the length of the ligand to yield $[Fe_4\mathbf{44}_6]$ (with $n=2$) expanded the cavity volume such that four tetrahydrofuran molecules could be enclathrated (Figure 10.14).

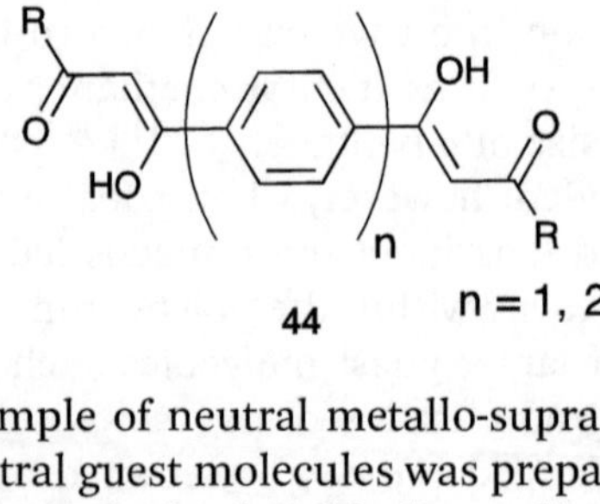

44 n = 1, 2

A similarly rare example of neutral metallo-supramolecular host capable of binding charge-neutral guest molecules was prepared by the group of Cui. By using a ligand with resolved axial chirality, a diastereoselective synthesis

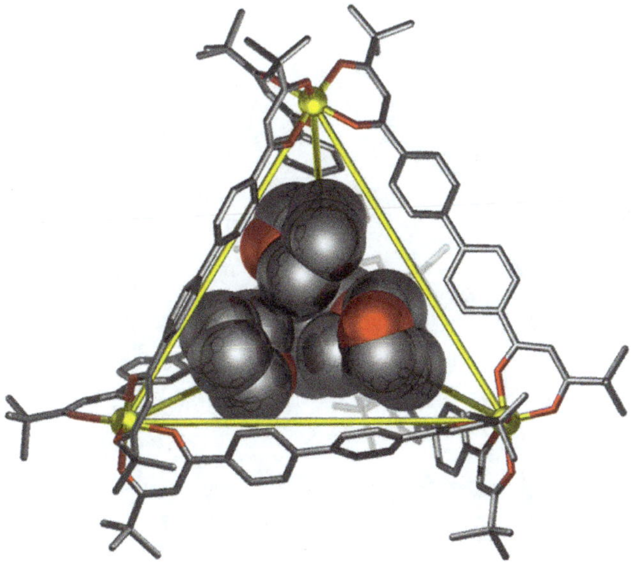

Figure 10.14 Crystal structure of the [$thf_4 \subset Fe_4$**44**$_6$] (where $n = 2$) tetrahedron. The four encapsulated thf molecules are shown as space-filling models. The smaller [Fe_4**44**$_6$] (where $n = 1$) tetrahedron encapsulates only one thf molecule.

of a M_4L_6 tetrahedral cage was achieved.[117] The di-acetylacetonato ligand with biphenylene backbone 5,5′,6,6′-tetramethyl-3,3′-diketone-2,2′-bis(methoxymethoxy)-biphenyl, **45**, forms metallo-supramolecular tetrahedra with Fe(III) and Ga(III) with high enantiopurity at the metal centres (Figure 10.15).

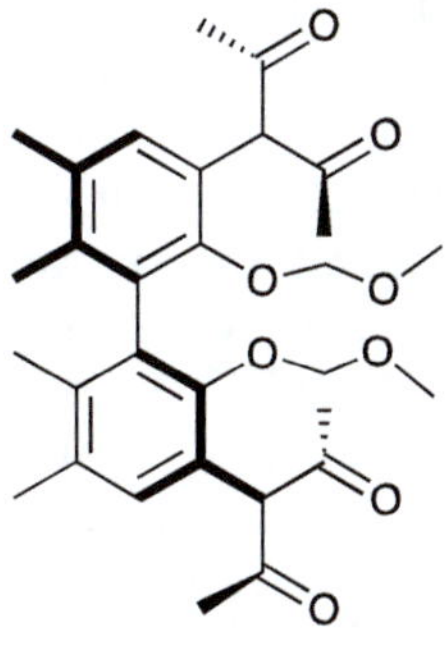

(*S*)-**45**

The authors synthesised both *R* and *S* forms of the ligand from commercially available enantiopure starting materials. The *S* hand of the ligand induced the tetrahedron to form with Λ stereochemistry at all four metal centres. Conversely, when the *R* hand of the ligand was used the other

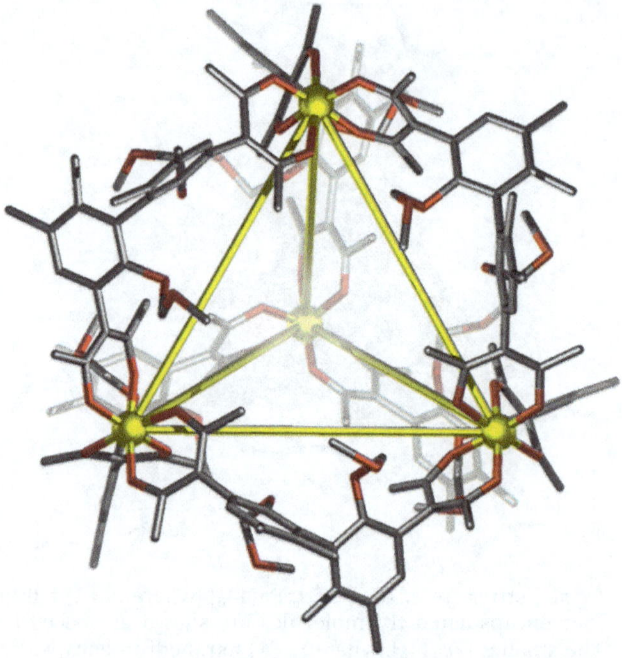

Figure 10.15 The crystal structure of the ΛΛΛΛ-[Fe_4((*S*)-**45**)$_6$] metallo-supramolecular tetrahedron. The ligand (*S*)-**45** exclusively forms the ΛΛΛΛ-enantiomer of the [M_4((*S*)-**27**)$_6$] (M = Fe or Ga), while the *R* form of the ligand produces the ΔΔΔΔ enantiomer.

enantiomer was produced. The chirality of the cage cavity could be exploited to perform enantioselective separations. 2-butanol could be bound with 98.8% *ee* and 3-methyl-2-butanol with 99.5% *ee*.

Nitschke *et al.* investigated the effects of varying cavity and pores sizes on guest-binding behaviour in a series of $[Fe_4L_4]^{8+}$ face-capped tetrahedra.[118] Guest binding in MeCN solution was observed only when ligand **46** was used. This observation was rationalised as only this cage combined a sufficiently large cavity with the sufficiently small pores required for guest binding. The $[Fe_4\mathbf{46}_4]^{8+}$ capsule was shown to bind a wide variety of guest species in size- and shape-selective fashion. Carbon tetrachloride, cyclopentene, cyclohexane, cyclopentane, cyclohexene, pyridine, and benzene were all encapsulated inside the cavity of cage. Interestingly, if the polarity of the solvent used for the self-assembly reaction that produced $[Fe_4\mathbf{46}_4]^{8+}$ is increased through the use of a MeOH–MeCN mixture, an $[Fe_{12}\mathbf{46}_{12}]^{24+}$ icosahedron is produced, as opposed to the previously seen tetrahedron. This larger structure does not bind any organic guest molecules, but was found to incorporate a $[B_{12}F_{12}]^{2-}$ anion.[119]

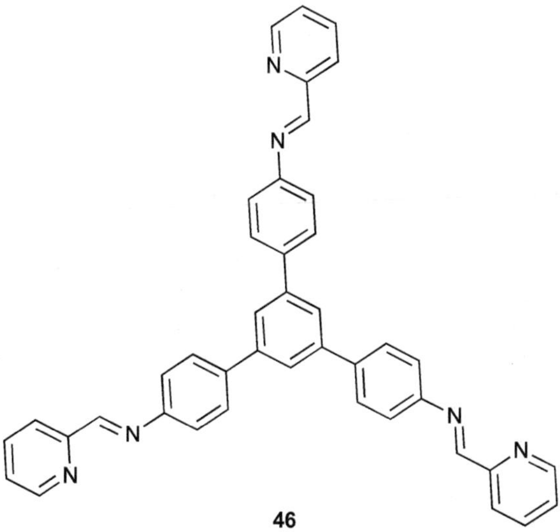

46

The reaction of ligand **47**, which has two coordination vectors that diverge by 120°, with iron(II) was expected to form an $[Fe_8\mathbf{47}_{12}]^{16+}$ cube with octahedral metal centres located on the vertices and the ligands defining the edges. The ^1H NMR spectrum was consistent with the formation of a high symmetry polyhedron with only seven signals present (one for each symmetry independent proton on the ligand) and the X-ray structure confirmed the formation of the cubic architecture. Compared to the cubic hosts prepared by Ward *et al.*, the $[Fe_8\mathbf{47}_{12}]^{16+}$ assembly was expected to have open pores. Guest binding studies were carried out in both MeCN (R = Me) and cyclohexane (R = $C_{10}H_{12}$) solvents. Only ferrocene was found to interact with the cube in acetonitrile while in cyclohexane 9-acetylanthracene was encapsulated, presumably due to enhanced π–π stacking in the less polar cyclohexane.[120]

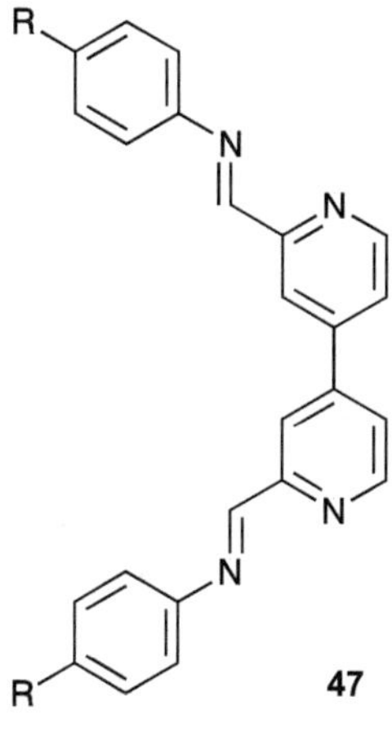

47

Taking inspiration from nature, the group of Nitschke designed and synthesised a metallo-supramolecular cube for the selective encapsulation of large aromatic guests. It was reasoned that to achieve such a goal, the cube would require the following features: (i) large regions of the cage walls rich in π-electron density to provide favourable interactions with targeted guests (ii) a cavity size of sufficient size to ensure the guests of interest could be enclosed and (iii) small pore sizes to isolate guests from the external environment. A design incorporating the C_4-symmetric *tetrakis*-bidentate ligand **48**, which when combined with iron(II), was expected to form a *O*-symmetric cubic structure of general formula $[Fe_8\mathbf{48}_6]^{8+}$, in which the corners of the cube are defined by the metal centres and the faces by the ligands, was chosen. The porphyrin groups were expected to both tightly close off the interior of the cage from the outside environment and also provide favourable π-electron density for guest recognition. The crystal structure of the resulting complex is shown in Figure 10.16. The solid-state structure showed that the metallo-supramolecular cube encloses a volume of 1340 $Å^3$ which was well isolated; the largest sphere that could enter or exit the cavity through a pore window has a radius of only 1 Å.

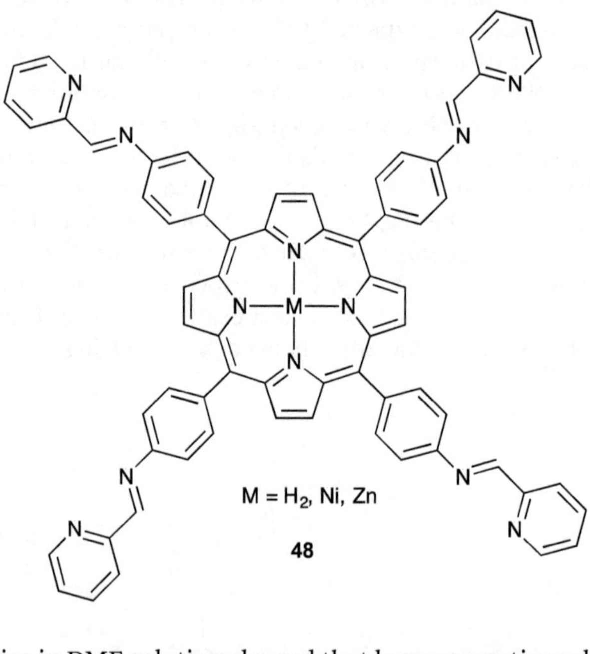

48

NMR studies in DMF solution showed that large aromatic molecules could indeed by selectively bound within the cage cavity. Three coronene molecules could be encapsulated while both C_{60} and C_{70} were observed to form 1:1 host–guest complexes with cage. Mass spectroscopy showed that the cage could bind higher-order fullerenes from fullerene soot including C_{76}, C_{78} and C_{82}. A notable feature of the host–guest complexes is that through encapsulation the maximum concentration of C_{60} in DMF solution can be

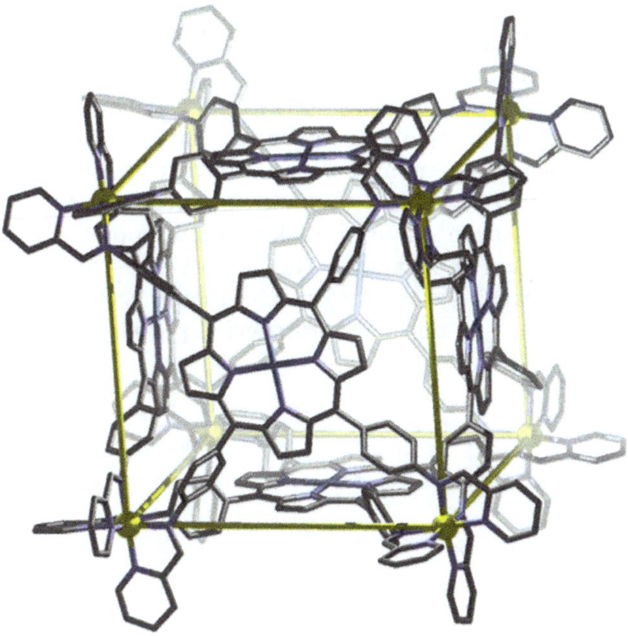

Figure 10.16 Crystal structure of the $[Fe_8\mathbf{48}_6]^{8+}$ (M = Ni) metallo-supramolecular cube. The π-rich walls of the cage allow a series of large aromatic guest molecules to be selectively encapsulated.

increased sixfold.[121] A triangular bipyramidal cage formed from a similar ligand was shown to bind two 2,7-dinitro-9-fluorenone molecules.[122]

Following a related design strategy, the incorporation of large polyaromatic panels such as perylene bisimide[123] or pyrene derivatives[124] into suitable difunctional ligands has allowed for the encapsulation of similar large aromatic guests including fullerenes, polycyclic aromatic hydrocarbons, and structurally complex natural products such as steroids into $[M_4L_6]$ tetrahedra.[123,124]

10.4.3.3 Gas Uptake in the Solid State

Following a 'secondary building unit'[125] approach that has been extensively used in the design and preparation of polymeric metal-organic framework materials,[126] a number of groups have successfully prepared charge-neutral metallo-supramolecular polyhedra (which are often referred to as "Metal-Organic Polyhedra" or "MOPs" (*sic*), in this sub-class of the literature) in which small metal clusters or paddlewheels are featured as design elements.[12] When these clusters are combined with ligands (generally di- or tricarboxylates such as **49–51**) of the appropriate geometry, discrete polyhedral assemblies result, which often have the potentially useful features of both coordinatively unsaturated metal centres and permanent porosity. Metal clusters of this type are not only robust design elements, but also have been demonstrated to provide sites for gases such as hydrogen or carbon dioxide to bind.[127] Once a suitable structural motif has been identified it is possible to prepare a series of polyhedra by substituting the chosen ligand with an extended analogue in a reticular fashion; even mixed ligand complexes can be prepared. Remarkably, many materials formed in this way display permanent porosity, that is, voids within the materials remain open after any residual solvent has been removed, making them potentially useful in gas storage and separation applications. The relative ease with which the metal ions and ligands can be exchanged means that it is possible to tune the binding behaviour of the polyhedra for specific purposes.[128]

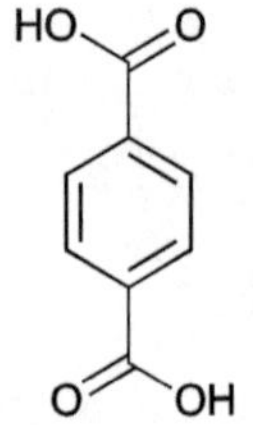

49

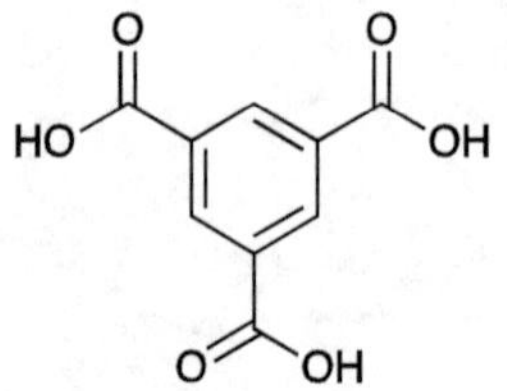

50

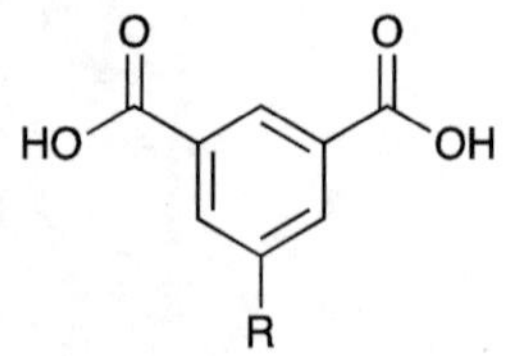

51

Following this approach a wide range of polyhedral structures have been prepared, including examples of tetrahedra, octahedra, cuboctahedra, truncated tetrahedra, truncated heterocubes and rhombic dodecahedra.[129] For example, the reaction of **51** (R = ^{t}Bu) with copper(II) acetate produced a cuboctahedral [Cu$_{24}$**51**$_{24}$] (where **51** is doubly deprotonated) complex (Figure 10.17). After suitable activation this material was found to display selective gas adsorption; at 77 K significantly more $O_{2(g)}$ and $H_{2(g)}$ adsorbed than either of $N_{2(g)}$ or $Ar_{(g)}$, while at 195 K carbon dioxide could be adsorbed seven times more effectively than methane.[130] It must be noted that while the cavity inside the polyhedron certainly plays a role in enhancing the surface area of these materials, it is not yet clear that the gas sorption properties arise from the void inside the polyhedron or, if instead, guests are merely bound in cavities in the crystal structure which exist between the polyhedra in the solid state, although some X-ray structural studies have shown that solvent molecules are often located within the central cavity of the polyhedra.[131]

Closely related structures in which sulfonylcalix[4]arene and thiacalix[4]-arene macrocycles have been incorporated into the polyhedral design have also been prepared.[132] The macrocyclic ligands act as caps for small clusters of metal ions which are further bridged by carboxylate ligands. The calix[4]arene units provide distinct guest-binding sites from the octahedral cavity encapsulated by the cage superstructure. While these cages display similar gas binding properties in the solid state to the MOPs just discussed due to their permanent porosity, and show some interesting CO_2/N_2 selectivity, in the solution phase they are also able to encapsulate a variety of small drug (*i.e.* aspirin, ibuprofen) and dye molecules (*e.g.* methylene blue). The release profiles of the guest molecules suggested that they were encapsulated in both the cage cavity and the calix[4]arene binding sites.[133]

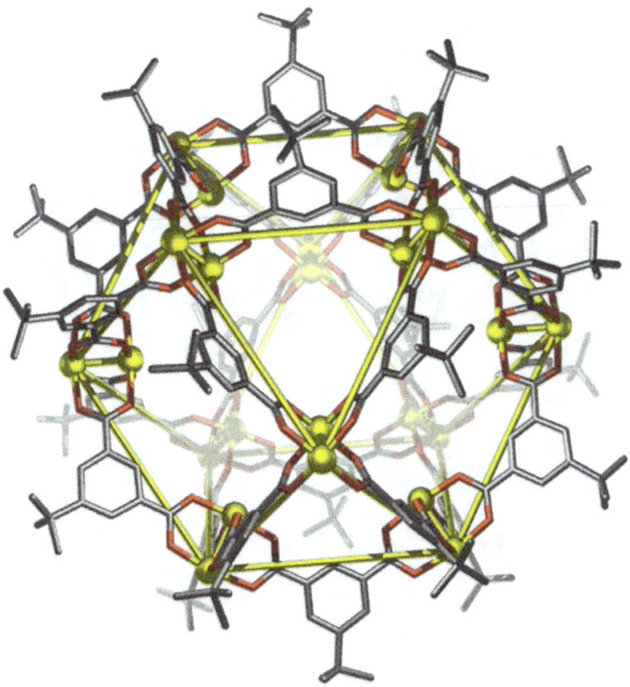

Figure 10.17 Crystal structure of the $[Cu_{24}\mathbf{51}_{24}]$ (R = ^{t}Bu) cuboctahedron. This material displays selective gas sorption behaviour.

The group of Batten designed a series of metallo-supramolecular cubes for hydrogen storage applications. They combined eight equivalents of the Cu(I) scorpionate ligand **52** with six equivalents of divalent transition metals to form $[M_6\mathbf{52}_8]^{16+}$ cubic architectures (Figure 10.18) of approximately 2.8 nm in diameter. Hydrogen gas adsorption and desorption isotherms that up to 1.2 wt% of hydrogen could be adsorbed, equating to approximately 34 hydrogen molecules per cage. The calculated enthalpy of adsorption (8.75 kJ mol^{-1}) indicated a strong binding affinity to the walls of the cage, probably on the exterior.[134] Further studies showed that a large range of solvent molecules could be enclathrated within cavities in between the cages in the crystal lattice; the identity of the solvent subtly affected the spin-crossover (and other magnetic) properties of the cage materials allowing them to be employed as nano scale molecular switches.[135]

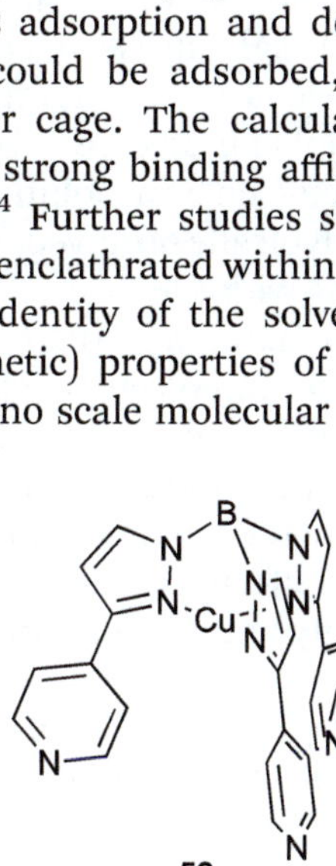

52

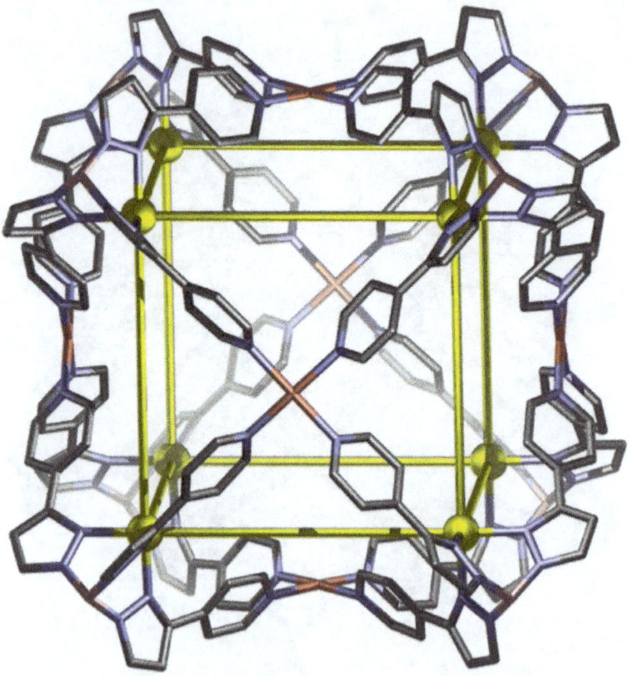

Figure 10.18 Crystal structure of the $[Cu_6\mathbf{5}2_8]^{16+}$ cube. The iron(II) analogue of this polyhedron acts as a sensor for a wide range of guest solvent molecules, signalling their presence through changes in the spin-crossover behaviour.

10.5 Reactivity Modulation

Once guest molecules are encapsulated within metallo-supramolecular polyhedra, they are confined in tight spaces in close proximity to the walls of the cage and other guest molecules. Such confinement means that the behaviour of the isolated guest(s) can diverge substantially from what would normally be observed of free molecules in solution. as the random collisions with other molecules that are typically observed in bulk solution are replaced by a smaller set of interactions with the host or other guest molecules. This in turn can lead to significant changes in the reactivity of the encapsulated species. Accordingly, metallo-supramolecular cages can have a wide range of reactivity-modulating functions, from accelerating reaction rates to promoting the formation of unusual products that are not formed under non-encapsulated conditions or when using non-supramolecular metal catalysts, or in unusual solvents. These effects can arise from a variety of factors, including the variation of polarity and/or electron density in the internal cage void space compared to the bulk solution, higher effective concentrations of the substrate in within the cage and the confinement of the guest within close proximity to a catalytically-active metal centre. The size and shape of the internal cavity can also have an effect in an 'enzyme-like' manner, where substrates are bound in such an orientation as to encourage the formation of a certain product. Notably, the entrapment of a guest can also stabilise otherwise unstable molecules effectively inhibiting the rates of reactions or preventing them from occurring.

10.5.1 Catalytic Polyhedra

Not unexpectedly, the cage molecules that display the most diverse guest-binding behaviour also have shown diverse catalytic properties. Fujita's $[Pd_6\mathbf{2}_4]^{12+}$ octahedron is no exception.[7] The four open faces of the assembly are a contributing feature allowing rapid catalytic turnover as the open faces facilitate effective substrate entry and product departure from the hydrophobic interior of the cage. A wide range of reactions has been shown to occur within the octahedron, including Wacker oxidations,[136] photodimerisation of alkenes,[137] alkane oxidation[138] and a number of other photocatalysed reactions.[139] Notably, due to the ability of the cage to simultaneously encapsulate more than one guest molecule, this system has had great success in the catalysis of Diels–Alder and other pericyclic reactions, including on inert aromatics that would not undergo cycloaddition under typical conditions due to the high stability of the aromatic system. For example, the combination of $[Pd_6\mathbf{2}_4]^{12+}$ with *N*-cyclohexylmaleimide and 2,3-diethylnaphthalene in water produced the Diels–Alder adduct in a 60% yield, whereas the reaction did not proceed without the presence of the cage.[140] Similar reactions between *N*-cyclohexylmaleimide and other normally unreactive aromatic molecules **53** and **54** could also be induced to occur. The regioselectivity of these reactions could be controlled through the use of either photoirradiation, which produced [2 + 2] cycloaddition adducts, or heating which produced the [2 + 4] adducts giving products with high regio- and stereoselectivity (Figure 10.19).[141] It was been proposed that the promotion and acceleration of these reactions is due to the cage holding both substrates in a specific, defined orientation that readily facilitates the reaction to occur. The encapsulation of a number of Diels–Alder products within the cage was confirmed by X-ray crystallography (Figure 10.20).[140]

The catalytic performance of the octahedron could be tuned by exchanging the ethylenediamine capping ligand for a different capping group. If the chiral diaminocyclohexane capped palladium derivative **38** was used instead of **1** in the self-assembly reaction, a chiral analogue of the $[Pd_6\mathbf{2}_4]^{12+}$ octahedron results. When the same [2 + 2] and [2 + 4] cycloadditions just discussed were repeated in the chiral cage, the formation of one enantiomer was favoured, with enantiomeric excess ranging from 6 to 50%.[142] The chiral capping ligands induce chirality in encapsulated void, which in turn preorganises the substrates in such a way as to promote the formation of one enantiomeric product over the other.

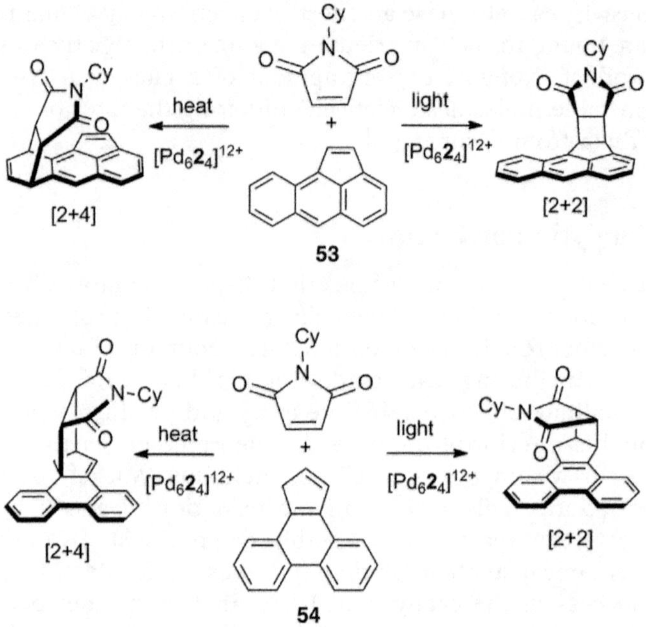

Figure 10.19 [2 + 2] and [2 + 4] pericyclic reactions between *N*-cyclohexylmaleimide and two inert aromatic molecules (**53** and **54**) occur with very high regio- and stereoselectivity when the substrates are encapsulated inside the $[Pd_6\mathbf{2}_4]^{12+}$ octahedron.

Further, modifications to capping ligands can be used to better encapsulate particular substrates, which in turn effects their reactivity. The use of the bulky capping group **37** reduces the void space within the cage through steric crowding by up to 20%.[143] This reduced volume allows for a 'tighter' fit for the host–guest complex of the substrates naphthalene and an N-substituted maleimide. This arrangement promotes a Diels–Alder reaction which does not occur in the cage prepared without the bulky mesitylene substituents, where the 'looser' fit of the substrates does not allow for the rigid pre-organisation required for the reaction of such an unreactive substrate. Interactions with the electron-deficient panels of the cage walls or stabilisation of anionic intermediates due to the cage's significant overall change can be used to catalyse a range of other reactions or lead to unusual regio-selectivities. The scope of reactions that are effected in this manner include nucleophilic substitution reactions,[144] oxidations[145] and Knoevenagel condensations[146] to name but a few.

Mukherjee *et al.* have prepared a similar polyhedron that has also been shown to catalyse the Knoevenagel condensation. The reaction of the tri(imidazole) ligand **55** with the *cis*-capped palladium(II) complex **1** results in the formation of a $[Pd_6\mathbf{55}_4]^{12+}$ octahedron analogous to Fujita's.[147] This cage catalysed the condensation reaction in aqueous solution by stabilising the anionic transition state inside the hydrophobic interior of the cage.

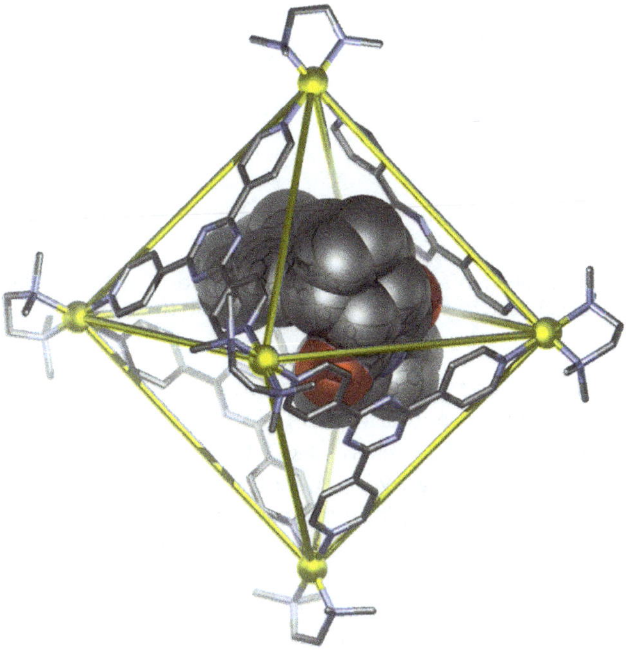

Figure 10.20 Crystal structure of the Diels–Alder adduct of *N*-cyclohexylmaleimide and 2,3-diethylnaphthalene encapsulated within the $[Pd_6\mathbf{2}_4]^{12+}$ octahedron. Guest molecule shown as a space-filling model.

$[Pd_6\mathbf{55}_4]^{12+}$ was also reported to catalyse the Diels–Alder cycloaddition between N-substituted maleimides and 9-hydroxymethyl anthracene, leading to much higher yields than in the non-encapsulated reaction.

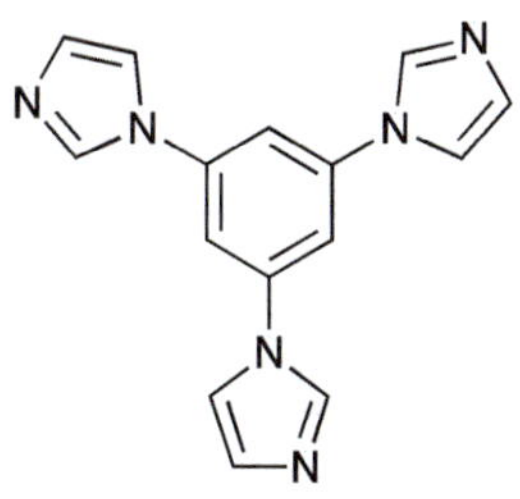

55

The $[Co_8\mathbf{42}_{12}]^{16+17}$ cube of Ward *et al.* catalyses the Kemp elimination of benzisoxazole (Figure 10.21) in water, accelerating the reaction rate by a factor of 2×10^5. It was proposed that the highly positive cage causes the aggregation of negatively charged hydroxide ions in its local area. These hydroxide ions then deprotonate the carbon adjacent to the benzisoxazole nitrogen, the crucial step in this elimination reaction. The product is not strongly bound within the cage, freely diffusing into solution and allowing catalytic turnover to occur.

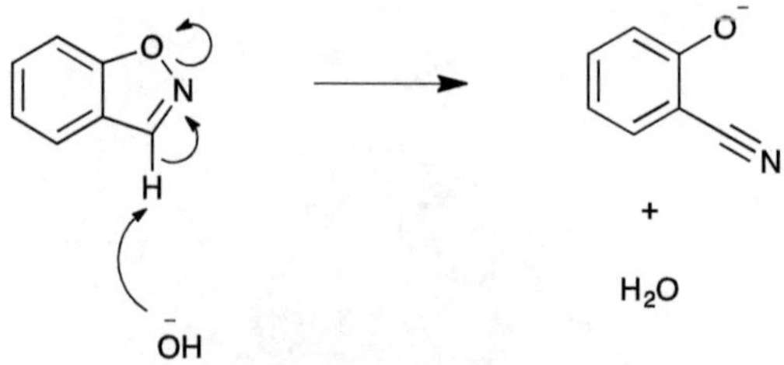

Figure 10.21 The mechanism of the Kemp elimination of benzisoxazole, which is accelerated by a factor of 2×10^5 inside the $[Co_8\mathbf{42}_{12}]^{16+}$ cube.

A number of cages constructed from multi-bidentate pyridyl-imine ligands have shown reaction modulation functionality. For example, Nitschke's $[Fe_4\mathbf{34}_6]^{4-}$ tetrahedron has been used to control the progress of the Diels–Alder reaction of furan with maleimide through a competitive host–guest inhibition process.[113] When the cage, furan and maleimide are combined in stoichiometric quantities, furan is encapsulated and sequestered keeping it separate from the maleimide, preventing the cycloaddition reaction from taking place. Upon addition of benzene, the benzene is bound within the cage, ejecting furan into the bulk solution where it can freely react with maleimide, forming the Diels–Alder adduct. A related, chiral, enantiopure cage from ligand **56** has been shown to have catalytic activity in the hydrolysis of organophosphates, demonstrated with the pesticide dichlorvos.[148] It is proposed that this is due to the interaction of the pendant hydroxyl groups with the substrate or due to polarisation of the substrate by the highly cationic cage, encouraging nucleophilic substitution at the phosphorous.

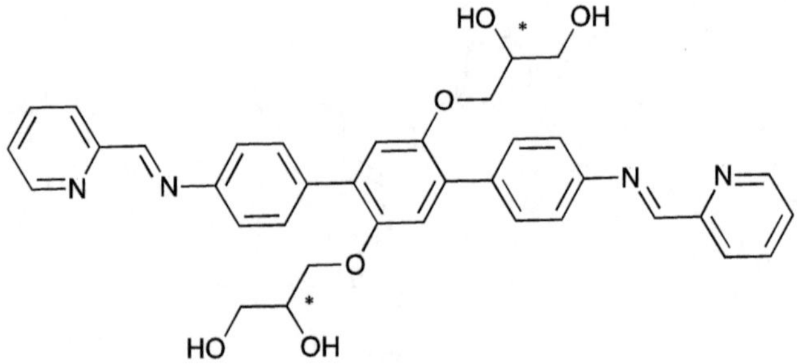

56

A third related tetrahedron, this time containing a BODIPY chromophore (**25**), has also been reported by Nitschke and co-workers.[149] When the terminal anilines had an iodine-containing substituent in the 4-position, photogeneration of singlet oxygen by the BODIPY unit was enhanced by up to a factor of two over non-iodine-containing residues. By using a mixed-system where the initial ligand is the 4-methylthio derivative and 4-iodoaniline is added to the solution, an autocatalytic process could be triggered. Small quantities of the iodoaniline replace the thiomethyl substituent due to the dynamic covalent chemistry of the imine bond. This leads to accelerated photogeneration of singlet oxygen, which oxidises the thiomethyl groups to methylsulfinyl moieties. These electron-poor methylsulfinyl groups are more readily replaced by the iodoaniline, leading to increased production of singlet oxygen, which in turn increases the rate of oxidation of the thiomethylaniline in a continuous cycle until oxidation of all thiomethylaniline is complete. The coordination of iron(ɪɪ) to form the cage stabilises the imine bonds to a sufficient extent to allow more than transient ligand formation, but still allows exchange of the terminal anilines.

A cubic, face-capped cage has been reported by de Bruin,[150] as an expanded form of a cage reported earlier by Nitschke.[121] The ligand (**57**) contains four bidentate 2,2′-bipyridyl coordinating groups linked by a zinc porphyrin core. The void space was sufficient to encapsulate a cobalt tetra(4-pyridyl) porphyrin, which coordinates through the four 4-pyridyl groups to fours of the zinc atoms on the cage faces. This cage was shown to be effective in catalysing a range of Co(porphyrin)-catalysed radical reactions with diazo compounds, such as cyclopropanation and olefin formation. The cage was shown to be similar in performance to the best free Co(porphyrin) catalysts for cyclopropanation and superior for olefin formation reactions.

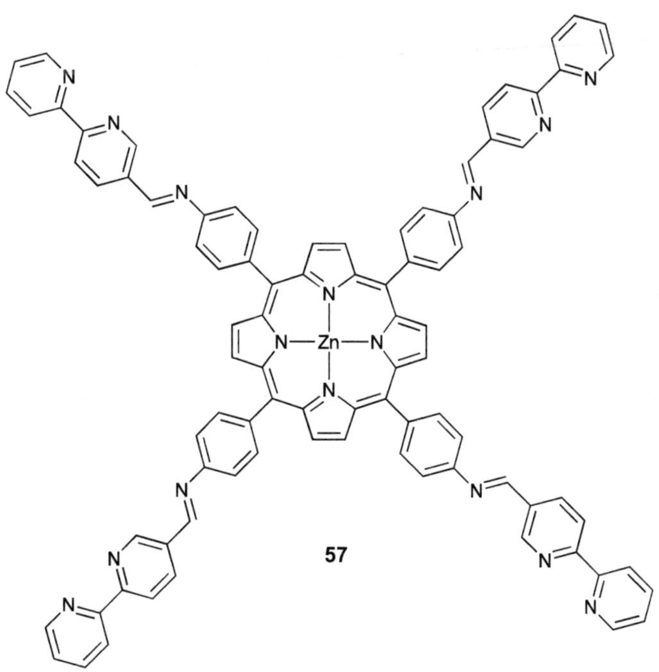

57

As discussed in Section 10.4.2 the $[Ga_4\mathbf{32}_6]^{12-}$ tetrahedron prepared by the group of Raymond readily binds a variety of cations in aqueous solvent. Indeed, the capsule has been shown to encapsulate protonated guest molecules at neutral or even basic pH, allowing reactions that are typically acid-catalysed to proceed under unusual conditions.[3] This has allowed reactions like acetal hydrolysis[151] and orthoformate hydrolysis[152] to proceed *via* acid-catalysed mechanisms within the cage cavity, even when at pH 10. The significant negative charge of the assembly catalyses a number of reactions that proceed *via* cationic transition states or begin from cationic substrates. For example, this cage has been shown to catalyse the aza-Cope rearrangement of allyl eneammonium cations. This reaction takes place *via* a [3,3] sigmatropic arrangement to form a cationic iminium product which is also stabilised by the cage. The iminium cation is then hydrolysed to give the corresponding aldehyde product which is ejected from the cage.[153] When an enantiopure tetrahedron was used, the aza-Cope rearrangement induced enantiomeric excesses ranging from 6 to 78% in the products.[154]

Other cyclisation reactions that proceed *via* cationic intermediates or transition states can also be catalysed by the $[Ga_4\mathbf{32}_6]^{12-}$ tetrahedron. The Nazarov cyclisation, which converts 1,4-pentadien-3-ols to cyclopentadiene derivatives, proceeds by a series of cationic intermediates and transition states, each of which are stabilised by the cage.[155] By reacting the product with maleimide in a Diels–Alder reaction, to remove it from the cage, 2.1–10^6-fold acceleration over the uncatalysed reaction was achieved. This compares favourably with reaction accelerations seen in enzymes. The cage was also seen to impart regioselectivity to the Nazarov cyclisation that is not seen in uncatalysed systems (Figure 10.22).[156]

The same material has been shown to catalyse the intramolecular Prins cyclisation of a molecule containing both and aldehyde and an alkene.[157] A library of four cages containing $[Ga_4\mathbf{32}_6]^{12-}$, its chiral analogue $[Ga_4\mathbf{58}_6]^{12-}$ and two larger cages ($[Ga_4\mathbf{59}_6]^{12-}$ and $[Ga_4\mathbf{60}_6]^{12-}$) in which the naphthalene spacer was replaced with pyrene, was prepared and the effect of cage volume and chirality upon the reaction was probed. It was found that enantioselectivity was more pronounced in the smaller naphthalene-spaced cages, most likely because the substrate is held more tightly in a single orientation, favouring the formation of one stereoisomer over the other three potential products. Reaction rates of the catalysed reaction were observed to be on the order of 10^5–10^6 times greater than the unencapsulated reaction. A related reaction, the aza-Prins cyclisation, where the aldehyde is replaced by a secondary amine in the presence of formaldehyde, was also reported to be catalysed.[158] For this reaction, a transient iminium ion intermediate is stabilised in the anionic cage, causing the selective formation of only a single one of two potential products, again due to orientation within the cage cavity.

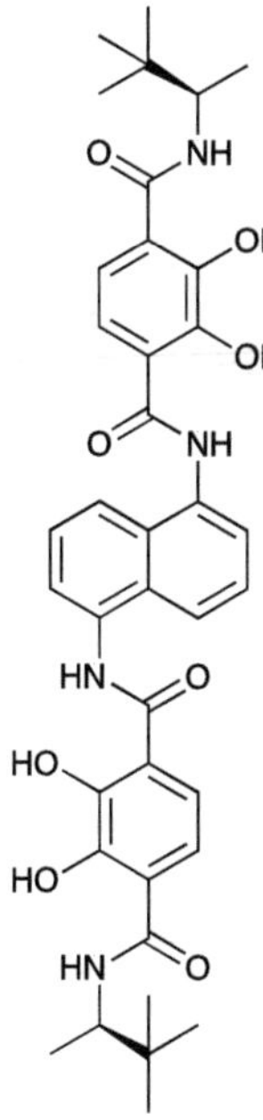

58

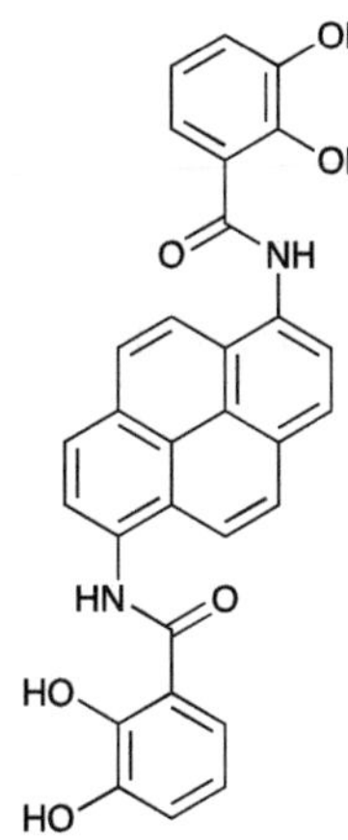

59

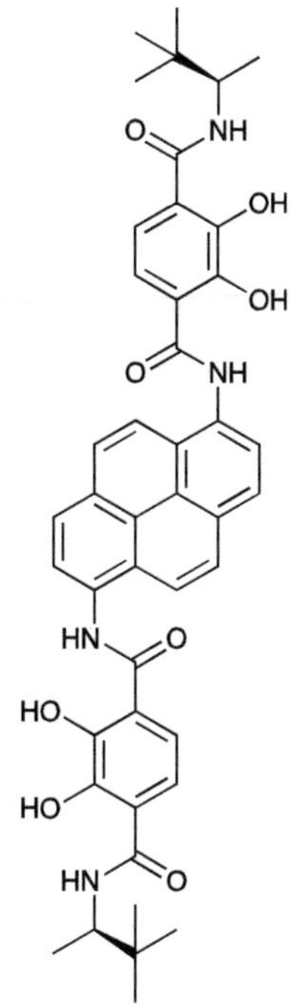

60

The incorporation of chiral amides into the cages formed from **58** allowed enantioselective reactions to be catalysed. For example, the Prins cyclisation could be catalysed with enantiomeric excesses 60–70%, with the major product dictated by the stereochemistry of the chiral amide.[159] This same chiral cage was also shown to catalyse asymmetric nucleophilic substitutions – namely the solvolytic substitution of imidates or alkyl chlorides to form ethers. In the presence of the chiral cage, retention of stereochemistry from substrate to product was observed to be 88% or greater, as opposed to the use of no catalyst, where only 30–40% stereochemical retention was observed.[77]

Raymond's tetrahedron has also been reported to act as a photocatalyst for the 1,3 rearrangement of cinnamylammonium guests. Unencapsulated, this reaction does not occur as it is highly thermodynamically disfavoured. When the substrate is encapsulated and UV light is applied, photoinduced electron transfer (PET) from the cage to the substrate occurs. This causes cleavage of the C–N bond, resulting in a free tertiary amine and an allyl radical cation, which then recombine to give the product.[160]

Neutral cages that catalyse reactions are less common. Duan *et al.* have reported two charge-neutral cerium-containing face-capped tetrahedra based on phenol-hydrazone coordination containing internal hydrogen bonding groups to encourage guest encapsulation.[161] These cages, [Ce_4**61**$_4$]

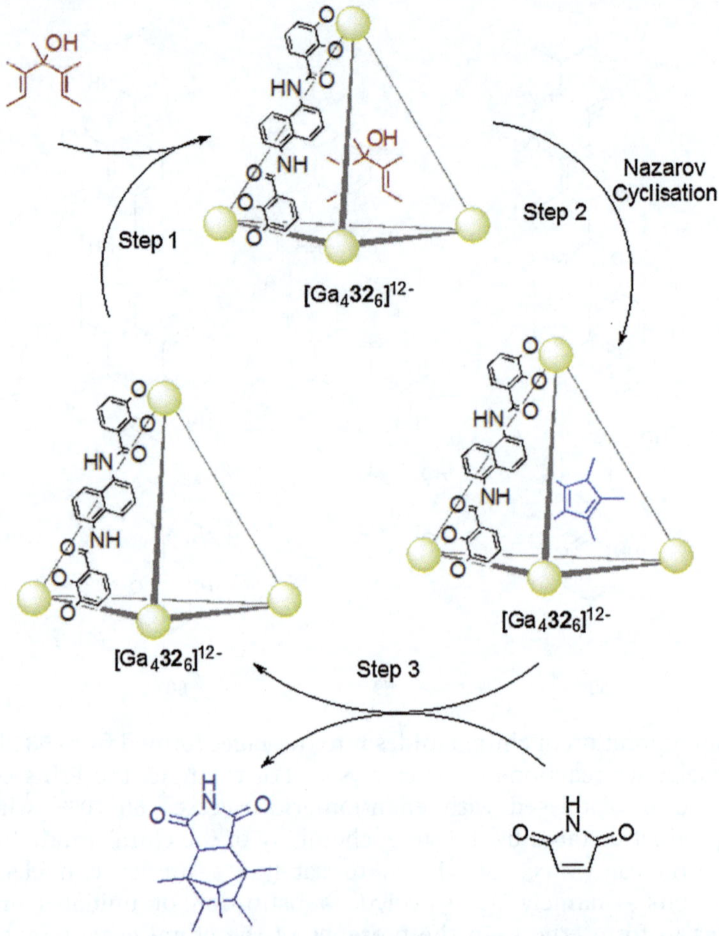

Figure 10.22 The conversion of 1,4-pentadien-3-ol to pentamethylcyclopentadiene *via* a Nazarov cyclisation is catalysed by the $[Ga_4\mathbf{32}_6]^{12-}$ metallo-supramolecular tetrahedron. Step 1: the 1,4-pentadien-3-ol substrate is encapsulated with in the hydrophobic cavity of the cage. Step 2: the cage catalyses conversion of the substrate to the pentamethylcyclopentadiene product *via* Nazarov cyclisation by stabilising the cationic transitions states. Step 3: the pentamethylcyclopentadiene is removed from the tetrahedron by a Diels–Alder reaction with malemide regenerating the empty host.

and $[Ce_4\mathbf{62}_4]$, have been shown to catalyse Knoevenagel condensations of salicylaldehyde derivatives and cyanosilylations of aryl aldehydes. The amine and amide hydrogens act as hydrogen bond donors to lock the substrates into a preferred orientation, as well as acting as the weak base catalyst that is required for these reactions to take place.

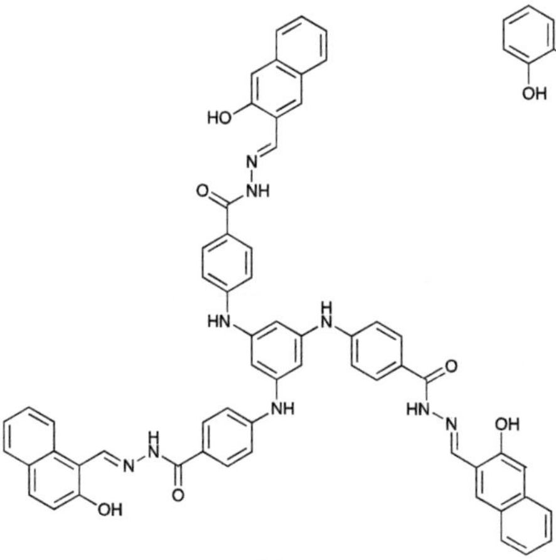

61

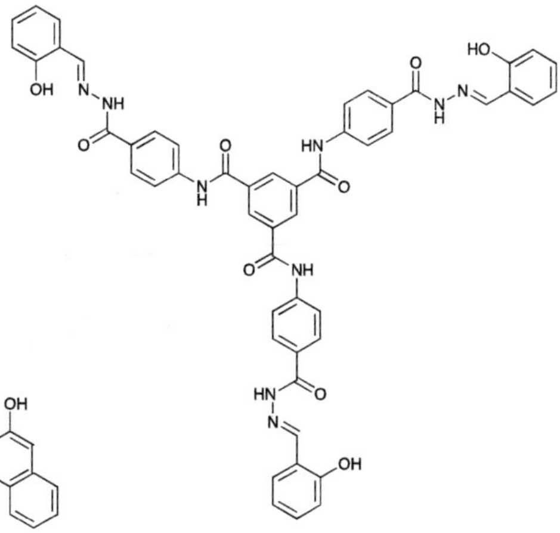

62

10.5.2 Reaction Inhibition

By applying many of the same principles used to impart catalytic properties on metallo-supramolecular polyhedra, these systems can also be used to stabilise highly reactive molecules or intermediates and/or inhibit reactivity. For example, Nitschke's $[Fe_4\mathbf{34}_6]^{4-}$ cage mentioned above has been shown to encapsulate white phosphorus, P_4, rendering it air-stable under atmospheric and aqueous conditions.[111] Without the cage, P_4 would spontaneously combust under these conditions. However, encapsulation within the hydrophobic interior of the cage stabilises P_4 by a steric method – the oxidation product and intermediates in the oxidation would be too large to fit within the cage (Figure 10.23).

A number of other highly reactive molecules, including intermediates, have been shown since then to be encapsulated, inhibiting reactions that normally occur under ambient conditions. Fujita's $[M_6\mathbf{2}_4]^{12+}$ octahedron has been shown encapsulate a number of reactive species. The palladium(II) version of the cage has been shown to encapsulate an unstable ruthenocene, containing a Ru–Ru bond, thereby creating a metastable host–guest complex. It has been proposed that this stability to the normally rapid photocleavage of the Ru–Ru bond is due to tight packing of the complex within the cage, forcing the ruthenium ions to remain in close proximity.[162] Encapsulated complexes of this type can then undergo transformations within the cage, reacting with alkynes to form diruthenacyclopentenones that maintain the Ru–Ru bond after ejection from the cage.[163] A range of other reactive metal species can be encapsulated and stabilised by this cage[164] and within Raymond's anionic tetrahedron[165] through similar mechanisms.

Earlier work by Fujita[166] showed that the platinum(II) version of this octahedral cage could be used to prepare and encapsulate small, highly reactive cyclic trimers of trialkylsiloxanes. These trimers are generally rapidly

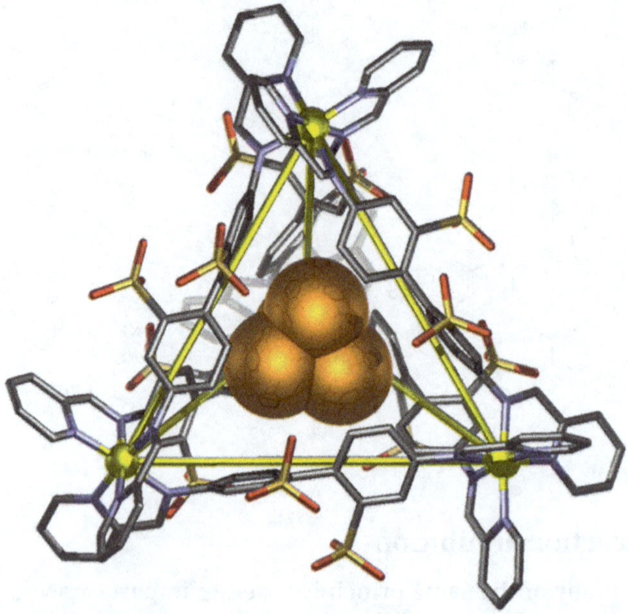

Figure 10.23 Crystal structure of white phosphorous encapsulated within the $[Fe_4\mathbf{34}_6]^{4-}$ metallo-supramolecular tetrahedron. The guest molecule is shown as a space-filling model. Encapsulation of the P_4 renders it air-stable as the tight-binding within the cavity prevents the formation of oxidation products or transition states.

consumed intermediates in the formation of silica nanoparticles *via* sol–gel condensations, but reaction in the presence of the cage led to the encapsulation of the trimeric species in 92% yield. If the cage was added after the condensation reaction was begun, no encapsulation of the trimer was observed. Again, this inhibition is due to tight packing of the siloxane trimer within the hydrophobic cavity of the octahedron.

Another cationic cage that inhibits photodegradation of encapsulated guests has been reported by Su and co-workers.[167] These rhombododecahedral $[Pd_6\mathbf{63}_8]^{28+}$ cages are assembled from the tritopic metalloligand **63**. This cage has been shown to prevent degradation of photosensitive organics such as 2,2-dimethoxy-2-phenylacetophenone, a light-curing agent found in some paints. The origin of this inhibition appears to be a combination of guest encapsulation and absorbance of UV light by the ruthenium metalloligand, as the cage can inhibit degradation for up to 120 h of exposure compared to 24 h for the free metalloligand.

Some cages have also been shown to regulate the formation of larger structures, finely controlling the size and morphology of the final product. For example, Fujita's group have reported the synthesis of a number of monodisperse nanoparticles by forcing the formation of these to occur within $[Pd_{12}L_{24}]^{48+}$ 'spherical' cages. The metallo-supramolecular assemblies essentially work as a mould, *endo*-templating the nano-particle formation. By attaching sugar residues to the inner face of the $[Pd_{12}\mathbf{64}_{24}]$[48] a hydrophilic surface was created. This surface proved to suitable for the sol–gel condensation of tetramethoxysilane producing silica nanoparticles of very narrow polydispersity. PdO nanospheres and TiO_2 nanoparticles have been prepared in a similar approach.[168]

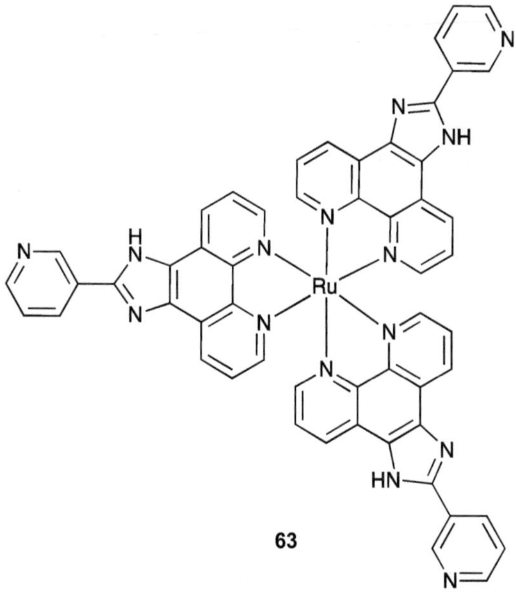

63

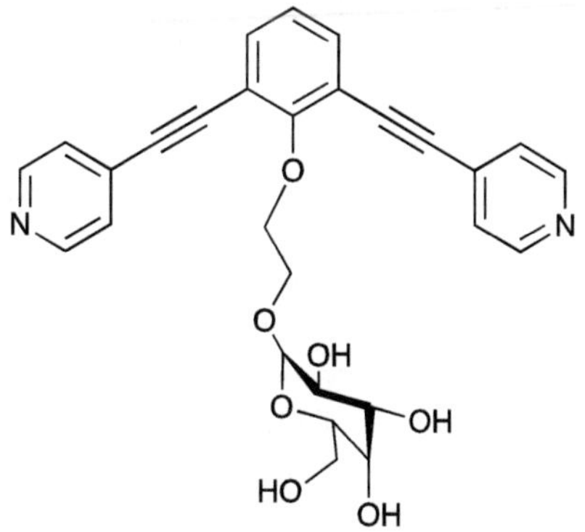

64

10.6 Conclusions and Outlook

There is enormous scope for the creation of functional metallo-supramolecular polyhedra. This is in large part because the polyhedra, by their very nature, are constructed of multiple components (metals, bridging ligands, counter-ions, guests) and each of these is a potential source of functionality.

Polyhedra are fascinating for their beauty and symmetry and this in its own right has provided significant motivation for many researchers interested in the creation of supramolecular structures. The deliberate preparation of functional systems, while challenging, serves to further amplify their beauty. Perhaps more importantly, efforts to create such species underpin the development and understanding of the fundamental principles of supramolecular design, allowing the creation of more structurally sophisticated examples, or those with more remarkable functions including stimuli responsive behaviour. While many of the endearing features of metallo-supramolecular capsules and cages could be viewed as esoteric, it is clear that these remarkable materials have great potential for commercial applications. Recent advances in the field are beginning to focus on the applications of metallo-supramolecular polyhedra for asymmetric synthesis and catalysis, or for drug delivery. It is likely that their use for these purposes will expand significantly over the coming years, particularly in light of the need for efficient synthesis of pharmaceuticals and biologically active agents. Whatever the future holds for these materials, one thing is certain – their design, construction and functionality has been at the forefront of supramolecular chemistry, where it will stay while there are chemists fascinated by the challenges of their creation and the beauty of their structures.

References

1. R. W. Saalfrank, A. Stark, K. Peters and H. G. von Schnering, *Angew. Chem., Int. Ed.*, 1988, **27**, 851–853.
2. R. Custelcean, *Chem. Soc. Rev.*, 2014, **43**, 1813–1824.
3. M. D. Pluth, R. G. Bergman and K. N. Raymond, *Acc. Chem. Res.*, 2009, **42**, 1650–1659.
4. D. Samanta and P. S. Mukherjee, *Dalton Trans.*, 2013, **42**, 16784–16795.
5. A. J. McConnell, C. S. Wood, P. P. Neelakandan and J. R. Nitschke, *Chem. Rev.*, 2015, **115**, 7729–7793.
6. Y. R. Hristova, M. M. J. Smulders, J. K. Clegg, B. Breiner and J. R. Nitschke, *Chem. Sci.*, 2011, **2**, 638–641; K. Suzuki, J. Iida, S. Sato, M. Kawano and M. Fujita, *Angew. Chem., Int. Ed.*, 2008, **47**, 5780–5782; R. W. Saalfrank, H. Glaser, B. Demleitner, F. Hampel, M. M. Chowdhry, V. Schunemann, A. X. Trautwein, G. B. M. Vaughan, R. Yeh, A. V. Davis and K. N. Raymond, *Chem. – Eur. J.*, 2002, **8**, 493–497.
7. M. Fujita, D. Oguro, M. Miyazawa, H. Oka, K. Yamaguchi and K. Ogura, *Nature*, 1995, **378**, 469–471.
8. W. J. Ramsay, T. K. Ronson, J. K. Clegg and J. R. Nitschke, *Angew. Chem., Int. Ed.*, 2013, **52**, 13439–13443.
9. L. F. Lindoy and I. Atkinson, *Self-Assembly in Supramolecular Systems*, Royal Society of Chemistry, Cambridge, UK., 2000.
10. R. W. Saalfrank, H. Maid and A. Scheurer, *Angew. Chem., Int. Ed.*, 2008, **47**, 8794–8824; D. L. Caulder and K. N. Raymond, *Acc. Chem. Res.*, 1999, **32**, 975–982; R. Chakrabarty, P. S. Mukherjee and P. J. Stang, *Chem.*

Rev., 2011, **111**, 6810–6918; M. D. Ward, *Chem. Commun.*, 2009, 4487–4499; S. J. Dalgarno, N. P. Power and J. L. Atwood, *Coord. Chem. Rev.*, 2008, **252**, 825–841; M. Yoshizawa, J. K. Klosterman and M. Fujita, *Angew. Chem., Int. Ed.*, 2009, **48**, 3418–3438; M. M. J. Smulders, I. A. Riddell, C. Browne and J. R. Nitschke, *Chem. Soc. Rev.*, 2013, **42**, 1728–1754; M. Fujita, *Chem. Soc. Rev.*, 1998, **27**, 417; T. R. Cook and P. J. Stang, *Chem. Rev.*, 2015, **115**, 7001–7045; C. Richardson, J. K. Clegg and L. F. Lindoy, in *Bioinspiration and Biomimicry in Chemistry*, ed. G. F. Swiegers, Wiley-VCH, Weinheim, 2012, pp. 17–46.

11. T. R. Cook, Y.-R. Zheng and P. J. Stang, *Chem. Rev.*, 2013, **113**, 734–777; N. J. Young and B. P. Hay, *Chem. Commun.*, 2013, **49**, 1354–1379.
12. D. J. Tranchemontagne, Z. Ni, M. O'Keeffe and O. M. Yaghi, *Angew. Chem., Int. Ed.*, 2008, **47**, 5136–5147.
13. D. L. Caulder and K. N. Raymond, *J. Chem. Soc., Dalton Trans.*, 1999, 1185–1200; L. Li, D. J. Fanna, N. D. Shepherd, L. F. Lindoy and F. Li, *J. Inclusion Phenom. Macrocyclic Chem.*, 2015, **82**, 3–12.
14. S. Leininger, B. Olenyuk and P. J. Stang, *Chem. Rev.*, 2000, **100**, 853–908; R. Chakrabarty, P. S. Mukherjee and P. J. Stang, *Chem. Rev.*, 2011, **111**, 6810–6918.
15. Y. Ozaki, M. Kawano and M. Fujita, *Chem. Commun.*, 2009, 4245–4247.
16. T. Sawada, H. Hisada and M. Fujita, *J. Am. Chem. Soc.*, 2014, **136**, 4449–4451.
17. W. Cullen, M. C. Misuraca, C. A. Hunter, N. H. Williams and M. D. Ward, *Nat. Chem.*, 2016, **8**, 231–236.
18. L. Tosheva and V. P. Valtchev, *Chem. Mat.*, 2005, **17**, 2494–2513.
19. S. Kitagawa, R. Kitaura and S. Noro, *Angew. Chem., Int. Ed.*, 2004, **43**, 2334–2375.
20. M. M. Conn and J. Rebek, *Chem. Rev.*, 1997, **97**, 1647–1668.
21. R. Pinalli, M. Suman and E. Dalcanale, *Eur. J. Inorg. Chem.*, 2004, 451–462.
22. G. Wenz, B. H. Han and A. Muller, *Chem. Rev.*, 2006, **106**, 782–817.
23. J. Lagona, P. Mukhopadhyay, S. Chakrabarti and L. Isaacs, *Angew. Chem., Int. Ed.*, 2005, **44**, 4844–4870.
24. U. K. Gautam, P. Costa, Y. Bando, X. S. Fang, L. Li, M. Imura and D. Golberg, *Sci. Tech. Adv. Mat.*, 2010, **11**, 1–9.
25. A. Rodriguez-Fortea, A. L. Balch and J. M. Poblet, *Chem. Soc. Rev.*, 2011, **40**, 3551–3563.
26. J. L. Sessler, P. A. Gale and W.-S. Cho, *Anion Receptor Chemistry*, Royal Society of Chemistry, 2006; *Supramolecular Chemistry of Anions*, ed. K. Bowman-James, A. Bianchi and E. García-Espana, Wiley-VCH, New York, 1997.
27. *Anion Coordination Chemistry*, ed. K. Bowman-James, A. Bianchi and E. García-Espana, Wiley-VCH Verlag GmbH & Co., Weinheim, 2012.
28. M. Frank, J. M. Dieterich, S. Freye, R. A. Mata and G. H. Clever, *Dalton Trans.*, 2013, **42**, 15906–15910; M. Han, R. Michel, B. He, Y.-S. Chen, D. Stalke, M. John and G. H. Clever, *Angew. Chem., Int. Ed.*, 2013, **52**,

1319–1323; J. E. M. Lewis and J. D. Crowley, *Aust. J. Chem.*, 2013, **66**, 1447–1454; R. Custelcean, *Chem. Commun.*, 2013, **49**, 2173–2182.
29. J. S. Fleming, K. L. V. Mann, C.-A. Carraz, E. Psillakis, J. C. Jeffery, J. A. McCleverty and M. D. Ward, *Angew. Chem., Int. Ed.*, 1998, **37**, 1279–1281.
30. S. Mann, G. Huttner, L. Zsolnai and K. Heinze, *Angew. Chem., Int. Ed.*, 1996, **35**, 2808–2809.
31. Z. R. Bell, J. C. Jeffery, J. A. McCleverty and M. D. Ward, *Angew. Chem., Int. Ed.*, 2002, **41**, 2515–2518.
32. R. L. Paul, S. P. Argent, J. C. Jeffery, L. P. Harding, J. M. Lynam and M. D. Ward, *Dalton Trans.*, 2004, 3453–3458.
33. F. Li, N. F. Sciortino, J. K. Clegg, S. M. Neville and C. J. Kepert, *Aust. J. Chem.*, 2014, **67**, 1625–1628.
34. C. R. K. Glasson, G. V. Meehan, J. K. Clegg, L. F. Lindoy, P. Turner, M. B. Duriska and R. Willis, *Chem. Commun.*, 2008, 1190–1192.
35. C. R. K. Glasson, J. K. Clegg, J. C. McMurtrie, G. V. Meehan, L. F. Lindoy, C. A. Motti, B. Moubaraki, K. S. Murray and J. D. Cashion, *Chem. Sci.*, 2011, **2**, 540–543.
36. C. R. K. Glasson, G. V. Meehan, C. A. Motti, J. K. Clegg, P. Turner, P. Jensen and L. F. Lindoy, *Dalton. Trans.*, 2011, **40**, 10481–10490.
37. A. Ferguson, M. A. Squire, D. Siretanu, D. Mitcov, C. Mathoniere, R. Clerac and P. E. Kruger, *Chem. Commun.*, 2013, **49**, 1597–1599.
38. B. El Aroussi, L. Guenee, P. Pal and J. Hamacek, *Inorg. Chem.*, 2011, **50**, 8588–8597.
39. J. Hamacek, D. Poggiali, S. Zebret, B. E. Aroussi, M. W. Schneider and M. Mastalerz, *Chem. Commun.*, 2012, **48**, 1281–1283.
40. C. J. Sumby and M. J. Hardie, *Angew. Chem., Int. Ed.*, 2005, **44**, 6395–6399; T. K. Ronson, J. Fisher, L. P. Harding and M. J. Hardie, *Angew. Chem., Int. Ed.*, 2007, **46**, 9086–9088; T. K. Ronson, J. Fisher, L. P. Harding, P. J. Rizkallah, J. E. Warren and M. J. Hardie, *Nat. Chem.*, 2009, **1**, 212–216.
41. N. J. Cookson, J. J. Henkelis, R. J. Ansell, C. W. G. Fishwick, M. J. Hardie and J. Fisher, *Dalton Trans.*, 2014, **43**, 5657–5661.
42. R. Custelcean, J. Bosano, P. V. Bonnesen, V. Kertesz and B. P. Hay, *Angew. Chem., Int. Ed.*, 2009, **48**, 4025–4029.
43. R. Custelcean, P. V. Bonnesen, N. C. Duncan, X. Zhang, L. A. Watson, G. Van Berkel, W. B. Parson and B. P. Hay, *J. Am. Chem. Soc.*, 2012, **134**, 8525–8534.
44. S. Yi, V. Brega, B. Captain and A. E. Kaifer, *Chem. Commun.*, 2012, **48**, 10295–10297.
45. Y. R. Hristova, M. M. J. Smulders, J. K. Clegg, B. Breiner and J. R. Nitschke, *Chem. Sci.*, 2011, **2**, 638–641.
46. S. Ma, M. M. J. Smulders, Y. R. Hristova, J. K. Clegg, T. K. Ronson, S. Zarra and J. R. Nitschke, *J. Am. Chem. Soc.*, 2013, **135**, 5678–5684.
47. I. A. Riddell, M. M. J. Smulders, J. K. Clegg, Y. R. Hristova, B. Breiner, J. D. Thoburn and J. R. Nitschke, *Nat. Chem.*, 2012, **4**, 751–756.

48. I. A. Riddell, Y. R. Hristova, J. K. Clegg, C. S. Wood, B. Breiner and J. R. Nitschke, *J. Am. Chem. Soc.*, 2013, **135**, 2723–2733; W. Meng, T. K. Ronson, J. K. Clegg and J. R. Nitschke, *Angew. Chem., Int. Ed.*, 2013, **52**, 1017–1021.
49. I. A. Riddell, T. K. Ronson, J. K. Clegg, C. S. Wood, R. A. Bilbeisi and J. R. Nitschke, *J. Am. Chem. Soc.*, 2014, **136**, 9491–9498.
50. I. A. Riddell, T. K. Ronson and J. R. Nitschke, *Chem. Sci.*, 2015, **6**, 3533–3537.
51. J. K. Clegg, J. Cremers, A. J. Hogben, B. Breiner, M. M. J. Smulders, J. D. Thoburn and J. R. Nitschke, *Chem. Sci.*, 2013, **4**, 68–76.
52. P. P. Neelakandan, A. Jimenez and J. R. Nitschke, *Chem. Sci.*, 2014, **5**, 908–915.
53. W. Meng, J. K. Clegg and J. R. Nitschke, *Angew. Chem., Int. Ed.*, 2012, **51**, 1881–1884.
54. W. Meng, A. B. League, T. K. Ronson, J. K. Clegg, W. C. Isley, D. Semrouni, L. Gagliardi, C. J. Cramer and J. R. Nitschke, *J. Am. Chem. Soc.*, 2014, **136**, 3972–3980.
55. M. M. J. Smulders, A. Jiménez and J. R. Nitschke, *Angew. Chem., Int. Ed.*, 2012, **51**, 6681–6685.
56. W. J. Ramsay, F. T. Szczypiński, H. Weissman, T. K. Ronson, M. M. J. Smulders, B. Rybtchinski and J. R. Nitschke, *Angew. Chem., Int. Ed.*, 2015, 5636–5640.
57. W. J. Ramsay and J. R. Nitschke, *J. Am. Chem. Soc.*, 2014, **136**, 7038–7043.
58. X. Hang, S. Wang, X. Zhu, H. Han and W. Liao, *CrystEngComm*, 2016.
59. R. W. Saalfrank, A. Stark, M. Brenner and H. U. Hummel, *Angew. Chem., Int. Ed.*, 1990, **29**, 311–315; R. M. Yeh, J. Xu, G. Seeber and K. N. Raymond, *Inorg. Chem.*, 2005, **44**, 6228–6239; C. Bruckner, R. E. Powers and K. N. Raymond, *Angew. Chem., Int. Ed.*, 1998, **37**, 1837–1839; M. Ziegler, J. J. Miranda, U. N. Andersen, D. W. Johnson, J. A. Leary and K. N. Raymond, *Angew. Chem., Int. Ed.*, 2001, **40**, 733–736.
60. R. W. Saalfrank, R. Burak, A. Breit, D. Stalke, R. Herbst-Irmer, J. Daub, M. Porsch, E. Bill, M. Muether and A. X. Trautwein, *Angew. Chem., Int. Ed. Engl.*, 1994, **33**, 1621–1623.
61. M. D. Pluth, D. W. Johnson, G. Szigethy, A. V. Davis, S. J. Teat, A. G. Oliver, R. G. Bergman and K. N. Raymond, *Inorg. Chem.*, 2009, **48**, 111–120.
62. M. Scherer, D. L. Caulder, D. W. Johnson and K. N. Raymond, *Angew. Chem., Int. Ed.*, 1999, **38**, 1588–1592.
63. M. D. Pluth, R. G. Bergman and K. N. Raymond, *Acc. Chem. Res.*, 2009, **42**, 1650–1659; T. N. Parac, D. L. Caulder and K. E. Raymond, *J. Am. Chem. Soc.*, 1998, **120**, 8003; D. L. Caulder, R. E. Powers, T. N. Parac and K. N. Raymond, *Angew. Chem., Int. Ed.*, 1998, **37**, 1840–1843.
64. M. D. Pluth, D. Fiedler, J. S. Mugridge, R. G. Bergman and K. N. Raymond, *Proc. Natl. Acad. Sci. U. S. A.*, 2009, **1–6**, 6; M. D. Pluth, R. G. Bergman and K. N. Raymond, *J. Am. Chem. Soc.*, 2008, **130**,

6362–6366; V. M. Dong, D. Fiedler, B. Carl, R. G. Bergman and K. N. Raymond, *J. Am. Chem. Soc.*, 2006, **128**, 14464–14465; T. N. Parac, D. L. Caulder and K. N. Raymond, *J. Am. Chem. Soc.*, 1998, **120**, 8003–8004.
65. T. N. Parac, M. Scherer and K. N. Raymond, *Angew. Chem., Int. Ed.*, 2000, **39**, 1239–1242.
66. D. Fiedler, D. Pagliero, J. L. Brumaghim, R. G. Bergman and K. N. Raymond, *Inorg. Chem.*, 2004, **43**, 846–848.
67. B. E. F. Tiedemann and K. N. Raymond, *Angew. Chem., Int. Ed.*, 2006, **45**, 83–86.
68. A. V. Davis, D. Fiedler, G. Seeber, A. Zahl, R. Van Eldik and K. N. Raymond, *J. Am. Chem. Soc.*, 2006, **128**, 1324–1333.
69. V. M. Dong, D. Fiedler, B. Carl, R. G. Bergman and K. N. Raymond, *J. Am. Chem. Soc.*, 2006, **128**, 14464–14465.
70. M. Ziegler, J. L. Brumaghim and K. N. Raymond, *Angew. Chem., Int. Ed.*, 2000, **39**, 4119.
71. D. Fiedler, R. G. Bergman and K. N. Raymond, *Angew. Chem., Int. Ed.*, 2006, **45**, 745–748.
72. C. Sgarlata, J. S. Mugridge, M. D. Pluth, B. E. F. Tiedemann, V. Zito, G. Arena and K. N. Raymond, *J. Am. Chem. Soc.*, 2010, **132**, 1005–1009.
73. A. J. Terpin, M. Ziegler, D. W. Johnson and K. N. Raymond, *Angew. Chem., Int. Ed.*, 2001, **40**, 157–160; A. V. Davis, D. Fiedler, M. Ziegler, A. Terpin and K. N. Raymond, *J. Am. Chem. Soc.*, 2007, **129**, 15354–15363.
74. D. Fiedler, D. H. Leung, R. G. Bergman and K. N. Raymond, *J. Am. Chem. Soc.*, 2004, **126**, 3674–3675.
75. C. J. Brown, R. G. Bergman and K. N. Raymond, *J. Am. Chem. Soc.*, 2009, **131**, 17530–17531.
76. C. Zhao, Q.-F. Sun, W. M. Hart-Cooper, A. G. Di Pasquale, F. D. Toste, R. G. Bergman and K. N. Raymond, *J. Am. Chem. Soc.*, 2013, **135**, 18802–18805.
77. C. Zhao, F. D. Toste, K. N. Raymond and R. G. Bergman, *J. Am. Chem. Soc.*, 2014, **136**, 14409–14412.
78. B. F. Abrahams, N. J. FitzGerald and R. Robson, *Angew. Chem., Int. Ed.*, 2010, **49**, 2896–2899; B. F. Abrahams, B. A. Boughton, N. J. FitzGerald, J. L. Holmes and R. Robson, *Chem. Commun.*, 2011, **47**, 7404–7406.
79. P. Mal, D. Schultz, K. Beyeh, K. Rissanen and J. R. Nitschke, *Angew. Chem., Int. Ed.*, 2008, **47**, 8297–8301.
80. S. Zarra, M. M. J. Smulders, Q. Lefebvre, J. K. Clegg and J. R. Nitschke, *Angew. Chem., Int. Ed.*, 2012, **51**, 6882–6885.
81. C. Reichardt and T. Welton, *Solvents and Solvent Effects in Organic Chemistry*, VCH, New York, 2011; N. T. Southall, K. A. Dill and A. D. J. Haymet, *J. Phys. Chem. B.*, 2002, **106**, 521–533; D. B. Smithrud, E. M. Sanford, I. Chao, S. B. Ferguson, D. R. Carcanague, J. D. Evanseck, K. N. Houk and F. Diederich, *Pure Appl. Chem.*, 1990, **62**, 2227–2236.
82. J. H. Jordan and B. C. Gibb, *Chem. Soc. Rev.*, 2015, **44**, 547–585; J. Rebek Jr, *Hydrogen-bonded Capsules: Molecular Behavior in Small Spaces*, World Scientific, Singapore, 2015.

83. Y. Fang, T. Murase and M. Fujita, *Chem. – Asian J.*, 2014, **9**, 1321–1328; Y. Fang, T. Murase, S. Sato and M. Fujita, *J. Am. Chem. Soc.*, 2013, **135**, 613–615.
84. S. Tashiro, M. Kobayashi and M. Fujita, *J. Am. Chem. Soc.*, 2006, **128**, 9280–9281; B. Roy, A. K. Ghosh, S. Srivastava, P. D'Silva and P. S. Mukherjee, *J. Am. Chem. Soc.*, 2015, **137**, 11916–11919; S. Bivaud, J. Y. Balandier, M. Chas, M. Allain, S. Goeb and M. Salle, *J. Am. Chem. Soc.*, 2012, **134**, 11968–11970.
85. K. Ono, J. K. Klosterman, M. Yoshizawa, K. Sekiguchi, T. Tahara and M. Fujita, *J. Am. Chem. Soc.*, 2009, **131**, 12526–12527.
86. H. Takezawa, T. Murase, G. Resnati, P. Metrangolo and M. Fujita, *J. Am. Chem. Soc.*, 2014, **136**, 1786–1788.
87. T. Sawada, M. Yoshizawa, S. Sato and M. Fujita, *Nat. Chem.*, 2009, **1**, 53–56.
88. H. Takezawa, T. Murase, G. Resnati, P. Metrangolo and M. Fujita, *Angew. Chem., Int. Ed.*, 2015, **54**, 8411–8414.
89. T. Kusukawa and M. Fujita, *J. Am. Chem. Soc.*, 2002, **124**, 13576–13582; Y. Kohyama, T. Murase and M. Fujita, *Angew. Chem., Int. Ed.*, 2014, **53**, 11510–11513.
90. S. Tashiro and M. Fujita, *Bull. Chem. Soc. Jpn.*, 2006, **79**, 833–837.
91. S. Tashiro, M. Tominaga, M. Kawano, B. Therrien, T. Ozeki and M. Fujita, *J. Am. Chem. Soc.*, 2005, **127**, 4546–4547.
92. K. Kumazawa, K. Biradha, T. Kusukawa, T. Okano and M. Fujita, *Angew. Chem., Int. Ed.*, 2003, **42**, 3909–3913; M. Yoshizawa, K. Kumazawa and M. Fujita, *J. Am. Chem. Soc.*, 2005, **127**, 13456–13457.
93. G. Gupta, E. Denoyelle-Di-Muro, J. P. Mbakidi, S. Leroy-Lhez, V. Sol and B. Therrien, *J. Organomet. Chem.*, 2015, **787**, 44–50; B. Kilbas, S. Mirtschin, T. Riis-Johannessen, R. Scopelliti and K. Severin, *Inorg. Chem.*, 2012, **51**, 5795–5804; Y. Y. Zhang, L. Zhang, Y. J. Lin and G. X. Jin, *Chem. – Eur. J.*, 2015, **21**, 14893–14900.
94. M. Yoshizawa, J. Nakagawa, K. Kumazawa, M. Nagao, M. Kawano, T. Ozeki and M. Fujita, *Angew. Chem., Int. Ed.*, 2005, **44**, 1810–1813.
95. Y. Yamauchi, M. Yoshizawa, M. Akita and M. Fujita, *J. Am. Chem. Soc.*, 2010, **132**, 960–966.
96. M. Kiguchi, T. Takahashi, Y. Takahashi, Y. Yamauchi, T. Murase, M. Fujita, T. Tada and S. Watanabe, *Angew. Chem., Int. Ed.*, 2011, **50**, 5708–5711.
97. T. Osuga, T. Murase and M. Fujita, *Angew. Chem., Int. Ed.*, 2012, **51**, 12199–12201.
98. P. C. Duan, Z. Y. Wang, J. H. Chen, G. Yang and R. G. Raptis, *Dalton Trans.*, 2013, **42**, 14951–14954.
99. S. Turega, M. Whitehead, B. R. Hall, M. F. Haddow, C. A. Hunter and M. D. Ward, *Chem. Commun.*, 2012, **48**, 2752–2754.
100. S. Turega, M. Whitehead, B. R. Hall, A. J. H. M. Meijer, C. A. Hunter and M. D. Ward, *Inorg. Chem.*, 2013, **52**, 1122–1132.
101. M. Whitehead, S. Turega, A. Stephenson, C. A. Hunter and M. D. Ward, *Chem. Sci.*, 2013, **4**, 2744–2751.

102. S. Turega, W. Cullen, M. Whitehead, C. A. Hunter and M. D. Ward, *J. Am. Chem. Soc.*, 2014, **136**, 8475–8483.
103. W. Cullen, S. Turega, C. A. Hunter and M. D. Ward, *Chem. Sci.*, 2015, **6**, 2790–2794.
104. W. Cullen, S. Turega, C. A. Hunter and M. D. Ward, *Chem. Sci.*, 2015, **6**, 625–631.
105. W. Cullen, K. A. Thomas, C. A. Hunter and M. D. Ward, *Chem. Sci.*, 2015, **6**, 4025–4028.
106. C. J. Hastings, M. D. Pluth, S. M. Biros, R. G. Bergman and K. N. Raymond, *Tetrahedron*, 2008, **64**, 8362–8367.
107. B. E. F. Tiedemann and K. N. Raymond, *Angew. Chem., Int. Ed.*, 2007, **46**, 4976–4978.
108. M. D. Pluth, R. G. Bergman and K. N. Raymond, *J. Org. Chem.*, 2008, **73**, 7132–7136.
109. J. S. Mugridge, R. G. Bergman and K. N. Raymond, *Angew. Chem., Int. Ed.*, 2010, **49**, 3635–3637.
110. T. K. Ronson, C. Giri, N. Kodiah Beyeh, A. Minkkinen, F. Topić, J. J. Holstein, K. Rissanen and J. R. Nitschke, *Chem. – Eur. J.*, 2013, **19**, 3374–3382.
111. P. Mal, B. Breiner, K. Rissanen and J. R. Nitschke, *Science*, 2009, **324**, 1697–1699.
112. P. Mal and J. R. Nitschke, *Chem. Commun.*, 2010, **46**, 2417–2419; I. A. Riddell, M. M. J. Smulders, J. K. Clegg and J. R. Nitschke, *Chem. Commun.*, 2011, **47**, 457–459; J. A. Foster, R. M. Parker, A. M. Belenguer, N. Kishi, S. Sutton, C. Abell and J. R. Nitschke, *J. Am. Chem. Soc.*, 2015, **137**, 9722–9729; J. Roukala, J. F. Zhu, C. Giri, K. Rissanen, P. Lantto and V. V. Telkki, , *J. Am. Chem. Soc.*, 2015, **137**, 2464–2467.
113. M. M. J. Smulders and J. R. Nitschke, *Chem. Sci.*, 2012, **3**, 785–788.
114. A. B. Grommet, J. L. Bolliger, C. Browne and J. R. Nitschke, *Angew. Chem., Int. Ed.*, 2015, **54**, 15100–15104.
115. J. L. Bolliger, T. K. Ronson, M. Ogawa and J. R. Nitschke, *J. Am. Chem. Soc.*, 2014, **136**, 14545–14553.
116. J. K. Clegg, L. F. Lindoy, B. Moubaraki, K. S. Murray and J. C. McMurtrie, *Dalton. Trans.*, 2004, 2417–2423; J. K. Clegg, F. Li, K. A. Jolliffe, G. V. Meehan and L. F. Lindoy, *Chem. Commun.*, 2011, **47**, 6042–6044.
117. T. Liu, Y. Liu, W. Xuan and Y. Cui, *Angew. Chem., Int. Ed.*, 2010, **49**, 4121–4124.
118. R. A. Bilbeisi, J. K. Clegg, N. Elgrishi, X. D. Hatten, M. Devillard, B. Breiner, P. Mal and J. R. Nitschke, *J. Am. Chem. Soc.*, 2012, **134**, 5110–5119.
119. R. A. Bilbeisi, T. K. Ronson and J. R. Nitschke, *Angew. Chem., Int. Ed.*, 2013, **52**, 9027–9030.
120. C. Browne, S. Brenet, J. K. Clegg and J. R. Nitschke, *Angew. Chem., Int. Ed.*, 2013, **52**, 1944–1948.
121. W. Meng, B. Breiner, K. Rissanen, J. D. Thoburn, J. K. Clegg and J. R. Nitschke, *Angew. Chem., Int. Ed.*, 2011, **50**, 3479–3483.

122. T. Nakamura, H. Ube, M. Shiro and M. Shionoya, *Angew. Chem., Int. Ed.*, 2013, **52**, 720–723.
123. P. D. Frischmann, V. Kunz and F. Wurthner, *Angew. Chem., Int. Ed.*, 2015, **54**, 7285–7289.
124. T. K. Ronson, A. B. League, L. Gagliardi, C. J. Cramer and J. R. Nitschke, *J. Am. Chem. Soc.*, 2014, **136**, 15615–15624.
125. D. J. Tranchemontagne, J. L. Mendoza-Cortes, M. O'Keeffe and O. M. Yaghi, *Chem. Soc. Rev.*, 2009, **38**, 1257–1283.
126. M. Eddaoudi, D. B. Moler, H. L. Li, B. L. Chen, T. M. Reineke, M. O'Keeffe and O. M. Yaghi, *Acc. Chem. Res.*, 2001, **34**, 319–330.
127. N. Ahmad, A. H. Chughtai, H. A. Younus and F. Verpoort, *Coord. Chem. Rev.*, 2014, **280**, 1–27.
128. A. Mallick, B. Garai, D. D. Díaz and R. Banerjee, *Angew. Chem., Int. Ed.*, 2013, **52**, 13755–13759.
129. D. J. L. Tranchemontagne, Z. Ni, M. O'Keeffe and O. M. Yaghi, *Angew. Chem., Int. Ed.*, 2008, **47**, 5136–5147.
130. J.-R. Li and H.-C. Zhou, *Nat. Chem.*, 2010, **2**, 893–898.
131. A. K. Gupta, A. Yadav, A. K. Srivastava, K. R. Ramya, H. Paithankar, S. Nandi, J. Chugh and R. Boomishankar, *Inorg. Chem.*, 2015, **54**, 3196–3202; J.-R. Li, J. Yu, W. Lu, L.-B. Sun, J. Sculley, P. B. Balbuena and H.-C. Zhou, *Nat. Commun.*, 2013, **4**, 1538.
132. S. Pasquale, S. Sattin, E. C. Escudero-Adan, M. Martinez-Belmonte and J. de Mendoza, *Nat. Commun.*, 2012, **3**, 785.
133. F. R. Dai and Z. Q. Wang, *J. Am. Chem. Soc.*, 2012, **134**, 8002–8005; F. R. Dai, U. Sambasivam, A. J. Hammerstrom and Z. Q. Wang, *J. Am. Chem. Soc.*, 2014, **136**, 7480–7491; S. Du, T.-Q. Yu, W. Liao and C. Hu, *Dalton Trans.*, 2015, **44**, 14394–14402.
134. M. B. Duriska, S. M. Neville, J. Lu, S. S. Iremonger, J. F. Boas, C. J. Kepert and S. R. Batten, *Angew. Chem., Int. Ed.*, 2009, **48**, 8919–8922.
135. M. B. Duriska, S. M. Neville, B. Moubaraki, J. D. Cashion, G. J. Halder, K. W. Chapman, C. Balde, J.-F. Letard, K. S. Murray, C. J. Kepert and S. R. Batten, *Angew. Chem., Int. Ed.*, 2009, **48**, 2549–2552; M. B. Duriska, S. M. Neville, B. Moubaraki, K. S. Murray, C. Balde, J.-F. Letard, C. J. Kepert and S. R. Batten, *ChemPlusChem*, 2012, **77**, 616–623.
136. H. Ito, T. Kusukawa and M. Fujita, *Chem. Lett.*, 2000, 598–599; M. Yoshizawa, N. Sato and M. Fujita, *Chem. Lett.*, 2005, **34**, 1392–1393.
137. M. Yoshizawa, Y. Takeyama, T. Kusukawa and M. Fujita, *Angew. Chem., Int. Ed.*, 2002, **41**, 1347–1349; M. Yoshizawa, Y. Takeyama, T. Okano and M. Fujita, *J. Am. Chem. Soc.*, 2003, **125**, 3243–3247.
138. M. Yoshizawa, S. Miyagi, M. Kawano, K. Ishiguro and M. Fujita, *J. Am. Chem. Soc.*, 2004, **126**, 9172–9173.
139. Y. Nishioka, T. Yamaguchi, M. Kawano and M. Fujita, *J. Am. Chem. Soc.*, 2008, **130**, 8160–8161; T. Yamaguchi and M. Fujita, *Angew. Chem., Int. Ed.*, 2008, **47**, 2067–2069.
140. T. Murase, S. Horiuchi and M. Fujita, *J. Am. Chem. Soc.*, 2010, **132**, 2866–2867.

141. S. Horiuchi, Y. Nishioka, T. Murase and M. Fujita, *Chem. Commun.*, 2010, **46**, 3460–3462.
142. T. Murase, S. Peschard, S. Horiuchi, Y. Nishioka and M. Fujita, *Supramol. Chem.*, 2011, **23**, 199–208.
143. Y. Fang, T. Murase and M. Fujita, *Chem. Lett.*, 2015, **44**, 1095–1097.
144. Y. Kohyama, T. Murase and M. Fujita, *Chem. Commun.*, 2012, **48**, 7811–7813.
145. X. Qiu, W. Zhong, C. Bai and Y. Li, *J. Am. Chem. Soc.*, 2016, **138**, 1138–1141.
146. T. Murase, Y. Nishijima and M. Fujita, *J. Am. Chem. Soc.*, 2012, **134**, 162–164.
147. D. Samanta, S. Mukherjee, Y. P. Patil and P. S. Mukherjee, *Chem. – Eur. J.*, 2012, **18**, 12322–12329; D. Samanta and P. S. Mukherjee, *Chem. Commun.*, 2013, **49**, 4307–4309.
148. J. L. Bolliger, A. M. Belenguer and J. R. Nitschke, *Angew. Chem., Int. Ed.*, 2013, **52**, 7958–7962.
149. P. P. Neelakandan, A. Jimenez, J. D. Thoburn and J. R. Nitschke, *Angew. Chem., Int. Ed.*, 2015, **54**, 14378–14382.
150. M. Otte, P. F. Kuijpers, O. Troeppner, I. Ivanovic-Burmazovic, J. N. H. Reek and B. de Bruin, *Chem. – Eur. J.*, 2013, **19**, 10170–10178.
151. M. D. Pluth, R. G. Bergman and K. N. Raymond, *J. Org. Chem.*, 2009, **74**, 58–63.
152. M. D. Pluth, R. G. Bergman and K. N. Raymond, *J. Am. Chem. Soc.*, 2008, **130**, 11423–11429.
153. D. Fiedler, H. van Halbeek, R. G. Bergman and K. N. Raymond, *J. Am. Chem. Soc.*, 2006, **128**, 10240–10252.
154. C. J. Brown, R. G. Bergman and K. N. Raymond, *J. Am. Chem. Soc.*, 2009, **131**, 17530–17531.
155. C. J. Hastings, M. D. Pluth, R. G. Bergman and K. N. Raymond, *J. Am. Chem. Soc.*, 2010, **132**, 6938–6940.
156. C. J. Hastings, M. P. Backlund, R. G. Bergman and K. N. Raymond, *Angew. Chem., Int. Ed.*, 2011, **50**, 10570–10573.
157. W. M. Hart-Cooper, C. Zhao, R. M. Triano, P. Yaghoubi, H. L. Ozores, K. N. Burford, F. D. Toste, R. G. Bergman and K. N. Raymond, *Chem. Sci.*, 2015, **6**, 1383–1393.
158. D. M. Kaphan, F. D. Toste, R. G. Bergman and K. N. Raymond, *J. Am. Chem. Soc.*, 2015, **137**, 9202–9205.
159. C. Zhao, Q.-F. Sun, W. M. Hart-Cooper, A. G. Di Pasquale, F. D. Toste, R. G. Bergman and K. N. Raymond, *J. Am. Chem. Soc.*, 2013, **135**, 18802–18805.
160. D. M. Dalton, S. R. Ellis, E. M. Nichols, R. A. Mathies, F. D. Toste, R. G. Bergman and K. N. Raymond, *J. Am. Chem. Soc.*, 2015, **137**, 10128–10131.
161. J. Zhang, H. Yu, C. X. Zhang, C. He and C. Y. Duan, *New J. Chem.*, 2014, **38**, 3137–3145.
162. S. Horiuchi, T. Murase and M. Fujita, *J. Am. Chem. Soc.*, 2011, **133**, 12445–12447.

163. S. Horiuchi, T. Murase and M. Fujita, *Angew. Chem., Int. Ed.*, 2012, **51**, 12029–12031.
164. M. Kawano, Y. Kobayashi, T. Ozeki and M. Fujita, *J. Am. Chem. Soc.*, 2006, **128**, 6558–6559.
165. C. J. Brown, G. M. Miller, M. W. Johnson, R. G. Bergman and K. N. Raymond, *J. Am. Chem. Soc.*, 2011, **133**, 11964–11966; D. H. Leung, R. G. Bergman and K. N. Raymond, *J. Am. Chem. Soc.*, 2006, **128**, 9781–9797.
166. M. Yoshizawa, T. Kusukawa, M. Fujita and K. Yamaguchi, *J. Am. Chem. Soc.*, 2000, **122**, 6311–6312.
167. K. Li, L. Y. Zhang, C. Yan, S. C. Wei, M. Pan, L. Zhang and C. Y. Su, *J. Am. Chem. Soc.*, 2014, **136**, 4456–4459.
168. K. Takao, K. Suzuki, T. Ichijo, S. Sato, H. Asakura, K. Teramura, K. Kato, T. Ohba, T. Morita and M. Fujita, *Angew. Chem., Int. Ed.*, 2012, **51**, 5893–5896; T. Ichijo, S. Sato and M. Fujita, *J. Am. Chem. Soc.*, 2013, **135**, 6786–6789; K. Harris, D. Fujita and M. Fujita, *Chem. Commun.*, 2013, **49**, 6703–6712.

CHAPTER 11

Designing Functional Porous Organic Frameworks for Gas Storage and Separation

PEZHMAN ARAB,[a] OUSSAMA M. EL-KADRI[b] AND HANI M. EL-KADERI*[a]

[a] Department of Chemistry, Virginia Commonwealth University, Richmond, Virginia 23284-2006, USA; [b] Department of Biology, Chemistry, and Environmental Sciences, American University of Sharjah, PO Box 26666, Sharjah, United Arab Emirates
*Email: helkaderi@vcu.edu

11.1 Introduction

Owing to their large surface area and accessible pores, porous materials play important roles in a wide variety of fields, such as catalysis, energy conversion, sensing, gas separation, and gas storage.[1–5] The accessibility of pores and the surface of porous materials allows for their physical and chemical interactions with molecules and ions in both gas phase and liquid phase.[6] The nature and the degree of such interactions can be governed by the chemical nature of the pores as well as size, shape, and volume of the pores.[6,7] Conventional porous materials such as zeolites and porous carbons have been widely used in industry due to their high surface area and high physicochemical stability.[4,8] However, the surface chemistry of the pores in these materials cannot be easily controlled as it is mainly determined by the nature of the material.[8] Although the surface chemistry of zeolites and

Monographs in Supramolecular Chemistry No. 22
Functional Supramolecular Materials: From Surfaces to MOFs
Edited by Rahul Banerjee

Published by the Royal Society of Chemistry, www.rsc.org

porous carbons can be tuned to some extent *via* post-synthetic modification methods, this low degree of control over surface chemistry of the pores can limit their advanced applications.[9] As a result, developing new synthetic strategies to prepare novel porous materials with desired physicochemical properties have attracted considerable attention by scientific and industrial communities.[5,9] Very recently, porous organic frameworks (POFs) have been introduced as a new class of materials with high surface area, physicochemical stability, and chemical tunability.[4,10] Porous organic frameworks are synthesized *via* a bottom-up approach upon polymerization of rigid organic monomers having star-shaped topology.[11] In this approach, the pores are created *in situ* during polymerization due to the topology of monomers used.[11] The rigidity of the monomers prevent these polymeric frameworks from packing into a dense structure which leads to permanent porosity of POFs.[11] Owing to availability of a wide variety of organic monomers and polymerization reactions, several families of POFs such as porous aromatic frameworks (PAFs),[10] conjugated microporous polymers (CMPs),[12] covalent organic frameworks (COFs),[13] hyper-cross-linked polymers (HCPs),[14] and porous organic polymers (POPs)[15] with high porosity levels have been successfully synthesized. The versatility of organic chemistry and polymerization reactions allow for high degree of control over the chemical nature of the pores in POFs.[4] In addition, by judicious choice of synthetic conditions, crystalline POFs can be synthesized, and thereby the physical nature of the pores such as pore size and pore shape can also be controlled.[13] Accordingly, due to their physical and chemical tunability, POFs have garnered a lot of attention for a plethora of applications in catalysis, sensing, gas storage, gas separation, *etc.*[4,8,16]

11.2 Synthesis of POFs

11.2.1 General Strategy for Synthesis of Porous Organic Polymers (POPs)

Porous organic polymers (POPs) can be synthesized *via* a topology-directed approach using rigid star-shaped monomers.[11] In this methodology, atoms and bonds are considered as vertices and edges, respectively; and therefore, the structure of the porous organic framework can be predicted based on the configuration of the monomers (Figure 11.1).[11,13] By judicious choice of rigid monomers having star-shaped topology, a porous organic framework can be synthesized where the pores are formed upon polymerization due to the topology of the monomers.[12,13] In this approach, the respective position of bonds and atoms in the framework can be predicted before synthesis of the POP.[11] This topology-directed approach have been developed as a bottom-up strategy for construction of POPs with predetermined structures.[11] The use of rigid monomers having well-defined structures allow for construction of POPs with high stability and prevents the

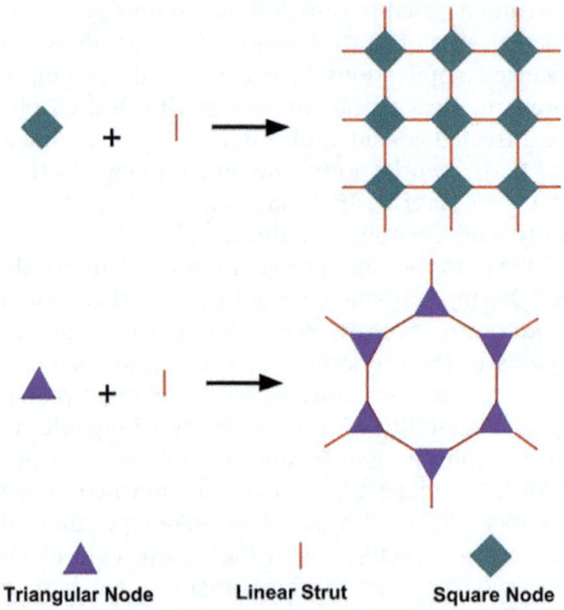

Figure 11.1 Schematic representation of topology-directed approach for synthesis of POPs using planar monomers.

framework from collapsing after removal of the solvent from the pores.[4] Accordingly, POPs are usually synthesized by linking aromatic monomers having well-defined and rigid structures as nodes with rigid struts such as alkynes, alkenes, as well as aromatic and heterocyclic rings.[4,12]

POPs synthesized using the topology-directed strategy might have crystalline or amorphous structures.[4,13] The crystallinity of porous organic polymers is mainly governed by the nature of polymerization reactions.[4,13] Irreversible polymerization reactions usually leads to formation of kinetically-controlled amorphous POPs.[4,13] On the other hand, the use of reversible reactions for construction of porous polymers can allow for synthesis of thermodynamically-controlled crystalline frameworks.[4,13] Due to the dynamic nature of reversible polymerization reactions, the linkages can be broken and reformed to afford the most thermodynamically stable crystalline structure. This allows for a self-healing feedback with "error checking" and "proof-reading".[13] Crystalline POPs are known as covalent organic frameworks (COFs)[13] which are discussed in details in another chapter of this book. The synthetic route for synthesis of COF-5[17] is schematically shown in Figure 11.2.

It is worth noting that the crystallinity is not a prerequisite for achieving a high surface area.[4] For example, PAF-1, an amorphous porous aromatic framework, exhibits the Brunauer–Emmett–Teller (BET) surface area of 6460 $m^2 g^{-1}$ which is among the most porous materials.[18] The use of irreversible polymerization reactions results in formation of amorphous POPs

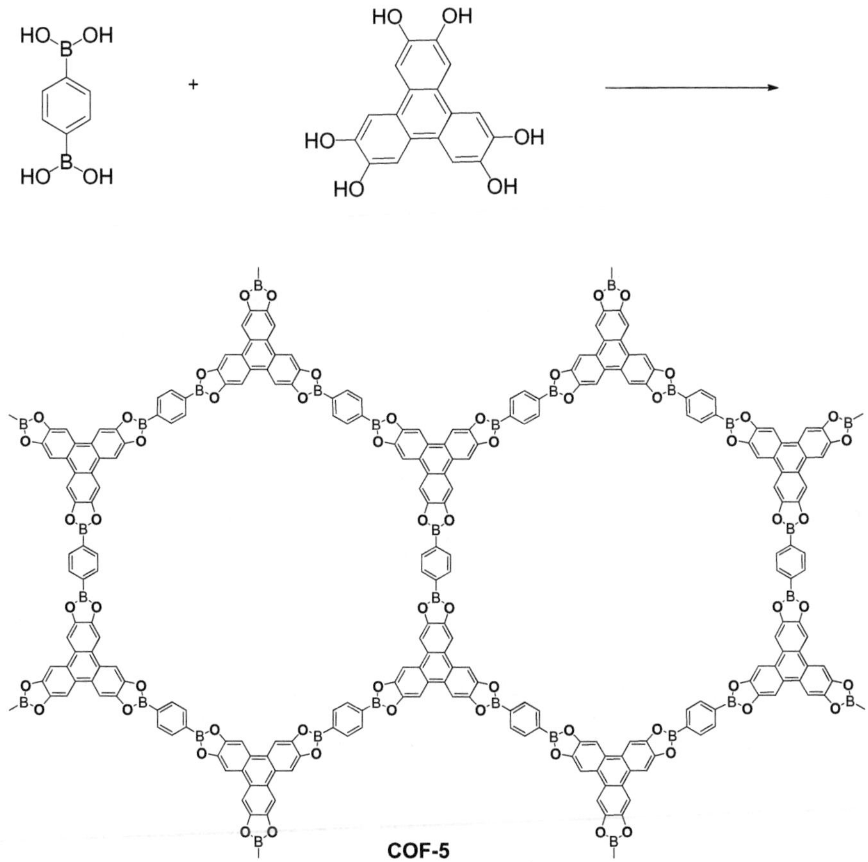

Figure 11.2 Synthesis of a porous covalent organic framework, COF-5, *via* a topology-directed approach.

and limits the control over the porosity parameters (such as pore size and pore shape) of such POPs.[4] However, the irreversibility of the linkage formation in such POPs provide them with a high physicochemical stability. This is especially advantageous for practical applications of POPs as sorbents for gas storage and separation where humidity and acidic gases are typically involved.[19] In addition, the high physicochemical stability of POPs allows for their post-synthetic functionalization under harsh reaction conditions.[20] For example, several POPs have been successfully functionalized with $-NO_2$ groups upon treatment with nitric acid.[20,21] Thanks to versatility of organic chemistry, a wide variety of polymerization reactions have been successfully employed for synthesis of POPs functionalized with diverse polar groups.[4,22] Typical monomers used for construction of POPs are shown in Figure 11.3. Several classes of POPs, such as azo-linked,[19,23] benzimidazole-linked,[15] benzoxazole-linked,[24] thiazolothiazole-linked,[25] imide-linked,[26] acetal-linked,[27] ethynylene-linked, triazole-linked,[28] urea-linked,[29] porphyrin-based,[30]

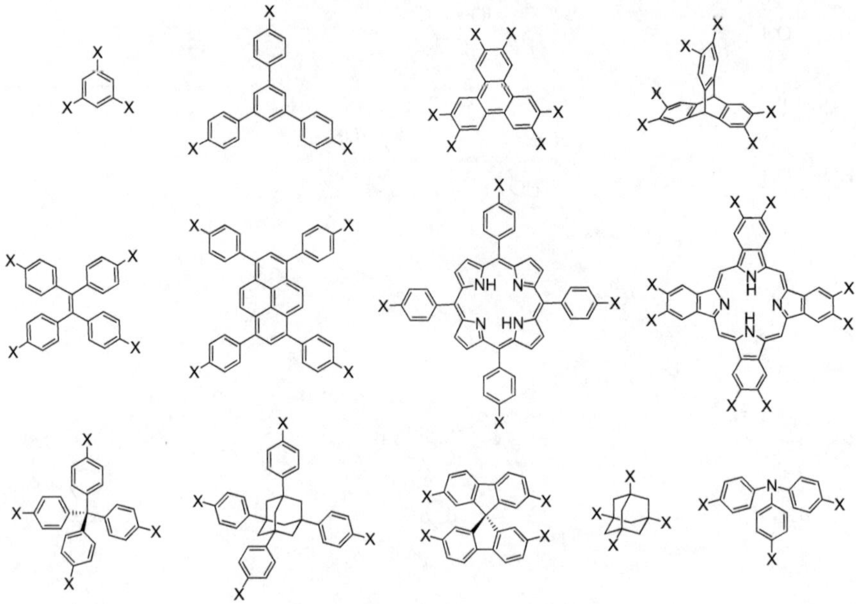

Figure 11.3 Typical monomers for synthesis of POFs *via* topology-directed approach. X represents the reacting groups of monomers which can be tuned according to polymerization reactions.

and diaminotriazine-based[31] porous polymers have been synthesized and used for gas storage and/or separation applications.

11.2.2 Different Approaches for Synthesis of HCPs

HCPs are amorphous microporous organic polymers in which the pores are formed due to extensive cross-linking of polymer chains.[14] Due to the high degree of cross-linking, the polymer chains form a rigid porous framework and cannot pack into a dense nonporous structure (Figure 11.4).[14] Because of their high surface area, high physicochemical stability, low cost, and ease of preparation, HCPs have recently received a great interest for gas separation.[14] Three strategies can be used for preparation of HCPs: (1) post-cross-linking of preformed linear polymers, (2) polycondensation of appropriate monomers, (3) knitting aromatic compounds using an external cross-linker.[14] In order to synthesize a HCP from a linear polymer, the polymer is first dissolved or swollen in a solvent to make the polymer chains separated from each other.[14] Then, the polymer chains are crosslinked to lock the polymer chains in their expanded form for creating a porous network.[14] This crosslinked network can preserve its structure after removing the solvent and therefore it can exhibit high permanent porosity.[14]

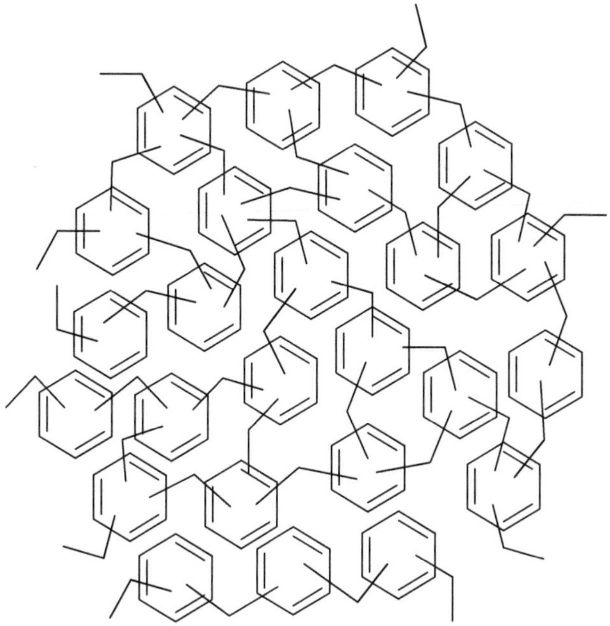

Figure 11.4 Schematic representation of a hyper-cross-linked porous polymer.

HCPs with high surface area of up to 2090 $m^2\ g^{-1}$ have been synthesized *via* post-cross-linking of poly(divinylbenzene-*co*-vinylbenzyl chloride) using Friedel–Crafts reactions of CH_2Cl moieties.[32] Moreover, Friedel–Crafts reaction has been used to synthesis HCPs through polycondensation of bis(chloromethyl) aromatic monomers without the need to prepare a preformed linear polymer precursor.[14] Friedel–Crafts alkylation of bis(chloromethyl)-based monomers such as dichloroxylene (DCX), bis(chloromethyl)biphenyl (BCMBP), and bis(chloromethyl)anthracene (BCMA) resulted in the synthesis of HCPs with high BET surface areas of up to 1900 $m^2\ g^{-1}$.[33] Very recently, Li *et al.* introduced a novel synthetic approach for synthesis of HCPs *via* the Friedel–Crafts reaction of aromatic monomers with an external cross-linker, formaldehyde dimethyl acetal (FDA), as shown in Figure 11.5.[34] This approach allows for knitting aromatic compounds using an external linker to form porous frameworks without the need for the presence of specific polymerizable groups on the monomers, which can be considered as the main advantage of this method.[14,34] By varying the ratio of FDA to monomer, the degree of cross-linking can be changed, and HCPs with different porosity parameters can be constructed. By using different monomers and FDA : monomer ratios, various HCPs with surface areas of 530–1900 $m^2\ g^{-1}$ were synthesized by Li *et al.*[34] The synthesis of HCPs featuring functional groups such as OH, NH_2, and Cl can be easily achieved by using suitable monomers such as phenol, aniline, and chlorobenzene.[34,35] Moreover, heteroatom functionalized HCPs can be synthesized from heteroatomic

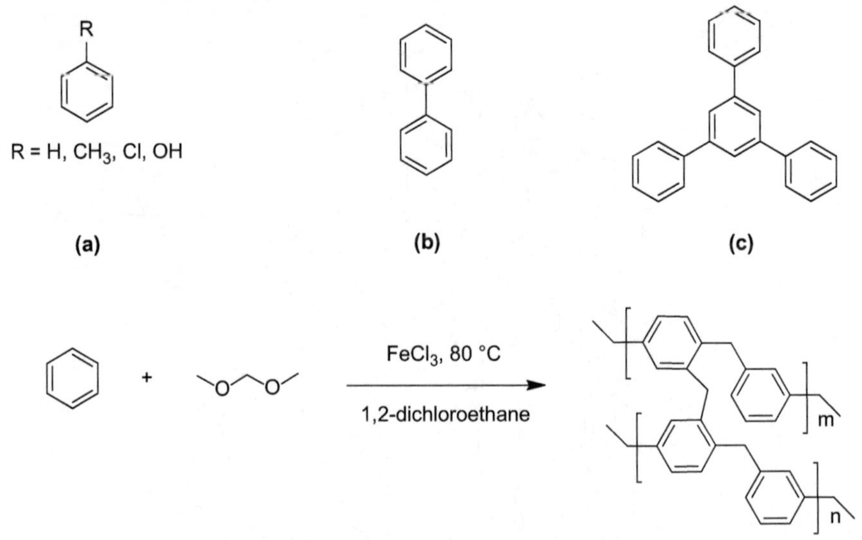

Figure 11.5 Building blocks (a, b, c), and a typical synthetic pathway for preparation of HCPs using FDA as an external cross-linker.
Adapted with permission from B. Li, R. Gong, W. Wang, X. Huang, W. Zhang, H. Li, C. Hu and B. Tan, A New Strategy to Microporous Polymers: Knitting Rigid Aromatic Building Blocks by External Cross-Linker. *Macromolecules*, 2011, **44**, 2410. Copyright 2011 American Chemical Society.

aromatic building blocks such as carbazole, dibenzothiophene, dibenzofuran, indole, benzothiophene, benzofuran, thiophene, pyrrole, and furan.[36,37] Due to their high surface area, microporous structure, and high physicochemical stability, HCPs are among promising sorbents for CO_2 capture and separation. However, due to the amorphous nature of this class of materials, their porosity parameters cannot be predicted.[14] Therefore, new synthetic routes are needed to design HCPs with long-range-order and well-defined structures.[14]

11.3 Design and Synthesis of POFs for Hydrogen Storage

The use of hydrogen as a clean energy source for fuel-cell-powered vehicles has attracted the attention of scientific and industrial communities.[3,38] Carbon dioxide emissions and air pollution in general can be significantly decreased *via* replacement of fossil fuels by hydrogen as a green energy source for transportation applications.[38] Hydrogen has a high energy density (142 MJ kg^{-1}) which is more than three times that of fossil fuels, and it has a clean combustion by-product (water).[4] As a result, significant research efforts have been aimed at achieving an efficient system for the utilization of hydrogen as energy source for on-board applications.[39] However, the efficient storage of hydrogen at operational conditions for fuel-cell-powered vehicles has remained as a great challenge to be addressed.[3] A typical compact vehicle powered by fuel cell needs ~4 kg of hydrogen for 300 miles driven.[4] The liquid or highly pressurized hydrogen tanks are neither safe nor practical for on-board applications. An efficient storage technology should be able to store large amount of hydrogen at ambient temperature and maximum pressure of 100 atm.[40] In addition to high hydrogen delivery capacity, an ideal storage system must have a fast kinetics for storage and release of hydrogen.[40] A wide range of materials such as metal hydrides, organic materials, and porous materials have been investigated as hydrogen storage media for on-board applications; however none of them could meet the criteria needed for an efficient hydrogen storage system.[3,39] Chemisorption-based storage materials such as metal hydrides have high hydrogen storage capacities; however, they need to be heated to release hydrogen which limits their applications as hydrogen storage media.[39] On the other hand, physisorption-based storage materials such as porous sorbents exhibit low binding affinity for hydrogen; and therefore, the physisorbed hydrogen can be easily released at ambient temperature.[4,41] However, physisorption-based materials usually show very low binding affinities of below 10 kJ mol^{-1} for hydrogen, which leads to their poor hydrogen storage capacity.[3]

Due to their high surface area, large pore volume, adjustable pore size, and chemical tunability, POFs have recently garnered a lot of attention for hydrogen storage.[18,40,42] For practical applications, the ultimate target for hydrogen storage capacity of a physisorptive material is 70 g L^{-1} at room temperature and pressure of <100 atm.[40] However, this target has not been achieved by any sorbent so far, mainly due to the low adsorption enthalpy of porous sorbents for hydrogen (<10 kJ mol^{-1}). The hydrogen adsorption capacity of non-functionalized POFs[4] are similar to that of other high-surface-area materials such as porous carbons[43] and metal-organic frameworks.[3] Yaghi and co-workers have studied the use of boronate-based and boraxine-based COFs having high surface areas of up to 3620 $m^2 g^{-1}$ for hydrogen storage.[42] Their results suggest that three-dimensional COFs outperform two-dimensional COFs due to their higher surface area and larger pore volumes.[42] Boron-based COFs show low binding affinity for hydrogen (6–8 kJ mol^{-1}) and they show 7.2 wt% hydrogen uptake capacity at $T = 77$ K and $P = 40$ bar.[42] Very recently, Zhou and co-workers reported the synthesis of a porous aromatic framework (PPN-4) with exceptionally high surface area of 6460 $m^2 g^{-1}$.[18] Although PPN-1 exhibits very high surface area and large pore volume, its low binding affinity for hydrogen (4 kJ mol^{-1}) leads to a poor hydrogen uptake capacity of 8.34 wt% at $T = 77$ K and $P = 55$ bar.[18] Theoretical studies have shown that the binding affinity of porous sorbents for hydrogen should be more than 15 kJ mol^{-1} to be promising for the physisorptive storage of hydrogen at ambient temperature.[44] However, the binding energy of POPs for hydrogen is usually below 10 kJ mol^{-1} which leads to their poor performance in hydrogen storage. For example, the hydrogen uptake capacity of PS4TH, a thiophene-based POP at 298 K and 70 bar is only 0.45 wt%.[45] This shows how far POFs are from meeting the U.S. Department of Energy (DOE) target of 5.5 wt% for 2015 at room temperature and 100 atm.[40] Theoretical studies have shown that functionalization of POFs with metal cations can lead to enhancement in binding affinity for H_2 which leads to high H_2 uptake capacity.[46] Simulation studies by Klontzas *et al.* have shown that the incorporation of lithium alkoxide into a 3D covalent organic framework, COF-105, can lead to a high H_2 uptake capacity of more than 6 wt% at $T = 300$ K and $P = 100$ bar.[46] Their results suggest that H_2 molecules can interact with lithium cations through charge–quadrupole and charge–induced dipole interactions.[46] A different research group experimentally studied the effect of post-synthetic functionalization of a CMP with lithium on its hydrogen storage capacity.[47] Upon lithiation, the hydrogen uptake capacity at $T = 77$ and $P = 1$ bar increased from 1.6 wt% to more than 6 wt%.[47] Despite its exceptional hydrogen uptake capacity at 77 K, Li-doped CMP exhibits a very low hydrogen uptake capacity of 0.1 wt% at 273 K.[47] Another study by Lu and co-workers has shown that post-synthetic functionalization of POFs can lead to enhancement in their hydrogen storage capacity.[48] They synthesized an alkyne-based porous polymer, COP-1, which was functionalized with lithium carboxylate groups using lithium naphthalenide and CO_2, as shown in Figure 11.6.[48] Upon post-synthetic functionalization, the surface area of the parent polymers usually decrease due to addition of mass to networks.[48] As a result, the surface area of COP-1 decreased from 830 $m^2 g^{-1}$ to 570 $m^2 g^{-1}$, although only some of alkyne moieties were functionalized.[48] Despite the decrease in surface area upon functionalization, the incorporation of lithium carboxylate groups into COP-1 resulted in more than 70% enhancement in hydrogen uptake capacity at $T = 77$ K and $P = 1$ bar.[48] This enhancement in H_2 uptake capacity was explained by an increase in the binding affinity of the framework from 6.7 kJ mol^{-1} to 7.56 kJ mol^{-1} upon lithiation.[48] Although theoretical[46] studies have shown that Li-functionalized POFs can be promising sorbents

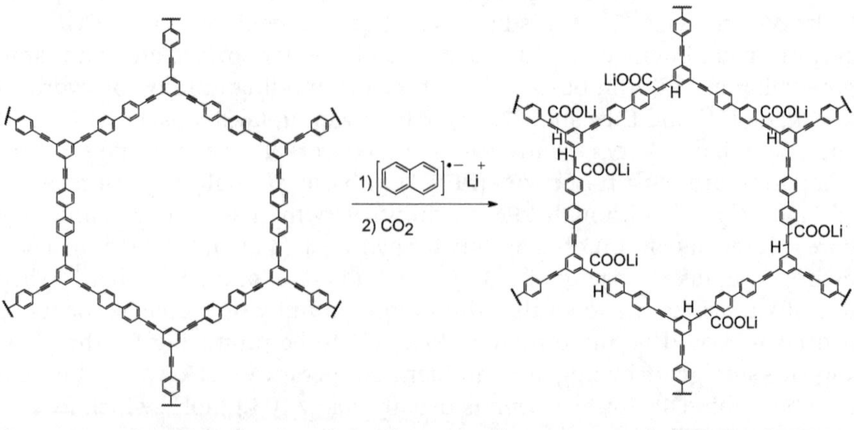

Figure 11.6 Functionalization of an alkyne-based porous organic framework, COP-1, with lithium carboxylate groups.

for hydrogen storage at ambient temperature, they have not been supported by experimental[47,48] studies. This might be due to the low surface area of experimentally synthesized Li-functionalized POFs (<800 $m^2 g^{-1}$) and also their low Li content (<3 wt%).[47,48]

Very recently, Weston *et al.* studied the impact of post-synthetic functionalization of a catechol-containing POP with open metal sites on its hydrogen storage capacity.[49] The synthesis of this catechol-containing POP was achieved *via* a cobalt-catalysed acetylene trimerization.[49] In order to prevent the coordination of catechol moiety to the cobalt atoms of the catalyst, the catechol moieties of the monomer were protected by an acid-labile orthoester group (Figure 11.7).[49] After polymerization, the labile orthoester group was cleaved by washing the polymer with concentrated HCl to obtain a catechol-decorated POP.[49] The polymer was successfully metalated upon treatment with different metal precursors such as $MgMe_2$, $Cu(OAc)_2$, or $Mn(OAc)_2$.[49] The metalation resulted in enhancement in binding affinity values for hydrogen (up to 1.5 kJ mol^{-1}); however, it did not lead to increases in hydrogen uptake capacity of the POP due to the decrease in the surface area of the polymer.[49] The enhancement of binding affinity for hydrogen in metalated catechol-based POPs was mainly attributed to

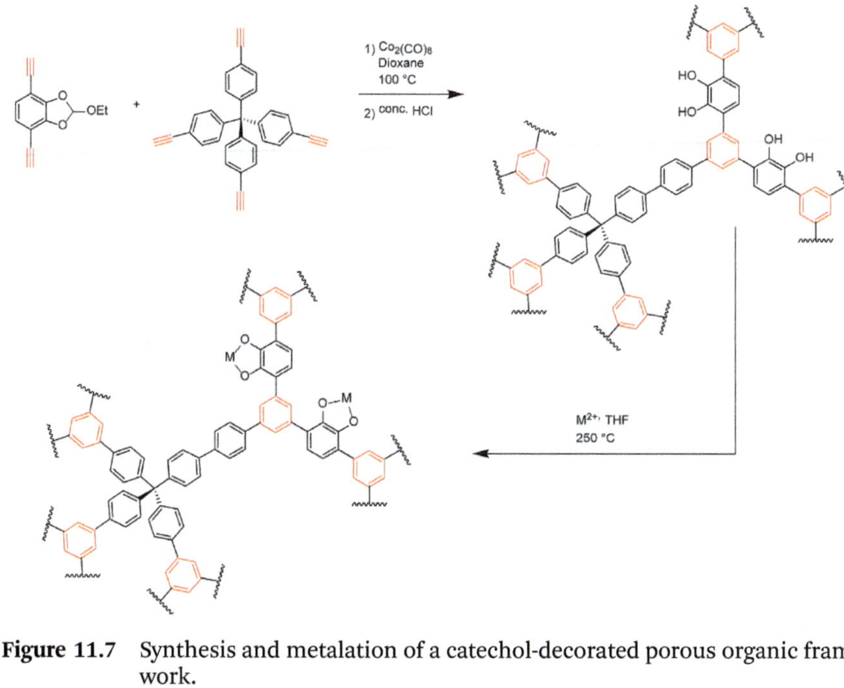

Figure 11.7 Synthesis and metalation of a catechol-decorated porous organic framework.
Adapted with permission from M. H. Weston, O. K. Farha, B. G. Hauser, J. T. Hupp and S. T. Nguyen, Synthesis and Metalation of Catechol-Functionalized POPs. *Chem. Mater.*, 2012, **24**, 1292–1296. Copyright 2012 American Chemical Society.

charge–quadrupole and charge–induced-dipole interactions between metal ions and hydrogen molecules.[49] Moreover, metalation of catechol-based POPs resulted in decrease in the pore size of the framework, which leads to stronger framework-hydrogen interactions due to multiple interactions of the adsorbed hydrogen molecule with pore walls.[49] Metalated catechol-based POPs exhibit low hydrogen uptake capacities (<1.2 wt% at 77 K and 1 bar) due to their low binding affinities for hydrogen and low surface areas.[49]

A recent theoretical study by Yaghi and co-workers suggests that the incorporation of $PdCl_2$ into a 3D porous frameworks can lead to exceptionally high hydrogen uptake capacity (Figure 11.8).[40] Their results suggest that the $PdCl_2$-fucntionalized porous framework shows a high binding affinity of 24.5 kJ mol^{-1} at 1 bar which slightly decreases at high pressures (23 kJ mol^{-1} at 100 bar). The simulation studies have shown that the strong interactions between the d orbitals of transition metals and hydrogen molecules can be a promising strategy for developing sorbents for hydrogen storage. The strength of such interactions depends on the type of the transition metal and its ligands. The effect of functionalization with $PdCl_2$ on hydrogen uptake capacity was studied for a 3D COF due to its high surface area and large pore volume.[40] Upon metalation the surface area was predicted to decrease from 3700 m^2 g^{-1} to 1100 m^2 g^{-1}.[40] While the non-metalated COF has a poor performance for hydrogen uptake (1 wt%) at room temperature and 100 bar, the $PdCl_2$-fucntionalized exhibit a very high hydrogen storage capacity

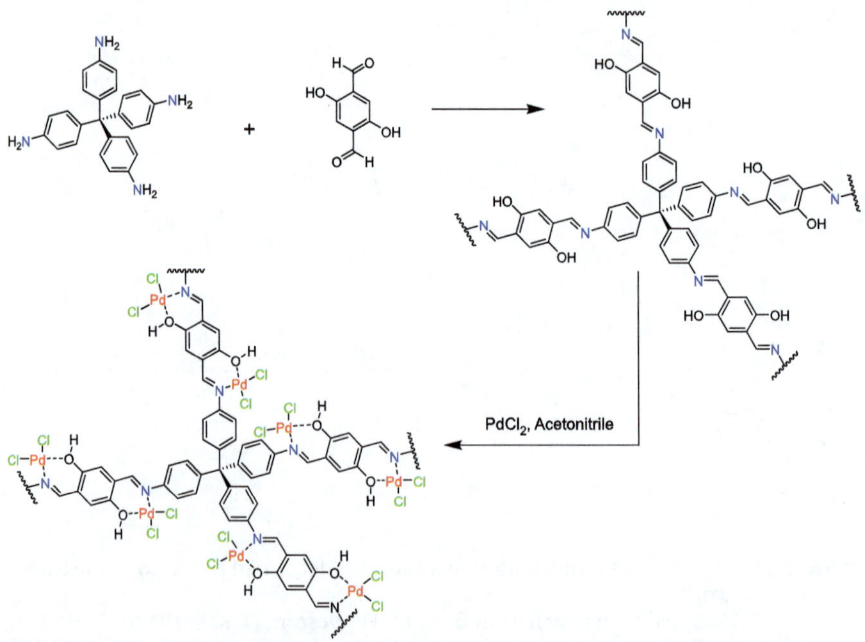

Figure 11.8 The proposed synthetic route for preparation and post-synthetic functionalization of COF-301 with $PdCl_2$.

of 4.2 wt% at 298 K and 100 bar.[40] This significant enhancement in hydrogen uptake capacity is attributed to the high binding energy of $PdCl_2$-functionalized COF for hydrogen over the entire adsorption pressure. Although the theoretical studies suggest that this proposed $PdCl_2$-COF could be a promising hydrogen sorbent, experimental attempts for synthesis of this COF was unsuccessful.[40]

In summary, POFs can have high hydrogen uptake capacities at cryogenic temperatures;[18] however, their low binding affinity for hydrogen leads to their poor performance for hydrogen storage at ambient temperature for practical applications. Due to this low binding energy, even POFs with ultra-high surface areas cannot meet the ultimate target for hydrogen storage capacities for practical applications.[40] Theoretical studies suggest that functionalization of POFs with lithium[46] or transition metals[40] can significantly enhance their binding affinity for hydrogen which can lead to high hydrogen uptake capacity at ambient temperature. Synthesis of such simulated metalated POFs are challenging and are yet to be achieved by experimental chemists. Accordingly, the use of POFs as hydrogen sorbents is still promising albeit challenging.

11.4 Methane Storage by POFs

The use of methane (natural gas) as a clean source of energy for transportation has attracted a lot of attention due to concerns about ground-level air quality and also climate change.[3] Methane is a promising fuel for mobile applications since the technology needed for commercialization of methane-based vehicles has been well-developed. Compared to gasoline, methane provides much more energy with much lower carbon dioxide emission.[3] Although methane is considered as a promising alternative clean fuel for cars, the technologies for efficient storage of methane for on-board applications have not been well-developed.[3] Currently, for on-board applications, methane is sorted at a very high pressure of 250 bar in compressed natural gas tanks which are heavy and potentially explosive.[42] In order for methane to be considered as a more efficient fuel for cars, methane tanks of small size but capable of storing a high amount of methane at relatively low pressures are required. One strategy to achieve a more safe and efficient methane storage technology is to use porous materials in tanks as sorbents to store a high amount of methane at relatively low pressures.[50] The storage of methane using porous sorbents is much easier than the storage of hydrogen due to its higher boiling point and stronger interactions with common sorbents.[3] The binding affinity of porous sorbents for methane can reach more than 20 $kJ\,mol^{-1}$ while that for hydrogen is usually below 10 $kJ\,mol^{-1}$.[3] Ultra-high-surface-area materials have been shown to be promising methane sorbents for mobile applications since the methane storage capacity of porous materials increases with their surface area and pore volume.[42] Indeed, some of metal-organic frameworks with high porosity levels have shown exceptional volumetric methane storage capacities

(>200 v/v at 298 K and 65 bar).[3,50] These values are approaching the target of 263 v/v at ambient temperature which is set by the DOE.[50] For on-board applications, the methane storage working capacity, which is defined as the difference between the methane uptake capacity at 65 bar and 5 bar, is more important than the storage capacity since the working capacity shows the deliverable amount of methane for the engine.[50] Accordingly, novel porous sorbents with high working capacity and physiochemical stability should be designed and investigated for storage of methane for on-board applications.[50]

Due to their high surface area, large pore volume, and high physicochemical stability, porous organic frameworks have garnered a lot of attention for methane storage.[41] Due to their microporous nature and high aromatic moiety content, POFs usually exhibit high binding affinity values (up to 20 kJ mol^{-1}) for methane.[19] The binding energy of POFs for CH_4 falls within the desirable range since the optimum binding affinity of an efficient sorbent for methane storage has been estimated to be 18.8 kJ mol^{-1} to achieve a fast rate of uptake and release.[4] The methane uptake of POFs increases with their surface area and pore volume. As a result, POFs with high porosity levels have been investigated for methane storage.[18,41] Zhou and co-workers synthesized a series of highly porous polymers (PPNs) using homocoupling of tetrahedral monomers using Yamamoto reaction (Figure 11.9).[18]

PPNs show exceptionally high surface areas of 4220–6460 m^2 g^{-1} and are among the most porous materials.[18] This high level of porosity can be attributed to the combined effect of several factors: (1) due to tetrahedral geometry of the monomers, PPNs have a diamond framework topology with widely open and interconnected pores; (2) high robustness of C–C bonds connecting the whole framework results in high physicochemical stability of the framework which prevents it from collapsing upon removal of the

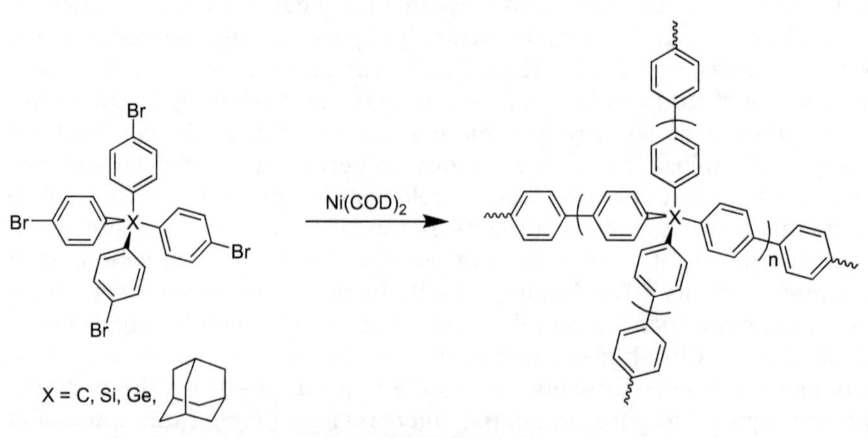

Figure 11.9 Synthesis of 3D POPs (PPNs) using tetrahedral monomers *via* the Yamamoto homocoupling reaction.

solvent; (3) the high efficiency of the Yamamoto reaction leads to reaction of all sites of the monomers which results in formation of a highly connected network without unreacted terminal groups.[41] It is noteworthy that the synthesis of PPNs was carried out at room temperature to prevent undesired side reactions and to prepare PPNs with higher molecular weight and less degree of network interpenetration, leading to higher surface area levels.[18] The potential applications of PPN-4 for methane storage was investigated due to its high surface area (6460 $m^2 g^{-1}$).[18] The methane uptake capacity of PPN-4 at 55 bar and 297 K was found to be 269 $mg g^{-1}$ which is the highest value among all classes of POPs.[18] Yaghi and co-workers synthesized a series of COFs and studied their methane storage capacities.[42] Their results showed that 3D COFs outperform 2D COFs for methane storage because of their higher surface area and larger pore volume.[42] They have shown that COF-102, a 3D covalent organic framework with surface area of 4650 $m^2 g^{-1}$, has a high methane uptake capacity (229 $mg g^{-1}$ at 298 K and 85 bar) which is the highest value among COFs.[42] Other POPs with moderate surface areas (<1900 $m^2 g^{-1}$) such as azo-linked porous polymers (ALPs)[19] and HCPs[51] exhibit low gravimetric methane uptake capacity values (<110 $mg g^{-1}$ at 298 K, up to 55 bar). Likewise, the volumetric methane uptake capacity of ALPs (75 v/v at 70 bar and 298 K) is much lower than that of PPN-4 (120 v/v at 75 bar and 298 K).[19] These values are much lower than the target of 263 v/v at ambient temperature which is set by DOE.[50] Accordingly, due to their high physicochemical stability, high binding affinity for CH_4, and ultra-high surface areas, POFs are attractive candidates for on-board methane storage applications. However, for practical applications, POFs with higher surface area and pore volume should be synthesized to achieve high CH_4 uptake capacity values. Moreover, for commercialization of POFs as methane sorbents, other criteria such as working capacity, scalability, and cost should also be taken into account.

11.5 Synthesis of Functionalized POFs for CO_2 Capture

Over the past century, carbon dioxide emission to the atmosphere has significantly increased due to anthropogenic activities for energy production from fossil fuels.[52] This increase in atmospheric CO_2 level is one of the main causes of global warming and climate change.[52] Since fossil fuels have remained the main sources of energy, carbon dioxide capture and sequestration has been considered as a promising solution for mitigating climate change until renewable green energy sources become widely available.[22,53] From a practical point of view, in order to reduce CO_2 emissions to atmosphere, CO_2 capture and separation should take place at stationary point sources such as fossil fuel-based power plants.[54] Due to the high concentration of CO_2 at stationary point sources, CO_2 can be efficiently captured by CO_2 sorbents.[54] On the other hand, CO_2 capture from air is very challenging due to the relatively low concentration of CO_2 in the air.[54] The captured CO_2 can be permanently sequestrated by different scenarios such as conversion to fuel and useful chemicals, as well as injection to underground geological formations such as empty oil wells.[54] Regardless of the sequestration strategy, carbon dioxide needs to be first separated from gas mixtures at stationary point sources before sequestration.[54] Currently, chemical absorption of CO_2 by aqueous amine solutions is the most widely-used technology in industry for the separation of CO_2 from flue gas.[53] However, the use of amine solutions for CO_2 separation can be limited due to the high energy input required for liberation of the captured CO_2 to regenerate the amine solution for the next cycles.[53] In addition, amine solutions are corrosive and degrade over time which leads to decrease in their CO_2 uptake capacity.[53] As a result, to address these limitations, physisorption of CO_2 by porous materials has attracted a lot of attention.[22,52,55] The low binding affinity of porous sorbents for CO_2 leads to physisorption of CO_2; and therefore, the captured CO_2 can be desorbed with a low energy cost by lowering the pressure without applying heat.[22] However, the low binding energy of porous sorbents for CO_2 results in their low CO_2 uptake capacity values.[22] The CO_2 binding affinity of porous sorbents can be increased by functionalization of the pores with CO_2-philic groups to achieve an optimal binding energy for CO_2.[22] Recently, POFs have emerged as promising sorbents for selective CO_2 capture due to their high surface area, high physicochemical stability, and chemical tunability.[22,56] Thanks to versatility of organic chemistry, POFs functionalized with a wide variety of CO_2-philic groups can be synthesized *via* pre-synthetic or post-synthetic modification strategies.[2,14,22] Both theoretical[57] and experimental[22] studies have shown that incorporation of polar groups such as $-OH$, $-NH_2$, $-NO_2$, and $-SO_3H$ into POFs' structure can lead to stronger CO_2-framework interactions through hydrogen bonding or dipole–quadrupole interactions between CO_2 and polar functional groups. Due to its large quadrupole moment, CO_2 molecule can act as a Lewis acid; and therefore, POPs functionalized with Lewis basic sites exhibit high binding affinities and uptake capacities for CO_2.[19,58] Three main approaches have been sought for synthesis of POFs functionalized with CO_2-philic groups: (1) polymerization of functionalized monomers;[7] (2) formation of functional groups during polymerization by reaction of polymerizable sites of the monomers;[15] (3) introduction of functional groups into the pores after synthesis of the POF.[58] Cooper and co-workers synthesized a series of functionalized porous polymers (CMPs) by Sonogashira cross-coupling reaction of the monomers functionalized with $-NH_2$, $-OH$, and $-COOH$ groups.[7] They have shown that the functionalized CMPs have higher biding affinities than that of the non-functionalized CMP.[7] Likewise, Yang *et al.* synthesized a series of fluorinated porous polymers (MOPs) *via* polymerization of fluorinated monomers by Sonogashira cross-coupling reaction.[59] They have shown that fluorinated MOPs have much higher biding affinities for CO_2 (34–49 $kJ\ mol^{-1}$) than non-functionalized MOPs (22–26 $kJ\ mol^{-1}$).[59] The high binding affinities of fluorinated MOPs can be attributed to dipole–quadrupole interactions between CO_2 and fluorine groups which leads to their high CO_2 uptake capacities (up to 5.1 mmol g^{-1} at 1 bar and 273 K).[59] Recently, heteroatomic aromatic monomers such as pyrrole, furan, thiophene, dibenzofuran, dibenzothiophene, indole, benzothiophene, and benzofuran were used for synthesis of a series of HCPs functionalized with heteroatoms such as N, O, and S.[36,37] Due to Lewis acid–base interactions of CO_2 with their heteroatoms, HCPs exhibit high biding affinity values of up to 49 kJ mol^{-1} for CO_2.[37] Some of the reactions used for synthesis of POFs from functionalized monomers are schematically shown in Figure 11.10.

Synthesis of POFs from functionalized monomers can be challenging due to difficulties in preparation of functionalized monomers. Moreover, the reactivity of functional groups of the monomers during the polymerization process can impose synthetic challenges for synthesis of functionalized POFs. As a result, formation and incorporation of functional groups into POFs' structure during the polymerization process has been widely used as a useful strategy for synthesis of functionalized POFs.[56] In this approach, functional groups are formed *in situ* during the polymerization process upon

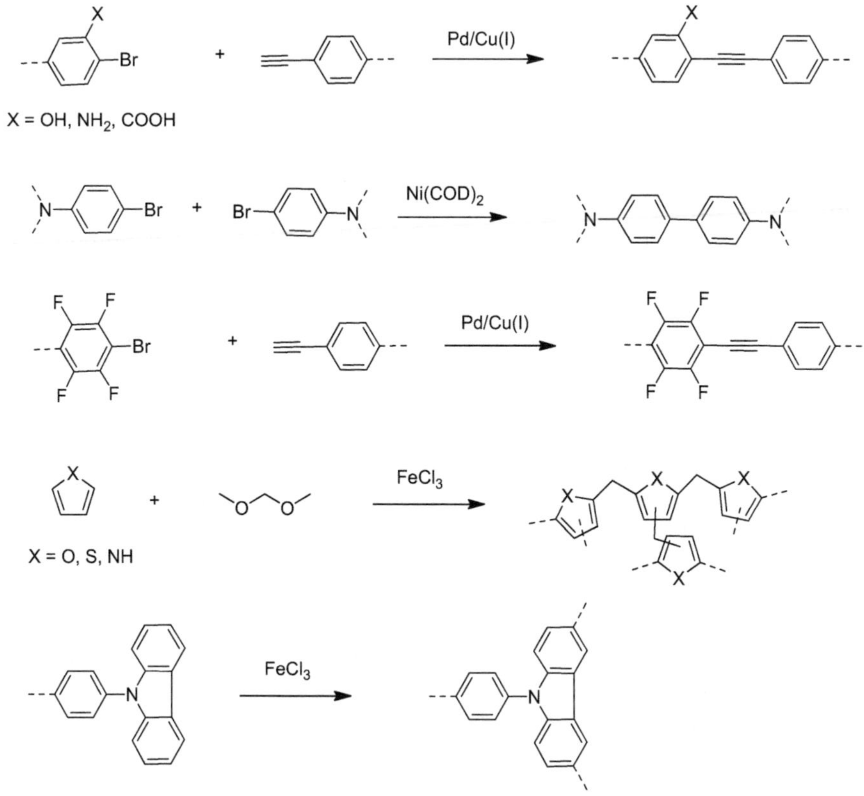

Figure 11.10 Polymerization reaction examples for synthesis of POFs from functionalized monomers.

reaction of polymeriseable sites of the monomers.[56] These incorporated functional groups are part of the framework and can act as linkers or nodes of the network. By judicious choice of nitrogen-rich or oxygen-rich monomers, a wide variety of functionalized POFs have been synthesized using this strategy.[22,56] El-Kaderi and co-workers synthesized a series of benzimidazole-linked porous polymers (BILPs) by condensation reactions between aryl-*o*-diamines and aryl-aldehyde monomers.[15] BILPs are mainly microporous and have moderate surface areas (<1500 $m^2 g^{-1}$).[60] Because of Lewis acid–base interactions between CO_2 and nitrogen atoms of benzimidazole groups, BILPs exhibit high binding affinities for CO_2 (27–38 kJ mol^{-1}) which lead to their high CO_2 uptake capacities (up to 5.34 mmol g^{-1} and at 273 K and 1 bar).[60] Azo-linked porous polymers (ALPs) have been recently synthesized by homocoupling reaction of aryl-amines using Cu(I) as a catalyst.[19,23] ALPs exhibit high binding energy values for CO_2 (~30 kJ mol^{-1}) because of Lewis acid–base interactions between CO_2 and nitrogen atoms of azo moieties.[23] Due to their high surface areas and high binding affinities for CO_2, ALPs show high CO_2 uptake capacities of up to 5.37 mmol g^{-1} and at 273 K and 1 bar).[23] A diverse range of functional groups such as thiazolothiazole,[25] triazole,[28,61] imide,[26] benzoxazole,[24] diaminotriazine,[31] triazine,[62] urea,[29] acetal,[27] and porphyrin[30] have been incorporated into structure of POFs through *in situ* formation of functional groups during the polymerization process. Figure 11.11 depicts selected polymerization reactions reported for formation and incorporation of functional groups into POFs.

Post-synthetic functionalization of POFs is a promising strategy to modulate gas–framework interactions to achieve a high CO_2 uptake capacity and selectivity.[20,58] After synthesis of frameworks, CO_2-philic functional groups can be introduced to the pores of POFs by organic or inorganic reactions (Figure 11.12).[20,58,63] In addition to tailoring the chemical nature of the pores, the size of the pores can be tuned by post-synthetic modification strategies to increase the binding affinity of frameworks for CO_2.[58] The size of the pores is usually decreased by addition of functional groups into the pores.[58] This is desirable for achieving higher binding affinity for CO_2 since the number of multiple interactions between an adsorbed CO_2 molecule and pore walls of the framework increases as the size of the pores decreases.[58] Zhou and co-workers showed that sulfonation of PPN-6, a porous aromatic framework with surface area of 4023 $m^2 g^{-1}$, resulted in a significant increase in CO_2 binding affinity from 17 to 30 kJ mol^{-1}.[63] As a result, upon sulfonation, the CO_2 uptake capacity of the framework dramatically increased from 1.16 to 3.60 mmol g^{-1} at 295 K and 1 bar.[63] Likewise, a significant enhancement in CO_2 binding affinity and CO_2 uptake capacity was observed upon post-synthetic functionalization of PPN-6 with amine functionalities.[64] For example, PPN-6-CH_2DETA exhibits the CO_2 uptake capacity of 4.3 mmol g^{-1} at 295 K and 1 bar which is the highest value reported for any POPs under similar conditions.[64] Very recently, El-Kaderi and co-workers have shown that post-synthetic functionalization of a porous bis(imino)-pyridine-linked polymer (BIPLP-1) with $Cu(BF_4)_2$ can lead to a 19% increase

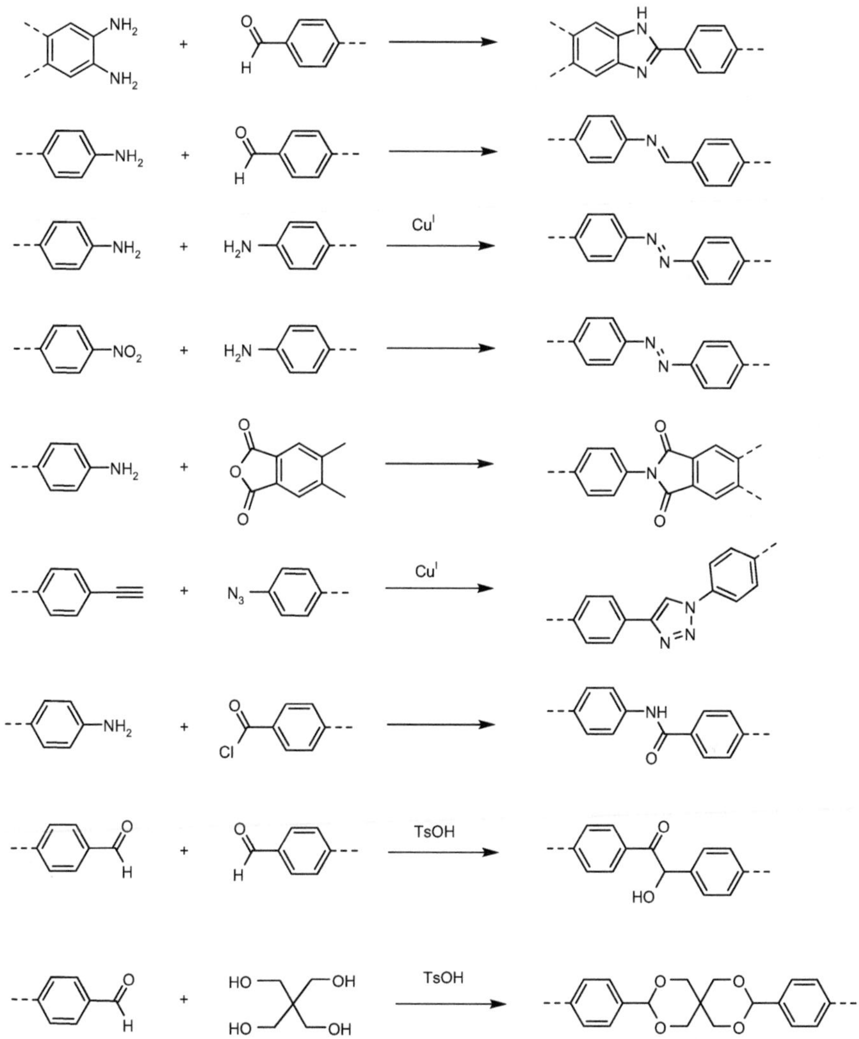

Figure 11.11 Typical polymerization reactions for *in situ* formation and incorporation of functional groups into POFs.

in CO_2 binding affinity.[58] This resulted in 50% enhancement in CO_2 uptake capacity at 298 K and 1 bar.[58] It is noteworthy that the $Cu(BF_4)_2$ functionalized polymer showed a 200% enhancement in CO_2 uptake capacity at 0.15 bar which is the partial pressure of CO_2 in flue gas.[58] The enhanced CO_2 binding affinity of the $Cu(BF_4)_2$ functionalized polymer can be mainly attributed to Lewis acid–base interactions between CO_2 molecules and BF_4^- ions.[58] Moreover, a 50% decrease in pore size was observed upon functionalization which can result in stronger CO_2-farmework interactions due to higher number of interactions between an adsorbed CO_2 molecule and pore walls of the network.[58] A recent study has shown that post-synthetic

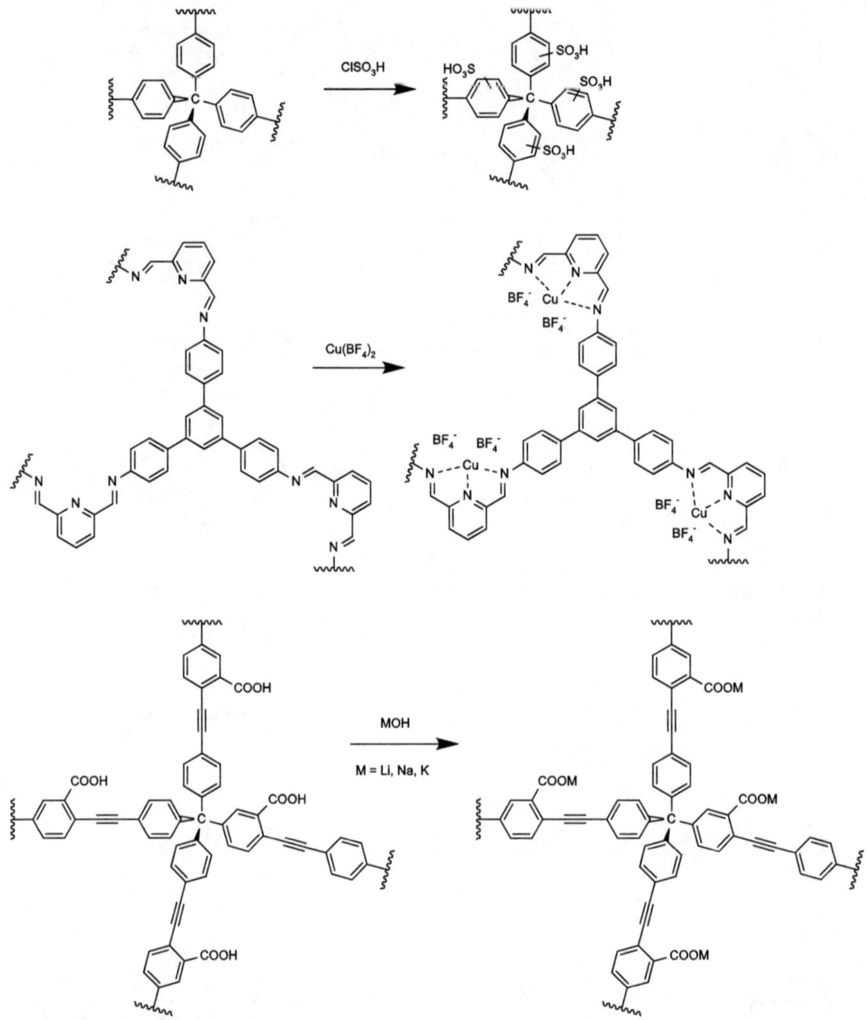

Figure 11.12 Different representative example synthetic strategies for post-synthetic functionalization of POFs.

functionalization of a POPs with Na^+ ions can increase its binding affinity for CO_2 from 28.1 kJ mol^{-1} to 35.0 kJ mol^{-1}.[65] Figure 11.12 shows three representative example synthetic strategies for post-synthetic functionalization of POFs.

11.6 Conclusion

In the proceeding sections, we have summarized recent advances in the targeted synthesis of highly porous POFs and their potential in small gas storage and separation. POFs have many desired textural and physical properties that make them very promising for these applications, however, future research in this field should focus on viable synthetic methodologies that rely on none-expensive solvent, building units, and catalysts especially for applications that demand large scale use of adsorbents like CO_2 capture and sequestration. Unlike MOFs, POFs are amorphous and therefore full understanding of their structure–function relationship remains a challenge. More efforts should be directed toward the synthesis of structurally more ordered POFs and their integration into membranes for gas separation.

References

1. Y. Zhang and S. N. Riduan, Functional Porous Organic Polymers for Heterogeneous Catalysis, *Chem. Soc. Rev.*, 2012, **41**, 2083–2094.
2. J.-X. Jiang and A. Cooper, Microporous Organic Polymers: Design, Synthesis, and Function, In *Functional Metal-Organic Frameworks: Gas Storage, Separation and Catalysis*, ed. M. Schröder, Springer, Berlin Heidelberg, 2010, vol. 293, pp. 1–33.
3. S. Ma and H.-C. Zhou, Gas Storage in Porous Metal-Organic Frameworks for Clean Energy Applications, *Chem. Commun.*, 2010, **46**, 44–53.
4. R. Dawson, A. I. Cooper and D. J. Adams, Nanoporous Organic Polymer Networks, *Prog. Polym. Sci.*, 2012, **37**, 530–563.
5. L. E. Kreno, K. Leong, O. K. Farha, M. Allendorf, R. P. Van Duyne and J. T. Hupp, Metal–Organic Framework Materials as Chemical Sensors, *Chem. Rev.*, 2012, **112**, 1105–1125.
6. M. E. Davis, Ordered Porous Materials for Emerging Applications, *Nature*, 2002, **417**, 813–821.
7. R. Dawson, D. J. Adams and A. I. Cooper, Chemical Tuning of CO_2 Sorption in Robust Nanoporous Organic Polymers, *Chem. Sci.*, 2011, **2**, 1173–1177.
8. P. Kaur, J. T. Hupp and S. T. Nguyen, Porous Organic Polymers in Catalysis: Opportunities and Challenges, *ACS Catal.*, 2011, **1**, 819–835.
9. J. Lee, O. K. Farha, J. Roberts, K. A. Scheidt, S. T. Nguyen and J. T. Hupp, Metal-organic framework materials as catalysts, *Chem. Soc. Rev.*, 2009, **38**, 1450–1459.
10. T. Ben and S. Qiu, Porous aromatic frameworks: Synthesis, structure and functions, *CrystEngComm*, 2013, **15**, 17–26.
11. X. Zou, H. Ren and G. Zhu, Topology-Directed Design of Porous Organic Frameworks and their Advanced Applications, *Chem. Commun.*, 2013, **49**, 3925–3936.
12. Y. Xu, S. Jin, H. Xu, A. Nagai and D. Jiang, Conjugated Microporous Polymers: Design, Synthesis and Application, *Chem. Soc. Rev.*, 2013, **42**, 8012–8031.
13. X. Feng, X. Ding and D. Jiang, Covalent Organic Frameworks, *Chem. Soc. Rev.*, 2012, **41**, 6010–6022.
14. S. Xu, Y. Luo and B. Tan, Recent Development of Hypercrosslinked Microporous Organic Polymers, *Macromol. Rapid Commun.*, 2013, **34**, 471–484.

15. M. G. Rabbani and H. M. El-Kaderi, Synthesis and Characterization of Porous Benzimidazole-Linked Polymers and Their Performance in Small Gas Storage and Selective Uptake, *Chem. Mater.*, 2012, **24**, 1511–1517.
16. M. Rose, Nanoporous Polymers: Bridging the Gap between Molecular and Solid Catalysts? *ChemCatChem*, 2014, **6**, 1166–1182.
17. A. P. Côté, A. I. Benin, N. W. Ockwig, M. O'Keeffe, A. J. Matzger and O. M. Yaghi, Porous, Crystalline, Covalent Organic Frameworks, *Science*, 2005, **310**, 1166–1170.
18. D. Yuan, W. Lu, D. Zhao and H.-C. Zhou, Highly Stable Porous Polymer Networks with Exceptionally High Gas-Uptake Capacities, *Adv. Mater.*, 2011, **23**, 3723–3725.
19. P. Arab, M. G. Rabbani, A. K. Sekizkardes, T. İslamoğlu and H. M. El-Kaderi, Copper(I)-Catalyzed Synthesis of Nanoporous Azo-Linked Polymers: Impact of Textural Properties on Gas Storage and Selective Carbon Dioxide Capture, *Chem. Mater.*, 2014, **26**, 1385–1392.
20. T. Islamoglu, M. Gulam Rabbani and H. M. El-Kaderi, Impact of Post-Synthesis Modification of Nanoporous Organic Frameworks on Small Gas Uptake and Selective CO_2 Capture, *J. Mater. Chem. A*, 2013, **1**, 10259–10266.
21. C. Shen and Z. Wang, Tetraphenyladamantane-Based Microporous Polyimide and Its Nitro-Functionalization for Highly Efficient CO_2 Capture, *J. Phys. Chem. C*, 2014, **118**, 17585–17593.
22. R. Dawson, A. I. Cooper and D. J. Adams, Chemical Functionalization Strategies for Carbon Dioxide Capture in Microporous Organic Polymers, *Polym. Int.*, 2013, **62**, 345–352.
23. P. Arab, E. Parrish, T. Islamoglu and H. M. El-Kaderi, Synthesis and Evaluation of Porous Azo-Linked Polymers for Carbon Dioxide Capture and Separation, *J. Mater. Chem. A*, 2015, **3**, 20586–20594.
24. H. A. Patel, D. Ko and C. T. Yavuz, Nanoporous Benzoxazole Networks by Silylated Monomers, Their Exceptional Thermal Stability, and Carbon Dioxide Capture Capacity, *Chem. Mater.*, 2014, **26**, 6729–6733.
25. X. Zhu *et al.*, Thiazolothiazole-linked porous organic polymers, *Chem. Commun.*, 2014, **50**, 15055–15058.
26. O. K. Farha, A. M. Spokoyny, B. G. Hauser, Y.-S. Bae, S. E. Brown, R. Q. Snurr, C. A. Mirkin and J. T. Hupp, Synthesis, Properties, and Gas Separation Studies of a Robust Diimide-Based Microporous Organic Polymer, *Chem. Mater.*, 2009, **21**, 3033–3035.
27. Y.-C. Zhao, L.-M. Zhang, T. Wang and B.-H. Han, Microporous Organic Polymers with Acetal Linkages: Synthesis, Characterization, and Gas Sorption Properties, *Polym. Chem.*, 2014, **5**, 614–621.
28. J. R. Holst, E. Stöckel, D. J. Adams and A. I. Cooper, High Surface Area Networks from Tetrahedral Monomers: Metal-Catalyzed Coupling, Thermal Polymerization, and "Click" Chemistry, *Macromolecules*, 2010, **43**, 8531–8538.

29. S.-Y. Moon, J.-S. Bae, E. Jeon and J.-W. Park, Organic Sol–Gel Synthesis: Solution-Processable Microporous Organic Networks, *Angew. Chem., Int. Ed.*, 2010, **49**, 9504–9508.
30. A. Modak, M. Nandi, J. Mondal and A. Bhaumik, Porphyrin based porous organic polymers: novel synthetic strategy and exceptionally high CO_2 adsorption capacity, *Chem. Commun.*, 2012, **48**, 248–250.
31. W.-C. Song, X.-K. Xu, Q. Chen, Z.-Z. Zhuang and X.-H. Bu, Nitrogen-Rich Diaminotriazine-Based Porous Organic Polymers for Small Gas Storage and Selective Uptake, *Polym. Chem.*, 2013, **4**, 4690–4696.
32. J.-H. Ahn, J.-E. Jang, C.-G. Oh, S.-K. Ihm, J. Cortez and D. C. Sherrington, Rapid Generation and Control of Microporosity, Bimodal Pore Size Distribution, and Surface Area in Davankov-Type Hyper-Cross-Linked Resins, *Macromolecules*, 2006, **39**, 627–632.
33. C. D. Wood *et al.*, Hydrogen Storage in Microporous Hypercrosslinked Organic Polymer Networks, *Chem. Mater.*, 2007, **19**, 2034–2048.
34. B. Li, R. Gong, W. Wang, X. Huang, W. Zhang, H. Li, C. Hu, B. Tan and A. New, Strategy to Microporous Polymers: Knitting Rigid Aromatic Building Blocks by External Cross-Linker, *M*acromolecules, 2011, **44**, 2410–2414.
35. R. Dawson, T. Ratvijitvech, M. Corker, A. Laybourn, Y. Z. Khimyak, A. I. Cooper and D. J. Adams, Microporous Copolymers for Increased Gas Selectivity, *Polym. Chem.*, 2012, **3**, 2034–2038.
36. Y. Luo, B. Li, W. Wang, K. Wu and B. Tan, Hypercrosslinked Aromatic Heterocyclic Microporous Polymers: A New Class of Highly Selective CO_2 Capturing Materials, *Adv. Mater.*, 2012, **24**, 5703–5707.
37. M. Saleh, H. M. Lee, K. C. Kemp and K. S. Kim, Highly Stable CO_2/N_2 and CO_2/CH_4 Selectivity in Hyper-Cross-Linked Heterocyclic Porous Polymers, *ACS Appl. Mater. Interfaces*, 2014, **6**, 7325–7333.
38. J. Lan, D. Cao, W. Wang, T. Ben and G. Zhu, High-Capacity Hydrogen Storage in Porous Aromatic Frameworks with Diamond-like Structure, *J. Phys. Chem. Lett.*, 2010, **1**, 978–981.
39. L. Schlapbach and A. Zuttel, Hydrogen-storage materials for mobile applications, *Nature*, 2001, **414**, 353–358.
40. J. L. Mendoza-Cortes, W. A. Goddard, H. Furukawa and O. M. Yaghi, A Covalent Organic Framework that Exceeds the DOE 2015 Volumetric Target for H_2 Uptake at 298 K, *J. Phys. Chem. Lett.*, 2012, **3**, 2671–2675.
41. W. Lu *et al.*, Porous Polymer Networks: Synthesis, Porosity, and Applications in Gas Storage/Separation, *Chem. Mater.*, 2010, **22**, 5964–5972.
42. H. Furukawa and O. M. Yaghi, Storage of Hydrogen, Methane, and Carbon Dioxide in Highly Porous Covalent Organic Frameworks for Clean Energy Applications, *J. Am. Chem. Soc.*, 2009, **131**, 8875–8883.
43. Y. Xia, Z. Yang and Y. Zhu, Porous carbon-based materials for hydrogen storage: advancement and challenges, *J. Mater. Chem. A*, 2013, **1**, 9365–9381.

44. S. K. Bhatia and A. L. Myers, Optimum Conditions for Adsorptive Storage, *Langmuir*, 2006, **22**, 1688–1700.
45. S. Yuan, S. Kirklin, B. Dorney, D.-J. Liu and L. Yu, Nanoporous Polymers Containing Stereocontorted Cores for Hydrogen Storage, *Macromolecules*, 2009, **42**, 1554–1559.
46. E. Klontzas, E. Tylianakis and G. E. Froudakis, Hydrogen Storage in Lithium-Functionalized 3-D Covalent-Organic Framework Materials, *J. Phys. Chem. C*, 2009, **113**, 21253–21257.
47. A. Li, R.-F. Lu, Y. Wang, X. Wang, K.-L. Han and W.-Q. Deng, Lithium-Doped Conjugated Microporous Polymers for Reversible Hydrogen Storage, *Angew. Chem., Int. Ed.*, 2010, **49**, 3330–3333.
48. Z. Xiang, D. Cao, W. Wang, W. Yang, B. Han and J. Lu, Postsynthetic Lithium Modification of Covalent-Organic Polymers for Enhancing Hydrogen and Carbon Dioxide Storage, *J. Phys. Chem. C*, 2012, **116**, 5974–5980.
49. M. H. Weston, O. K. Farha, B. G. Hauser, J. T. Hupp and S. T. Nguyen, Synthesis and Metalation of Catechol-Functionalized Porous Organic Polymers, *Chem. Mater.*, 2012, **24**, 1292–1296.
50. Y. Peng, V. Krungleviciute, I. Eryazici, J. T. Hupp, O. K. Farha and T. Yildirim, Methane Storage in Metal–Organic Frameworks: Current Records, Surprise Findings, and Challenges, *J. Am. Chem. Soc.*, 2013, **135**, 11887–11894.
51. C. D. Wood, B. Tan, A. Trewin, F. Su, M. J. Rosseinsky, D. Bradshaw, Y. Sun, L. Zhou and A. I. Cooper, Microporous Organic Polymers for Methane Storage, *Adv. Mater.*, 2008, **20**, 1916–1921.
52. J.-R. Li, Y. Ma, M. C. McCarthy, J. Sculley, J. Yu, H.-K. Jeong, P. B. Balbuena and H.-C. Zhou, Carbon Dioxide Capture-Related Gas Adsorption and Separation in Metal-Organic Frameworks, *Coord. Chem. Rev.*, 2011, **255**, 1791–1823.
53. G. T. Rochelle, Amine Scrubbing for CO_2 Capture, *Science*, 2009, **325**, 1652–1654.
54. K. Sumida, D. L. Rogow, J. A. Mason, T. M. McDonald, E. D. Bloch, Z. R. Herm, T.-H. Bae and J. R. Long, Carbon Dioxide Capture in Metal–Organic Frameworks, *Chem. Rev.*, 2011, **112**, 724–781.
55. A.-H. Lu, G.-P. Hao and X.-Q. Zhang, Porous Carbons for Carbon Dioxide Capture. In *Porous Materials for Carbon Dioxide Capture*, ed. A.-H. Lu and S. Dai, Springer, Berlin Heidelberg, 2014, pp. 15–77.
56. C. Xu and N. Hedin, Microporous Adsorbents for CO_2 Capture – a Case for Microporous Polymers? *Mater. Today*, 2014, **17**, 397–403.
57. H. M. Lee, I. S. Youn, M. Saleh, J. W. Lee and K. S. Kim, Interactions of CO2 with various functional molecules, *Phys. Chem. Chem. Phys.*, 2015, **17**, 10925–10933.
58. P. Arab, A. Verlander and H. M. El-Kaderi, Synthesis of a Highly Porous Bis(imino)pyridine-Linked Polymer and Its Postsynthetic Modification with Inorganic Fluorinated Ions for Selective CO_2 Capture, *J. Phys. Chem. C*, 2015, **119**, 8174–8182.

59. Z.-Z. Yang, Y. Zhao, H. Zhang, B. Yu, Z. Ma, G. Ji and Z. Liu, Fluorinated Microporous Organic Polymers: Design and Applications in CO_2 Adsorption and Conversion, *Chem. Commun.*, 2014, **50**, 13910–13913.
60. M. G. Rabbani, A. K. Sekizkardes, O. M. El-Kadri, B. R. Kaafarani and H. M. El-Kaderi, Pyrene-Directed Growth of Nanoporous Benzimidazole-Linked Nanofibers and their Application to Selective CO_2 Capture and Separation, *J. Mater. Chem.*, 2012, **22**, 25409–25417.
61. P. Pandey, O. K. Farha, A. M. Spokoyny, C. A. Mirkin, M. G. Kanatzidis, J. T. Hupp and S. T. Nguyen, A "Click-Based" Porous Organic Polymer from Tetrahedral Building Blocks, *J. Mater. Chem.*, 2011, **21**, 1700–1703.
62. S. Ren, M. J. Bojdys, R. Dawson, A. Laybourn, Y. Z. Khimyak, D. J. Adams and A. I. Cooper, Porous, Fluorescent, Covalent Triazine-Based Frameworks Via Room-Temperature and Microwave-Assisted Synthesis, *Adv. Mater.*, 2012, **24**, 2357–2361.
63. W. Lu, D. Yuan, J. Sculley, D. Zhao, R. Krishna and H.-C. Zhou, Sulfonate-Grafted Porous Polymer Networks for Preferential CO_2 Adsorption at Low Pressure, *J. Am. Chem. Soc.*, 2011, **133**, 18126–18129.
64. W. Lu, J. P. Sculley, D. Yuan, R. Krishna, Z. Wei and H.-C. Zhou, Polyamine-Tethered Porous Polymer Networks for Carbon Dioxide Capture from Flue Gas, *Angew. Chem., Int. Ed.*, 2012, **51**, 7480–7484.
65. H. Ma, H. Ren, X. Zou, S. Meng, F. Sun and G. Zhu, Post-Metalation of Porous Aromatic Frameworks for Highly Efficient Carbon Capture from $CO_2 + N_2$ and $CH_4 + N_2$ Mixtures, *Polym. Chem.*, 2014, **5**, 144–152.

CHAPTER 12

Gated and Stepwise Sorption Processes in Functional Metal-organic Frameworks

PRAKASH KANOO,[a,†] RITESH HALDAR,[b] PAPRI SUTAR,[a] ANINDITA CHAKRABORTY[a] AND TAPAS KUMAR MAJI*[a,b]

[a] Chemistry and Physics of Materials Unit, Jawaharlal Nehru Centre for Advanced Scientific Research, Jakkur, Bangalore 560 064, India; [b] New Chemistry Unit, Jawaharlal Nehru Centre for Advanced Scientific Research, Jakkur, Bangalore 560 064, India
*Email: tmaji@jncasr.ac.in

12.1 Introduction

The chemistry of metal-organic frameworks (MOFs) or porous coordination polymers (PCPs) has received enormous attention because of their widespread and possible applications in the field of storage and separation, catalysis, sensing, drug delivery *etc.*[1–8] The advantage of MOFs over conventional porous materials such as zeolites or porous carbons is its structural modularity and tunability for tailor-made applications. This enables one to elucidate molecular level structure with the aid of single crystal X-ray diffraction which becomes a very good starting point to explore the properties of the material thereafter. The competent reticular chemistry and the enormous possibility of functionalizing the organic backbones are the

[†]Current Address: Department of Chemistry, School of Chemical Sciences, Central University of Haryana, Jant-Pali, Mahendergarh 123031, Haryana, India.

Monographs in Supramolecular Chemistry No. 22
Functional Supramolecular Materials: From Surfaces to MOFs
Edited by Rahul Banerjee

Published by the Royal Society of Chemistry, www.rsc.org

strongest tools associated with the framework materials for generating wide variety of functionalized structures.[9–13] In 1998, MOFs were first classified into three categories based on their structural response: (a) Frameworks which collapse their structure upon guest removal were termed as **First-Generation MOFs**; (b) There are rigid frameworks which retain their original structure even if they undergo loss of guest molecules and were termed as **Second-Generation MOFs**; (c) In the third case, there might be some structural rearrangement or reorganization when the framework loses guest molecules and were named as **Third-Generation MOFs**.[14]

Third-Generation MOFs have the characteristics of structural dynamism or flexibility wherein the framework undergoes definite change in the global or at least local structure in response to external stimuli such as guest, temperature, pressure, light *etc.*[15–19] In a broad sense, a flexible or soft MOF can be defined as one that possess both a highly ordered network and structural transformability.[4,20–23] These can have more than one intermediate states, both kinetically as well as thermodynamically stable, with long range ordering, a reversible transformability between the states, and permanent porosity. The transformations which these solids encounter can be of single-crystal-to-single-crystal type or may be from a single crystal to a less ordered or periodic crystalline state.[24–30] The dynamic nature of such flexible or soft frameworks has remarkable impact on the sorption processes of gases and vapours. Unlike the rigid frameworks, the sorption isotherms obtained with flexible MOFs are difficult to categorize according to IUPAC classification. The flexible structure of MOFs could be compared with biological systems wherein the protein or enzymes undergo conformational changes in response to the external guest molecules.

12.2 Rigid and Flexible MOFs

One of the important divergences between rigid and flexible MOFs lies in the predictability of their crystalline state properties. While the properties arising from rigid frameworks are more or less predictable, it is not the case with flexible frameworks. Rigid porous frameworks have highly regular and open channel structure and hence can accommodate a range of guest molecules making itself less competent for selective sorption and separation.[28,31–35] However, frameworks having channel sizes smaller than the kinetic diameter of the adsorbates would not allow the molecules to easily pass through, and in such cases selectivity may arise. This is very important from the view point of separation and selectivity as only certain molecules would be able to interact with the solid while it excludes other molecules.[17,18,36,37]

On the other hand, the flexible frameworks can adopt its structures according to the incoming guests through strong non-covalent interactions. Such responses to the incoming guest molecules can bring large structural changes which can alter its adsorption, magnetic or photophysical properties.[17–19] Thus flexible frameworks can be extremely useful for molecular

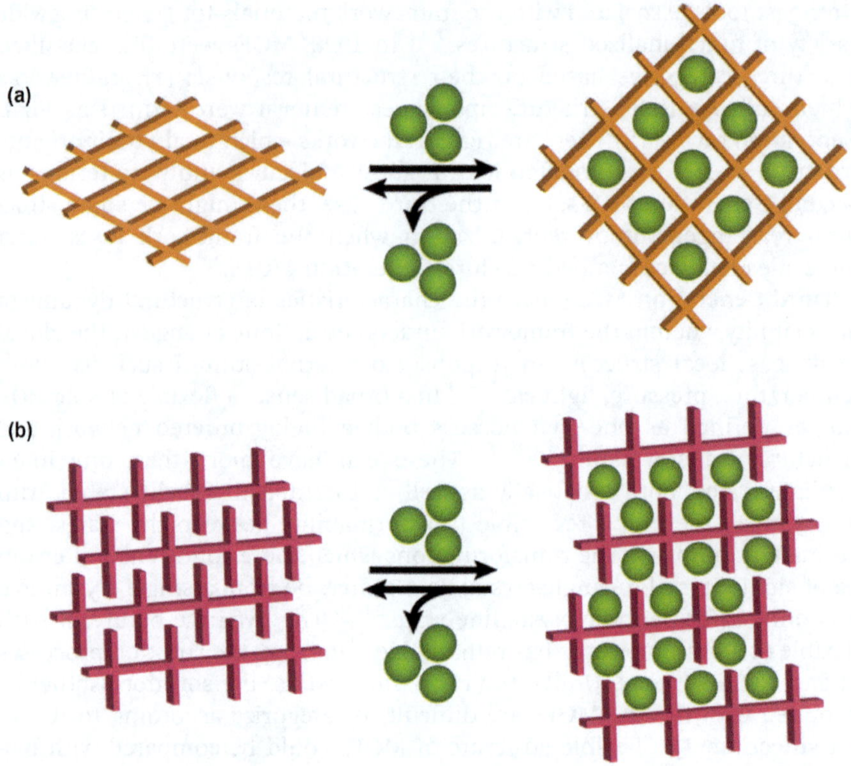

Scheme 12.1 Schematic of guest responsive nature in flexible metal-organic frameworks, (a) 3D frameworks; and (b) interdigitated MOFs.

recognition applications.[19,38,39] The behaviour of the flexible MOFs towards incoming guest molecules become dependent on several parameters, such as: (a) chemical nature of the incoming guest molecule; (b) chemical nature and size of the pore window; (c) pressure; (d) temperature *etc.* Guest molecules having size comparable to the pore window and with significant dipole or quadrupole moment can bring enhanced interaction with the pore surfaces and ultimately diffuse into the channel at a particular onset pressure (Scheme 12.1).

12.3 Sorption in Flexible MOFs: Gated and Stepwise Adsorption

Kondo *et al.* investigated adsorption using a 3D MOF for the first time, while in the following year the presence of permanent porosity was demonstrated in MOF-5.[40] After those initial studies there has been an upsurge in the area of adsorption of MOFs which is reflected in the exponential increase of research papers published. For practical applications, selective adsorption and separation of guest molecules by MOFs is very important.[41–51] In case of an open porous structure, it is difficult to control the diffusion of guest molecules selectively. Soft porous MOFs usually undergo change in the structure from an open phase to a close phase upon guest removal. This alters its channel structure and pore volume thus making the diffusion of incoming guest molecules difficult. Due to structural change two cases may arise: either the pore volume of new closed phase is not at all accessible or it may be available partially. In the first the case, the solid does not show any guest uptake at low pressures, however as the pressure increases there is a sudden rise in the uptake (at threshold pressure or gate-opening pressure) and the desorption process is accompanied by a large hysteresis. This results in gate-opening type sorption accompanied by a sudden structural rearrangement.[4,44,52–54] The unusual sorption behaviour in such inorganic–organic hybrid compounds can be modulated by the adsorbate and adsorbent interactions such as H-bonding, π–π interactions.[55,56] Both inert and polar gas molecules such as N_2 and CO_2 were found to display this unusual behaviour suggesting both kinetic and thermodynamic factors controlling the sorption processes.

In the second case, the partially-available space is first filled at low guest loading (low pressure) and as the pressure increases gradually the framework undergoes further changes in the structure displaying additional uptake of guest molecules inside the pore. In general, such sorption processes are selective as the incoming guest molecule has to interact with the framework in a specific manner to bring upon the framework rearrangement. The framework rearrangement can occur in various manners depending upon the host structure and the incoming guest molecules. Specific cases are described below in different coordination polymers having different structural aspects.

12.4 Structural Flexibility in a 1D Coordination Polymer

A recent report from our group showed dynamic behaviour of a 1D coordination polymer, {[Zn(*o*-phen)(2,6-ndc)] · DMF}$_n$.[19] The dynamics were observed upon removal of guest DMF molecules and during CO_2 adsorption. The coordination polymer is formed *via* tethering of Zn(II) through 2,6-ndc and the *o*-phen chelating linker restricts the structure to 1D. Through C–H···O hydrogen bonding and C–H···π interactions these 1D coordination chains self-assemble to form 1D supramolecular pores which are occupied by the guest DMF molecules (Figure 12.1a–d). The guest-removed structure is different from the as-synthesized structure as realized from the PXRD studies. Such structural rearrangement is probably due to the presence of non-covalent interactions that stabilize the structure by reducing the void space of the system (reduction in cell volume). Interestingly, the desolvated framework does not adsorb N_2 (77 K) but shows stepwise uptake of CO_2 at 195 K with large hysteresis (Figure 12.1e). The uptake rises slowly till

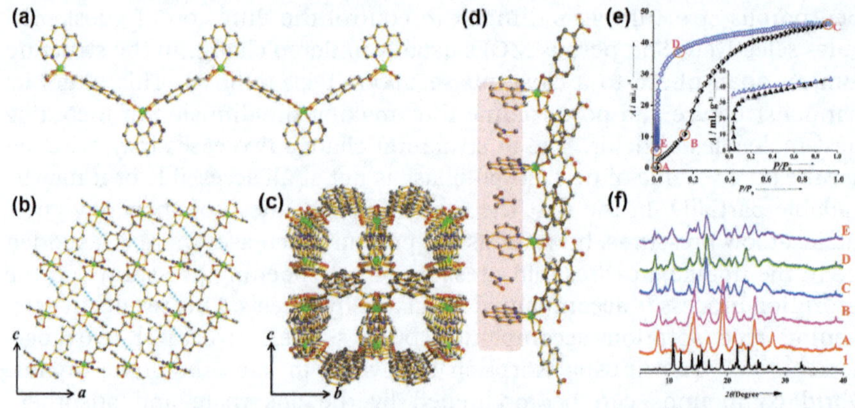

Figure 12.1 (a) 1D zigzag chain of compound {[Zn(*o*-phen)(2,6-ndc)] · DMF}$_n$; (b) 2D corrugated sheet; (c) view of the rectangular channels along the *a* axis formed by the supramolecular organization of 1D chains; (d) columnar stacking of the [Zn(*o*-phen)] subunit and the guest DMF molecules; (e) CO_2 adsorption (half-filled diamonds) and desorption (black dot centered diamonds) isotherms of desolvated framework at 195 K. Inset: C_6H_6 vapor adsorption (half-filled triangles) and desorption (triangles) isotherms of desolvated framework at 298 K; (f) PXRD patterns of desolvated framework at different stages of CO_2 adsorption–desorption. (Adapted with permission from R. Haldar, R. Matsuda, S. Kitagawa, S. J. George and T. K. Maji, Amine-Responsive Adaptable Nanospaces: Fluorescent Porous Coordination Polymer for Molecular Recognition, *Angew. Chem., Int. Ed.*, 2014, **126**, 11966. Copyright © 2014 WILEY-VCH Verlag GmbH & Co. KGaA, Weinheim.)

pressure P/P_0 ~0.18 and after that with a steep uptake it reaches the final amount, 48 mL g^{-1} (0.9 molecules of CO_2 per formula). A PXRD coupled adsorption experiment was carried out and at each point of adsorption and desorption, PXRD patterns were recorded which provided insightful observations (Figure 12.1f). Indexing of the PXRD patterns collected during adsorption process show that from the desolvated state to fully CO_2-loaded state the cell volume increases from 2143 to 2963 Å^3. During desorption the CO_2 molecules are not easily removed until very low pressures are reached, suggesting the strong adsorbate–adsorbent interactions. The increase in cell volume during the adsorption process might be referred to as the opening of the pores to permit the diffusion of CO_2 and this process is triggered by framework π surface and CO_2 quadrupole interaction. This sort of structural change and the manipulation of the desolvated structure are quite hard tasks to carry out but the PXRD patterns clearly suggest that CO_2 does open up the voids of the framework.

12.5 Flexibility in 2D MOFs

Coordination polymers made up of 2D structure exhibit greater degree of flexibility because of the possible movement of the 2D networks in addition to local bond length and angle shift or change during guest removal or accommodation.[57–60] Both stepwise and gated adsorptions are reported in the 2D MOFs and we will be discussing some of such fascinating examples.

12.5.1 Stepwise Sorption in 2D MOFs

In most cases, the pore surfaces of the MOFs are made up of π-electron-rich aromatic rings, polar organic functional groups, unsaturated metal sites (UMSs), *etc.* which are responsible for creating sorption potential in the nanochannel. As a consequence, polar guest molecules that are easily polarizable and possess significant quadrupole moment can interact efficiently with the pore surface of the MOFs and diffuse through the channel.[61,62] When the pore window or the pore surfaces of MOFs are highly polar such guest molecules can bring upon structural transformation resulting in steps in the sorption isotherm. Kanoo *et al.* reported a mixed-ligand MOF, $\{[Ni(1,3\text{-adc})(bpp)(H_2O)_2]\cdot(H_2O)\cdot(EtOH)\}_n$ which was constructed from flexible organic linkers and has a 2D triangular corrugated structure (Figure 12.2a–b).[63] The desolvated framework, $\{[Ni(1,3\text{-adc})(bpp)]\}_n$ undergoes structural contraction and was found to display double-step hysteretic profiles for CO_2, H_2O, MeOH and a single-step gate-opening behaviour with EtOH (Figure 12.2c–d). The double-step sorption process in the case of solvent vapours was interpreted by analysing the PXRD patterns recorded with the various adsorbed states which revealed structural transformation as the second step commences (Figure 12.2e). Since the MOF is composed of 2D layers, the most expected structural rearrangement was believed to be sliding of the layers. The transformation was realized only in case of small molecules as sorption measurement with EtOH revealed a one-step profile. It was concluded from the PXRD patterns of different adsorbed states that the compound adsorbs H_2O molecules maintaining narrow pores, in contrast to MeOH molecules which are adsorbed with opening of pores. Framework contraction upon H_2O sorption was interpreted on the basis of strong coordinating ability of the corresponding adsorbate with an unsaturated Ni(II) centre and hydrogen-bonding interaction between the framework and adsorbed H_2O molecules in the pore.

In continuation of the work on 2D systems, a Cd(II) MOF $\{[Cd_4(azpy)_2(pyrdc)_4(H_2O)_2]\cdot 9H_2O\}_n$ with two different type of spaces, 1D channel and 2D interlayer space occupied by water molecules, has been reported where dynamic motion of 2D bilayers was observed during the H_2O and MeOH sorption process.[64] Careful analysis of the PXRD pattern shows shifting of the (400) peak to a higher angle, from 2θ ~8.85° in the as-synthesized MOF to ~9.19° in the desolvated MOF suggesting a decrease in interlayer spacing with concomitant reduction of 2D space (Scheme 12.1). This is evident as it contains guest and coordinated water molecules in (400) plane, the removal of which causes the neighbouring bilayers to come closer and hence close the 2D space. PXRD patterns recorded at P/P_0 ~0.1 shows the (400) peak appears at 2θ ~9.01° (Figure 12.3b). This sudden decrease in 2θ

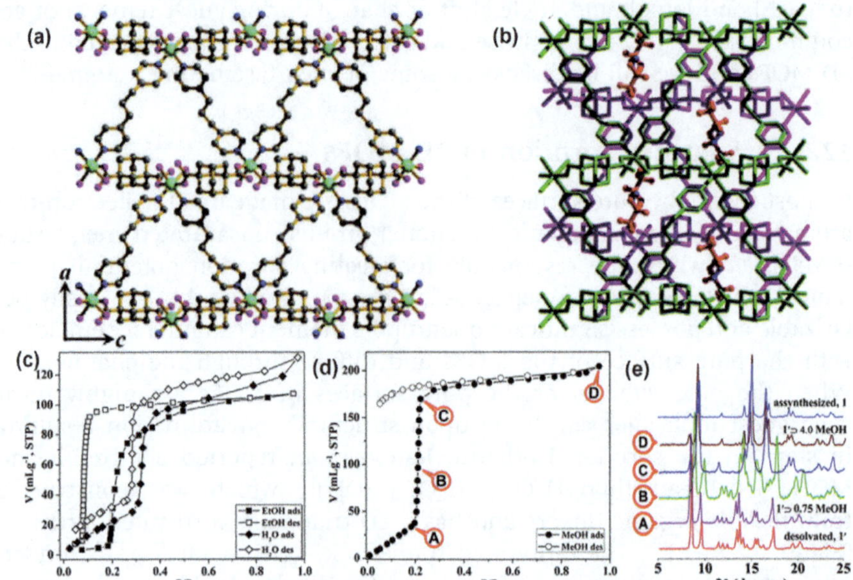

Figure 12.2 (a) 2D triangular sheet of $\{[Ni(1,3\text{-adc})(bpp)(H_2O)_2]\cdot(H_2O)\cdot(EtOH)\}_n$ composed of 1D $[Ni(1,3\text{-adc})]_n$ chains connected by the bpp linkers; (b) 2D sheets stack along the *b* axis in an AB fashion, forming 1D channels occupied by H_2O (blue) and EtOH (red) molecules. Vapour adsorption isotherms of desolvated framework: (c) H_2O and EtOH at 298 K; (d) MeOH at 293 K. (e) PXRD patterns of as-synthesized and desolvated framework at different states of MeOH adsorption. (Adapted with permission from P. Kanoo, R. Sambhu and T. K. Maji, Guest-Specific Double- or Single-Step Adsorption in a Flexible Porous Framework Based on a Mixed-Ligand System, *Inorg. Chem.*, 2011, **50**, 400. Copyright 2011 American Chemical Society.)

value (from 9.19° in the desolvated MOF) suggests that even at very low pressure the structure expands and the bilayers start moving with respect to each other. The 2θ value remains almost constant up to first step sorption, 2θ ~9.03° at P/P_0 ~0.7. Once the 1D channels are covered by the adsorbate molecules, the next available sorption sites are 2D interlayer spaces the dimension of which is smaller than the 1D channel. During the sorption process, at high pressure and P/P_0 ~0.7 the bilayers move away because of enhanced interaction of adsorbate molecules with unsaturated metal sites (coordinative) and –COO groups (*via* hydrogen bonding) and suddenly, the molecules starts diffusing into the 2D spaces and hence a rapid uptake is observed. This is evidenced by a further decrease in the 2θ value after second step sorption, 2θ ~9.01°→8.94°. Hence, it implies that a certain critical pressure forces the bilayers to move up to an extent whereby MeOH molecules start diffusing in the 2D interlayer space.

That 2D MOFs are more flexible than their 3D counterparts is best exemplified by $[Cu(bpy)_2(OTf)]_n$ (Figure 12.4).[65] With the same chemical

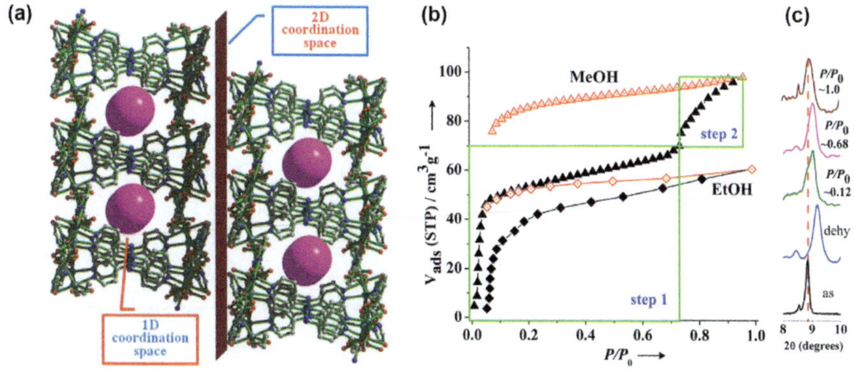

Figure 12.3 (a) The pillared-bilayer 2D networks of $\{[Cd_4(azpy)_2(pyrdc)_4(H_2O)_2]\cdot 9H_2O)\}_n$ show 1D channels along the crystallographic *c* axis and the 2D interlayer space in the *ab* plane, the latter separates the bilayers; (b) MeOH and EtOH adsorption isotherms at 293 and 298 K, respectively. Black curves indicate adsorption and red curves desorption. (c) PXRD pattern showing shifting of the (400) peak during MeOH adsorption. The dashed red line points peak shifting during MeOH adsorption. (Adapted from ref. 64 with permission from the Royal Society of Chemistry.)

composition, two variances, a 2D and another 3D MOF, have been synthesized which have similar Cu(II) coordination environments and exactly the same framework structures. The 3D compound exhibits typical type I sorption profiles with N_2, CO_2 and Ar with no definite steps suggesting no structural change during the sorption process (Figure 12.4). In contrast, the 2D variance shows two-step sorption profiles in all the cases suggesting greater degree of flexibility which the 3D variance does not show. The 2D MOF has an open channel that allows rapid uptake at low pressure, however, as the pressure increases the 2D layers move away from each other causing an expansion in the structure and a second step uptake. Synchrotron X-ray diffraction studies after second step sorption suggest shifting of some peaks to low angles indicating expansion of the structure. The sorption mechanism in the 2D MOF is believed to occur *via* a concerted mechanism involving expansion/shrinkage of the layers and the breathing transformation.

12.5.2 Gated Sorption in 2D MOFs

Kitaura *et al.* reported gated sorption behaviour of various gases, N_2, CO_2, O_2 and CH_4 at high pressures in a mixed-ligand based Cu(II)-compound, $\{[Cu(dhbc)_2(4,4'\text{-bpy})]\cdot H_2O\}_n$ with interdigitated structure (Figure 12.5).[66] In the interdigitated 2D network, the copper ions are connected by 4,4′-bpy to produce linear 1D chains, which are further linked by dhbc linkers to give the 2D sheet (Figure 12.5a–b). The π–π stacking interactions between nearest-neighbour dhbc ligands create mutually interdigitated 2D nets forming

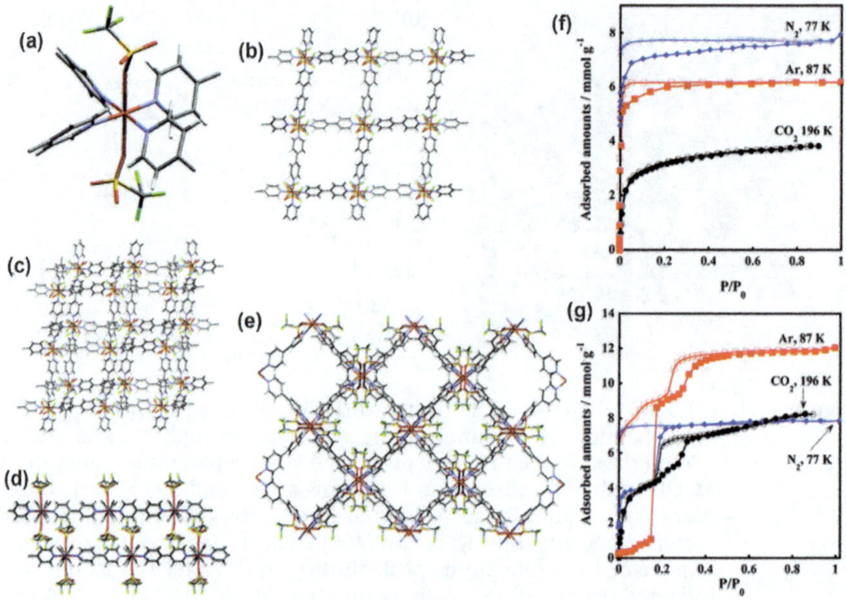

Figure 12.4 (a) Local coordination structure of the Cu(II) ion in 2D and 3D MOFs; (b) The 2D sheet in 2D MOF; (c) the top view and (d) the side view of stacking structure of the square grid sheets; (e) the view of 3D MOF with hydrophilic and hydrophobic pores along the *c* axis. Adsorption isotherms of N_2 (77 K), CO_2 (196 and 273 K), and Ar (87.3 K) on (f) 3D and (g) 2D MOFs. Solid and open symbols represent adsorption and desorption branches, respectively.
(Adapted with permission from A. Kondo, H. Kajiro, H. Noguchi, L. Carlucci, D. M. Proserpio, G. Ciani, K. Kato, M. Takata, H. Seki, M. Sakamoto, Y. Hattori, F. Okino, K. Maeda, T. Ohba, K. Kaneko and H. Kanoh, Super Flexibility of a 2D Cu-Based Porous Coordination Framework on Gas Adsorption in Comparison with a 3D Framework of Identical Composition: Framework Dimensionality-Dependent Gas Adsorptivities, *J. Am. Chem. Soc.*, 2011, **133**, 10512. Copyright 2011 American Chemical Society.)

1D channels along the *a* axis (Figure 12.5c). In dehydrated MOFs, peaks corresponding to the relative position of the 2D layer motifs show an apparent shift to the higher angle region between 15 and 20°. This result clearly indicated shrinking of the layer gap along the *c* axis, which is attributable to a gliding motion of the two π-stacked rings, which decreases the channel cross section. Structural re-expansion occurs with guest-inclusion at high pressure, which was reflected in the gated adsorption isotherms of various supercritical gases (Figure 12.5d–e). The most crucial outcome of this report is that it was realized that MOFs may have flexible or dynamic character and they can respond to external stimuli, in this case guest molecules (gas and solvent vapours).

In the subsequent years, a 2D bilayer open framework of Cu(II), $\{[Cu(pyrdc)(bpp)]\cdot 5H_2O\}_n$, was synthesized that was found to display gated

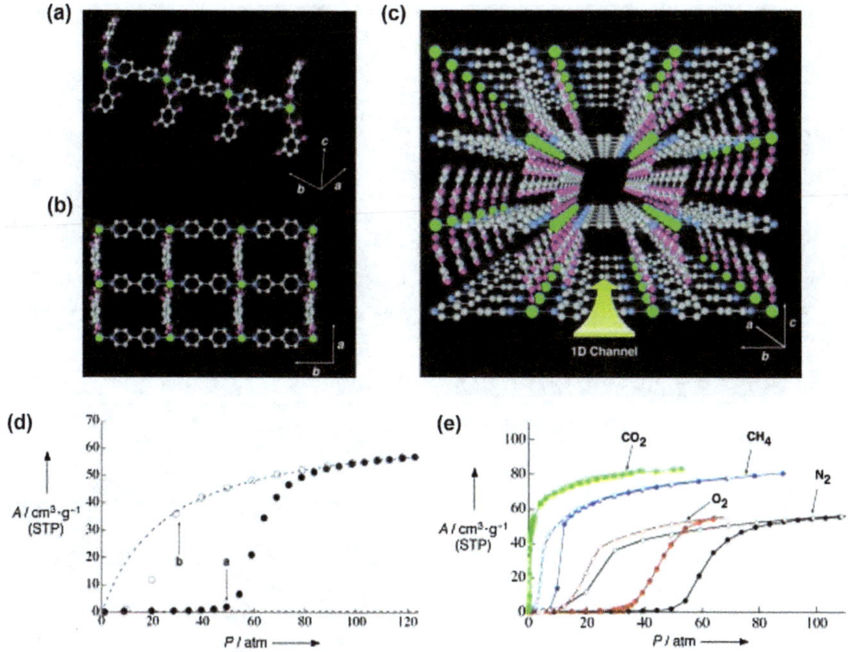

Figure 12.5 (a) 1D chain structure constructed by Cu^{II}, 4,4′-bpy, and dhbc along the b axis of $\{[Cu(dhbc)_2(4,4'\text{-bpy})]\cdot H_2O\}_n$ (Cu, green; O, red; C, grey; N, blue); (b) 2D layer structure along the ab plane; (c) 3D π-stacked pillared layer structure of $\{[Cu(dhbc)_2(4,4'\text{-bpy})]\cdot H_2O\}_n$; (d) Nitrogen adsorption (filled circles) and desorption isotherms (open circles) at 298 K; (e) Adsorption (filled circles) and desorption (open circles) isotherms of N_2, CH_4, CO_2 and O_2 at 298 K.
(Adapted with permission from R. Kitaura, K. Seki, G. Akiyama and S. Kitagawa, Porous Coordination-Polymer Crystals with Gated Channels Specific for Supercritical Gases, *Angew. Chem., Int. Ed.*, 2003, **42**, 428. Copyright 2002 WILEY-VCH Verlag GmbH & Co. KGaA, Weinheim.)

CO_2 sorption at 195 K.[30] Structure determination revealed that, upon dehydration, the cell volume decreases by 12.62%, indicating a drastic contraction of the framework (Figure 12.6a–d). This was the first observation of porous to non-porous phase transition of a dynamic framework, with retention of single crystallinity and introducing asymmetry into the framework upon dehydration. The selective gate-opening phenomenon with certain adsorbates is completely realized by the hydrogen-bonding interaction with apohost framework, that is, guests are permitted to pass the gate at specific gate-opening pressures that depend on the strength of the intermolecular interaction (Figure 12.6e–g).

Kaneko *et al.* reported crystal-to-crystal transformation of a 3D interpenetrated-type MOF $\{[Cu(BF_4)_2(bpy)(H_2O)_2]\cdot(bpy)\}_n$ to a 2D square-grid-type $[Cu(BF_4)_2(bpy)_2]_n$.[67] $[Cu(BF_4)_2(bpy)_2]_n$ consists of the quasi-square grid 2D layered stacking structure and the layers stack upon each other to block the interlayer voids, thus leaving no open spaces for CO_2 gas

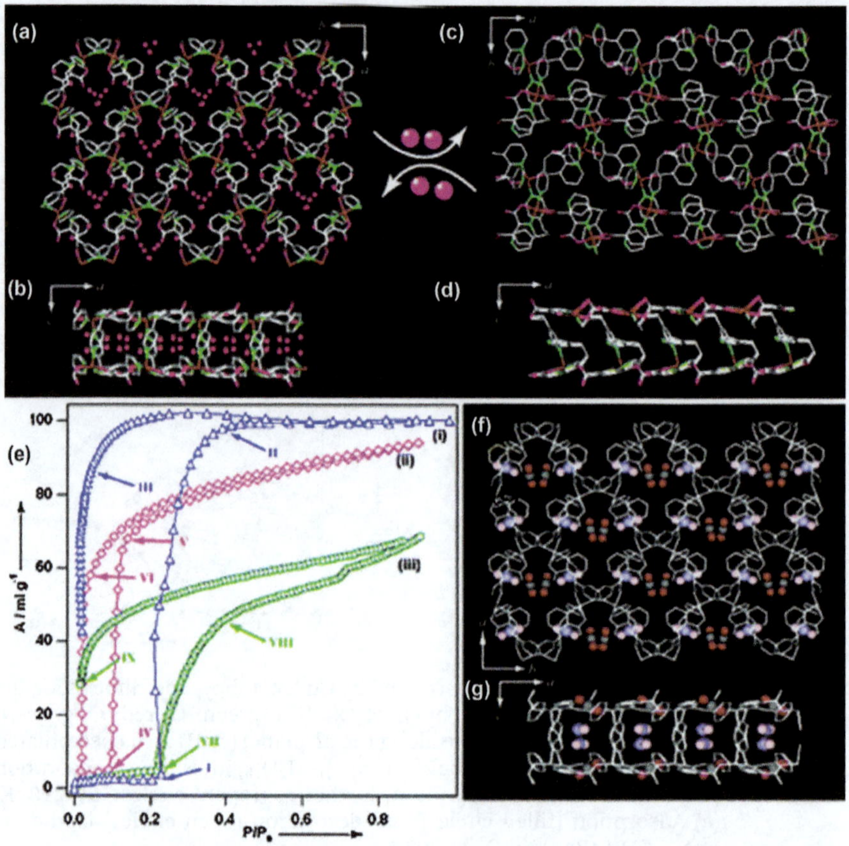

Figure 12.6 (a) Honeycomb-like 2D layers of $\{[Cu(pyrdc)(bpp)]\cdot 5H_2O\}_n$ showing a water-filled channel along the *c* axis; (b) Pillared-bilayer network showing the 1D channel along the *b* axis; (c) Honeycomb-like 2D layers of dehydrated framework along the *c* axis; (d) Pillared-bilayer network along the *b* axis (the pendent bpp ligand is occupied in the channel); (e) Sorption isotherm for different adsorbate: (i) CO_2 (195K); (ii) MeOH; and (iii) EtOH (298 K); (f) Honeycomb-like 2D layer with CO_2 molecules along the *c* axis; (g) Pillared-bilayer network along the *b* axis with different CO_2 molecules.
(Adapted with permission from T. K. Maji, G. Mostafa, R. Matsuda and S. Kitagawa, Guest-Induced Asymmetry in a Metal-organic Porous Solid with Reversible Single-Crystal-to-Single-Crystal Structural Transformation, *J. Am. Chem. Soc.*, 2005, **127**, 17152. Copyright 2005 American Chemical Society.)

adsorption (Figure 12.7a–d). During guest (CO_2) adsorption, an expansion of the interlayer structure occurs concomitantly inducing the formation of a structure having many open pores, and thus CO_2 molecules can be accommodated in the open structure (Figure 12.7e–f). This structural change of the 2D flexible framework caused by the interaction between the pore wall and

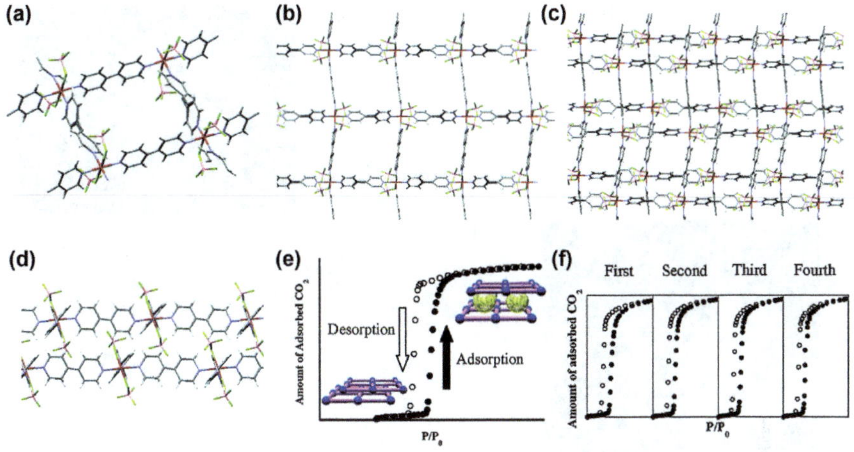

Figure 12.7 (a) Fundamental framework structure of$\{[Cu(BF_4)_2(bpy)(H_2O)_2]\cdot(bpy)\}_n$; (b) 2D sheet structure; (c) stacking structure; (d) side view of 2D sheets (orange, Cu; grey, C; blue, N; pink, B; yellow green, F; white, H); (e) Schematic representation of the gate adsorption phenomenon of CO_2 associated with structural transformation and clathrate formation; (f) The high successive reproducibility over four cycles.
(Adapted with permission from A. Kondo, H. Noguchi, S. Ohnishi, H. Kajiro, A. Tohdoh, Y. Hattori, W.-C. Xu, H. Tanaka, H. Kanoh and K. Kaneko, Novel Expansion/Shrinkage Modulation of 2D Layered MOF Triggered by Clathrate Formulation with CO_2 Molecules, *Nano Lett.*, 2006, **6**, 2581. Copyright 2006 American Chemical Society.)

gas molecules is triggered by the clathrate formation. The expansion–shrinkage reversible modulation of the flexible 2D framework results in novel "gated adsorption" at 273 K and the sorption isotherm has a predominantly rectangular shaped hysteresis, which is quite unique.

In an attempt to understand the gate-opening process, a kinetic study was carried on the O_2, N_2 and Ar sorption isotherm on a flexible mixed-ligand Cd(II)MOF, $\{[Cd(bpndc)(bpy)]\}_n$.[68] The single crystal structure reveals that the Cd(II) ions are connected by bpndc to produce 1D double-chain structures of $\{[Cd(bpndc)]\}_n$ along the *c* axis, which are linked by bpy along the *b* axis to give a 2D motif (Figure 12.8a–b). The 2D layers are mutually interdigitated to create a 3D assembled framework. The framework shows a large difference in the gate-opening pressures, P_{go} between O_2, Ar, and N_2 (Figure 12.8e). Although the volumetric uptake of three gases is comparable, the gate-opening pressures, P_{go} for the gases are different which means the MOF is able to adsorb the gases at different pressures. Kinetic studies revealed that formation of intermediates between the adsorbate and adsorbent is crucial in the gate-opening process which eventually determines P_{go}. Comparison of the kinetic data showed the rate constant k_{go} for the same pressure is $O_2 > Ar > N_2$ which was found to be consistent with the guest dependency of P_{go}. This trend is consistent with the boiling point of the gases (O_2, 90.2 K;

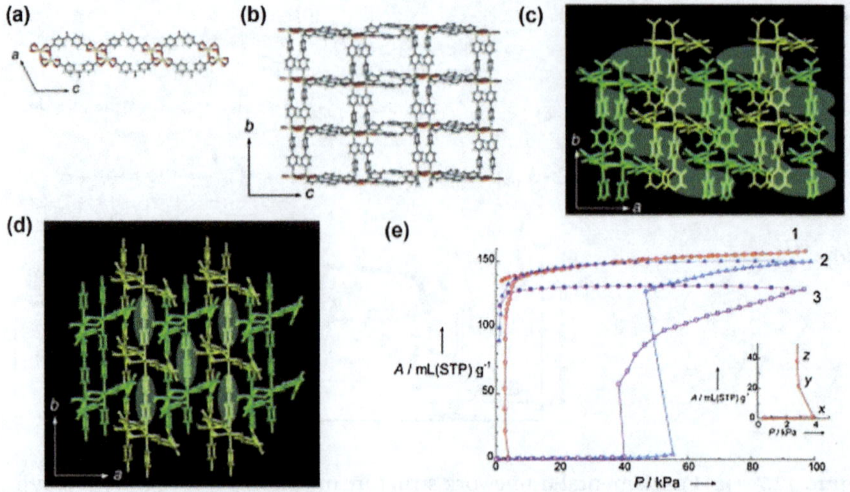

Figure 12.8 (a) Crystal structures of {[Cd(bpndc)(bpy)]}$_n$ showing 1D double chain of {[Cd(bpndc)]}$_n$; (b) 2D layer structure of {[Cd(bpndc)(bpy)]}$_n$ (grey, purple, red, and yellow are C, N, O, and Cd, respectively); (c) 3D assembled structure of {[Cd(bpndc)(bpy)]}$_n$ with solvents; (d) 3D assembled structure of {[Cd(bpndc)(bpy)]}$_n$. Cavities and windows are highlighted; (e) Sorption isotherms for adsorbates in {[Cd(bpndc)(bpy)]}$_n$ at 90 K; (1) O_2 adsorption (red open circles) and desorption (red filled circles); (2) N_2 adsorption (blue open triangles) and desorption (blue filled triangles); (3) Ar adsorption (purple open squares) and desorption (purple filled squares). Inset: O_2 adsorption isotherm at low pressure. (Adapted with permission from D. Tanaka, K. Nakagawa, M. Higuchi, S. Horike, Y. Kubota, T. C. Kobayashi, M. Takata and S. Kitagawa, Kinetic Gate-Opening Process in a Flexible Porous Coordination Polymer, *Angew. Chem., Int. Ed.*, 2008, **120**, 3978. Copyright 2008 WILEY-VCH Verlag GmbH & Co. KGaA, Weinheim.)

Ar, 87.3 K; N_2, 77.4 K) suggesting intermolecular interaction plays vital role in the sorption process.

12.6 Flexibility in 3D MOFs

12.6.1 Stepwise Sorption in 3D MOFs

12.6.1.1 Stepwise Sorption in 3D Interpenetrated MOFs

Interpenetration in MOFs either leads to a completely non-porous dense structure or a porous structure with smaller channels compare to an individual 3D net.[69–74] The porous interpenetrated structures can be either rigid or flexible which depends upon several factors. Structural rigidity is realized when two interpenetrated nets are hold tightly by non-covalent π–π or H-bonding interactions or closely packed nets in one side which does not allow to move. The structural dynamics of entangled flexible nets are quite interesting as they can show guest responsive deformation and concomitant shearing of nets. Due to such structural transformation, the desolvated phase may transform to a completely non-porous phase or to a semi-open phase based on the degree of net movement. The effect of such structural changes is quite significant in adsorption properties or other physical properties like magnetism and optical properties.[75–78] Particularly, such desolvated frameworks show stepwise or gated adsorption profile which have implication for a selective uptake and separation of a particular adsorbent molecule.[79,80] We will discuss here some of such structural transformations.

A partially double-interpenetrated MOF NOTT-202 where one net is fully occupied while the other net has an occupancy of 0.75 was found to display a three step CO_2 sorption at 195 K with a large hysteresis (Figure 12.9).[71] The three steps were observed at P/P_0 0.07, 0.52 and 0.95 with uptake capacities of ~7, 14 and 20 mmol g^{-1}, respectively. Interestingly, the sorption profiles measured between 205–213 K, below the triple point of CO_2, show similar behaviour but with the second and third steps slowly merging. However, the isotherms measured between 221–303 K, above the triple point of CO_2 do not

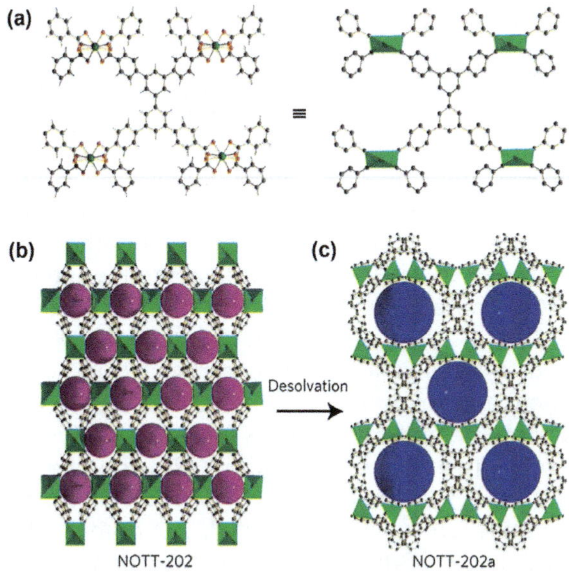

Figure 12.9 (a) View of the coordination environment at the $[In(O_2CR)_4]$ nodes and ligands L^{4-}. The $[In(O_2CR)_4]$ nodes are represented by green tetrahedra. Schematics of pore framework structure for NOTT-202 (b) and NOTT-202a (c).
(Adapted by permission from Macmillan Publishers Ltd: Nature Materials (S. Yang, X. Lin, W. Lewis, M. Suyetin, E. Bichoutskaia, J. E. Parker, C. C. Tang, D. R. Allan, P. J. Rizkallah, P. Hubberstey, N. R. Champness, K. Mark Thomas, A. J. Blake and M. Schröder, *Nat. Mater.*, 2012, **11**, 710), copyright (2012).)

display any step, suggesting highly temperature dependent behaviour of the sorption process (Figure 12.10). *In situ* powder diffraction studies at various pressures of CO_2 show changes in the diffraction pattern at 195 K, while that measured at 273 K did not show any changes in the peak positions (Figure 12.10). It has been mentioned that at temperatures below the triple point, the sorption equilibria can be built up by adsorbed CO_2 molecules in a thermodynamically more ordered state than in the fluid phase above the triple point (Figure 12.10). The sorption–desorption hysteresis in this MOF is exclusively coupled with CO_2-host ordering in the microporous interactions below the triple point.

In another report from our group, we have shown that even the secondary building unit (SBU) can critically affect the flexibility.[81] Two supramolecular isomers were synthesized with same molecular formula, $\{[Cd(NH_2\text{-bdc})(bphz)_{0.5}]\cdot DMF\}_n$ but different structural features. Both the structures are two-fold interpenetrated; however one structure has a paddlewheel type SBU and the other one is μ-oxo bridged (Figure 12.11a,b). Due to requisite available space with paddlewheel type SBU, it shows flexible behaviour which is reflected in the CO_2 adsorption profile. The 1D channels created due to entanglement are filled with guest DMF molecules; removal of those at 160 °C shows a distinct change in the PXRD pattern compared to that of the as-synthesized suggesting structural rearrangement after desolvation (Figure 12.11c). Such a flexible nature is also reflected in the CO_2 adsorption profile which shows stepwise uptake at 195 K (Figure 12.11d). The desorption curve creates a distinct hysteresis indicating strong CO_2-framework interaction. We conjectured that the strong interaction between the CO_2 molecules and the $-NH_2$ functional groups of the framework rearrange the interpenetrated nets resulting in the stepwise adsorption.

Another 3D interpenetrated framework, $\{[Zn_2(bdc)_2(bpNDI)]\cdot 4DMF\}_n$ with a pore surface decorated with polar functional groups and dense aromatic π clouds, shows CO_2 capture, storage and separation of small hydrocarbons in ambient conditions.[18] The ability to encapsulate both polar and non-polar aromatic solvent molecules as reported earlier led us to believe that this could be a potential framework for this purpose.[39] MOFs with UMSs have already proved to be potential candidates for storage and separation but lack of efficiency in the presence of moisture makes the process adverse. Utilization of functional linkers in framework could be also an alternative pathway where this problem can be avoided. The framework is composed of $\{Zn_2(COO)_4\}$ paddlewheel SBUs which are connected through bdc linkers to form 2D square grid along the *bc* plane. Those square grids are further extended by bpNDI pillars along [101] to generate an extended 3D framework (Figure 12.12a–d). Sufficient void space in a single net allows growth of another leading to a twofold interpenetration with two different types of channels (site A and site B) occupied by four DMF molecules (Figure 12.12a–c). Interestingly, upon removal of three DMF molecules, the framework changes to a different phase, $\{[Zn_2(bdc)_2(bpNDI)]\}_n$, with contracted

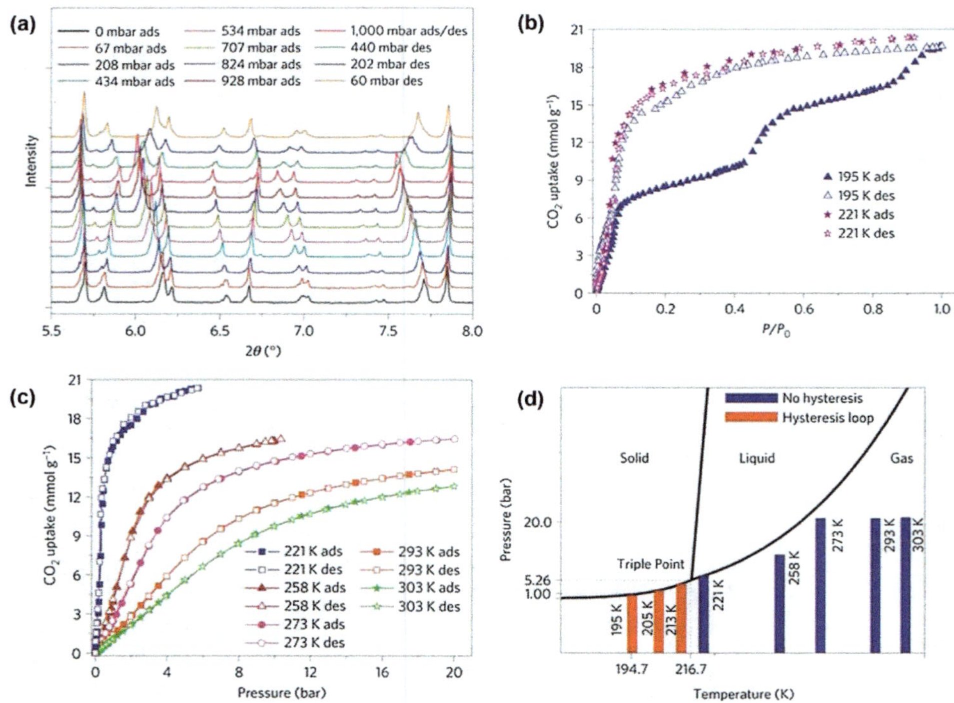

Figure 12.10 (a) Powder X-ray diffraction patterns at different CO_2 adsorption and desorption loadings at 195 K (the powder patterns of NOTT-202a were recorded at 195 K at 0, 67, 208, 434, 534, 707, 824, 928 and 1000 mbar of CO_2 adsorption loadings and at 440, 202 and 60 mbar of CO_2 desorption loadings); (b) CO_2 sorption isotherms at 195 and 221 K; (c) CO_2 sorption isotherms at various temperatures between 221 and 303 K; (d) CO_2 phase diagram with CO_2 sorption isotherms for NOTT-202a at different temperatures.
(Adapted by permission from Macmillan Publishers Ltd: Nature Materials (S. Yang, X. Lin, W. Lewis, M. Suyetin, E. Bichoutskaia, J. E. Parker, C. C. Tang, D. R. Allan, P. J. Rizkallah, P. Hubberstey, N. R. Champness, K. Mark Thomas, A. J. Blake and M. Schröder, *Nat. Mater.*, 2012, **11**, 710), copyright (2012).)

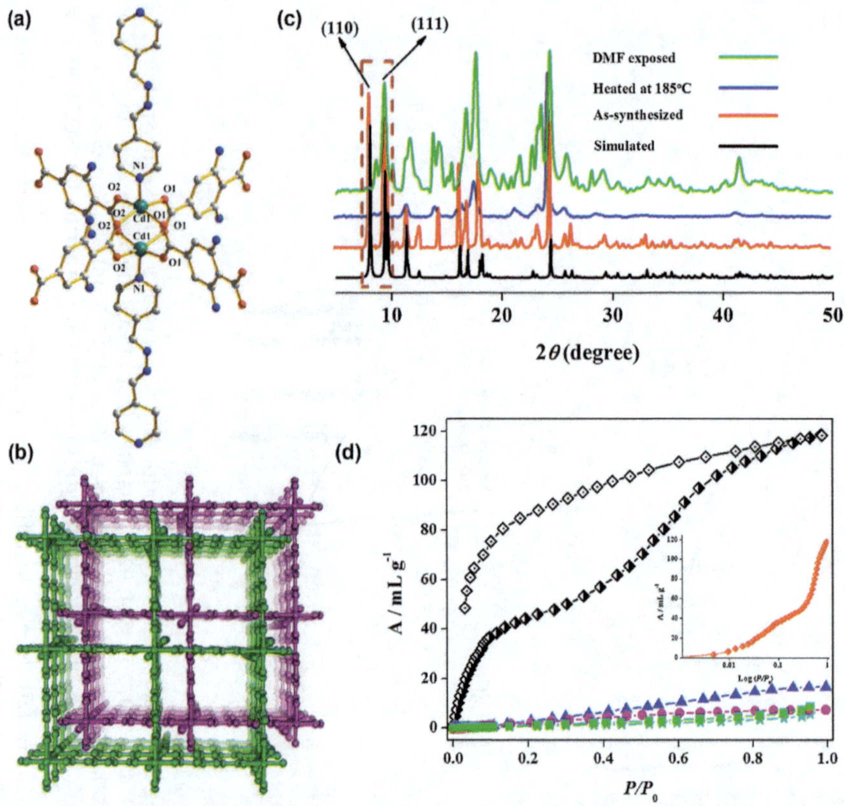

Figure 12.11 (a) Coordination environment of Cd(II) in $\{[Cd(NH_2\text{-}bdc)(bphz)_{0.5}]\cdot DMF\}_n$ with a paddle wheel SBU; (b) Two fold interpenetrated 3D framework of $\{[Cd(NH_2\text{-}bdc)(bphz)_{0.5}]\cdot DMF\}_n$; (c) PXRD patterns of $\{[Cd(NH_2\text{-}bdc)(bphz)_{0.5}]\cdot DMF\}_n$ at different conditions; (d) Adsorption isotherms for desolvated framework at 195 K: CO_2 adsorbtion (half-filled diamonds), CO_2 desorption (open diamonds), N_2 (stars), H_2 (circles), Ar (triangles), O_2 (squares); Inset: CO_2 adsorption profile at 195 K (log scale)
(Adapted with permission from R. Haldar, S. K. Reddy, V. M. Suresh, S. Mohapatra, S. Balasubramanian and T. K. Maji, Flexible and Rigid Amine-Functionalized Microporous Frameworks Based on Different Secondary Building Units: Supramolecular Isomerism, Selective CO_2 Capture, and Catalysis, *Chem. – Eur. J.*, 2014, **20**, 4347. Copyright 2014 WILEY-VCH Verlag GmbH & Co. KGaA, Weinheim.)

pores and one of the adsorption sites (Site B) remaining open. Even after complete removal of the guest molecules, the structure resembles $\{[Zn_2(bdc)_2(bpNDI)]\cdot DMF\}_n$. Our group has exploited this framework for CO_2 capture, hydrocarbon storage and separation of the gas mixtures CO_2–CH_4, C_2H_2–CH_4, C_2H_6–CH_4, and C_2H_4–CH_4 at 298 K and 1 bar (Figure 12.12e–i). It shows double-step hysteric CO_2 and C_2H_2 profiles at

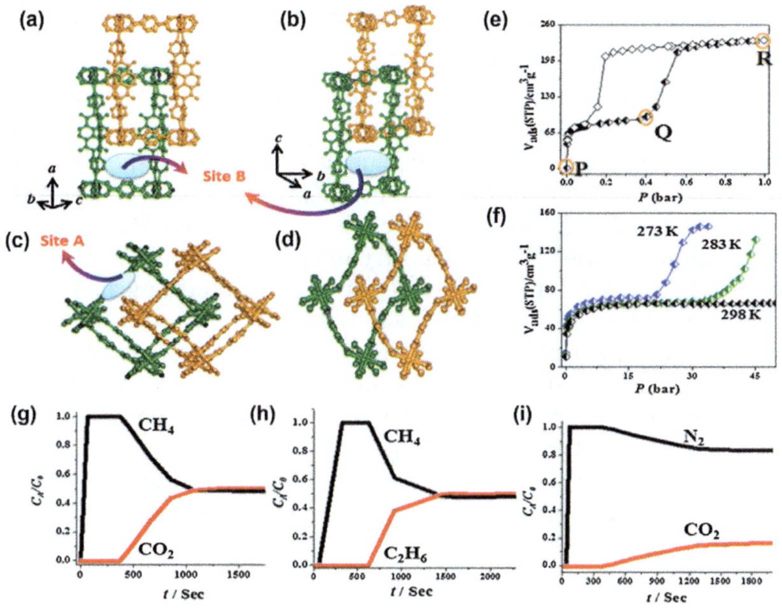

Figure 12.12 Crystal structure of $\{[Zn_2(bdc)_2(bpNDI)]\cdot 4DMF\}_n$ and $\{[Zn_2(bdc)_2(bpNDI)]\}_n$: (a and b) view of two-fold entangled frameworks of $\{[Zn_2(bdc)_2(bpNDI)]\cdot 4DMF\}_n$ and $\{[Zn_2(bdc)_2(bpNDI)]\}_n$ showing the similar adsorption sites; (c and d) two-fold entangled nets along the *a* axis in $\{[Zn_2(bdc)_2(bpNDI)]\cdot 4DMF\}_n$ showing site A which is blocked in $\{[Zn_2(bdc)_2(bpNDI)]\}_n$ due to the shearing of nets; (e) CO_2 adsorption studies of 1a in different conditions at 195 K [P is starting point of adsorption, Q and R are at the end of first step and second step, respectively]; (f) High pressure isotherms of CO_2 at 273 (33 bar), 283 (45 bar) and 298 K (50 bar). Breakthrough curves for binary mixture of (g) CH_4–CO_2 (50 : 50) and (h) CH_4–C_2H_6 (50 : 50) and (i) N_2–CO_2 (85 : 15) at 298 K. $p = 5$ kPa.
(Adapted with permission from N. Sikdar, S. Bonakala, R. Haldar, S. Balasubramanian and T. K. Maji, Dynamic Entangled Porous Framework for Hydrocarbon (C2–C3) Storage, CO_2 Capture, and Separation, *Chem. – Eur. J.*, 2016, **22**, 6059. Copyright 2016 WILEY-VCH Verlag GmbH & Co. KGaA, Weinheim.)

different temperatures which indicates the highly dynamic nature of the framework. The CO_2 uptake at 298 K is 12.7 wt% and shows high IAST CO_2–CH_4 selectivity of 598 in ambient conditions. In addition to this, this framework also shows good storage capacity of small chain hydrocarbons such as propane, ethane, ethylene and acetylene over methane at ambient condition and hence, high IAST selectivity. Density functional theory (DFT) calculations also infer the imide and aromatic π electrons of the framework to be preferable interaction sites for CO_2 and hydrocarbons. The most significant outcome of this work is the real separation of gas mixtures demonstrated by breakthrough column experiments (Figure 12.12g–i).

12.6.1.2 Stepwise Sorption in a Post-synthetically Modified MOF

Wang *et al.* have demonstrated a unique way to make a MOF breathe *via* postsynthetic modification at the organic backbone.[82] DMOF-1-NH_2 is an amino derivative of DMOF-1, a 3D porous MOF constructed from Zn(II)-based paddle wheel SBUs, 1,4-benzenedicarboxylate, and pillaring 1,4-diazabicyclo[2.2.2]octane (DABCO) ligands (Figure 12.13a). DMOF-1-AM3 is a MOF decorated with flexible amide chains in the nanochannel which is obtained from DMOF-1-NH_2 *via* postsynthetic modification at the –NH_2 group (Figure 12.13a). Interestingly, N_2 sorption isotherm measurements at 77 K demonstrate a two-step profile with the first step starting at very low P, P/P_0 ~0.015 and then at high pressure there is a change in the structure resulting in a second step uptake (Figure 12.13b). Similar behaviour was observed with Ar isotherm measured at 87 K and a more pronounced effect was seen in CO_2 isotherm at 196 K (Figure 12.13c). It has been proposed steric, weak chain–chain interactions, and/or chain flexibility may contribute to the observed breathing behaviour that helps to switch structure between two different forms.

12.6.1.3 Stepwise Sorption via Breathing Mechanism

MIL-53, a well-known MOF shows a large guest-dependent "breathing" effect.[83,84] Structure determination using X-ray powder diffraction data revealed a fully reversible transition between the hydrated form (MIL-53lt) and anhydrous solid (MIL-53ht) and this is accompanied by a high breathing effect (more than 5 Å), the pores being clipped in the presence of water molecules (MIL-53lt) and reopened when the channels are empty (MIL-53ht). Moreover, in MIL-53lt, water is easily exchanged by DMF, showing the great affinity of DMF for this type of structure. The resulting product has cell parameters intermediate between MIL-53as (as-synthesized MIL-53 with neutral terepthalic acid as the guest) and MIL-53lt, proving the adaptability of the breathing network to the shape of the chemical stimulus. This breathing is the result of strong guest–skeleton interaction and this has been verified by *in situ* solid state ^{13}C, ^{1}H and ^{27}Al NMR experiments of the Al-homologue. During rehydration water molecules show three types of hydrogen bonds; one corresponds to guest–guest interaction of the water molecules along the axis of the tunnels and the other two show host–guest interactions. Structural change in flexible MOF of Fe-MIL-88b was also investigated recently by an *in situ* XRD study.[85] Surface-grown, oriented crystal films and in bulk form, and its structural changes upon adsorption and desorption of water were recorded. Due to the orientation of the grown crystals on the gold substrates, structural changes in [001] direction could be observed. For the randomly oriented bulk crystals the structural changes in all crystallographic directions were observable and the changes of the lattice constants *a* and *c* and the cell volume could be determined quantitatively by indexing the complete diffraction patterns. The MOF crystals show two

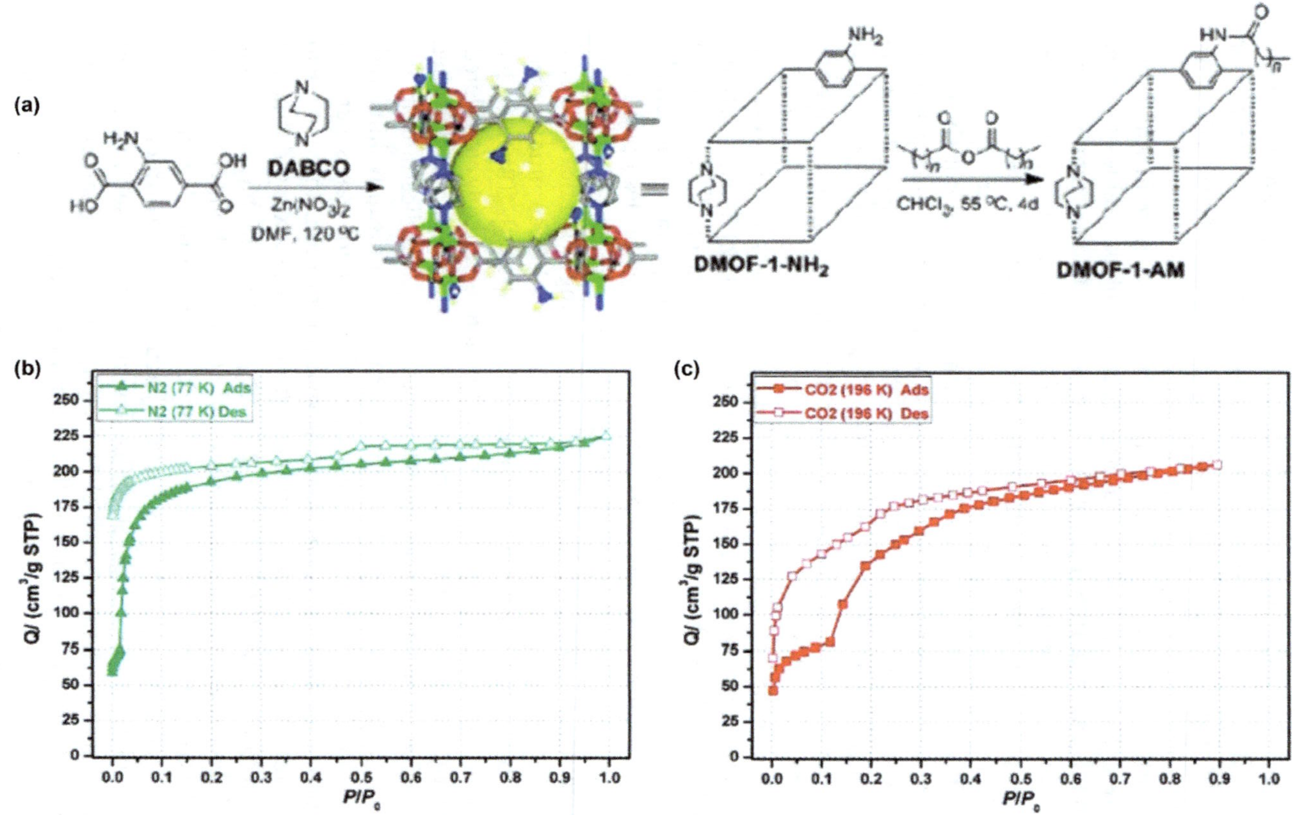

Figure 12.13 (a) Postsynthetic modification of DMOF-1-NH$_2$ with linear alkyl anhydrides; (b) N_2 sorption isotherm at 77 K; and (c) CO_2 sorption isotherm at 196 K for DMOF-1-AM3.
(Adapted with permission from Z. Wang and S. M. Cohen, Modulating Metal-Organic Frameworks to Breath: A Postsynthetic Covalent Modification Approach, *J. Am. Chem. Soc.*, 2009, **131**, 16675. Copyright 2009 American Chemical Society.)

distinct steps of structural changes during water sorption. During sorption of water the lattice parameter *c* is slightly decreased, whereas the lattice parameter *a* shows a prominent two-step increase resulting in an increase of the cell volume by about 40%. The conventional volumetric water sorption measurement also shows two distinct steps in the uptake of water vapour that can be correlated with the structural changes observed in X-ray diffraction.

MIL-53 (Al) is a very interesting example of soft porous solid which has two different and distinct phases, a large pore (lp) phase stable at higher temperatures and a narrow pore phase (np) stable at lower temperatures (Figure 12.14a–e).[86] The sorption isotherms of gases such as CO_2, CH_4, Xe at high temperatures display reversible sorption–desorption process owing to the sorption in lp phase (Figure 12.15a). The CO_2 isotherm at 343 K, CH_4 isotherms between 250–298 K and Xe isotherms at 323 K do not show any step (Figure 12.15a). However, the isotherms measured at low temperatures clearly show np to lp transitions in all cases. It has been proposed that the np to lp transition is very general in these cases and regardless of guest molecules (Figure 12.15b). It is expected that the affinity of np phase towards guest molecule would be higher than the lp phase.

In another work *in situ* synchrotron-based XRD analysis was utilized to demonstrate structural changes in $\{[Zn_3(AmTAZ)_3S](NO_3)\cdot(H_2O)\}_n$ and $\{[Zn_7(AmTAZ)_8(CO_3)_2(OH)_2]\cdot 2(EtOH)\}_n$ (Figure 12.16a,b).[87] No nitrogen uptake was observed for $\{[Zn_3(AmTAZ)_3S](NO_3)\cdot(H_2O)\}_n$, but nanoporosity was confirmed through *in situ* X-ray diffraction studies of desorption and resorption of the guest water molecules. The cubic symmetry of the system was maintained throughout these processes revealing a subtle, concerted "breathing" of the host framework. The inclusion of three unique types of anionic bridging ligands in $\{[Zn_7(AmTAZ)_8(CO_3)_2(OH)_2]\cdot 2(EtOH)\}_n$ has generated an intricate 3D network structure that defines a highly convoluted, and primarily 1D, pore system. Nitrogen sorption studies of the desorbed host material, $\{[Zn_7(AmTAZ)_8(CO_3)_2(OH)_2]\cdot 2(EtOH)\}_n$, indicate that the passage of nitrogen through the pore system is significantly hindered. This suggests that this material may be well suited for gas separation applications, especially as related to hydrogen purification. In contrast to $\{[Zn_3(AmTAZ)_3S](NO_3)\cdot(H_2O)\}_n$, *in situ* synchrotron-based X-ray diffraction studies of $\{[Zn_7(AmTAZ)_8(CO_3)_2(OH)_2]\cdot 2(EtOH)\}_n$ revealed highly anisotropic structural flexing during the desorption and sorption of both ethanol and water molecules, due in part to the increased degrees of freedom imparted by its complex, lower symmetry structure (Figure 12.16c,d). Interestingly, the uptake of the different guest molecules proceeds *via* subtly different structural flexing mechanisms, which leads to a larger overall lattice volume increase for the sorption of the smaller water molecules.

$[Co(BDP)]_n$ (BDP^{2-} = 1,4-benzenedipyrozolate) is an interesting material that exhibits remarkable multistep N_2 "breathing" at 77 K in the low pressure region of P/P_0 <0.05 (Figure 12.17a,b).[88] The 3D pseudotetragonal

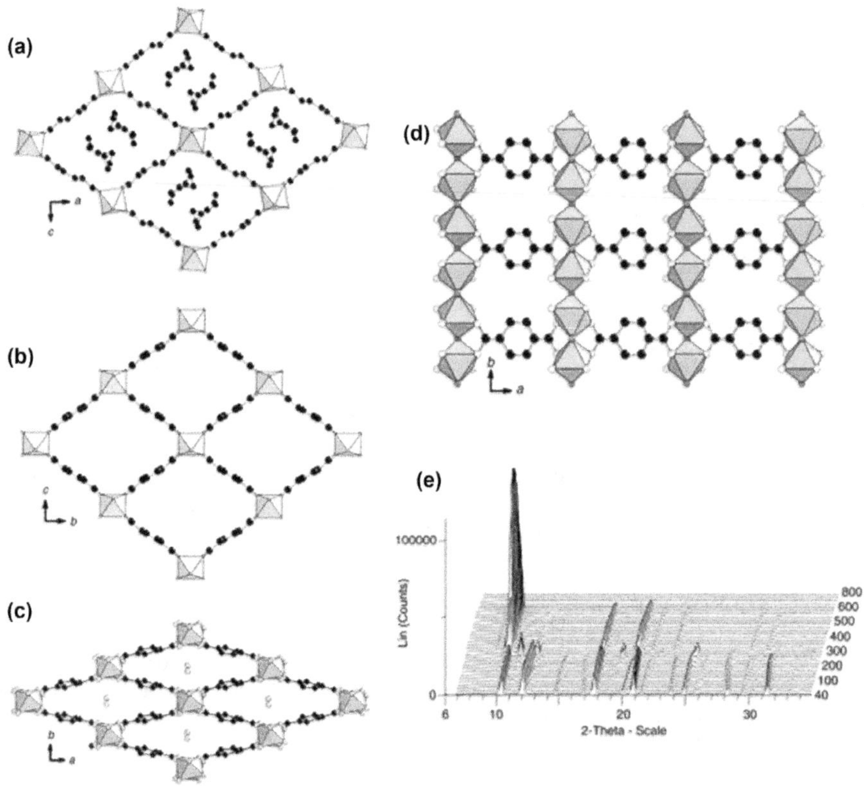

Figure 12.14 (a) Views of the 3D structure of MIL-53 (Al) showing the channel system $Al(OH)[O_2C\text{–}C_6H_4\text{–}CO_2]\cdot[HO_2C\text{–}C_6H_4\text{–}CO_2H]_{0.70}$ or MIL-53*as* (Al), in which the channels are occupied by free disordered 1,4-benzenedicarboxylic acid molecules; (b) calcined form, MIL-53ht (Al) or $Al(OH)[O_2C\text{–}C_6H_4\text{–}CO_2]$ with empty channels; (c) room temperature form, MIL-53lt (Al) or $Al(OH)[O_2C\text{–}C_6H_4\text{–}CO_2]\cdot H_2O$, in which a water molecule is located at the centre of the channels. Grey octahedra: $AlO_4(OH)_2$; black circles: carbon; grey circles: oxygen; (d) View of the infinite chains of corner-sharing octahedral $AlO_4(OH)_2$ units connected through the 1,4-benzenedicarboxylate ligands, grey circles indicate the hydroxyl groups bridging the aluminium atoms; (e) X-ray thermodiffractogram of MIL-53*as*(Al) in air (40–800 °C). (Adapted with permission from T. Loiseau, C. Serre, C. Huguenard, G. Fink, F. Taulelle, M. Henry, T. Bataille and G. Férey, A Rationale for the Large Breathing of the Porous Aluminum Terephthalate (MIL-53) Upon Hydration, *Chem. – Eur. J*, 2004, **10**, 1373. Copyright 2004 WILEY-VCH Verlag GmbH & Co. KGaA, Weinheim.)

structure of the as-synthesized solid, $\{[Co(BDP)]\cdot 2DEF\cdot H_2O\}_n$, can be described as a 4^4 net composed of chains of tetrahedrally coordinated Co^{II} centres along [001] linked in the two orthogonal directions by bridging BDP^{2-} ligands (Figure 12.17a). The framework shows an N_2 adsorption

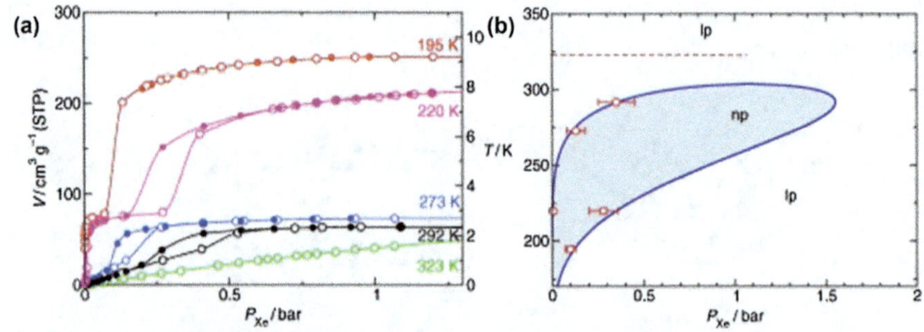

Figure 12.15 (a) Experimental xenon adsorption isotherms in MIL-53(Al). Open symbols: adsorption branch; filled symbols: desorption branch. Left scale: measured adsorbed volumes; right scale: corresponding number of adsorbed molecules per unit cell. The low-pressure desorption branch at 195 K (hysteresis loop) was not recorded for technical reasons; (b) Temperature–xenon pressure phase diagram of MIL-53(Al). Solid line: osmotic thermodynamic model; open symbols: experimental transition steps in the isotherms. The dashed line corresponds to the experimental temperature (323 K) for which no transition took place.
(Adapted with permission from A. Boutin, M.-A. Springuel-Huet, A. Nossov, A. Gédéon, T. Loiseau, C. Volkringer, G. Férey, F.-X. Coudert and A. H. Fuchs, Breathing Transitions in MIL-53(Al) Metal–Organic Framework Upon Xenon Adsorption, *Angew. Chem., Int. Ed.*, 2009, **48**, 8314. Copyright 2009 WILEY-VCH Verlag GmbH & Co. KGaA, Weinheim.)

isotherm with five steps and probably such five-step isotherm for a non-polar gas N_2 is the only isotherm reported in MOFs (Figure 12.17b). A similar behaviour like N_2 adsorption was observed in the sorption measured at 87 K but the steps appear at higher pressures and are less distinct. Interestingly, in both the cases desorption isotherms show hysteresis, but reversible at low pressures suggesting a pore-opening and pore-closing process. Such step-wise processes have been attributed to multilayer sorption on energetically homogeneous surfaces, multiple adsorbent–adsorbate interactions that substantially differ in energy, or various structural phase transitions occurring under different adsorbate pressures (Figure 12.17c). For the material under consideration, it is possibly that the last of these mechanisms is operational, likely in conjunction with either or both of the other effects. On the contrary, with H_2 as an adsorbate no such step was observed at 77 K and low pressures, suggesting the guest-dependent behaviour of $[Co(BDP)]_n$. Interestingly, as the pressure is increased to *ca.* 20 bar, a sudden uptake of H_2 was observed which is reminiscent of gate-opening behaviour. The computationally assisted structure determinations suggest structural changes involving the orientation of the benzene rings relative to the pyrazolate rings, the dihedral angles for the pyrazolate rings bound at the metal centres, and a change in the metal coordination geometry from square planar to tetrahedral.

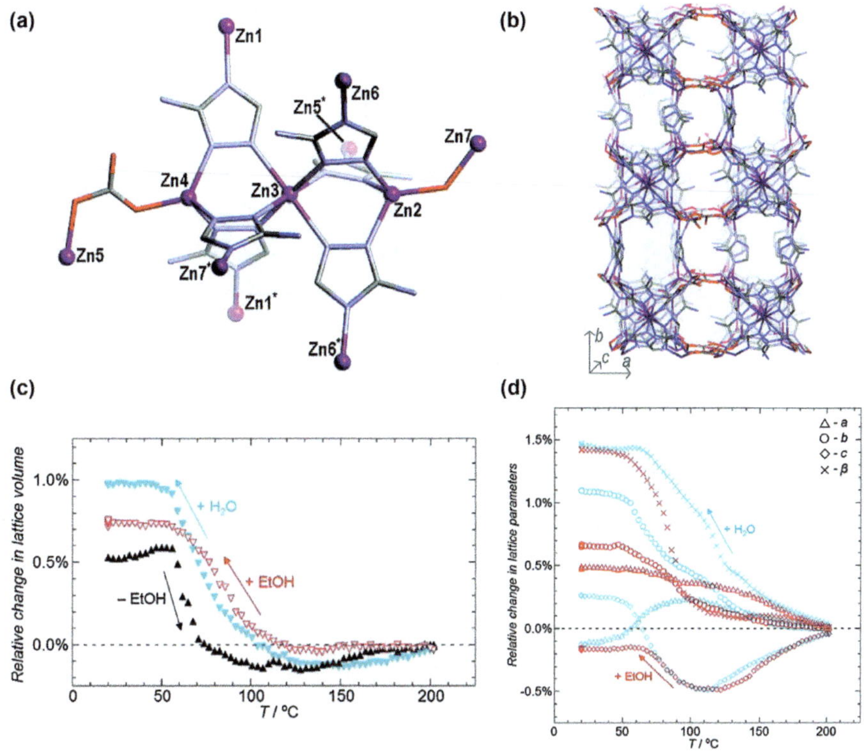

Figure 12.16 (a) The side view of the eight connecting $Zn_3(AmTAZ)_6$ clusters of $\{[Zn_7(AmTAZ)_8(CO_3)_2(OH)_2]\cdot 2(EtOH)\}_n$; (b) Representation of the highly-connected 3D network structure of $\{[Zn_7(AmTAZ)_8(CO_3)_2(OH)_2]\cdot 2(EtOH)\}_n$; (c) Relative change of lattice volume for $\{[Zn_7(AmTAZ)_8(CO_3)_2(OH)_2]\cdot 2(EtOH)\}_n$ from Le Bail analysis of S-PXRD data collected during heating from 20 to 202 °C under a flow of helium gas (black), and subsequently with cooling to 20 °C under a flow of either ethanol- (red) or water-saturated (blue) helium gas; (d) Relative changes in the monoclinic lattice parameters for $\{[Zn_7(AmTAZ)_8(CO_3)_2(OH)_2]\cdot 2(EtOH)\}_n$ upon sorption of ethanol (red) and water (blue). Lattice parameters were determined by Le Bail analysis of S-PXRD data collected during cooling from 202 to 20 °C under a flow of either ethanol- or water-saturated helium gas.
(Adapted with permission from G. J. Halder, H. Park, R. J. Funk, K. W. Chapman, L. K. Engerer, U. Geiser and J. A. Schlueter, In-Situ X-ray Diffraction Studies of Host-Guest Properties in Nanoporous Zinc-Triazolate-Based Framework Materials, *Cryst. Growth Des.*, 2009, **9**, 3609. Copyright 2009 American Chemical Society.)

12.6.2 Gated Sorption in 3D MOFs

Kitaura *et al.* for the first time observed gated and hysteretic sorption in a 3D pillared-layer MOF of Cu(II), $\{[Cu_2(pzdc)_2(dpyg)]\cdot 8H_2O\}_n$ with MeOH and H_2O vapour (Figure 12.18a–c).[89] MeOH sorption profile shows a sudden

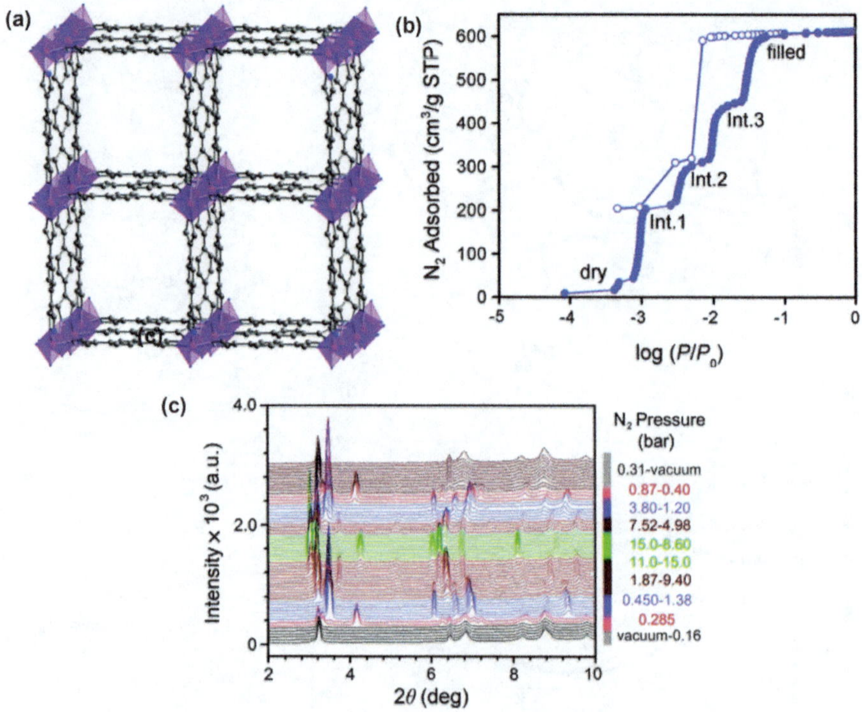

Figure 12.17 (a) Pseudotetragonal unit cell of $[Co(BDP)]_n$, with purple, grey, and blue spheres representing Co, C, and N atoms, respectively; (b) N_2 adsorption isotherm measured at 77 K, indicating the five pressure-dependent phases. Filled and empty symbols represent adsorption and desorption, respectively; (c) Layered plots of X-ray powder diffraction data measured at 100 K, where each pattern was measured under a constant N_2 pressure indicated as ranges.
(Adapted with permission from F. Salles, G. Maurin, C. Serre, P. L. Llewellyn, C. Knöfel, H. J. Choi, Y. Filinchuk, L. Oliviero, A. Vimont, J. R. Long and G. Férey, Multistep N2 Breathing in the Metal-Organic Framework Co(1,4-benzenedipyrazolate), *J. Am. Chem. Soc.*, 2010, **132**, 13782. Copyright 2010 American Chemical Society.)

uptake at 298 K and P/P_0 ~0.23 and then reaches saturation at higher pressures (Figure 12.18d). A similar sorption isotherm was recorded in the case of H_2O vapour, however no uptake of CH_4 gas observed for the same compound. This observation was chiefly associated with the hydrogen-bonding interaction between the H_2O and MeOH molecules and the –OH groups of the dpyg ligands; the attractive force should be strong enough to transform the channel structure to allow the diffusion of the guest molecules.

Among the various flexible MOFs, the zeolitic imidazolate framework (ZIF) series are most commonly known for their gated sorption behaviour.[90–92] A prototypical structure with a sodalite (SOD) topology, ZIF-8 have relatively small window dimensions (~3.40 Å) with large cavity size (~11.6 Å).[93–96]

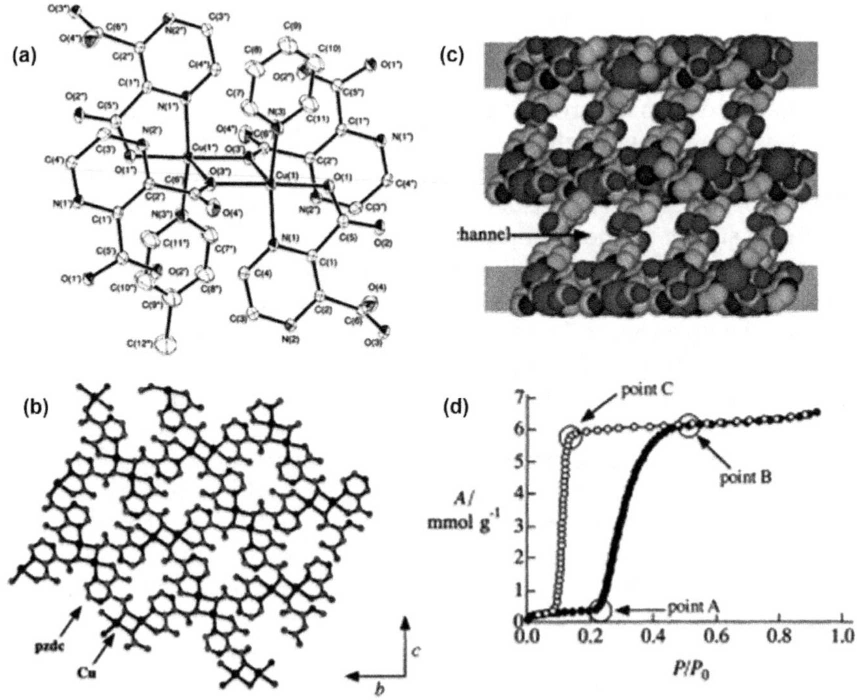

Figure 12.18 (a) Molecular structure of the dicopper sites of $\{[Cu_2(pzdc)_2(dpyg)]\cdot 8H_2O\}_n$. View of the two-dimensional sheet (b) and channel structure (c); (d) Methanol adsorption (filled circles) and desorption (open circles) isotherms at 298 K. A = absolute absorption of $\{[Cu_2(pzdc)_2(dpyg)]\cdot 8H_2O\}_n$. Solvent water and hydrogen atoms are omitted for clarity.
(Adapted with permission from R. Kitaura, K. Fujimoto, S.-I. Noro, M. Kondo and S. Kitagawa, A Pillared-Layer Coordination Polymer Network Displaying Hysteretic Sorption: $[Cu_2(pzdc)_2(dpyg)]_n$ (pzdc = Pyrazine-2,3-dicarboxylate; dpyg = 1,2-Di(4-pyridyl)glycol), *Angew. Chem., Int. Ed.*, 2002, **114**, 141. Copyright 2002 WILEY-VCH Verlag GmbH, Weinheim, Fed. Rep. of Germany.)

Cheetham *et al.* investigated first pressure-induced structural flexibility in ZIF-8 using diamond anvil cell experiments.[97] It is the first indication about the nature of the framework flexibility induced by sorption at much lower pressures. At 1.47 GPa, a new phase was observed which was referred to as ZIF-8HP. This high-pressure phase maintained the $I43m$ space-group symmetry of ZIF-8 but showed a reorientation of the imidazolate linkers, increasing the accessible pore volume and the size of the 6-ring windows. The structural change was found to be reversible after decreasing the pressure. The experimental sorption isotherm of N_2 at 77 K shows two steps occurring at 2×10^{-3} and 2×10^{-2} P/P_0 or 0.002 and 0.02 bar; the desorption process was completely reversible, and no hysteresis was found. The fitting of the Brunauer–Emmett–Teller (BET) equation to both steps revealed the uptake of N_2 for the first and second step, corresponding to 36.9 and 49.4 molecules per unit cell, respectively. This corresponds to an increase of 26% in uptake going from the first to the second step. This two-step isotherm, which has also been observed in two other ZIFs with a SOD topology (ZIF-90, ZIF-91), was previously interpreted as a result of a reorganization of the adsorbed nitrogen molecules during the sorption process, rejecting the existence of structural changes. A GCMS simulation study was carried out to investigate such a phenomenon and this study indicates a subtle change in structure and also suggests the important role of the linear shape of the N_2 molecule. *In situ* PXRD measurements clearly indicated the existence of structural change caused by the swing effect of imidazolate linker.

A 3D MOF, $[Zn(pydc)(dma)]_n$, based on a non-interpenetrating (10,3)-a topology was found to exhibit unusual gate behaviour with N_2 and Ar at 77 and 87 K, respectively (Figure 12.19a–d).[98] As the temperature increases, the breakthrough or gate pressure was found to increase (Figure 12.19e). Surprisingly, no H_2 uptake was recorded at 77 K and was reasoned on the basis of high kinetic energy of small hydrogen molecules (Figure 12.19e). The guest-specific sorption behaviour was not size-selective, and it was concluded that the interactions between gas molecules and the framework play more important role in deciding the gate pressure.

Although there are several reports on gated sorption in 2D coordination polymers, the gate effect in 3D systems at very low pressures is rare. Mulfort

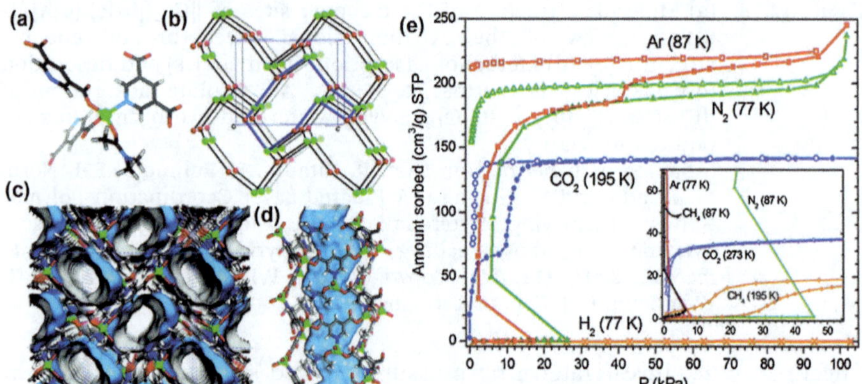

Figure 12.19 (a) Coordination environment of the Zn^{2+} ion; (b) Simplified net of $[Zn(pydc)(dma)]_n$. The green and red spheres represent the metal and ligand centres, respectively. One of the 10-gon circuits is emphasized by thick blue lines; (c) Perspective view along the *a* axis and the Connolly surface (1.4 Å probe); (d) Side view of the 1D channel; (e) Gas sorption isotherms at various temperatures. Closed and open symbols denote adsorption and desorption, respectively. Ar and N_2 are not adsorbed at all at 195 K.
(Adapted with permission from H. Chun and J. Seo, Discrimination of Small Gas Molecules through Adsorption: Reverse Selectivity for Hydrogen in a Flexible Metal-Organic Framework, *Inorg. Chem.*, 2009, **48**, 9980. Copyright 2009 American Chemical Society.)

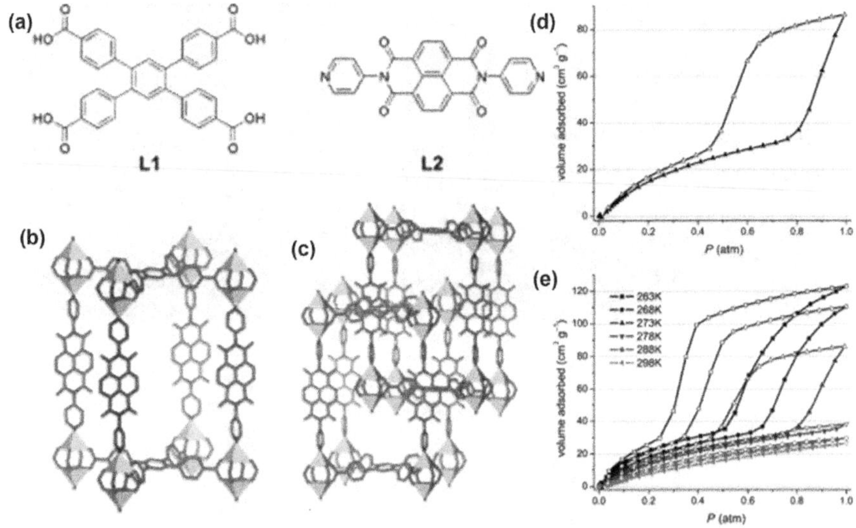

Figure 12.20 (a) Chemical structure of L1 and L2; (b) Crystal structure of $[Zn_2(L1)(L2)]_n$, one level of catenation is omitted to illustrate connectivity. Polyhedra represent zinc ions. Hydrogens omitted for clarity; (c) Catentation of $[Zn_2(L1)(L2)]_n$. Low-pressure CO_2 isotherms of $[Zn_2(L1)(L2)]_n$. (d) 273 K, and (e) temperatures from 263 K to 298 K. Closed symbols, adsorption; open symbols, desorption.
(Adapted with permission from K. L. Mulfort, O. K. Farha, C. D. Malliakas, M. G. Kanatzidis and J. T. Hupp, An Interpenetrate Framework Materials with Hysteretic CO2 Uptake, *Chem. – Eur. J*, 2010, **16**, 276. Copyright 2010 WILEY-VCH Verlag GmbH & Co. KGaA, Weinheim.)

et al. observed gate effect in N_2 and Ar sorption isotherms while a stepwise profile was recorded with CO_2 gas in a twofold interpenetrated 3D compound $[Zn_2(L1)(L2)]_n$ (Figure 12.20).[99] Interestingly, the sorption results, supported by pair distribution function (PDF) data, qualitatively point to dynamic framework behaviour upon both solvent removal (activation) and introduction of adsorbate molecules. The dynamic behaviour appeared to be dominated by inter-conversion between interpenetrated and interwoven geometries, with accompanying sizable changes in guest-accessible surface area.

Ghosh *et al.* reported two isomorphous compounds, $\{[Ce(tci)(H_2O)_2]\cdot 2H_2O\}_n$ and $\{[Pr(tci)(H_2O)_2]\cdot 2H_2O\}_n$.[100] In the crystal structure of $\{[Ce(tci)(H_2O)_2]\cdot 2H_2O\}_n$, Ce^{3+} and the carboxylate groups of tci formed 1D metal–carboxylate chains which were connected to the other metal–carboxylate chains through the flexible arms of tci linker and thereby formed 2D sheet (Figure 12.21a). Upon dehydration, $\{[Ce(tci)(H_2O)_2]\cdot 2H_2O\}_n$ converted to a non-porous 3D framework (Figure 12.21b). Crystal structure of this desolvated framework indicated that movement of flexible alkyl chains of tci linker was responsible for 2D to 3D structural transformation. However, upon rehydration, the 3D framework was transformed back to the original 2D structure. The water

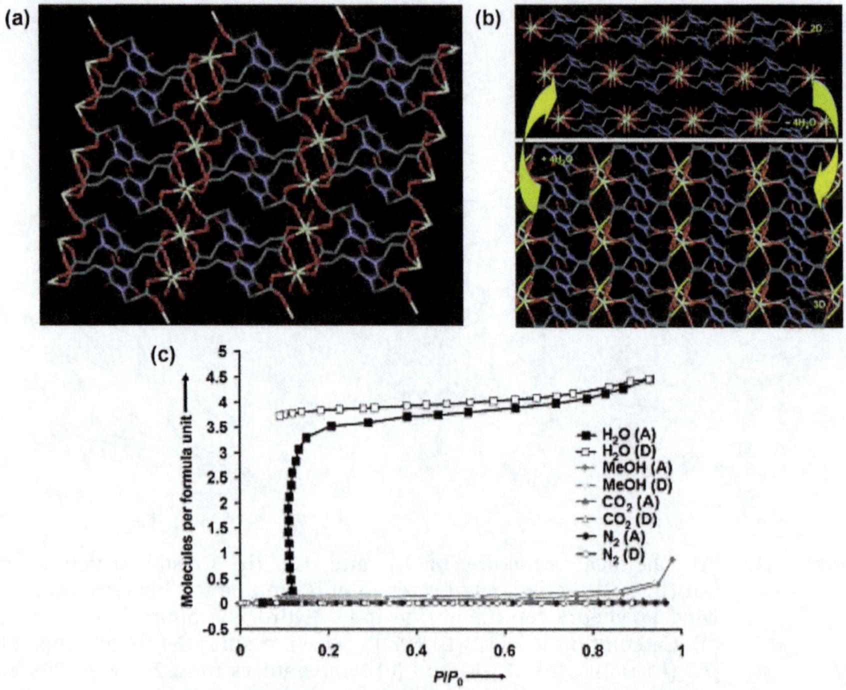

Figure 12.21 (a) 2D sheet structure of $\{[Ce(tci)(H_2O)_2]\cdot 2H_2O\}_n$; (b) Perspective view of the reversible transformation between the 2D layer structure of the as-synthesized framework and the 3D structure of dehydrated framework. New bonds between the sheets are shown in yellow; (c) Adsorption (A) and desorption (D) isotherms of H_2O (RT), MeOH (RT), CO_2 (195 K), and N_2 (77 K).
(Adapted with permission from S. K. Ghosh, J.-P. Zhang and S. Kitagawa, Reversible Topochemical Transformation of a Soft Crystal of a Coordination Polymer, *Angew. Chem., Int. Ed.*, 2007, **46**, 7965. Copyright 2007 WILEY-VCH Verlag GmbH & Co. KGaA, Weinheim.)

adsorption isotherm of the 3D framework exhibited gate-opening behaviour which indicated guest responsive structural transformation from one phase to another (Figure 12.21c).

12.6.2.1 Gate-opening Sorption for Separation of Small Molecules

Separation of small molecules such as hydrocarbons is an energy consuming process and hence, development of suitable materials for efficient separation, under suitable conditions is therefore has paramount importance.[101–104] There has been an ample amount of work done where gate-opening sorption plays an important role. Among this, separation of acetylene and ethylene *via* gate-opening selective sorption of acetylene by MOF, ethylene and ethane separation *via* gate-opening sorption of ethane by ZIF-7 and hydrocarbon separation by a flexible framework RPM3-Zn MOF are of special interest.[44,105]

ZIF-7 is the first example of microporous material that showed selective sorption of paraffins over olefins.[44] The interaction between the benzene ring and the hydrocarbons in the narrow window of ZIF-7 dominated the selectivity process by inducing a gate-opening type sorption that discriminate molecules of very similar size and shape.

$[Zn_2(bpdc)_2(bpee)]_n$, also known as RPM3-Zn, is capable of separating C1–C4 paraffins and two pairs of C2 isomers, C_2H_2–C_2H_4 and C_2H_4–C_2H_6 because of the differences in the gate-opening pressure of the gas molecules (Figure 12.22).[106] The overall 3D framework can be viewed as bpee ligands

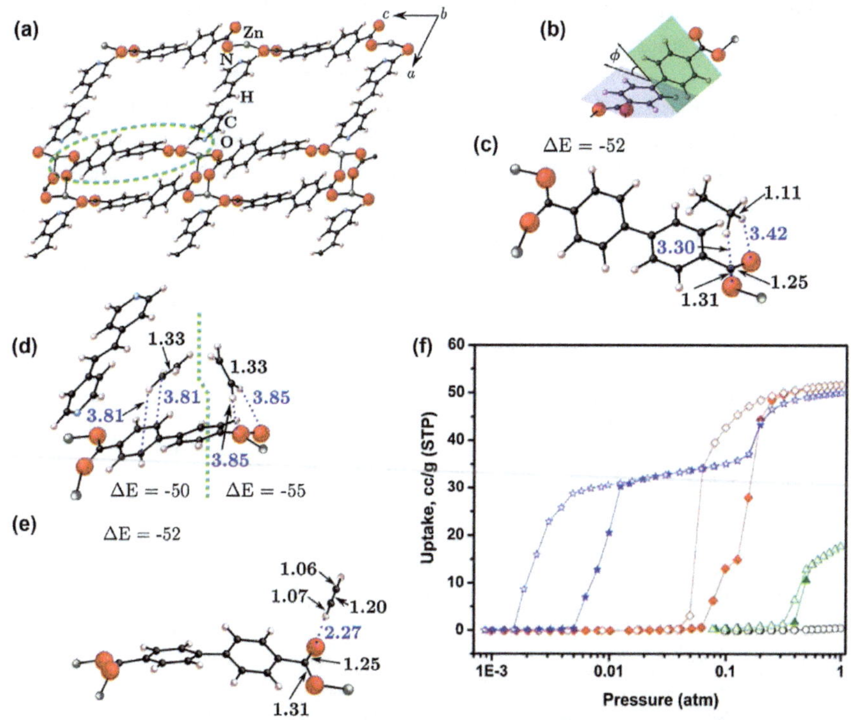

Figure 12.22 (a) Side view of the RPM3-Zn structure; (b) dihedral angle in the bpdc ligand; (c), (d), (e) local fragments of ethane, ethylene showing the interaction with the inter-ring C–C (left) and with the C–O (right), and acetylene respectively adsorbed in the RPM3-Zn. Bond lengths in blue are in Å, and ΔE in kJ mol^{-1}. For ethylene two possible adsorption sites are presented; (f) Adsorption–desorption isotherms of short alkanes in RPM3-Zn at rt (298 K) plotted as a function of relative pressure. Filled and open symbols represent adsorption and desorption branch, respectively. Colour schemes: black circles, CH_4; green triangles, C_2H_6; red diamonds, C_3H_8; blue asterisks, C_4H_{10}.
(Adapted with permission from N. Nijem, H. Wu, P. Canepa, A. Marti, K. J. Balkus, T. Thonhauser, J. Li and Y. J. Chabal, Tuning the Gate Opening Pressure of Metal-Organic Frameworks (MOFs) for the Selective Separation of Hydrocarbons, *J. Am. Chem. Soc.*, 2012, **134**, 15201. Copyright 2012 American Chemical Society).

pillaring the undulating charge-neutral $[Zn_2(bpdc)_2]_n$ layers (Figure 12.22a). Raman spectroscopy and *ab initio* DFT calculations suggest H-bonding interaction between the hydrcarbons (*via* terminal groups) and framework (*via* non-coordinating carboxylate group) play key roles for the differences in gate-opening pressure (Figure 12.22b–e). It has been concluded that stronger H-bonding interaction leads to a lower gate-opening pressure and suggests a pressure swing sorption type separation based on H-bonding (Figure 12.22f). Selective sorption of benzene over cyclohexane is very important from an application point of view and few MOFs were reported where such experiments were performed.

The flexible MOF $\{[Zn(\mu_4\text{-TCNQ-TCNQ})bpy]\cdot 1.5benzene\}_n$ was found to exhibit highly selective and gated sorption of benzene over cyclohexane.[106] In this MOF, the Zn ions are octahedrally coordinated to the four cyanide nitrogen atoms of TCNQ in the equatorial plane and the two nitrogen atoms of the bpy ligand at the axial sites (Figure 12.23a). The Zn ions are connected by bpy to give a 1D chain along the *a* axis and $[TCNQ\text{-}TCNQ]^{2-}$ acts as cross-linker connecting the four 1D chains of Zn ions and bpy to form a 3D open framework (Figure 12.23b). The guest benzene molecule is accommodated strongly in the cavity and the H–π interaction with the host framework, which facilitates strong interaction between benzene and TCNQ ligand

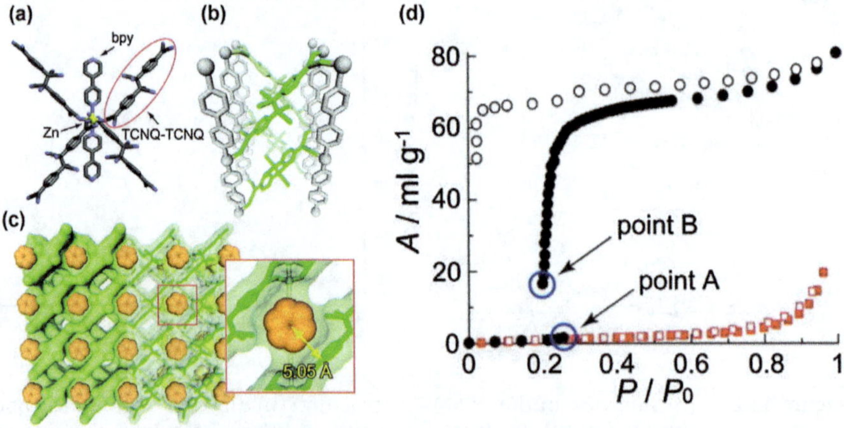

Figure 12.23 (a) Coordination environment of Zn(II) ion of $\{[Zn(\mu_4\text{-TCNQ-TCNQ})bpy]\cdot 1.5benzene\}_n$; (b) TCNQ dimer (green) connected to four 1D chains of Zn and bpy (grey); (c) Benzene arranged in the cage of the undulating channel of $\{[Zn(\mu_4\text{-TCNQ-TCNQ})bpy]\cdot 1.5benzene\}_n$; (d) Sorption isotherm for each adsorbate in guest-free crystal at 298 K – Benzene adsorption (filled circles) and desorption (open circles); Cyclohexane adsorption (filled squares) and desorption (open squares).
(Adapted with permission from S. Shimomura, S. Horike, R. Matsuda and S. Kitagawa, Guest-Specific Function of a Flexible Undulating Channel in a 7,7,8,8-Tetracyano-p-quinodimethane Dimer-Based Porous Coordination Polymer, *J. Am. Chem. Soc.*, 2007, **129**, 10990.

resulting in selective sorption of benzene over cyclohexane (Figure 12.23c–d). Synergy of the undulating channel formed gives the MOF structural flexibility to adapt to the target molecular shape and, with the interaction site located in the right position of the pore surface, makes MOFs an effective separation system.

12.6.2.2 Methane Storage in Flexible MOFs

Long *et al.* reported two flexible MOFs Fe(bdp) and Co(bdp) which undergo a structural phase transition in response to specific CH_4 pressures, resulting in adsorption and desorption isotherms that feature a sharp 'step' (Figure 12.24).[88] Such behaviour enabled greater storage capacities than have been achieved for classical adsorbents, while also reducing the amount of heat released during adsorption and the impact of cooling during desorption. Co(bdp) was selected as a potential responsive adsorbent for methane storage due to its large internal surface area and high degree of flexibility. In its solvated form, Co(bdp) exhibited one-dimensional chains of tetrahedral Co^{2+} cations bridged by μ_2-pyrazolates to form a structure with square channels. As mentioned earlier, evacuated Co(bdp) showed a five-stepped N_2 adsorption isotherm at 77 K which has been attributed to four structural transitions. High-pressure CH_4 adsorption isotherm, measured at 25 °C showed a minimal CH_4 uptake at low pressures and a sharp step in the adsorption isotherm at 16 bar. The desorption isotherm showed hysteresis which was closed by 7 bar. The step in CH_4 isotherm was attributed to reversible structural phase transition between a collapsed, non-porous phase and an expanded, porous phase at transition pressures (Figure 12.24d). To determine the specific structural changes responsible for the stepped CH_4 adsorption isotherm of Co(bdp), *in situ* powder X-ray diffraction experiments were performed under various pressures of CH_4 at 25 °C (Figure 12.24a,b). At low pressure, the central benzene ring of the bdp^{2-} ligand twisted out of the plane of the two pyrazolates by 25° resulting in edge-to-face π–π interactions with four neighbouring benzene rings which provided most of the thermodynamic driving force for the collapse of Co(bdp) (Figure 12.24e). The close contacts between neighbouring bdp^{2-} ligands lead to no accessible porosity and therefore no CH_4 adsorption was observed in the collapsed phase. Moreover, Co(bdp) showed intrinsic heat management within the material by using the enthalpy change of phase transition (Figure 12.25c,d). The expansion of Co(bdp) during adsorption is endothermic, because energy is needed to overcome the greater thermodynamic stability of the collapsed phase. On the other hand, during the desorption process the transition to the collapsed phase is exothermic. They also prepared an isostructural iron equivalent, Fe(bdp), which showed a stepped high-pressure CH_4 isotherm at 25 °C, suggesting that Fe(bdp) also undergoes a reversible phase transition between a collapsed and expanded framework. Although the total CH_4 uptake is comparable to that of Co(bdp), the adsorption and desorption steps occurred at the considerably higher pressures of 24 bar and 10 bar,

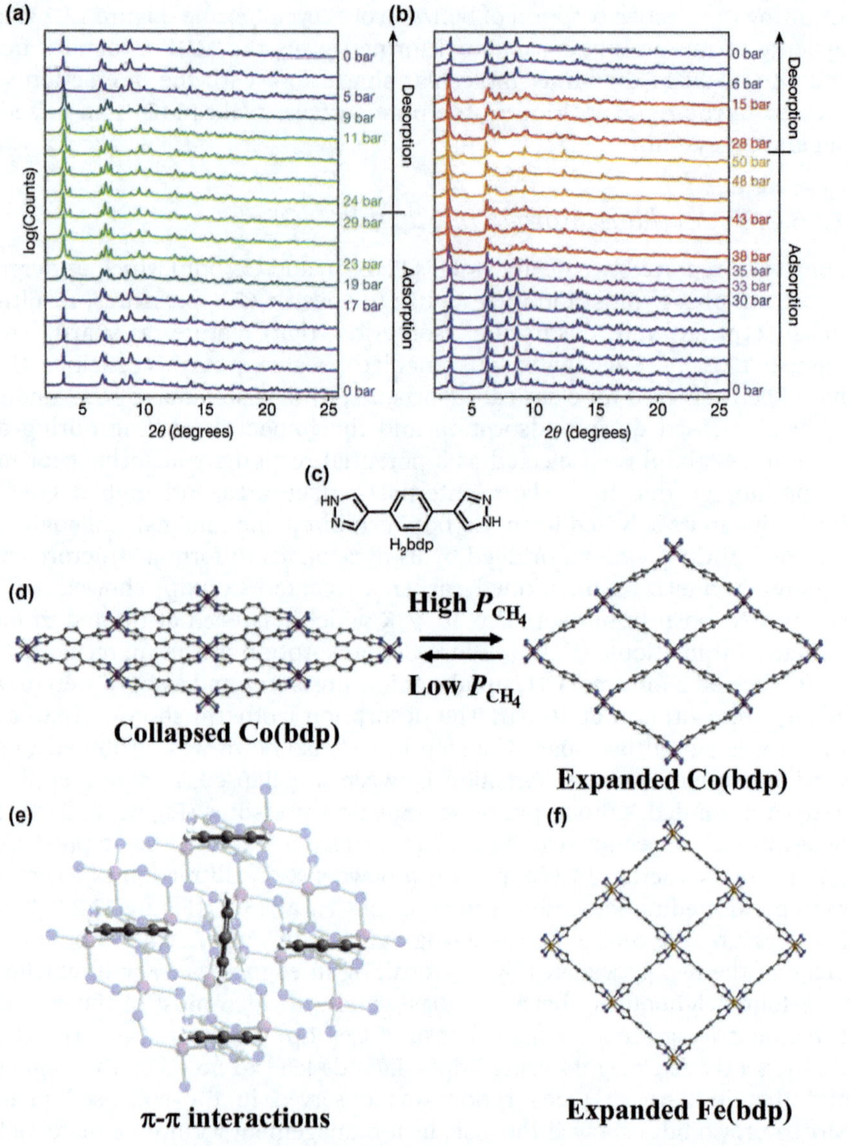

Figure 12.24 (a), (b) powder X-ray diffraction patterns of Co(bdp) and Fe(bdp) respectively at 25 °C and variable CH_4 pressures; (c) The bridging ligand precursor H_2bdp; (d) Crystal structure of the collapsed and CH_4 expanded phase of Co(bdp); (e) Each benzene ring in the collapsed phase of Co(bdp) has four edge-to-face π–π interactions with neighbouring benzene rings; (f) Crystal structure of the CH_4-expanded (40 bar) phase of Fe(bdp).
(Adapted by permission from Macmillan Publishers Ltd: Nature (J. A. Mason, J. Oktawiec, M. K. Taylor, M. R. Hudson, J. Rodriguez, J. E. Bachman, M. I. Gonzalez, A. Cervellino, A. Guagliardi, C. M. Brown, P. L. Llewellyn, N. Masciocchi and J. R. Long, *Nature*, 2015, **527**, 357), copyright (2015).)

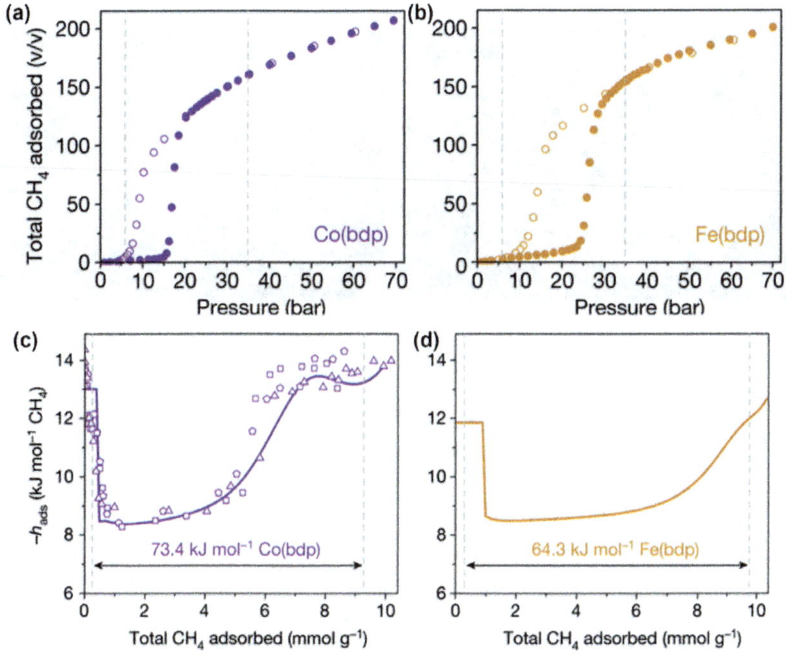

Figure 12.25 (a), (b) Total CH_4 adsorption isotherms for Co(bdp) and Fe(bdp) respectively at 25 °C; (c), (d) Differential enthalpies of CH_4 adsorption (hads) for Co(bdp) and Fe(bdp) respectively as determined from variable-temperature adsorption isotherms.
(Adapted by permission from Macmillan Publishers Ltd: Nature (J. A. Mason, J. Oktawiec, M. K. Taylor, M. R. Hudson, J. Rodriguez, J. E. Bachman, M. I. Gonzalez, A. Cervellino, A. Guagliardi, C. M. Brown, P. L. Llewellyn, N. Masciocchi and J. R. Long, *Nature*, 2015, **527**, 357), copyright (2015).)

respectively, suggesting that replacing Co with Fe increases the energy of the phase transition.

12.6.2.3 Temperature Dependency of Gate-opening

In general, the gate-opening process is accompanied by either a global transformation, in the majority of the cases or by local transformation of the framework structure. However, there are rare cases when the framework does not change the structure and it is the kinetic factor that leads to gate-opening sorption. Microporous manganese formate $[Mn(HCO_2)_2 \cdot \frac{1}{3}(C_4H_8O_2)]_n$ is one such example where the gate-opening is triggered by temperature which is presumably due to the dynamic opening of the pore structure and/or sufficient kinetic energy of the adsorbates that overcome the diffusion barrier above the critical temperature of the gases.[107] The 3D framework is built with corner- and edge-shared MnO_6 octahedra and contains 1D zigzag channel (Figure 12.26a,b). The compound does not show any uptake for N_2

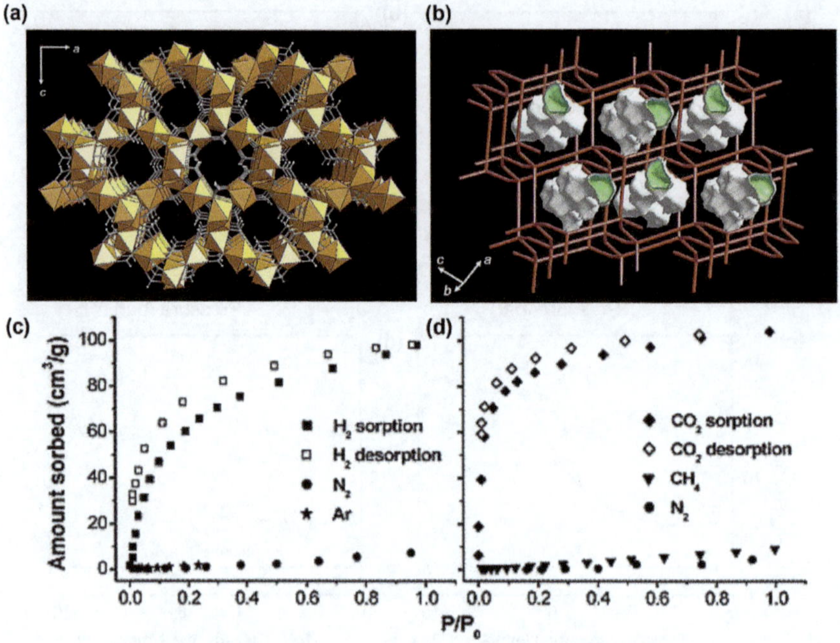

Figure 12.26 (a) X-ray crystal structure of $[Mn(HCO_2)_2 \cdot \frac{1}{3}(C_4H_8O_2)]_n$ showing channels along the *b* axis. Guest molecules are omitted. Each octahedron represents Mn(II) cation coordinated by six formate ligands; (b) A simplified presentation of the network in $[Mn(HCO_2)_2 \cdot \frac{1}{3}(C_4H_8O_2)]_n$. Only Mn(II) centres are connected (red wire) to show the framework topology. Accessible zigzag open channels along the *b* axis are shown as white surfaces with green entrances. BET gas sorption isotherms at 78 K (c) and 195 K (d).
(Adapted from ref. 108 with permission from the Royal Society of Chemistry).

at 77 K, however, surprisingly at higher temperatures above 110 K there is a significant uptake of the gas *via* a gate-opening process (Figure 12.26c,d). The single crystal structure of the MOF determined at 90 and 196 K revealed no significant change in the structure, including cell parameter and pore size, suggesting the compound has a "kinetically closed" pore towards N_2 at low temperatures which then becomes open at higher temperatures.

12.7 Future Outlook

From the foregoing discussion it can be understood that MOFs which have flexible or soft structures are amenable to structural change when they interact with various incoming guest molecules. For MOFs, the entire framework is supported by coordination bonds and/or other weak co-operative interactions such as H-bonding, π–π stacking, and van der Waals interaction. The flexibility is thus expected even under mild conditions that alter the cooperative interactions. In most cases the structural change is associated with a phase change from open to close form upon guest removal. This leads to the deformation of the pore structure and closure of pore aperture. The diffusion of the guest molecules into the nanochannels of such pore structures is highly restricted which leads to unusual gated and stepwise sorption phenomena. At the same time it creates opportunity to load the molecules of interest into the nanochannels by altering various parameters such as pressure, temperature, size and polarity of the adsorbates *etc.* In a way one can controllably and selectively deliver guest molecules in the micropores of these soft MOFs which is of paramount importance from application point of view. Since the pore structure of the MOFs can be suitably predesigned in the micropore region small molecules can be separated, which is not possible with the traditional porous materials such as porous carbon and zeolites. Depending on this phenomena gas separation is also possible and there are plenty of examples of such kind.

Over the last decade, a number of soft or flexible MOFs have been synthesized and characterized and their sorption behaviour reported. It is now strongly believed that these materials have the remarkable property of dynamism controlled by external stimuli and admittedly unmatchable with traditional porous materials such as zeolites and porous carbon. This has made these materials highly promising for various applications such as selective sorption and separation. The few reports on the separation of lower hydrocarbons and gas mixtures such as $CO–N_2$ are quite encouraging. With the availability of advanced *in situ* tools such as X-ray-sorption and IR-sorption it has become possible to characterize the intermediates associated with such unusual sorption processes. However, it is quite evident that more advancement is required to design and synthesize more sophisticated MOFs suitable for practical applications.

Abbreviations

PCP	porous coordination polymer
MOF	metal-organic framework
IUPAC	International Union of Pure and Applied Chemistry
DMF	*N*,*N*′-dimethyl formamide
PXRD	powder X-ray diffraction
SBU	secondary building unit
1D	one dimensional
2D	two dimensional
3D	three dimensional
o-phen	*ortho*-phenanthroline
2,6-ndc	2,6-napthalene dicarboxylic acid
1,3-adc	1,3-adamantanedicarboxylate
bpp	1,3-bis(4-pyridyl) propane
azpy	4,4′-azopyridine

pyrdc	pyridine-2,3-dicarboxylate
aip	5-azidoisophthalate
4,4′-bpy, bpy	4,4′ bipyridine
OTf	trifluoromethanesulfonate
Hdhbc	2,5-dihydroxybenzoic acid
bpndc	benzophenone-4,4′-dicarboxylate
NH_2-bdc	2-aminobenzenedicarboxylic acid
bphz	1,2-bis(4-pyridylmethylene) hydrazine
DABCO	1,4-diazabicyclo[2.2.2]octane
BDP^{2-}	1,4-benzenedipyrozolate
pzdc	pyrazine-2,3-dicarboxylate
dpyg	1,2-dipyridylglycol
dma	*N*,*N*′-dimethylacetamide
L1	4,4′,4″,4‴-benzene-1,2,4,5-tetrayl-tetra-benzoic acid
L2	*N*,*N*′-di-(4-pyridyl)-1,4,5,8-naphthalenetetracarboxydiimide
tci	tris(2-carboxyethyl) isocyanurate
bpdc	4,4′- biphenyldicarboxylate
bpee	1,2-bipyriylethylene
TCNQ	7,7,8,8-tetracyano-*p*-quinodimethane
bdc	1,4-benzene dicarboxylate
bpNDI	*N*,*N*′-bis-(4-pyrid-yl)-1,4,5,8-naphthalenediimide
AmTAZ	3-amino-1,2,4-triazole
UMSs	unsaturated metal sites
ZIF	zeolitic imidazolate framework

Acknowledgements

R. H. and A. C. acknowledge DST, Govt. of India and JNCASR, Bangalore for financial support.

P. S. acknowledges CSIR, India for fellowship. TKM gratefully acknowledges DST, Govt. of India for the financial support (Project No. MR-2015/001019).

References

1. L. J. Murray, M. Dincâ and J. R. Long, *Chem. Soc. Rev.*, 2009, **38**, 1294.
2. S. Chaemchuen, N. A. Kabir, K. Zhou and F. Verpoort, *Chem. Soc. Rev.*, 2013, **42**, 9304.
3. S. Furukawa, J. Reboul, S. Diring, K. Sumida and S. Kitagawa, *Chem. Soc. Rev.*, 2014, **43**, 5700.
4. A. Schneemann, V. Bon, I. Schwedler, I. Senkovska, S. Kaskel and R. A. Fischer, *Chem. Soc. Rev.*, 2014, **43**, 6062.

5. A. H. Chughtai, N. Ahmad, H. A. Younus, A. Laypkov and F. Verpoort, *Chem. Soc. Rev.*, 2015, **44**, 6804.
6. J. Liu, L. Chen, H. Cui, J. Zhang, L. Zhang and C.-Y. Su, *Chem. Soc. Rev.*, 2014, **43**, 6011.
7. Z.-J. Lin, J. Lu, M. Hong and R. Cao, *Chem. Soc. Rev.*, 2014, **43**, 5867.
8. A. U. Czaja, N. Trukhan and U. Muller, *Chem. Soc. Rev.*, 2009, **38**, 1284.
9. N. W. Ockwig, O. Delgado-Friedrichs, M. O'Keeffe and O. M. Yaghi, *Acc. Chem. Res.*, 2005, **38**, 176.
10. O. M. Yaghi and Q. Li, *MRS Bull.*, 2009, **34**, 682.
11. M. O'Keeffe, *Chem. Soc. Rev.*, 2009, **38**, 1215.
12. J. R. Long and O. M. Yaghi, *Chem. Soc. Rev.*, 2009, **38**, 1213.
13. M. O'Keeffe, M. A. Peskov, S. J. Ramsden and O. M. Yaghi, *Acc. Chem. Res.*, 2008, **41**, 1782.
14. J. Čejka, *Angew. Chem., Int. Ed.*, 2012, **51**, 4782.
15. M. Liu, H. S. Quah, S. Wen, Z. Yu, J. J. Vittal and W. Ji, *Chem. Mater.*, 2016, **28**, 3385.
16. S. Sanda, S. Parshamoni and S. Konar, *Inorg. Chem.*, 2013, **52**, 12866.
17. R. Haldar, M. Inukai, S. Horike, K. Uemura, S. Kitagawa and T. K. Maji, *Inorg. Chem.*, 2016, **55**, 4166.
18. N. Sikdar, S. Bonakala, R. Haldar, S. Balasubramanian and T. K. Maji, *Chem. – Eur. J.*, 2016, **22**, 6059.
19. R. Haldar, R. Matsuda, S. Kitagawa, S. J. George and T. K. Maji, *Angew. Chem., Int. Ed.*, 2014, **126**, 11966.
20. Z. Chang, D.-H. Yang, J. Xu, T.-L. Hu and X.-H. Bu, *Adv. Mater.*, 2015, **27**, 5432.
21. L. Sarkisov, R. L. Martin, M. Haranczyk and B. Smit, *J. Am. Chem. Soc.*, 2014, **136**, 2228.
22. Y. Yue, J. A. Rabone, H. Liu, S. M. Mahurin, M.-R. Li, H. Wang, Z. Lu, B. Chen, J. Wang, Y. Fang and S. Dai, *J. Phys. Chem. C*, 2015, **119**, 9442.
23. V. Haigis, F.-X. Coudert, R. Vuilleumier, A. Boutin and A. H. Fuchs, *J. Phys. Chem. Lett.*, 2015, **6**, 4365.
24. K. L. Gurunatha, S. Mohapatra, P. A. Suchetan and T. K. Maji, *Cryst. Growth Des.*, 2009, **9**, 3844.
25. M.-M. Liu, Y.-L. Bi, Q.-Q. Dang and X.-M. Zhang, *Dalton Trans.*, 2015, **44**, 19796.
26. T. K. Pal, D. De, S. Neogi and P. K. Bharadwaj, *Inorg. Chem. Front.*, 2015, **2**, 395.
27. L. Gurunatha, G. Mostafa, D. Ghoshal and T. K. Maji, *Cryst. Growth Des.*, 2010, **10**, 2483.
28. S. Mohapatra, H. Sato, R. Matsuda, S. Kitagawa and T. K. Maji, *CrystEngComm*, 2012, **14**, 4153.
29. R. Dey, R. Haldar, T. K. Maji and D. Ghoshal, *Cryst. Growth Des.*, 2011, **11**, 3905.
30. T. K. Maji, G. Mostafa, R. Matsuda and S. Kitagawa, *J. Am. Chem. Soc.*, 2005, **127**, 17152.

31. Y. He, J. Shang, Q. Gu, G. Li, J. Li, R. Singh, P. Xiao and P. A. Webley, *Chem. Commun.*, 2015, **51**, 14716.
32. P. Kanoo, R. Matsuda, R. Kitaura, S. Kitagawa and T. K. Maji, *Inorg. Chem.*, 2012, **51**, 9141.
33. P. Kanoo, K. L. Gurunatha and T. K. Maji, *J. Mater. Chem.*, 2010, **20**, 1322.
34. D. N. Dybtsev, H. Chun and K. Kim, *Angew. Chem., Int. Ed.*, 2004, **43**, 5033.
35. H. Li, M. Eddaoudi, M. O'Keeffe and O. M. Yaghi, *Nature*, 1999, **402**, 276.
36. J.-R. Li, R. J. Kuppler and H.-C. Zhou, *Chem. Soc. Rev.*, 2009, **38**, 1477.
37. A. Hazra, S. Bonakala, K. K. Bejagam, S. Balasubramanian and T. K. Maji, *Chem. – Eur. J.*, 2016, **22**, 7792.
38. B. Chen, S. Xiang and G. Qian, *Acc. Chem. Res.*, 2010, **43**, 1115.
39. Y. Takashima, V. M. Martínez, S. Furukawa, M. Kondo, S. Shimomura, H. Uehara, M. Nakahama, K. Sugimoto and S. Kitagawa, *Nat. Commun.*, 2011, **2**, 168.
40. M. Kondo, T. Yoshitomi, H. Matsuzaka, S. Kitagawa and K. Seki, *Angew. Chem., Int. Ed. Engl.*, 1997, **36**, 1725.
41. W. Huang, J. Jiang, D. Wu, J. Xu, B. Xue and A. M. Kirillov, *Inorg. Chem.*, 2015, **54**, 10524.
42. L. Wang, Y.-A. Li, F. Yang, Q.-K. Liu, J.-P. Ma and Y.-B. Dong, *Inorg. Chem.*, 2014, **53**, 9087.
43. J.-R. Li, Y. Tao, Q. Yu, X.-H. Bu, H. Sakamoto and S. Kitagawa, *Chem. – Eur. J.*, 2008, **14**, 2771.
44. C. Gücüyener, J. van den Bergh, J. Gascon and F. Kapteijn, *J. Am. Chem. Soc.*, 2010, **132**, 17704.
45. B. Chen, C. Liang, J. Yang, D. S. Contreras, Y. L. Clancy, E. B. Lobkovsky, O. M. Yaghi and S. Dai, *Angew. Chem., Int. Ed.*, 2006, **45**, 1390.
46. L. Alaerts, M. Maes, L. Giebeler, P. A. Jacobs, J. A. Martens, J. F. M. Denayer, C. E. A. Kirschhock and D. E. De Vos, *J. Am. Chem. Soc.*, 2008, **130**, 14170.
47. B. P. Biswal, A. Bhaskar, R. Banerjee and U. K. Kharul, *Nanoscale*, 2015, **7**, 7291.
48. B. Bhattacharya, R. Dey, P. Pachfule, R. Banerjee and D. Ghoshal, *Cryst. Growth Des.*, 2013, **13**, 731.
49. A. Mallick, S. Saha, P. Pachfule, S. Roy and R. Banerjee, *J. Am. Chem. Soc.*, 2010, **20**, 9073.
50. B. Manna, S. Mukherjee, A. V. Desai, S. Sharma, R. Krishna and S. K. Ghosh, *Chem. Commun.*, 2015, **51**, 15386.
51. S. Mukherjee, B. Manna, A. V. Desai, Y. Yin, R. Krishna, R. Babarao and S. K. Ghosh, *Chem. Commun.*, 2016, **52**, 8215.
52. F.-X. Coudert, C. Mellot-Draznieks, A. H. Fuchs and A. Boutin, *J. Am. Chem. Soc.*, 2009, **131**, 11329.
53. C. J. Sumby, *Nat. Chem.*, 2016, **8**, 294.

54. S. Sakaida, K. Otsubo, O. Sakata, C. Song, A. Fujiwara, M. Takata and H. Kitagawa, *Nat. Chem.*, 2016, **8**, 377.
55. X. Li, X. Chen, F. Jiang, L. Chen, S. Lu, Q. Chen, M. Wu, D. Yuan and M. Hong, *Chem. Commun.*, 2016, **52**, 2277.
56. S.-m. Hyun, J. H. Lee, G. Y. Jung, Y. K. Kim, T. K. Kim, S. Jeoung, S. K. Kwak, D. Moon and H. R. Moon, *Inorg. Chem.*, 2016, **55**, 1920.
57. R. Haldar and T. K. Maji, *Z. Anorg. Allg. Chem.*, 2014, **640**, 1102.
58. R. Haldar, K. L. Gurunatha, N. Sikdar and T. K. Maji, *Inorg. Chem. Front.*, 2015, **2**, 278.
59. M. Zhang, G. Feng, Z. Song, Y.-P. Zhou, H.-Y. Chao, D. Yuan, T. T. Y. Tan, Z. Guo, Z. Hu, B. Z. Tang, B. Liu and D. Zhao, *J. Am. Chem. Soc.*, 2014, **136**, 7241.
60. R. Koitz, M. Iannuzzi and J. Hutter, *J. Phys. Chem. C*, 2015, **119**, 4023.
61. M. Fischer, F. Hoffmann and M. Froba, *RSC Adv.*, 2012, **2**, 4382.
62. K. Sumida, D. L. Rogow, J. A. Mason, T. M. McDonald, E. D. Bloch, Z. R. Herm, T.-H. Bae and J. R. Long, *Chem. Rev.*, 2012, **112**, 724.
63. P. Kanoo, R. Sambhu and T. K. Maji, *Inorg. Chem.*, 2011, **50**, 400.
64. P. Kanoo, G. Mostafa, R. Matsuda, S. Kitagawa and T. K. Maji, *Chem. Commun.*, 2011, **47**, 8106.
65. A. Kondo, H. Kajiro, H. Noguchi, L. Carlucci, D. M. Proserpio, G. Ciani, K. Kato, M. Takata, H. Seki, M. Sakamoto, Y. Hattori, F. Okino, K. Maeda, T. Ohba, K. Kaneko and H. Kanoh, *J. Am. Chem. Soc.*, 2011, **133**, 10512.
66. R. Kitaura, K. Seki, G. Akiyama and S. Kitagawa, *Angew. Chem., Int. Ed.*, 2003, **42**, 428.
67. A. Kondo, H. Noguchi, S. Ohnishi, H. Kajiro, A. Tohdoh, Y. Hattori, W.-C. Xu, H. Tanaka, H. Kanoh and K. Kaneko, *Nano Lett.*, 2006, **6**, 2581.
68. D. Tanaka, K. Nakagawa, M. Higuchi, S. Horike, Y. Kubota, T. C. Kobayashi, M. Takata and S. Kitagawa, *Angew. Chem., Int. Ed.*, 2008, **120**, 3978.
69. R. Haldar, N. Sikdar and T. K. Maji, *Mater. Today*, 2015, **18**, 97.
70. Y. He, Z. Zhang, S. Xiang, F. R. Fronczek, R. Krishna and B. Chen, *Chem. Commun.*, 2012, **48**, 6493.
71. S. Yang, X. Lin, W. Lewis, M. Suyetin, E. Bichoutskaia, J. E. Parker, C. C. Tang, D. R. Allan, P. J. Rizkallah, P. Hubberstey, N. R. Champness, K. Mark Thomas, A. J. Blake and M. Schröder, *Nat. Mater.*, 2012, **11**, 710.
72. L. Qin, Z.-M. Ju, Z.-J. Wang, F.-D. Meng, H.-G. Zheng and J.-X. Chen, *Cryst. Growth Des.*, 2014, **14**, 2742.
73. M. J. Manos, M. S. Markoulides, C. D. Malliakas, G. S. Papaefstathiou, N. Chronakis, M. G. Kanatzidis, P. N. Trikalitis and A. J. Tasiopoulos, *Inorg. Chem.*, 2011, **50**, 11297.
74. J. Cai, J. Yu, H. Xu, Y. He, X. Duan, Y. Cui, C. Wu, B. Chen and G. Qian, *Cryst. Growth Des.*, 2013, **13**, 2094.

75. J. H. Lee, T. K. Kim, M. P. Suh and H. R. Moon, *CrystEngComm*, 2015, **17**, 8807.
76. Q. Yao, J. Su, O. Cheung, Q. Liu, N. Hedin and X. Zou, *J. Mater. Chem.*, 2012, **22**, 10345.
77. H. Kim, S. Das, M. G. Kim, D. N. Dybtsev, Y. Kim and K. Kim, *Inorg. Chem.*, 2011, **50**, 3691.
78. C.-Y. Sun, X.-L. Wang, X. Zhang, C. Qin, P. Li, Z.-M. Su, D.-X. Zhu, G.-G. Shan, K.-Z. Shao, H. Wu and J. Li, *Nat. Commun.*, 2013, **4**, 2717.
79. R. S. Pillai, M. L. Pinto, J. Pires, M. Jorge and J. R. B. Gomes, *ACS Appl. Mater. Interfaces*, 2015, **7**, 624.
80. O. Alduhaish, H. Wang, B. Li, T.-L. Hu, H. D. Arman, K. Alfooty and B. Chen, *ChemPlusChem*, 2016, **81**, 770.
81. R. Haldar, S. K. Reddy, V. M. Suresh, S. Mohapatra, S. Balasubramanian and T. K. Maji, *Chem. – Eur. J.*, 2014, **20**, 4347.
82. Z. Wang and S. M. Cohen, *J. Am. Chem. Soc.*, 2009, **131**, 16675.
83. T. Loiseau, C. Serre, C. Huguenard, G. Fink, F. Taulelle, M. Henry, T. Bataille and G. Férey, *Chem. – Eur. J.*, 2004, **10**, 1373.
84. C. Serre, F. Millange, C. Thouvenot, M. Noguès, G. Marsolier, D. Louër and G. Férey, *J. Am. Chem. Soc.*, 2002, **124**, 13519.
85. M. Ma, A. Bétard, I. Weber, N. S. Al-Hokbany, R. A. Fischer and N. Metzler-Nolte, *Cryst. Growth Des.*, 2013, **13**, 2286.
86. A. Boutin, M.-A. Springuel-Huet, A. Nossov, A. Gédéon, T. Loiseau, C. Volkringer, G. Férey, F.-X. Coudert and A. H. Fuchs, *Angew. Chem., Int. Ed.*, 2009, **48**, 8314.
87. G. J. Halder, H. Park, R. J. Funk, K. W. Chapman, L. K. Engerer, U. Geiser and J. A. Schlueter, *Cryst. Growth Des.*, 2009, **9**, 3609.
88. F. Salles, G. Maurin, C. Serre, P. L. Llewellyn, C. Knöfel, H. J. Choi, Y. Filinchuk, L. Oliviero, A. Vimont, J. R. Long and G. Férey, *J. Am. Chem. Soc.*, 2010, **132**, 13782.
89. R. Kitaura, K. Fujimoto, S.-I. Noro, M. Kondo and S. Kitagawa, *Angew. Chem., Int. Ed.*, 2002, **114**, 141.
90. B. Russell, J. Villaroel, K. Sapag and A. D. Migone, *J. Phys. Chem. C*, 2014, **118**, 28603.
91. S. Tanaka, K. Fujita, Y. Miyake, M. Miyamoto, Y. Hasegawa, T. Makino, S. Van, der Perre, J. Cousin Saint Remi, T. Van Assche, G. V. Baron and J. F. M. Denayer, *J. Phys. Chem. C*, 2015, **119**, 28430.
92. J. A. Gee, J. Chung, S. Nair and D. S. Sholl, *J. Phys. Chem. C*, 2013, **117**, 3169.
93. C.-H. Kuo, Y. Tang, L.-Y. Chou, B. T. Sneed, C. N. Brodsky, Z. Zhao and C.-K. Tsung, *J. Am. Chem. Soc.*, 2012, **134**, 14345.
94. Y. Dai, J. R. Johnson, O. Karvan, D. S. Sholl and W. J. Koros, *J. Membr. Sci.*, 2012, **401–402**, 76.
95. R. Kumar, K. Jayaramulu, T. K. Maji and C. N. R. Rao, *Chem. Commun.*, 2013, **49**, 4947.
96. Y. Hu, H. Kazemian, S. Rohani, Y. Huang and Y. Song, *Chem. Commun.*, 2011, **47**, 12694.

97. T. D. Bennett, S. Cao, J. C. Tan, D. A. Keen, E. G. Bithell, P. J. Beldon, T. Friscic and A. K. Cheetham, *J. Am. Chem. Soc.*, 2011, **133**, 14546.
98. H. Chun and J. Seo, *Inorg. Chem.*, 2009, **48**, 9980.
99. K. L. Mulfort, O. K. Farha, C. D. Malliakas, M. G. Kanatzidis and J. T. Hupp, *Chem. - Eur. J.*, 2010, **16**, 276.
100. S. K. Ghosh, J.-P. Zhang and S. Kitagawa, *Angew. Chem., Int. Ed.*, 2007, **46**, 7965.
101. Z. R. Herm, E. D. Bloch and J. R. Long, *Chem. Mater.*, 2014, **26**, 323.
102. Y. He, R. Krishna and B. Chen, *Energy Environ. Sci.*, 2012, **5**, 9107.
103. M. G. Plaza, A. M. Ribeiro, A. Ferreira, J. C. Santos, Y. K. Hwang, Y. K. Seo, U. H. Lee, J. S. Chang, J. M. Loureiro and A. E. Rodrigues, *Microporous Mesoporous Mater.*, 2012, **153**, 178.
104. S. Yang, A. J. Ramirez-Cuesta, R. Newby, V. Garcia-Sakai, P. Manuel, S. K. Callear, S. I. Campbell, C. C. Tang and M. Schröder, *Nat. Chem.*, 2015, **7**, 121.
105. N. Nijem, H. Wu, P. Canepa, A. Marti, K. J. Balkus, T. Thonhauser, J. Li and Y. J. Chabal, *J. Am. Chem. Soc.*, 2012, **134**, 15201.
106. S. Shimomura, S. Horike, R. Matsuda and S. Kitagawa, *J. Am. Chem. Soc.*, 2007, **129**, 10990.
107. J. A. Mason, J. Oktawiec, M. K. Taylor, M. R. Hudson, J. Rodriguez, J. E. Bachman, M. I. Gonzalez, A. Cervellino, A. Guagliardi, C. M. Brown, P. L. Llewellyn, N. Masciocchi and J. R. Long, *Nature*, 2015, **527**, 357.
108. H. Kim, D. G. Samsonenko, M. Yoon, J. W. Yoon, Y. K. Hwang, J.-S. Chang and K. Kim, *Chem. Commun.*, 2008, **39**, 4697.

Subject Index